Lecture Notes in Computer Science 9374

Commenced Publication in 1973
Founding and Former Series Editors:
Gerhard Goos, Juris Hartmanis, and Jan van Leeuwen

More information about this series at http://www.springer.com/series/7409

Ivan Lirkov · Svetozar D. Margenov
Jerzy Waśniewski (Eds.)

Large-Scale Scientific Computing

10th International Conference, LSSC 2015
Sozopol, Bulgaria, June 8–12, 2015
Revised Selected Papers

 Springer

Editors

Ivan Lirkov
Institute of Information and
    Communication Technologies
Bulgarian Academy of Sciences
Sofia
Bulgaria

Svetozar D. Margenov
Institute of Information and
    Communication Technologies
Bulgarian Academy of Sciences
Sofia
Bulgaria

Jerzy Waśniewski
Department of Informatics and Mathematical
    Modelling
Technical University of Denmark
Kongens Lyngby
Denmark

ISSN 0302-9743   ISSN 1611-3349 (electronic)
Lecture Notes in Computer Science
ISBN 978-3-319-26519-3   ISBN 978-3-319-26520-9 (eBook)
DOI 10.1007/978-3-319-26520-9

Library of Congress Control Number: 2015954587

LNCS Sublibrary: SL3 – Information Systems and Applications, incl. Internet/Web, and HCI

Springer International Publishing AG Switzerland is part of Springer Science+Business Media
(www.springer.com)

Preface

The 10th International Conference on Large-Scale Scientific Computations (LSSC 2015) was held in Sozopol, Bulgaria, June 8–12, 2015. The conference was organized by the Institute of Information and Communication Technologies at the Bulgarian Academy of Sciences in cooperation with Society for Industrial and Applied Mathematics (SIAM) and Sozopol municipality.

The following plenary invited speakers and lectures were hosted:

- Thierry Coupez, "Implicit Boundary in Multiphase Flows and Anisotropic Adaptive Meshing"
- David Keyes, "Algorithmic Adaptations to Extreme Scale"
- Johannes Kraus, "Combined Strategies in Algebraic Multilevel Preconditioning"
- Siegfried Selberherr, "Spin-Based CMOS-Compatible Devices"
- Ludmil Zikatanov, "Subspace Correction Methods: Theory, Practice, and Robustness"

The success of the conference and the present volume are the outcome of the joint efforts of many partners from various institutions and organizations. First, thanks to all the members of the Scientific Committee for their valuable contribution forming the scientific face of the conference, as well as for their help in reviewing contributed papers. We especially thank the organizers of the special sessions. We are also grateful to the staff involved in the local organization.

Traditionally, the purpose of the conference is to bring together scientists working with large-scale computational models in natural sciences and environmental and industrial applications, and specialists in the field of numerical methods and algorithms for modern high-performance computers. The invited lectures reviewed some of the most advanced achievements in the field of numerical methods and their efficient applications. The conference talks were presented by researchers from academic institutions and practical industry engineers including applied mathematicians, numerical analysts, and computer experts. The general theme of LSSC 2015 was "Large-Scale Scientific Computing" with a particular focus on the organized special sessions.

The special sessions and organizers are as follows:

- "A Posteriori Error Control and Iterative Methods for Maxwell Type Problems" — D. Pauly, J. Kraus
- "Multilevel Methods on Graphs" — P. Vassilevski, L. Zikatanov
- "Mathematical Modeling and Analysis of PDEs Describing Physical Problems" — O. Iliev
- "Numerical Methods for Multiphysics Problems" — J. Adler, X. Hu, R. Lazarov
- "Control and Uncertain Systems" — M. Krastanov, V. Veliov
- "Enabling Exascale Computation" — O. Iliev, D. Keyes

- "Efficient Algorithms for Hybrid HPC Systems" — A. Karaivanova, E. Atanassov, T. Gurov, M. Mascagni
- "Applications of Metaheuristics to Large-Scale Problems" — S. Fidanova, G. Luque
- "Computational Microelectronics — From Monte Carlo to Deterministic Approaches" — I. Dimov, M. Nedjalkov, J. Weinbub
- "Large-Scale Models: Numerical Methods, Paralel Computations and Applications" — K. Georgiev, Z. Zlatev

More than 120 participants from all over the world attended the conference representing some of the strongest research groups in the field of advanced large-scale scientific computing. This volume contains 49 papers by authors from 15 countries.

The next international LSSC conference will be organized in June 2017.

October 2015

Ivan Lirkov
Svetozar D. Margenov
Jerzy Waśniewski

Organization

Scientific Committee

James Adler	Tufts University, USA
Emanouil Atanassov	Institute of Information and Communication Technologies, BAS, Bulgaria
Ivan Dimov	Institute of Information and Communication Technologies, BAS, Bulgaria
Stefka Dimova	Sofia University, Bulgaria
Stefka Fidanova	Institute of Information and Communication Technologies, BAS, Bulgaria
Krassimir Georgiev	Institute of Information and Communication Technologies, BAS, Bulgaria
Todor Gurov	Institute of Information and Communication Technologies, BAS, Bulgaria
Xiaozhe Hu	Tufts University, USA
Oleg Iliev	ITWM, Germany
Aneta Karaivanova	Institute of Information and Communication Technologies, BAS, Bulgaria
David Keyes	KAUST, Saudi Arabia
Mikhail Krastanov	Sofia University, Bulgaria
Johannes Kraus	University of Duisburg-Essen, Germany
Ulrich Langer	Johannes Kepler University Linz, Austria
Raytcho Lazarov	Texas A&M University, USA
Ivan Lirkov	Institute of Information and Communication Technologies, BAS, Bulgaria
Gabriel Luque	University of Málaga, Spain
Svetozar Margenov	Institute of Information and Communication Technologies, BAS, Bulgaria
Marcin Paprzycki	Systems Research Institute, PAS, Poland
Siegfried Selberherr	Vienna University of Technology, Austria
Panayot Vassilevski	Lawrence Livermore National Laboratory, USA
Vladimir Veliov	TU-Vienna, Austria
Jerzy Waśniewski	Technical University of Denmark, Denmark
Zahari Zlatev	Aarhus University, Denmark
Ludmil Zikatanov	Pennsylvania State University, USA

Contents

Applications of Metaheuristics to Large-Scale Problems

Computational Microelectronics — From Monte Carlo to Deterministic Approaches

**Large-Scale Models: Numerical Methods, Paralel Computations
and Applications**

Contributed Papers

Invited Papers

Preconditioners for Mixed FEM Solution of Stationary and Nonstationary Porous Media Flow Problems

Owe Axelsson, Radim Blaheta[✉], and Tomáš Luber

Institute of Geonics AS CR,
Studentska 1768, 70800 Poruba, Ostrava, Czech Republic
blaheta@ugn.cas.cz

Abstract. The paper concerns porous media flow in rigid or deformable matrix. It starts with stationary Darcy flow, but the main interest is in extending Darcy problem to involve time dependent behaviour and deformation of the matrix. The considered problems are discretized by mixed FEM in space and stable time discretization methods as backward Euler and second order Radau methods. The discretization leads to time stepping methods which involve solution of a linear system within each time step. The main focus of the paper is then devoted to the construction of suitable preconditioners for these Euler and Radau systems. The paper presents also numerical experiments for illustration of efficiency of the suggested numerical algorithms.

Keywords: Darcy flow · Poroelasticity · Saddle point systems · Preconditioners

1 Introduction

The porous media flow in rigid or deformable matrix are basically described by Darcy flow and Biot poroelasticity models, respectively. The stationary Darcy problem can be written in the form

$$
\begin{aligned}
K^{-1}v + \nabla p &= 0, \\
\text{div}(v) &= Q
\end{aligned}
\tag{1}
$$

with two physical fields, the Darcy velocity v and the fluid (pore) pressure p, which have to be determined in a domain Ω. Here $v = \phi v_f$ where ϕ is the porosity and v_f is the fluid velocity. The parameter K is the matrix of permeabilities divided by fluid viscosity (effective permeability, $K_{ij} = \kappa_{ij}/\nu_f$) and Q stands for the fluid source/sink term. The introduced Darcy flow model can be formulated variationally and discretized by a mixed finite element method, which leads to saddle point systems. The solution of these systems can be done by use of iterative methods with preconditioners based on the natural block structure. Efficient preconditioners can be based on regularization of the zero block and

© Springer International Publishing Switzerland 2015
I. Lirkov et al. (Eds.): LSSC 2015, LNCS 9374, pp. 3–14, 2015.
DOI: 10.1007/978-3-319-26520-9_1

subsequent formulation with augmented blocks. For stationary problems, the regularization implies necessity of strong augmentation and possible difficulties in solving the augmented block system.

The source term and accordingly also the velocity and pressure can be time dependent. In this case the Darcy model usually involves a flow retardation mechanism, which is provided by a small compressibility of fluid and/or deformation of the porous matrix. The time dependent Darcy model then has the following form

$$
\begin{aligned}
K^{-1}v &+ \nabla p = 0, \\
\mathrm{div}(v) &+ c_{pp}\tfrac{\partial}{\partial t}p = Q.
\end{aligned}
\tag{2}
$$

The time dependent Darcy model can be also discretized by the mixed finite elements in space and a suitable method in time. For the time discretization, we shall use stable methods such as the first order backward Euler or higher order Radau methods. After discretization, we solve the evolution problems by a time stepping procedure with solving saddle point systems within each time step. Compared with the stationary Darcy systems, the backward Euler systems are naturally regularized by the time derivative term, which influence the block preconditioners. For higher order Radau methods, we introduce additional preconditioners, which involve the solution of backward Euler type systems.

The porous media flow can be coupled with deformation of the porous matrix. The basic model in this respect is the Biot poroelasticity, which can be described by the equations

$$
\begin{aligned}
-\mathrm{div}(C_{el} : \varepsilon(u)) & &+ c_{up}\nabla p &= f, \\
& K^{-1}v &+ \nabla p &= 0, \\
c_{pu}\tfrac{\partial}{\partial t}\mathrm{div}(u) &+ \mathrm{div}(v) &+ c_{pp}\tfrac{\partial}{\partial t}p &= Q.
\end{aligned}
\tag{3}
$$

There are three physical fields in the domain Ω entering the above model, besides the velocity v and the fluid pressure p, it is the displacement u, which defines the small strain tensor $\varepsilon(u)$. Further, C_{el} is the elasticity tensor and $c_{up} = c_{pu} = \alpha$ are Biot-Willis coefficients. For simplicity, we assume $c_{up} = c_{pu} = 1$.

The organization of the paper is as follows. The next Section concerns discretization of the described porous media flow problems. The space discretization uses the lowest order Raviart-Thomas elements. The time dependent problems are then solved by time stepping methods with the solution of Euler and Radau systems in each step. The preconditioners for Euler and Radau systems are investigated in Sects. 3 and 4. Section 5 introduces a model problem and describes numerical experiments which illustrate the efficiency of the preconditioners.

2   Space and Time Discretization

The introduced problems can be formulated variationally and discretized by the Galerkin technique using proper function spaces. Namely, for $\Omega \subset R^d$ and decomposition of the boundary $\partial\Omega$ corresponding to different boundary conditions for flow $\partial\Omega = \Gamma_{v,p} \cup \Gamma_{v,v}$ and for mechanical response $\partial\Omega = \Gamma_{u,u} \cup \Gamma_{u,\sigma}$, we take

$$u_h \in U_h \subset U = \{u \in [H^1(\Omega)]^d, \ u = u_D \text{ on } \Gamma_{u,u}\},$$

where U_h corresponds to a finite element mesh division $\mathcal{T}_h$ of Ω into system of triangles for $d = 2$ or tetrahedra for $d = 3$. The functions from U_h are continuous on Ω and piecewise linear on elements of $\mathcal{T}_h$. Further,

$$v_h \in V_h \subset V = \{w \in H(\text{div}, \Omega), \ w \cdot \nu = q_n \text{ on } \Gamma_{p,v}\},$$

where V_h contains the lowest order Raviart-Thomas finite elements on the same division $\mathcal{T}_h$ as used for elasticity. Finally,

$$p_h \in P_h \subset P = L_2(\Omega),$$

where P_h contains functions piecewise constant on the same mesh $\mathcal{T}_h$.

After taking proper bases in U_h, V_h, P_h and establishing isomorphism between finite element functions and algebraic vectors $u_h \leftrightarrow u$, $v_h \leftrightarrow v$ and $p_h \leftrightarrow p$, we can introduce the finite element matrices,

$$\langle Au, w \rangle = \int_\Omega C\varepsilon(u_h) : \varepsilon(w_h)\, dx \ \forall u_h, w_h \in U_h,$$

$$\langle Mv, z \rangle = \int_\Omega K^{-1}v_h \cdot z_h\, dx \ \forall v_h, z_h \in V_h,$$

$$\langle M_p p, q \rangle = \int_\Omega p_h q_h\, dx \ \forall p_h, q_h \in P_h,$$

$$\langle B_u u, q \rangle = \int_\Omega \text{div}(u_h)q_h\, dx \ \forall u_h \in U_h, \ q_h \in P_h,$$

$$\langle B_v v, q \rangle = \int_\Omega \text{div}(v_h)q_h\, dx \ \forall v_h \in V_h, \ q_h \in P_h.$$

Note that A, M, M_p are symmetric, $C = c_{pp}M_p$, A is positive definite if the displacement is prescribed on a part $\Gamma_{u,u} \subset \partial\Omega$ with a positive measure and M, M_p are always positive definite.

The discretization of time dependent Darcy and poroelasticity problems by mixed finite element methods results in a differential-algebraic (DAE) system of a general form

$$\mathcal{A}_1 \frac{\partial}{\partial t}\mathcal{U} + \mathcal{A}_0\mathcal{U} = \mathcal{F},$$

where

$$\mathcal{A}_1 = \begin{bmatrix} 0 & 0 \\ 0 & -C \end{bmatrix}, \quad \mathcal{A}_0 = \begin{bmatrix} M & B^T \\ B & 0 \end{bmatrix}, \quad \mathcal{U} = \begin{bmatrix} v \\ p \end{bmatrix}$$

for the time dependent Darcy problem and

$$\mathcal{A}_1 = \begin{bmatrix} 0 & 0 & 0 \\ 0 & 0 & 0 \\ B_u & 0 & -C \end{bmatrix}, \quad \mathcal{A}_0 = \begin{bmatrix} A & 0 & B_u^T \\ 0 & M & B^T \\ 0 & B & 0 \end{bmatrix}, \quad \mathcal{U} = \begin{bmatrix} u \\ v \\ p \end{bmatrix}$$

for poroelasticity.

The time discretization is performed by a sequence of time steps

$$0 = t_0 < t_1 < \ldots < t_N,$$

where $\tau_i = t_{i+1} - t_i$ are provided apriori or computed adaptively. To simplify the presentation, we shall assume constant time steps, $\tau_i = \tau$. For each time interval $\langle t_i, t_{i+1}\rangle$, we have

$$\int\limits_{t_i}^{t_i+\tau} \mathcal{A}_1 \frac{\partial}{\partial t}\mathcal{U}\,dt + \int\limits_{t_i}^{t_i+\tau} (\mathcal{A}_0\mathcal{U} - \mathcal{F})\,dt = \mathcal{A}_1\left(\mathcal{U}^{i+1} - \mathcal{U}^i\right) + \int\limits_{t_i}^{t_i+\tau} (\mathcal{A}_0\mathcal{U} - \mathcal{F})\,dt = 0.$$

The integration $\int_{t_i}^{t_i+\tau} (\mathcal{A}_0\mathcal{U} - \mathcal{F})\,dt$ has to be performed by a suitable approximate integration scheme. The use of the simple right-hand rectangle approximate integration provides the backward Euler method

$$\mathcal{A}_E\mathcal{U}^{i+1} = (\mathcal{A}_1 + \tau\mathcal{A}_0)\,\mathcal{U}^{i+1} = \mathcal{A}_1\mathcal{U}^i + \tau\mathcal{F}^{i+1} \ \forall i = 0,\ldots,N-1. \qquad (4)$$

The backward Euler method is stable and suitable for the solution of stiff and DAE problems, but has only a first order time discretization error. As an example of higher order discretization methods, we use the third order L-stable method based on the second order Radau integration (RADAU IIA), see e.g. [1]. It uses two integration points $t_{i+1/3}$ and t_{i+1} in the interval $\langle t_i, t_{i+1}\rangle$. The position of $t_{i+1/3}$ and the weights are determined from the condition that the integration scheme should be exact for polynomials up to second order. It provides

$$\int\limits_{t_i}^{t_i+\tau} \phi\,dx = \frac{3}{4}\tau\phi(t_i + \tau/3) + \frac{1}{4}\tau\phi(t_{i+1}) \ \forall\phi(t) = \sum_{i=0}^{2} \alpha_i t^i.$$

The Radau method leads to the system

$$\mathcal{A}_R \begin{bmatrix}\mathcal{U}^{i+1/3} \\ \mathcal{U}^{i+1}\end{bmatrix} = \begin{bmatrix}\mathcal{A}_1 + \frac{5\tau}{12}\mathcal{A}_0 & -\frac{\tau}{12}\mathcal{A}_0 \\ \frac{3\tau}{4}\mathcal{A}_0 & \mathcal{A}_1 + \frac{\tau}{4}\mathcal{A}_0\end{bmatrix} \begin{bmatrix}\mathcal{U}^{i+1/3} \\ \mathcal{U}^{i+1}\end{bmatrix}$$

$$= \begin{bmatrix}\mathcal{A}_1\mathcal{U}^i + \frac{\tau}{12}(5\mathcal{F}^{i+1/3} - \mathcal{F}^{i+1}) \\ \mathcal{A}_1\mathcal{U}^i + \frac{\tau}{4}(3\mathcal{F}^{i+1/3} + \mathcal{F}^{i+1})\end{bmatrix}. \qquad (5)$$

Elimination of $\mathcal{U}^{i+1/3}$ provides a reduced system

$$\mathcal{A}_{RR}\mathcal{U}^{i+1} = \left(\mathcal{A}_1\mathcal{A}_0^{-1} - \frac{1}{3}\tau\right)\mathcal{A}_1\mathcal{U}^i + \frac{\tau}{4}\mathcal{A}_1\mathcal{A}_0^{-1}\left(3\mathcal{F}^{i+1/3} + \mathcal{F}^{i+1}\right) + \frac{1}{6}\tau^2\mathcal{F}^{i+1}$$

with the matrix

$$\mathcal{A}_{RR} = \frac{1}{6}\tau^2\mathcal{A}_0 + \frac{2}{3}\tau\mathcal{A}_1 + \mathcal{A}_1\mathcal{A}_0^{-1}\mathcal{A}_1. \qquad (6)$$

The space and time discretization provides possibility to solve the porous media flow by time stepping algorithms which compute the vector of all unknowns by solving the corresponding system in each time step. The systems to be solved are (4), (5) and (6), depending on the chosen time discretization technique.

3   Preconditioners for the Euler Systems

The Euler system for the time dependent Darcy model has the form

$$\mathcal{A}_E = \mathcal{A}_1 + \tau \mathcal{A}_0 = \tau \begin{bmatrix} M & B^T \\ B & -\frac{1}{\tau}C \end{bmatrix},$$

which contains a regularization term in the (2,2) block. The regularization is based on pressure mass matrix multiplied by compressibility parameter which is typically constant in the whole domain. For the lowest order Raviart-Thomas elements the pressure mass matrix is diagonal, which allows an easy inverse of the matrix C. All these facts indicate that suitable preconditioners can be found of the augmented type, i.e.

$$\mathcal{P}_{ET} = \begin{bmatrix} M_C & B^T \\ & -\frac{1}{\tau}C \end{bmatrix}, \quad \mathcal{P}_{ED} = \begin{bmatrix} M_C & \\ & -\frac{1}{\tau}C \end{bmatrix}, \quad \mathcal{P}_{EDP} = \begin{bmatrix} M_C & \\ & \frac{1}{\tau}C \end{bmatrix}.$$

In all cases $M_C = M + \tau B^T C^{-1} B$ is the augmented matrix. For the Raviart-Thomas finite elements, M_C can be assembled as a sparse matrix, which allows to solve the inner system with M_C by various direct or iterative solvers. Note that small time steps improve conditioning, which is favourable.

As concerns the preconditioned systems, an analysis for both exact (the ideal case) and inexact solution of subproblems can be found in the literature, see e.g. [5–8]. Other applicable preconditioning techniques can be found e.g. in [4,8].

The Euler system for the poroelasticity problems has the form

$$\mathcal{A}_E = \mathcal{A}_1 + \tau \mathcal{A}_0 = \tau \begin{bmatrix} A & 0 & B_u^T \\ 0 & M & B^T \\ \frac{1}{\tau}B_u & B & -\frac{1}{\tau}C \end{bmatrix}.$$

$\mathcal{A}_E$ is not symmetric but the corresponding system can be easily symmetrized by row scaling which provides new system with the matrix $\tilde{\mathcal{A}}_E$,

$$\tilde{\mathcal{A}}_E = \begin{bmatrix} \frac{1}{\tau} & & \\ & 1 & \\ & & 1 \end{bmatrix} \mathcal{A}_E = \begin{bmatrix} A & 0 & B_u^T \\ 0 & \tau M & \tau B^T \\ B_u & \tau B & -C \end{bmatrix}$$

and suitable preconditioners for $\tilde{\mathcal{A}}_E$ can be again found of the augmented type form, i.e.

$$\mathcal{P}_{ET} = \begin{bmatrix} S_C & \bar{B}^T \\ & -C \end{bmatrix}, \quad \mathcal{P}_{ED} = \begin{bmatrix} S_C & \\ & -C \end{bmatrix}, \quad \mathcal{P}_{EDP} = \begin{bmatrix} S_C & \\ & C \end{bmatrix},$$

where $\bar{B} = \begin{bmatrix} B_u & B \end{bmatrix}$ and the pivot block has now a 2×2 structure

$$S_C = \begin{bmatrix} A + B_u^T C^{-1} B_u & \tau B_u^T C^{-1} B \\ \tau B^T C^{-1} B_u & \tau M + \tau^2 B^T C^{-1} B \end{bmatrix} = \begin{bmatrix} S_{11} & S_{12} \\ S_{21} & S_{22} \end{bmatrix}.$$

As S_C is now more complicated, a question arises if it can be simplified by considering its block diagonal or block triangular part.

Proposition 1. *There is a constant $0 \leq \gamma < 1$ such that*

$$|\langle S_{12}v, u \rangle| \leq \gamma \sqrt{\langle S_{11}u, u \rangle} \sqrt{\langle S_{22}v, v \rangle} \ \forall u, v.$$

If c_{el} is a positive constant such that

$$c_{el} \|\text{div}(u_h)\|_{L_2}^2 \leq \langle Au, u \rangle$$

then $\gamma^2 \leq (1 + c_{pp}c_{el})^{-1}$. For isotropic elasticity with Lamè constants λ and μ, $c_{el} = \lambda$. With the constant γ, we have the spectral equivalence

$$(1-\gamma)\begin{bmatrix} S_{11} & \\ & S_{22} \end{bmatrix} \leq \begin{bmatrix} S_{11} & S_{12} \\ S_{21} & S_{22} \end{bmatrix} \leq (1+\gamma)\begin{bmatrix} S_{11} & \\ & S_{22} \end{bmatrix}$$

Proof. The proof is based on the strengthened Cauchy-Schwarz-Bunyakowski constant γ. To show the estimate, we apply the CBS inequality to get

$$|\langle S_{12}v, u \rangle| = |\langle \tau B_u^T C^{-1} B v, u \rangle| = \left|\left\langle \tau C^{-1/2} B v, C^{-1/2} B_u u \right\rangle\right|$$
$$\leq \sqrt{\langle \tau^2 B^T C^{-1} B v, v \rangle} \sqrt{\langle B_u^T C^{-1} B_u u, u \rangle}.$$

Then

$$|\langle S_{12}v, u \rangle| \leq \gamma \sqrt{\langle (\tau M + \tau^2 B^T C^{-1} B)v, v \rangle} \sqrt{\langle (A + B_u^T C^{-1} B_u)u, u \rangle}$$
$$= \gamma \sqrt{\langle S_{11}u, u \rangle} \sqrt{\langle S_{22}v, v \rangle}.$$

The estimate for γ comes from

$$\langle B_u u, C^{-1} B_u u \rangle \leq \frac{1}{c_{el}c_{pp}} \langle Au, u \rangle. \tag{7}$$

Details can be found in [10].   □

Note that the constant γ and the spectral equivalence are independent on discretization (represented by h and τ) and also does not depend on oscillations of permeability. On the other hand, γ depends on compressibility c_{pp} and mechanical stiffness of the porous matrix. A stiffer porous matrix will decrease the coupling between flow and deformation and provide a smaller value of γ.

Note that the estimate of γ could be improved by taking into account the contribution of M similarly as the contribution of A in (7). The contribution of M can be significant if the permeability is small. On the other hand, the qualitative result can be obtained without considering the contribution of M, which has two benefits - avoiding assumptions on oscillatory character of the permeability coefficients and avoiding the fact, that we should use h dependent inverse inequality to bound L_2-norm of $\text{div}(v_h)$ by L_2-norm of v_h.

In the case of having a solver for the system $M + \tau B^T C^{-1} B$, which is robust with respect to coefficient oscillations (see e.g. [9]), the spectral equivalence above provide also a possibility to construct a robust solver for the poroelasticity problem.

4 Preconditioners for the Radau Systems

The more complex systems (5) and (6) arising from the Radau discretization of both time dependent Darcy and poroelasticity problems can be solved iteratively with very efficient preconditioners based on the solution of simpler Euler type systems with matrices

$$\mathcal{A}_1 + \tau \mathcal{A}_0.$$

Proposition 2. *Let $\mathcal{A}_{RR}$ be the matrix of the reduced Radau system (6). Then a suitable preconditioner is found in the form*

$$\mathcal{P}_{RR} = \left(\mathcal{A}_1 + \frac{1}{\sqrt{6}}\tau\mathcal{A}_0\right)\mathcal{A}_0^{-1}\left(\mathcal{A}_1 + \frac{1}{\sqrt{6}}\tau\mathcal{A}_0\right). \tag{8}$$

The spectrum of the preconditioned matrix $\mathcal{P}_{RR}^{-1}\mathcal{A}_{RR}$ is real and lies in the interval $\left\langle 1 - \frac{1}{6+\sqrt{24}}, 1\right\rangle$.

Proposition 3. *Let $\mathcal{A}_R$ be the matrix of the full Radau system (5). Then a preconditioner can be taken in the triangular form*

$$\mathcal{P}_{R-T} = \begin{bmatrix} \mathcal{A}_1 + \frac{5\tau}{12}\mathcal{A}_0 & 0 \\ \frac{3\tau}{4}\mathcal{A}_0 & \mathcal{A}_1 + \frac{\tau}{4}\mathcal{A}_0 \end{bmatrix}. \tag{9}$$

The spectrum of the preconditioned matrix $\mathcal{P}_{R-T}^{-1}\mathcal{A}_R$ is real and lies in the interval $\left\langle 1, \frac{8}{5}\right\rangle$.

Proposition 4. *Let $\mathcal{A}_R$ be the matrix of the full Radau system (5). Then a preconditioner can be taken in the diagonal form*

$$\mathcal{P}_{R-D} = \begin{bmatrix} \mathcal{A}_1 + \frac{5\tau}{12}\mathcal{A}_0 & 0 \\ 0 & \mathcal{A}_1 + \frac{\tau}{4}\mathcal{A}_0 \end{bmatrix}. \tag{10}$$

The spectrum of the preconditioned matrix $\mathcal{P}_{R-D}^{-1}\mathcal{A}_R$ is complex and lies in the interval $\left\{z \in \mathbb{C}: \ Re(z) = 1 \ \& \ |Im(z)| \leq \sqrt{3/5}\right\}$.

Proof. The proof of Proposition 2 can be found in [2]. We shall show the proof of Proposition 3, the proof of Proposition 4 is similar.

A simple manipulation provides

$$\mathcal{P}_{R-T}^{-1}\mathcal{A}_R = I + \begin{bmatrix} 0 & E_{12} \\ 0 & E_{22} \end{bmatrix},$$

where

$$E_{12} = \left(\mathcal{A}_1 + \frac{5\tau}{12}\mathcal{A}_0\right)^{-1}\left(-\frac{\tau}{12}\right)\mathcal{A}_0 = -\frac{1}{5}\left(\frac{12}{5\tau}\tilde{\mathcal{A}}_1 + I\right)^{-1}$$

$$E_{22} = -\left(\mathcal{A}_1 + \frac{\tau}{4}\mathcal{A}_0\right)^{-1}\left(\frac{3\tau}{4}\right)\mathcal{A}_0 E_{12} = \frac{3}{5}\left(\frac{4}{\tau}\tilde{\mathcal{A}}_1 + I\right)^{-1}\left(\frac{12}{5\tau}\tilde{\mathcal{A}}_1 + I\right)^{-1}.$$

Above $\tilde{\mathcal{A}}_1 = \mathcal{A}_0^{-1}\mathcal{A}_1$.

If $\mu \in \sigma(\tilde{A}_1)$ then for the time dependent Darcy model there exists $z = (z_1, z_2)^T$ such that

$$\begin{aligned} \mu M z_1 + \mu B^T z_2 &= 0 \\ \mu B z_1 &= -C z_2 \end{aligned}$$

Thus $\mu = 0$ is the eigenvalue with the eigenvector $z = (z_1, 0)^T$, $z_1 \neq 0$. Moreover, if $\mu \in \sigma(\tilde{A}_1)$ and $\mu \neq 0$ then $z_1 = -M^{-1} B^T z_2$ and μ fulfils $\mu B M^{-1} B^T z_2 = C z_2$. Therefore $\mu > 0$. Similarly, it is possible to show that $\sigma(\tilde{A}_1) \subset \langle 0, \infty \rangle$ for poroelasticity problem, see [2].

Consequently, $\sigma(E_{22}) \subset \langle 0, \frac{3}{5} \rangle$ as $\lambda \in \sigma(E_{22})$ if $\lambda = \frac{3}{5}(\frac{4}{\tau}\mu + 1)^{-1}(\frac{12}{5\tau}\mu + 1)^{-1}$ for $\mu \in \sigma(\tilde{A}_1)$, and $\mu \geq 0$. Finally,

$$\sigma\left(\mathcal{P}_{R-T}^{-1}\mathcal{A}_R\right) = 1 + (\{0\} \cup \sigma(E_{22})) \subset \left\langle 1, \frac{8}{5} \right\rangle.$$

$\square$

5   Numerical Tests

The described preconditioners are tested on a poroelasticity model problem defined in the square domain $\Omega = \langle 0, 1 \rangle \times \langle 0, 1 \rangle$, see Fig. 1. We shall test the following material properties

(a) effective permeability $\log k(x) \in N(0, \tilde{\sigma})$, storativity $c_{pp} = 1$ (fast flow regime),
(b) effective permeability $\log 10^5 k(x) \in N(0, \tilde{\sigma})$, storativity $c_{pp} = 0.000165$ (modest flow regime like in sand).

Here $\xi \in N(\mu, \tilde{\sigma})$ denotes that the quantity ξ has a normal distribution with mean μ and variance $\tilde{\sigma}$. For $\tilde{\sigma} \neq 0$, we can therefore model the flow problem in heterogeneous porous medium, increasing $\tilde{\sigma}$ increases the contrast in coefficients. In practice there is a high range of possible effective permeability values. In geo-applications, $k = \kappa/\nu_f$ (permeability divided by viscosity) usually lies between 10^{-4} (highly fractured rock) and 10^{-16} (intact granite).

Other parameters are not oscillatory and we shall assume that they are constant in the whole domain Ω. The tests are performed for very soft elastic material Lamè constants $\lambda = \mu = 4$ ($E = 10$ and Poisson's ratio $\nu = 0.25$) as well as for stiffer material with $\lambda = 10^3$ and $\lambda = 10^6$ to confirm the behaviour described in Proposition 1. The Biot-Willis coefficient is also held constant $c_{up} = c_{pu} = 1$ in Ω.

For the elastic and flow part we use zero volume force $F_s = 0$ and zero volume source $Q = 0$. The boundary conditions are specified in Fig. 1. The problem uses zero initial conditions

$$u(x_1, x_2, 0) = 0, \quad p(x_1, x_2, 0) = 0 \text{ for } (x_1, x_2) \in \Omega.$$

The elastic part is discretized by a finite element method on a regular grid Ω_h created by a division of Ω into $1/h^2$ small congruent squares and subsequent

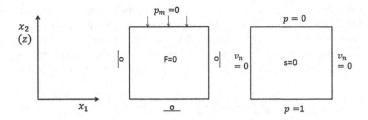

Fig. 1. The model problem in square Ω. Boundary conditions for elasticity are shown in the middle, boundary conditions for flow in the right square.

division of the squares into triangles. Then, the linear Courant elements are used. The flow part is discretized by a mixed finite element method on the same triangular grid with the lowest order Raviart-Thomas finite elements for the velocity and piecewise constant finite elements for the pressure. In this paper, we consider the time discretizations with fixed time steps, $\tau = 0.01$. The value of h is taken $h = 1/50$ in the reported experiments.

First, we test the efficiency of the iterative solution of the Euler system with the preconditioner

$$\mathcal{P}_{ED} = \begin{bmatrix} S_C & \\ & -C \end{bmatrix},$$

where

$$S_C = \begin{bmatrix} A + B_u^T C^{-1} B_u & \tau B_u^T C^{-1} B \\ \tau B^T C^{-1} B_u & \tau M + \tau^2 B^T C^{-1} B \end{bmatrix}$$

or S_C is replaced by its diagonal part $\tilde{S}_C$,

$$\tilde{S}_C = \begin{bmatrix} A + B_u^T C^{-1} B_u & \\ & \tau M + \tau^2 B^T C^{-1} B \end{bmatrix}.$$

Table 1. Numbers of GMRES iterations for solving Euler system with $\mathcal{A}_E$ using zero initial guess and relative residual accuracy $\varepsilon = 10^{-6}$. The numbers of iterations are averaged from the first ten time steps. The Euler system is preconditioned by $\mathcal{P}_{ED}$ with full and reduced Schur complements S_C and $\tilde{S}_C$, respectively. The systems with matrices S_C and $\tilde{S}_C$ are solved exactly. The model poroelasticity problem uses discretization $h = 1/50$, $\tau = 0.01$. The oscillations of effective permeability provide coefficient contrast $8.2 \cdot 10^5$ for $\tilde{\sigma} = 2$ and $6.8 \cdot 10^{11}$ for $\tilde{\sigma} = 4$.

	$\tilde{\sigma} = 0$				$\tilde{\sigma} = 2$				$\tilde{\sigma} = 4$			
λ	4	4	10^3	10^6	4	4	10^3	10^6	4	4	10^3	10^6
c_{pp}	1	10^{-3}	10^{-3}	10^{-3}	1	10^{-3}	10^{-3}	10^{-3}	1	10^{-3}	10^{-3}	10^{-3}
S_C	17	5	5	5	18	5	5	5	19	7	9	9
$\tilde{S}_C$	17	29	14	5	18	37	14	5	19	58	18	9

Table 2. The average numbers of GMRES iterations per one time step for Radau problem, block diagonal preconditioner $\mathcal{P}_{R-D}$, relative residual accuracy $\varepsilon = 10^{-6}$. The block systems are solved exactly. Column **Z** is for zero initial guess, **P** for initial guess taken from the previous time step. The model poroelasticity problem uses discretization $h = 1/50$, $\tau = 0.01$ and oscillations of effective permeability due to $\tilde{\sigma} = 2$.

$\log k \in N(0, \tilde{\sigma})$						$\log 10^5 k \in N(0, \tilde{\sigma})$					
$\tilde{\sigma} = 0$		$\tilde{\sigma} = 2$		$\tilde{\sigma} = 4$		$\tilde{\sigma} = 0$		$\tilde{\sigma} = 2$		$\tilde{\sigma} = 4$	
Z	**P**	**Z**	**P**	**Z**	**P**	**Z**	**P**	**Z**	**P**	**Z**	**P**
14	2.3	14	2.5	14	1.7	12	2.6	13	2.6	13	2.5

Table 3. The average numbers of GMRES iterations per one time step for Radau problem, block triangular preconditioner $\mathcal{P}_{R-T}$, relative residual accuracy $\varepsilon = 10^{-6}$. The block systems are solved exactly. Column **Z** is for zero initial guess, **P** for initial guess taken from the previous time step. The model problem and its discretization is the same as in Table 2.

$\log k \in N(0, \tilde{\sigma})$						$\log 10^5 k \in N(0, \tilde{\sigma})$					
$\tilde{\sigma} = 0$		$\tilde{\sigma} = 2$		$\tilde{\sigma} = 4$		$\tilde{\sigma} = 0$		$\tilde{\sigma} = 2$		$\tilde{\sigma} = 4$	
Z	**P**	**Z**	**P**	**Z**	**P**	**Z**	**P**	**Z**	**P**	**Z**	**P**
8	2.0	8	2.0	8	2.3	7	2.3	7.3	2.2	7	2.2

Table 4. The average numbers of GMRES iterations per one time step for the reduced Radau problem, preconditioner $\mathcal{P}_{RR}$, relative residual accuracy $\varepsilon = 10^{-6}$. The block systems are solved exactly. Column **Z** is for zero initial guess, **P** for initial guess taken from the previous time step. The model problem and its discretization is the same as in Table 2.

$\log k \in N(0, \tilde{\sigma})$						$\log 10^5 k \in N(0, \tilde{\sigma})$					
$\tilde{\sigma} = 0$		$\tilde{\sigma} = 2$		$\tilde{\sigma} = 4$		$\tilde{\sigma} = 0$		$\tilde{\sigma} = 2$		$\tilde{\sigma} = 4$	
Z	**P**	**Z**	**P**	**Z**	**P**	**Z**	**P**	**Z**	**P**	**Z**	**P**
5	1.4	5	1.5	5	1.7	4	2.2	4	2.1	4	2.1

We shall also examine the efficiency of preconditioners $\mathcal{P}_{R-D}$, $\mathcal{P}_{R-T}$, $\mathcal{P}_{RR}$ within Radau time steps, when the arising systems with Euler type matrices $\mathcal{A}_1 + c\tau\mathcal{A}_0$ are solved by a direct solution method (MATLAB backslash solver). The iterations are tested in two variants - with zero initial guess and with initial guess provided by the solution of system in the previous time step. As can be seen, the latter provides a significant reduction of the number of iterations. The case with zero initial guess somehow model the situation with adaptive time stepping, when the number of iterations are less reduced when the solution approaches the steady state. The results are summarized in Tables 2, 3 and 4.

6   Conclusions

The porous media flow problems are important in many applications. The numerical solution of these problems is not easy due to possible instabilities in the case of improper discretization and due to high heterogeneity and high contrast (oscillations) in the coefficients representing permeabilities, see also [3] and the references therein.

In this paper, we address both of the above mentioned aspects. A construction of preconditioner for Euler type systems in poroelasticity was shown, which is robust with respect to permeability oscillations and which can provide a fully robust and efficient solver if the subblock system corresponding to flow is solved by an inner robust solver, like solvers considered in [9].

The second focus is on solving still more complex systems arising in application of higher order Radau time integration method. It is shown that efficient preconditioning procedures to these systems can be created if solvers for the Euler systems are available. Preconditioning procedures can be applied to full or reduced Radau systems. The application to reduced system provides better convergence, the application to full systems brings more space for parallelization, which can be used for the whole matrix-vector multiplication as well as for the whole preconditioning in the case of diagonal $\mathcal{P}_{R-D}$ preconditioner. In this respect, we obtain efficiently preconditioned stable and accurate time discretization method.

Acknowledgement. This work was supported by the European Regional Development Fund in the IT4Innovations Centre of Excellence project (identification number CZ.1.05/1.1.00/02.0070).

References

1. Butcher, J.C.: Numerical Methods for Ordinary Differential Equations, 2nd edn. Wiley, Chichester (2008)
2. Axelsson, O., Blaheta, R., Kohut, R.: Preconditioning methods for high order strongly stable time integration methods with an application for a DAE problem. Numer. Linear Algebra Appl. (accepted)
3. Axelsson, O., Blaheta, R., Byczanski, P.: Stable discretization of poroelasticity problems and efficient preconditioners for arising saddle point type matrices. Comput. Vis. Sci. **15**(4), 191–207 (2012)
4. Axelsson, O., Blaheta, R., Hasal, M.: A comparison of preconditioning methods for saddle point problems with an application for porous media flow problems. Proceedings HPCSE 2015 (2015, submitted)
5. Axelsson, O., Blaheta, R.: Analysis of preconditioning methods based on augmentation. Math. Comput. Simul., special issue MODELLING 2014 **67**, 106–121 (2014). Kindly check and confirm the edit made in Ref.[5]
6. Axelsson, O.: Unified analysis of preconditioning methods for saddle point matrices. Numer. Linear Algebra Appl. **22**(2), 233–253 (2015)

7. Axelsson, O., Blaheta, R., Byczanski, P., Karatson, J., Ahmad, B.: Preconditioners for regularized saddle point problems with an application for heterogeneous Darcy flow problems. J. Comput. Appl. Math. **280**, 141–157 (2015)
8. Benzi, M., Golub, G.H., Liesen, J.: Numerical solution of saddle point problems. Acta Numerica **14**, 1–137 (2005)
9. Kraus, J., Lazarov, R., Lymbery, M., Margenov, S., Zikatanov, L.: Preconditioning of Weighted H(div)-Norm and Applications to Numerical Simulation of Highly Heterogeneous Media. Submitted, http://arxiv.org/pdf/1406.4455.pdf
10. Axelsson, O., Blaheta, Luber, T.: Robust preconditioners for flow problems in rigid and deformable porous media. In preparation

Fast Constrained Image Segmentation Using Optimal Spanning Trees

Stanislav Harizanov[1]([✉]), Svetozar Margenov[1], and Ludmil Zikatanov[2,3]

[1] Institute of Information and Communication Technologies,
Bulgarian Academy of Sciences, Sofia, Bulgaria
sharizanov@parallel.bas.bg
[2] Department of Mathematics, Pennsylvania State University,
University Park 16802, USA
[3] Institute of Mathematics and Informatics,
Bulgarian Academy of Sciences, Sofia, Bulgaria

Abstract. We propose a graph theoretical algorithm for image segmentation which preserves both the volume and the connectivity of the solid (non-void) phase of the image. The approach uses three stages. Each step optimizes the approximation error between the image intensity vector and piece-wise constant (indicator) vector characterizing the segmentation of the underlying image. The different norms in which this approximation can be measured give rise to different methods. The running time of our algorithm is $\mathcal{O}(N \log N)$ for an image with N voxels.

1 Introduction

We focus on the image segmentation problem, which is one of the challenging problems in Computer Vision. Throughout the last three-four decades numerous approaches have been proposed and developed in order to attack the problem. In 1989, Mumford and Shah [18] minimized a certain energy functional in order to compute the segmentation. The functional contains three terms: regularity term on the length of the inter-phase contours, regularity term on the smoothness of the intensity function v, and the data fidelity term which measures the L^2 distance between the input intensity u and the output intensity v. Since the Mumford-Shah functional is non-convex and non-smooth, the optimization problem is difficult to solve. Simplifications of the model, however, have been proposed. Among these is the piecewise smooth convex relaxation by functional lifting [19]. Another frequently applied strategy is to restrict v within the class of piece-wise constant functions, so that the second regularity term in the original functional is omitted. This piecewise constant model, combined with classical gradient based active contour models lead to the Chan-Vese model ([8] for 2-phase and [25] for multi-phase segmentation). There are many other approaches for 2-phase image segmentation based on [8] and its convex relaxation [7], e.g., [4,9,29]. In [6], the regularity term in the Mumford-Shah functional is replaced by the Rudin-Osher-Fatemi (ROF) functional [21]. A new multiphase segmentation model based on iteratively thresholding the minimizer of the ROF functional

© Springer International Publishing Switzerland 2015
I. Lirkov et al. (Eds.): LSSC 2015, LNCS 9374, pp. 15–29, 2015.
DOI: 10.1007/978-3-319-26520-9_2

of a convex relaxation of the Mumford-Shah functional is presented and relation between the solution in the 2-phase case and the one from the Chan-Vese model is established.

Graph-based image segmentation has been another active research field for the past 40 years. While the works we mentioned earlier use a continuous setup, in the graph based approaches, one studies a suitably constructed graphs whose vertices are the voxels describing the image. In [28], Zahn uses the minimum spanning tree (MST) of a weighted graph to obtain the phases. The edge weights are defined as the differences of the intensities of neighboring voxels. A "heavy" MST edges are then cut out, leaving the different connected components to be the different phases. Some shortcomings of the model have been overcome in [24], where the weights are normalized. A segmentation method, based on finding minimum cuts in a graph, is developed in [27]. The method appeared to be biased more towards finding small components. A normalized cut criterion, which takes into account self-similarity of regions, is developed in [22]. The latter leads to an NP-hard problem and in [22] the authors propose several polynomial approximating algorithms. The work of Weiss [26] relates such eigenvector-based approximations to more standard (spectral) partitioning methods on graphs. An efficient greedy algorithm for multiphase segmentation, based on edge detection, is developed in [10]. Although the algorithm uses local optimization procedures it runs in almost linear time with respect to the number of edges, and the output segmentation satisfies global properties.

Other graph-Laplacian-based segmentation models [13,16,17,23] can be viewed as an intersection of the general approaches we have just described. In these works, the image voxels are again part of a graph structure, and the corresponding 2-Laplacian functional [1,5] is used as the data fidelity term in an optimization problem. The data smoothing is not addressed via additional regularity terms in the functional, but by carefully choosing the edge weights.

The variety of segmentation techniques is huge and we cannot cover it all. There are other approaches, some of them considered classical (e.g., the K-means method and its modifications). We refer to [3,12], for the corresponding techniques and review of the literature.

In this paper we consider a constraint 2-phase segmentation problem, where one of the phases is simply connected and of fixed volume. The motivation comes from industry and more precisely from Computer Tomography (CT). Porous materials are of current interest within a wide range of applications and their properties strongly depend on various measurements such as absolute porosity, average pore size, size and shape of individual pores. Therefore, accurate segmentation of the 3D industrial CT reconstruction of the corresponding specimen is crucial for further numerical simulations. Due to the highly irregular structure of the segmentation phases and the presence of noise in the image, the methods, described earlier are not reliable and in some cases the results between different algorithms may differ drastically (even in 50 % of the voxels). To say the least, such a task is nontrivial.

An important constraint is the volume constraint. It is practical to assume that: (1) the volume (e.g. the number of "solid" voxels) of the solid phase is fixed (determined from the density and the weight of the material); and (2) the specimen consists of only one material piece (connected component).

In our work, we aim to design algorithms that give accurate segmentation, but also which respect the constraints given in (1) and (2). Recently, a promising step in this direction was made in [11], where, based on the techniques from [13] algorithms for accurate segmentation (with the volume constraint) were reported. Here, we introduce a different approach, based on MST properties. We propose a new class of algorithms, which give promising results and provide a framework for future research on the constrained image segmentation. Our mathematical model requires minimization of functionals measuring the approximation error with piece-wise constant functions. For the theory, we consider a discrete version of the fidelity term of the Chan-Vese model (called *fitting energy*) defined on the characteristic functions χ_S of simply-connected subsets $S \subset \Omega$ of cardinality $|S| = M$. Then, in the experimental part we add a regularity term to it for data smoothing.

The rest of the paper is organized as follows. We give preliminary notation and definitions in Sect. 2. Next, in Sect. 3 we formulate the problem relevant to the 2-phase constrained image segmentation. In Sect. 4 we describe in more detail the 3 stage algorithm and in Sect. 5 we test its performance.

2   Preliminaries

We introduce the notation needed to formulate a problem which we refer to in what follows as the *2-phase image segmentation problem*. We are given a volume Ω in 3D (2D) which is split in $n_x \times n_y \times n_z$ cubes (squares in 2D case when $n_z = 1$). The cubes are called voxels. The total number of voxels is $N = n_x n_y n_z$, and the set of voxels is $\mathcal{V}$. We thus have

$$\overline{\Omega} = \bigcup_{K \in \mathcal{V}} \overline{K}.$$

We assume that we are given a piece-wise constant (with respect to the partition of Ω in voxels) function called *intensity* u. Denoting by χ_S the characteristic function of a set S, we have that

$$u = \sum_{K \in \mathcal{V}} u_K \chi_K.$$

Here, $u_K = u\big|_K$. Since the space of such piece-wise constant functions is isomorphic to $\mathbb{R}^N$, we also denote by u the corresponding vector $\{u_K\}_{K \in \mathcal{V}}$ in $\mathbb{R}^N$, hoping that there is no ambiguity in such notation.

The formulation of the 2-phase image segmentation problem involves topological connectivity with respect to various graphs, thus we introduce the relevant notation next. We denote $G_1(\mathcal{V}, \mathcal{E}_1)$ and $G_\infty(\mathcal{V}, \mathcal{E}_\infty)$ to be the indirected graphs with vertices the set of voxels and edge sets $\mathcal{E}_p$, $p = 1, \infty$ defined as follows:

$$\mathcal{E}_p = \left\{ (i,j) \in \mathcal{V} \times \mathcal{V} \mid \|i - j\|_{\ell^p} = 1 \right\}.$$

Since we only consider undirected graphs, $(i,j) \in \mathcal{E}_p$ implies that $(j,i) \in \mathcal{E}_p$. The neighborhood $\mathcal{N}_p(i)$ of a voxel is defined as

$$\mathcal{N}_p(i) = \left\{ j \in \mathcal{V} \mid (i,j) \in \mathcal{E}_p \right\}.$$

For example, $\mathcal{N}_1(i)$ consists of the 6 voxels that share a common face with i, while $\mathcal{N}_\infty(i)$ consists of the 26 voxels that build the $3 \times 3 \times 3$ cube, centered at i.

The graph G is called connected if and only if for every pair of voxels i and j, there is a path formed by elements of $\mathcal{E}$ connecting them.

In the following definitions, we assume that we have fixed a connected graph $G = (\mathcal{V}, \mathcal{E})$ whose set of vertices is the set of voxels.

Definition 1. *Let $S \subset \mathcal{V}$ be a set of voxels. We call $G_S = (S, \mathcal{E}_S)$ the graph induced by S if G_S has as vertices the voxels in S and as edges all edges in $\mathcal{E}$ for which both ends are in S.*

Definition 2. *Let $S \subset \mathcal{V}$ be a set of voxels. We call S a G-connected set if the graph G_S induced by S is connected.*

When each edge $e = (i,j) \in \mathcal{E}$ has a weight $\omega_{ij} \geq 0$, the graph G is called weighted. The weights may have various meanings when the graph is related to real-life problems (e.g., gain, cost, penalty, etc.). In this paper, they will measure the dissimilarities between the edge endpoints (e.g., difference in intensities or/and gradient values of the corresponding voxels), so $\omega_{ij} \sim 0$ means that $i \sim j$ in a given sense. Every connected weighted graph possesses a minimum spanning tree. This tree is a computationally efficient way to store both connectivity and similarity information, and it plays a central role in our algorithm. Therefore, we briefly cover the MST theory we use in the paper. Let $G = G_\infty(\mathcal{V}, \mathcal{E})$.

Definition 3. *We say that the graph $T(\mathcal{V}_T, \mathcal{E}_T)$ is a minimum spanning tree (MST) of G, if $\mathcal{V}_T = \mathcal{V}_G$, T contains no cycles, and $\displaystyle\sum_{(i,j)\in\mathcal{E}_T} \omega_{ij} \to \min.$*

Important properties of T are stated below:

- **Cycle property:** For any cycle C in G, if $\bar{e} = \mathrm{argmax}_{e \in C} \omega_e$ is unique, then $\bar{e} \notin T$.
- **Cut property:** For any cut C in G, if $\bar{e} = \mathrm{argmin}_{e \in C} \omega_e$ is unique, then $\bar{e} \in T$.
- **Contraction:** If $\mathcal{T} \subset T$ is a tree, then we can contract it to a single vertex and maintain the MST property for the factor graph.

Definition 4. *We call $\mathcal{L}_T := \{ i \in \mathcal{V} : \exists! j \in \mathcal{V} \text{ s.t. } (i,j) \in \mathcal{E}_T \}$ the set of leaves of T. The <u>heaviest leaf of T</u> is $l_T := \mathrm{argmax}_{i \in \mathcal{L}_T} \{ \omega_{ij} : (i,j) \in \mathcal{E}_T \}.$*

To construct the MST, we apply the Kruskal's algorithm [15], which is linear in the number of edges (thus of $O(N \log N)$ complexity). Possible accelerations using local techniques and parallel realizations (the so called approximate Kruskal algorithm is found on pp.600–602 in Kraus [14]) are available, but since the purpose of this paper is mainly to address the constraint segmentation problem, we do not pursue this avenue here, and we use the classical algorithm in [15].

3   Problem Formulation

Here we state in a precise fashion the mathematical problem for determining a 2-phase segmentation of an image with intensity $u \in [0, 1]^N$.

We begin by giving the definition of an *admissible 2-phase (image) segmentation*.

Definition 5. *Given an intensity u and a graph G, we say that S is an admissible 2-phase segmentation if the following properties are satisfied*

- **Connectedness Property (CP):** *S is G-connected.*
- **Approximate Dominating Property (ADP):** *There exists an intensity $v \approx u$ which provides the same solution S and satisfies Dominating Property (DP):*

$$\min_{K \in S} u_K > \max_{K \in \bar{S}} u_K \qquad \bar{S} = \mathcal{V} \setminus S.$$

When u itself satisfies (DP), we call S (DP)-admissible. In such a case, the image phases are well-separated and even direct segmentation methods such as hard thresholding will do the job. In this paper, we consider noisy and blurry images, where the boundary between S and $\bar{S}$ is not that sharp. We note that the notion of $v \approx u$ is a bit vague here. Intuitively, one may think of v as the original (denoised and deblurred version of u) image intensity, while the $\approx$ sign implies certain constraints on the magnitude of both the noise and blur levels of the image.

We consider a 2-phase segmentation problem, where the solid phase is connected and of fixed volume.

Problem 1. Given u and G, find 2-phase admissible segmentation S with a given cardinality $|S| = M > 1$.

We introduce the following family of functionals $J : 2^{\mathcal{V}} \mapsto \mathbb{R}$, where $2^{\mathcal{V}}$ is the set of all subsets of the set of vertices $\mathcal{V}$.

$$J(S) = \|u - \chi_S\|^2, \quad \chi_S \text{ is the characteristic function of } S. \tag{1}$$

The values of the functional depend on the norm that we take, and we are going to have two types of norm, thus two different functionals:

$$J_0(S) = \|u - \chi_S\|^2_{\ell^2(\mathcal{V})}, \quad J_1(S) = \|u - \chi_S\|^2_{\ell^2(\mathcal{V})} + \lambda \|\nabla (u - \chi_S)\|^2_\omega. \tag{2}$$

Here, $\lambda > 0$ is a parameter, $\nabla : \mathbb{R}^n \mapsto \mathbb{R}^{n_e}$ is the discrete gradient defined as

$$(\nabla v)_e = \delta_e v = v_i - v_j, \quad i < j, \quad (i,j) = e \in \mathcal{E}. \tag{3}$$

$$(\nabla v, \nabla w)_\omega = \sum_{e \in \mathcal{E}} \omega_e \delta_e v \delta_e w. \tag{4}$$

The non-negative weights $\{\omega_e\}_{e \in \mathcal{E}}$ may depend on the intensity u, and will be dealt with in Sect. 4. Finally, the weighted norm of the gradient is defined as

$$\|\nabla v\|_\omega^2 = (\nabla v, \nabla v)_\omega. \tag{5}$$

Note that J_0 can be seen as the discrete and simplified version of the 2D Chan-Vese fitting energy [8]

$$F_1(\mathcal{C}) + F_2(\mathcal{C}) = \int_{inside(\mathcal{C})} |u - c_0|^2 dx dy + \int_{outside(\mathcal{C})} |u - c_1|^2 dx dy,$$

where $\mathcal{C}$ is a 2D curve. The 2-phase segmentation there is obtained via minimizing the fitting energy with respect to $\mathcal{C}$, c_0, and c_1 together with regularity terms on the length of $\mathcal{C}$ and the area of its interior. Then, the interior and the exterior of $\mathcal{C}$ are the two phases. Here, we set $c_0 = \min_i u_i = 0$, and $c_1 = \max_i u_i = 1$. We have no regularity terms, but impose two additional constraints: the interior to be connected and of cardinality M. If no constraints are addressed, it is straightforward to check that the minimizer of J_0 corresponds to direct hard-threshold segmentation (e.g., $i \in S \Leftrightarrow u_i \geq 0.5$). The functional J_1 is J_0, penalized by a regularity term. The regularization depends on the choice of the weights ω.

We now have the following definition.

Definition 6. *We say that the set S provides an optimal 2-phase segmentation for u if and only if S is an admissible 2-phase segmentation of cardinality M and it minimizes the functional $J(S)$, namely,*

$$S = \arg\min \{ J(S) \mid |S| = M, \ S \text{ is connected} \}.$$

Such definition leads to a simple characterization of the minimizer for the norm choices (2). Indeed we have for all $S \in 2^\mathcal{V}$

$$J_0(S) = \sum_{j \in S} (u_j - 1)^2 + \sum_{j \notin S} u_j^2 = \|u\|_{\ell^2}^2 - 2 \sum_{j \in S} u_j + \|\chi_S\|_{\ell^2}^2$$

$$= -2 \sum_{j \in S} u_j + \|u\|_{\ell^2}^2 + M.$$

Thus, minimizing $J_0(S)$ is equivalent to finding a G-connected S, such that

$$S = \arg\max J_*(S), \quad J_*(S) := \sum_{j \in S} u_j.$$

Note that $J_*(\cdot)$ is a linear functional in u. Next, we look into the other norm. Denote the set of edges connecting S with the complement of S, denoted here by $\overline{S}$ by $\mathcal{E}_c$ (c stands for "cut"). We note that $\nabla \chi_S = 0$ for all edges interior to S or $\overline{S}$. We then compute

$$\|\nabla u - \nabla \chi_S\|_\omega^2 = \|\nabla u\|_\omega^2 - 2 \sum_{e \in \mathcal{E}_c} \omega_e \delta_e u + \sum_{e \in \mathcal{E}_c} \omega_e = \|\nabla u\|_\omega^2 + \sum_{e \in \mathcal{E}_c} (1 - 2\delta_e u)\omega_e$$

Note that, from the definition of the gradient, the above formula is correct if we have ordered first the vertices in S, so that $\delta_e \chi_S \geq 0$. Thus, we can obtain an optimal solution if we minimize

$$J_{**}(S) = -2 \sum_{j \in S} u_j + \lambda \sum_{e \in \mathcal{E}_c} (1 - 2\delta_e u)\omega_e$$

4   Constrained Segmentation

In this section, we propose a three stage segmentation algorithm. The steps are as follows: (1) smoothing step which removes the local extrema in the intensity vector; (2) Selecting M voxels and constructing a connected component in the graph containing all of these voxels; (3) trimming the connected component so that the approximation to the "solid" part of the image has exactly M voxels.

4.1   Step 1: Removing Local Maxima

For a fixed G, we say that u has a strict local maximum at $K \in \mathcal{V}$ if and only if

$$u_K > \max\{u_J, \quad J \in \mathcal{N}(K)\}.$$

Since by assumption we are looking for segmentation with $|S| > 1$, all strict local maxima are due to image artefacts (e.g., noise). We can now modify the intensity and remove them, still having an admissible solution S. We use the following algorithm for removal of local maxima:

Algorithm 1. (Removal of local max) *Input: u (a given intensity) and G (a graph). **Output:** v (modified intensity).*
***For** $i = 1, \ldots, N$.*

– ***If*** u_i *is a strict local maximum,* ***then***

$$v_i = \max\{u_j, \quad j \in \mathcal{N}(i)\};$$

– ***else*** $v_i = u_i$.

We have the following equivalence results about the segmentations corresponding to intensities u and v whose proof is straightforward.

Lemma 1. *Let v be obtained from u via Algorithm 1. Then*

- *If u satisfies (DP) then v also satisfies (DP).*
- *If $S(u)$ is a 2-phase admissible segmentation solving Problem 1, so is $S(v)$.*

4.2   Defining Edge Weights

From now on, we assume that the underlying graph G is fixed and in what follows, we take $G = G_\infty$. Furthermore, we assume $u \in [0, 1]^N$.

We aim at solving constraint segmentation problems, where the cardinality of the admissible 2-phase segmentation S is a priori known (i.e., $|S| = M$). For this purpose, we split $\mathcal{V}$ into three subsets:

$$\mathcal{V}_1 := \{i \in \mathcal{V} \mid u_i \sim 1 \,\&\, f(\nabla u_i) \le \varepsilon\},$$
$$\mathcal{V}_0 := \{i \in \mathcal{V} \mid u_i \sim 0 \,\&\, f(\nabla u_i) \le \varepsilon\},$$
$$\mathcal{V}_U := \mathcal{V} \setminus (\mathcal{V}_0 \cup \mathcal{V}_1).$$

Here, the maximal image intensity is 1 and the minimal is 0; ∇u_i is the G-gradient at i (i.e., $(\nabla u_i)_j = u_j - u_i$, $\forall j \in \mathcal{N}(i)$), $f : \mathbb{R}^{26} \to [0, +\infty)$, and ε is a small parameter, chosen by the user. The similarity relations $u_i \sim 1$ and $u_i \sim 0$ also need to be specified for the image. They should depend on the noise and blur levels. Typically, one uses $u_i \ge 1 - \eta$ and $u_i \le \eta$ for a suitable $\eta \in (0, 1/2)$.

The idea is that $\mathcal{V}_1 \subset S$, $\mathcal{V}_0 \subset \bar{S}$, while the origin of the voxels within $\mathcal{V}_U$ remains unclear and, depending on M, they should be distributed somehow between the two phases S and $\bar{S}$.

For the weights of the edges, we propose to add an "uncertainty penalizer", e.g.,

$$\omega_{ij} := |u_i - u_j| + \delta g\left(f(\nabla u_i), f(\nabla u_j)\right), \quad \forall(i, j) \in \mathcal{E}. \tag{6}$$

Here $g : [0, +\infty)^2 \to [0, +\infty)$, and δ is a small, positive parameter. Such weights need to favorize edges between $\mathcal{V}_U$ and $\mathcal{V}_0 \cup \mathcal{V}_1$ and penalize edges within $\mathcal{V}_U$. The latter will help us to "clarify" the origin of the unclear voxels, while in the same time it decouples them and they don't cluster. Hence, the elements of $\mathcal{V}_U$ can be treated individually, which is very important for our constraint problem.

To achieve that, we need to impose some assumptions on f, g:

Definition 7. *We say that f, g are admissible if they satisfy the following:*

- *Symmetry: $f(x_1, \ldots, x_{26}) = f(x_{\sigma(1)}, \ldots, x_{\sigma(26)})$, resp. $g(x_1, x_2) = g(x_2, x_1)$, where $\sigma : \{1, \ldots, 26\} \to \{1, \ldots, 26\}$ is an arbitrary permutation.*
- *Positivity: $f(x) = 0$, $g(x) = 0 \iff x = \mathbf{0}$.*
- *1-Homogeneity: $f(\lambda x) = \lambda f(x)$, $g(\lambda x) = \lambda g(x)$, $\forall \lambda \ge 0$.*
- *Monotonicity: $g(x_1 + \alpha, x_2) \ge g(x_1, x_2)$, $\forall \alpha > 0$.*

Examples: $\| \cdot \|_{\ell^p}$, $\forall p \ge 1$, $\min(\cdot)$, and many others.

Lemma 2 (Properties of ω_{ij} from (6)). *Let $G = G_\infty(\mathcal{V}, \mathcal{E})$, $i \in \mathcal{V}$, $j \in \mathcal{N}(i)$, f, g are admissible, and ω_{ij} is given by (6). Then*

(i) G is undirected graph ($\omega_{ij} = \omega_{ji}$) and ω_{ij} is invariant under rotations of Ω.

(ii) $\omega_{ij} = 0 \quad \Leftrightarrow \quad u|_{\mathcal{N}(i) \cup \mathcal{N}(j)} = const$. In particular, $\omega_{ij} \sim 0$ if $\{i, j\} \subset \mathcal{V}_0 \cup \mathcal{V}_1$.

(iii) Let $j \in \mathcal{V}_0 \cup \mathcal{V}_1$, $k \in \mathcal{V}_U \cap \mathcal{N}(i)$, and $u_i = u_j = u_k$. Then $\omega_{ij} \leq \omega_{ik}$. If f and g are strictly monotone, then $\omega_{ij} < \omega_{ik}$.

Proof. Since f is symmetric, G is Ω-rotation-invariant. Since g is symmetric, $\omega_{ij} = \omega_{ji}$. Hence (i) is verified. Property (ii) follows from the positivity of f and g. For (iii), w.l.o.g., let $j \in \mathcal{V}_1$. Then $u_j \sim 1$, thus $u_k \sim 1$, and $k \in \mathcal{V}_U$ iff $f(\nabla u_k) > \varepsilon \geq f(\nabla u_j)$. Finally,

$$\omega_{ik} = \delta g\left(f(\nabla u_i), f(\nabla u_k)\right) \geq \delta g\left(f(\nabla u_i), f(\nabla u_j)\right) = \omega_{ij},$$

due to monotonicity of g. If g is strictly monotone, the inequality is also strict. $\qquad \square$

4.3   Stage 2: Connecting Different Components

We now focus on the second stage in the image segmentation algorithm. Let $S \subset \mathcal{V}$ be the set of voxels whose intensities are the M largest components of the intensity vector. Since such hard thresholding does not guarantee any connectivity of S, we have several connected components $S_1 \ldots S_k$. Without loss of generality we may assume that S_1 has cardinality larger than the other components. Let $\widetilde{G}$ be the factor graph, where we consider two vertices equivalent if they lie in one and the same connected component S_j. We perform a *lexicographical breadth first search (LBFS)* [20] in $\widetilde{G}$ and construct the corresponding lexicographical BFS tree rooted at S_1. For the lexicographical BFS we need to introduce ordering of the vertices, which, in our case is by intensity values. This is aimed at minimizing J_0 (or alternatively maximizing J_*, because the BFS tree contains edges between vertices with high intensity. The final step of the algorithm connects each of S_j, $j = 2, \ldots, k$ with S_1 via the tree branches.

Other algorithms for choosing paths between S_1 and the rest of S_j, $j = 2, \ldots, k$ which maximizes J_* can also be used in place of what we propose here.

4.4   Stage 3: Cutting Heavy Leaves

We now discuss some theoretical aspects of the third phase of the algorithm aimed at trimming the connected component from the previous section in order to obtain an image segmentation that satisfies the volume constraint.

Let $u \in \{0, 1\}^N$ be discrete, and S be the (DP) admissible 2-phase segmentation with respect to u. The latter means that $u|_S = 1$, and $u|_{\bar{S}} = 0$. Let f, g be admissible and strictly monotone, and the graph $G = G_\infty$ is build w.r.t. the weights (6). Let S be G-connected, T be a MST of G_S, $l_T \in \mathcal{L}_T$ with $(l_T, j_T) \in \mathcal{E}_T$, and $\omega_{l_T j_T} > 0$. Then

$$l_T \in \partial S := \left\{ i \in S : \begin{array}{l} \mathcal{N}(i) \cap S \neq \emptyset \\ \mathcal{N}(i) \cap \bar{S} \neq \emptyset \end{array} \right\} = S \cap \mathcal{V}_U. \tag{7}$$

Indeed, first of all, $\mathcal{V}_0 = \text{int } \bar{S}$, $\mathcal{V}_1 = \text{int } S$, thus the set equality in (7) holds true.

Now, assume the contrary, i.e., $l_T \in \text{int } S$. Since $l_T \in \mathcal{L}_T \cap \text{int } S$, the MST cut property, applied for $(l_T, S \setminus \{l_T\})$ gives rise to $\min_{j \in \mathcal{N}(l_T)} \omega_{l_T j} = \omega_{l_T j_T} > 0$, and from Lemma 2(ii) it follows that $\mathcal{N}(l_T) \subset \partial S$. For every $j \in \mathcal{N}(l_T) \setminus \{j_T\}$, $(l_T, j) \notin \mathcal{E}_T$, because $l_T \in \mathcal{L}_T$. Due to the strict monotonicity of g and the positivity of f, we have for each $k \in \mathcal{N}(j) \cap (\text{int } S)^c$

$$\omega_{jk} = |u_j - u_k| + \delta g\left(f(\nabla u_j), f(\nabla u_k)\right) > \delta g\left(f(\nabla u_j), 0\right) = \omega_{j l_T}.$$

Applying again the MST cut property this time for $(j, S \setminus \{j\})$, we derive that there exists $k_j \in \mathcal{N}(j) \cap \text{int } S$, $k_j \neq l_T$, and $(j, k_j) \in \mathcal{E}_T$. Thus, we have a $3 \times 3 \times 3$ cube $\mathcal{N}(l_T)$, centered at l_T, all 26 boundary voxels of which are also from the G-boundary of the G-connected set S, while at least 25 of them are also G-connected to other interior points of S (different from l_T!) within the $5 \times 5 \times 5$ cube $\mathcal{N}^2(l_T)$, centered at l_T. It is straightforward to show that there should be at least 6 different external points (one for each 3×3 interior of a side of the cube), and at least 5 internal points (on the side of j_T there may not be one) each two of them at a distance at least 2 in $\| \cdot \|_{\ell^\infty}$. This is impossible. The rigorous proof of (7) is rather elaborate and is beyond the scope of the paper. What we need is a corollary from this result, which we state now.

Proposition 1. *Let $u \in \{0, 1\}^N$ be discrete, S be the admissible 2-phase segmentation with respect to u, f, g be admissible and strictly monotone, and the graph $G = G_\infty$ be build w.r.t. the weights (6). If S is G-connected, then for every MST T of G_S its heaviest leaf l_T belongs to $\mathcal{V}_U$.*

Proof. Let j_T be as before. If $\omega_{l_T j_T} > 0$ the result follows from the arguments above. Assume the contrary, i.e., $l_T \in \text{int } S$. Thus $\omega_{l_T j_T} = 0$ and Lemma 2(ii) implies $j_T \in \text{int } S$, and $\mathcal{L}_T \subset \text{int } S$. Now we aggregate l_T and j_T into a new (super) vertex/voxel l_T^1. We obtain $S^1 = S \cup \{l_T^1\} \setminus \{i \cup j\}$, and $\mathcal{E}_{S^1}$ can be straightforwardly derived from $\mathcal{E}_S$, since l_T and j_T agree on all the "doubled" edges (i.e., $\omega_{l_T k} = \omega_{j_T k}$, $\forall k \in \mathcal{N}(l_T) \cap \mathcal{N}(j_T)$). Due to the MST contraction property, the graph G_{S^1} is G-connected with MST $T^1(S^1, \mathcal{E}_T \setminus \{(l_T, j_T)\}$. Thus, all the leaves $\mathcal{L}_T \setminus \{l_T\}$ remain leaves in T^1 and their weight is preserved as zero, because $\omega_{l_T j_T} = 0$ was the heaviest leaf in T. $l_T^1 \in \text{int } S^1$ may or may not be a leaf in T^1, but since

$$|\text{int } S^1| = |\text{int } S| - 1 < |\text{int } S|,$$

after finitely many contractions m ($m \leq |\text{int } S|$) we will end up with a factor graph G_{S^m}, where $\mathcal{L}_{T^m} \subset \text{int } S^m$ and the heaviest leaf weight is strictly positive. This leaf can appear only after aggregation, thus belongs to $\text{int } S^m$. Contradiction with the arguments above.

4.5   An Algorithm for Constrained Image Segmentation

The steps of the algorithm described above are formally written as follows.

Algorithm 2 (Constraint segmentation) *Input: u (a given intensity), M (volume constraint), f, g (strictly monotone admissible functions), and δ (weight penalizer).*
Output: S (admissible 2-phase segmentation w.r.t. u).

1. ***Stage 1:*** *Compute G_∞ w.r.t. u and remove local maxima and local minima via Algorithm 1.*
2. *Sort the intensity vector u and take the M voxels of highest intensity to be S.*
3. *Find the connected components of S. If S is G-connected, **STOP**.*
4. *Else **Stage 2:** Attach all of the connected components of S to the one with largest cardinality as described in Sect. 4.3 until the union of S_j and the paths between them become G-connected. We denote the union of S_k by S again and move to the next step.*
5. ***Stage 3:*** *Calculate the weights for the graph G_S, using (6) with $f = g = \|\cdot\|_{\ell^1}$, and compute an MST T_S for G_S, using the Kruskal's algorithm [15].*
6. ***While*** *$|S| > M$: Cut the heaviest leaf of T_S.*

4.6   Properties of the Algorithm

Let u be the input intensity (input image). Compute $\mathcal{V}_0, \mathcal{V}_1, \mathcal{V}_U$ for it. Note that, removing local maxima and local minima is just denoising, so for the recomputed sets $\bar{\mathcal{V}}_0, \bar{\mathcal{V}}_1, \bar{\mathcal{V}}_U$ after stage 1 the following inclusions hold true $\mathcal{V}_0 \subseteq \bar{\mathcal{V}}_0, \mathcal{V}_1 \subseteq \bar{\mathcal{V}}_1$. Denote by C_1 the minimal G_∞-connected set that contains $\bar{\mathcal{V}}_1$. Let S_M be the set from step 3. We say that u is *admissible*, if $\mathcal{N}(\bar{\mathcal{V}}_1) \subseteq S_M$, $|\mathcal{N}(\bar{\mathcal{V}}_1) \cup C_1| \leq M$, and $\mathcal{N}(\bar{\mathcal{V}}_0) \subseteq \bar{S}_M$. If not, it is clear that either the parameter choices in $\mathcal{V}_1$ or $\mathcal{V}_0$ were poor or the constraint parameter M approximates badly the solid phase volume.

For admissible u, Proposition 1 implies that we cut only $\mathcal{V}_U$ voxels in stage 3, thus $\bar{\mathcal{V}}_1 \subseteq S$ at every moment. Moreover, due to Lemma 2(ii) the heaviest leaf weight is strictly positive. No $i \in \bar{\mathcal{V}}_0$ belongs to any shortest path between the S components, thus after stage 2 $\bar{\mathcal{V}}_0 \subseteq \bar{S}$. Since in step 6. We only cut, the inclusion remains true for the output image, as well. Finally, since we always cut out leaves from T_S, after stage 2 till the end S is always G_∞-connected. To summarize:

Theorem 3. *For any admissible input image u, Algorithm 2 terminates and produces an output 2-phase segmentation S that is G_∞-connected, has cardinality M, fully contains $\mathcal{V}_1$, and doesn't intersect with $\mathcal{V}_0$. The complexity of the algorithm is $\mathcal{O}(N \log N)$.*

5   A Numerical Test

In this section we assess the performance of our Algorithm 2 on a part of an image of a trabecular bone. The image is taken from [2], then convoluted with a Gaussian kernel with $\sigma = 2$ (i.e., the image is blurred), and $10\,\%$ white (Gaussian) noise is added to derive the input image u. The bone part image has size $64 \times 64 \times 64$. 50604 of its voxels are bone material (porosity $80.7\,\%$), thus $M = 50604$. Figure 1 summarizes the results. The most left image is of the original discretized bone. The second one is the result of direct segmentation, where the M voxels of highest intensity are taken as the solid phase. The third one is the output of our Algorithm 2. The last one is the output of the segmentation in [11], based on fully constrained convex ℓ^2-norm minimization.

Fig. 1. From left to right: Segmented bone part (binary image), direct M-segmentation of the noisy and blurry version u, connected M-segmentation via Algorithm 2, segmentation from [11].

The direct M-segmentation is quite noisy. It consists of lots of 1-element components, as well as other larger ones. Unlike it, our segmentation is G-connected. There are still some 1-voxel-wide branches, due to small noisy components in the set S_3 at step 3., which have been aggregated to the main component C_0 in stage 2 (see Fig. 2). The result of the segmentation in [11] lacks any noise, because of the smoothing role of the edge weights there, but is not G-connected and consists of three different components, thus it is not admissible with respect to Definition 5.

Fig. 2. From left to right: Segmented bone part (binary image), the set S after stage 2, the set of cut leaves in stage 3, and the final result of Algorithm 2.

Note that none of the S_3 connected components is of cardinality 1, due to the removing of local maximums. During the leaf cutting, some of those noisy branches have been erased, but some of them remain in the result. The reason is the usage of only voxels' intensity values throughout steps 1–4 of the algorithm, thus no regularization has been applied in the process, and the connected components of S_3 are not as "homogeneous" as they should be when the gradient is taken into account in the expanding process. Minimizing the functional J_1 instead of J_0 in step 4 should improve the quality of the result and is a subject of future work. We want to point out that out of 234 cut leaves in stage 3, only 5 of them belong to the actual bone, and the remaining 229 are indeed noise. The ℓ^1 difference of our result with the original bone is 20 844, which is larger than the 15 524 difference of the result in [11], but is by almost a thousand better than the difference of the direct M-segmentation (which is 21 796, as computed in [11]). The former means that there is plenty of room for improvement (e.g., possible combination of the two constraint algorithms, "thickening" the minimal paths, "homogenizing" the connected components, etc.), while the latter indicates that by simply removing local extrema and only replacing 234 candidate voxels with another, better group of 234 voxels, we already gain a lot.

6   Conclusions and Future Work

We proposed and tested a class of algorithms for *constrained image segmentation*. The algorithms are based on the minimization of suitable functionals measuring the best approximation of the input image within the space of step functions. For the output image, the approximate segmentation produced by the algorithm has a connected solid phase with fixed volume. Such type of algorithms, and especially multilevel versions of these algorithms, show potential to be robust tools in the image analysis.

Acknowledgments. The research is supported in part by the project AComIn "Advanced Computing for Innovation", grant 316087, funded by the FP7 Capacity Program. The research of Ludmil Zikatanov is supported in part by NSF DMS-1217142 and NSF DMS-1418843.

References

1. Amghibech, S.: Eigenvalues of the discrete p-Laplacian for graphs. Ars Combinatoria **67**, 283–302 (2003)
2. Beller, G., Burkhart, M., Felsenberg, D., Gowin, W., Hege, H.-C., Koller, B., Prohaska, S., Saparin, P., Thomsen, J.: Vertebral Body Data Set ESA29-99-L3. http://bone3d.zib.de/data/2005/ESA29-99-L3/
3. Bezdek, J., Ehrlich, R., Full, W.: FCM: the fuzzy c-means clustering algorithm. Comput. Geosci. **10**(2–3), 191–203 (1984)
4. Bresson, X., Esedoglu, S., Vandergheynst, P., Thiran, J., Osher, S.: Fast global minimization of the active contour/snake model. J. Math. Imaging Vis. **28**(2), 151–167 (2007)

5. Bühler, T., Hein, M.: Spectral clustering based on the graph p-Laplacian. In: Proceedings of the 26th Annual International Conference on Machine Learning, pp. 81–88 (2009)
6. Cai, X., Steidl, G.: Multiclass segmentation by iterated ROF thresholding. In: Heyden, A., Kahl, F., Olsson, C., Oskarsson, M., Tai, X.-C. (eds.) EMMCVPR 2013. LNCS, vol. 8081, pp. 237–250. Springer, Heidelberg (2013)
7. Chan, T., Esedolgu, S., Nikolova, M.: Algorithms for finding global minimizers of image segmentation and denoising models. SIAM J. Appl. Math. **66**(5), 1632–1648 (2006)
8. Chan, T., Vese, L.: Active contours without edges. IEEE Trans. Image Process. **10**(2), 266–277 (2001)
9. Dong, B., Chien, A., Shen, Z.: Frame based segmentation for medical images. Commun. Math. Sci. **32**, 1724–1739 (2010)
10. Felzenszwalb, P.F., Huttenlocher, D.P.: Efficient graph-based image segmentation. Int. J. Comput. Vis. **59**(2), 167–181 (2004). http://dx.doi.org/10.1023/B:VISI.0000022288.19776.77
11. Georgiev, I., Harizanov, S., Vutov, Y.: Supervised 2-phase segmentation of porous media with known porosity. In: 10th International Conference on Large-Scale Scientific Computations (2015, accepted)
12. He, Y., Shafei, B., Hussaini, M.Y., Ma, J., Steidl, G.: A new fuzzy c-means method with total variation regularization for segmentation of images with noisy and incomplete data. Pattern Recogn. **45**, 3436–3471 (2012). http://dx.doi.org/10.1016/j.patcog.2012.03.009
13. Kang, S.H., Shafei, B., Steidl, G.: Supervised and transductive multi-class segmentation using p-Laplacians and RKHS methods. J. Vis. Commun. Image Represent. **25**(5), 1136–1148 (2014)
14. Kraus, J.K.: An algebraic preconditioning method for M-matrices: linear versus non-linear multilevel iteration. Numer. Linear Algebra Appl. **9**(8), 599–618 (2002). http://dx.doi.org/10.1002/nla.281
15. Kruskal, J.: On the shortest spanning subtree of a graph and the traveling salesman problem. Proc. Am. Math. Soc. **7**(1), 48–50 (1956)
16. Law, N.Y., Lee, H.K., Ng, M.K., Yip, A.M.: A semisupervised segmentation model for collections of images. IEEE Trans. Image Proc. **21**(6), 2955–2968 (2012)
17. von Luxburg, U.: A tutorial on spectral clustering. Stat. Comput. **17**(4), 395–416 (2007)
18. Mumford, D., Shah, J.: Optimal approximations by piecewise smooth functions and associated variational problems. Commun. Pure Appl. Math. **42**(5), 577–685 (1989)
19. Pock, T., Cremers, D., Bischof, H., Chambolle, A.: An algorithm for minimizing the Mumford-Shah functional. In: Proceedings of IEEE 12th Conference Computer Vision, pp. 1133–1140 (2009)
20. Rose, D.J., Tarjan, R.E., Lueker, G.S.: Algorithmic aspects of vertex elimination on graphs. SIAM J. Comput. **5**(2), 266–283 (1976)
21. Rudin, L., Osher, S., Fatemi, E.: Nonlinear total variation based noise removal algorithms. Physica D **60**(1–4), 259–268 (1992)
22. Shi, J., Malik, J.: Normalized cuts and image segmentation. In: Proceedings of the IEEE Conference on Computer Vision and Pattern Recognition, pp. 731–737 (1997)
23. Shi, J., Szalam, J.: Normalized cuts and image segmentation. IEEE Trans. Pattern Anal. Mach. Intell. **22**(8), 888–905 (2000)

24. Urquhart, R.: Graph theoretical clustering based on limited neighborhood sets. Pattern Recogn. **15**(3), 173–187 (1982)
25. Vese, L., Chan, T.: A multiphase level set framework for image segmentation using the Mamford and Shah model. Int. J. Comput. Vis. **50**(3), 271–293 (2002)
26. Weiss, Y.: Segmentation using eigenvectors: a unifying view. Proc. Int. Conf. Comput. Vis. **2**, 975–982 (1999)
27. Wu, Z., Leahy, R.: An optimal graph theoretic approach to data clustering: theory and its application to image segmentation. IEEE Trans. Pattern Anal. Mach. Intell. **11**, 1101–1113 (1993)
28. Zahn, C.T.: Graph-theoretic methods for detecting and describing gestalt clusters. IEEE Trans. Comput. **20**, 68–86 (1971)
29. Zhang, Y., Matuszewski, B., Shark, L., Moore, C.: Medical image segmentation using new hybrid level-set method. In: BioMedical Visualization, MEDIVIS 2008, pp. 71–76 (2008)

On Computer Simulation of Fluid-Porous Structure Interaction Problems for a Class of Filtration Problems

Oleg Iliev$^{(\boxtimes)}$, Dimitar Iliev, and Ralf Kirsch

Fraunhofer ITWM, Fraunhofer-Platz 1, 67663 Kaiserslautern, Germany
`oleg.iliev@itwm.fraunhofer.de`

Abstract. Fluid-Porous Structure Interaction, FPSI, problem is first formulated and discussed in 3D in connection with modeling of flow processes in pleated filters. Solving the 3D problem is computationally expensive, therefore for a subclass of problems reduced model is considered, namely, the 3D poroelasticity problem is approximated by a poroelastic shell. Because resolving the geometry of a pleat is very important for obtaining accurate solution, interface fitted general quadrilateral grid is introduced. It is difficult to generate good quality grid in such complicated domains, therefore a discretization approach, which is robust on rough grids is selected, namely, multipoint flux approximation method. The coupled FPSI problem is solved with sequential approach, what allows to reuse an existing flow solver. Results from numerical simulations are presented and discussed.

Keywords: Fluid-porous structure interaction · Poroelasticity · Industrial filtration problem · Pleated filter · Thin porous media · Deflection

1 Introduction and Motivation

Fluid-structure interaction, FSI, is considered to be an essential class of multiphysics problems, the latter in general requiring significant computational resources and sophisticated algorithms. While a lot of research is dedicated in the last decade to developing and analyzing algorithms for FSI problems in the case of non-permeable structures, still very little is done in the case of porous structures. In [1] Richer describes in details the different ways to solve the multiphysics system of equations in the case of FSI. One of the main difficulties is that a part of the equations is formulated in Lagrangian coordinate system, while the other part is formulated in Eulerian one. Even if the equations describing the flow and the displacements are linear, the FSI problem is nonlinear due to the fact that the shape of the structure in the FSI problems is a part of the solution, and is not given a priori. For small size problems one can write a monolithic scheme and use a direct method at each iteration on nonlinearity. For larger problems and for cases when the coupling is not very strong, partitioned (sequential) approach might be preferable. One can in such a case reuse existing solvers for the

© Springer International Publishing Switzerland 2015
I. Lirkov et al. (Eds.): LSSC 2015, LNCS 9374, pp. 30–41, 2015.
DOI: 10.1007/978-3-319-26520-9_3

Fig. 1. Industrial motor oil filter for a car

elastic and for the flow subproblems, and impose the interface conditions via iterations between the subproblems. At the same time the sequential algorithms in general do not inherit the robustness of the monolithic approach. In general, the complexity of the FSI problems does not allow to develop algorithms which are at the same time very efficient, and applicable to a wide class of FSI problems. In this situation, our approach is to concentrate on a class of FSI problems and to develop a customized algorithm which will solve efficiently mathematical models describing this class of problems.

Such a mathematically challenging and practically relevant class of problems is filtration of solid particles out of fluid. Simulation of filtration processes helps designers and manufacturers of filter elements (e.g., like the one on Fig. 1) in the acceleration of the design process and in the optimization of the designed filters. In the case of non-deformable porous media, filtration-adapted algorithms and simulation software have proven its worth in industrial applications for years, see e.g. [2] and references therein. However, in many cases the deflection of the filtering medium can not be ignored.

Computer simulation of FPSI filtration problem requires the numerical solution of a complex multiphysics problem including flow through plain and porous regions as well as deformation of a porous structure. In this article we deal with simulation of pleated filters, in particular accounting for the pleat deflection.

2   Mathematical Model

Let us first discuss the 3D models for coupled fluid flow in a plain and in deformable porous media, coupled with equations describing the deformation of the porous media. Let us denote by Ω_p the region occupied by the filtering medium in non-deformed configuration, by Ω_f the region occupied by the pure fluid in non-deformed configuration, and by $\tilde{\Omega}_p$, $\tilde{\Omega}_f$ the respective domains in the deformed configuration. Further on, $\Omega = \Omega_p \cup \Omega_f = \tilde{\Omega}_p \cup \tilde{\Omega}_f$ is the total computational domain (which remains unchanged in our case), and $\partial\Omega$ is its (external) boundary. Finally, $\partial\tilde{\Omega}_{pf}$ stands for the interface between the plain and the porous media in deformed configuration. In order to write the interface conditions on $\partial\tilde{\Omega}_{pf}$ we will use Ω_f also as a reference domain. Further on, let $T : \Omega_f \to \tilde{\Omega}_f$ be a C^2-diffeomorphism mapping Ω_f to $\tilde{\Omega}_f$ and $F = \nabla T$.

In this paper we denote vectors with bold letters, unknowns defined in Lagrangian coordinates with tilde, unknowns defined in Eulerian coordinates without tilde. Unknowns defined in the plain fluid region and porous region are equipped with subscripts f and p, respectively. The following assumptions hold for the practical problem we are interested in, and are accounted for the model selection. The flow is single phase, incompressible and steady state. The filtering media is thin and it is an isotropic porous media (the permeability is just a scalar and not a tensor). No contact problems (e.g., no change in the topology of the computational domain) are considered at this stage. The so called dead end filtration is considered. In this case all the fluid to be filtered goes through the porous media (unlike the case of cross flow filtration when part of the flow is parallel to the filtering medium and never crosses it). The pore size is significantly smaller than the thickness of the media and thus homogenized (macroscale) models can be used to describe the flow through the porous media and its deformation. The flow is laminar and not very fast, so that Darcy law holds for the flow through porous media.

Flow Through Plain Region. In the plain fluid region the incompressible Navier-Stokes equations are used to describe the fluid motion. Because we are looking for an equilibrium steady state solution of the considered FPSI problem, the equations are written in deformed configuration.

$$(\tilde{\mathbf{v}}_f \cdot \nabla)\,\tilde{\mathbf{v}}_f - \nabla \cdot (\mu\nabla\tilde{\mathbf{v}}_f) = -\nabla\tilde{p}_f, \quad \nabla \cdot \tilde{\mathbf{v}}_f = 0 \ \text{ in } \tilde{\Omega}_f. \tag{1}$$

The notations $\tilde{\mathbf{v}}_f$ and $\tilde{p}_f$ stand for the fluid velocity and pressure in the region $\tilde{\Omega}_f$, respectively. With μ we denote the dynamic viscosity of the fluid.

Flow Through Rigid Porous Media. Before describing models for flow through deformable media, we first describe models for flow through rigid porous media ($\tilde{\Omega}_p \equiv \Omega_p$). Under the assumptions formulated above, the flow through saturated porous media is usually described by the Darcy equation coupled with continuity equation. In the case of highly porous filtering media and/or when Dirichlet boundary conditions for the velocity have to be imposed, the Brinkman system of equations (2) is used (for discussion on the Brinkman model in conjunction with simulation of filtration problems see [3]).

$$-\nabla \cdot (\mu\nabla\tilde{\mathbf{v}}_p) + \mu K^{-1}\tilde{\mathbf{v}}_p = -\nabla\tilde{p}_p, \quad \nabla \cdot \tilde{\mathbf{v}}_p = 0 \ \text{ in } \tilde{\Omega}_p \equiv \Omega_p, \tag{2}$$

The notations $\tilde{\mathbf{v}}_p$ and $\tilde{p}_p$ stand for the effective fluid velocity and effective pressure in the porous media, respectively. Note, that in some cases a value $\bar{\mu}$ different from μ (the fluid viscosity) is used for the effective viscosity of the fluid in the porous region. However our studies [3] show that the use of effective viscosity $\bar{\mu} = \mu$ is preferable for filtration problems. K stands for the permeability of the porous media. It should be noted that the term μK^{-1} usually is very large (the permeability is often of order 10^{-10}). Therefore the above Brinkman equation (2) can be considered as a perturbation of the Darcy equation, where the viscous terms are omitted.

Deformable Porous Media: Fluid Flow and Poroelasticity. Biot [4,5] derived the consolidation theory of a poromechanics coupling linear elasticity with the Darcy equations for a flow in a porous media. We have used this model in simulation of deflection of flat filter media, and the comparison with experiments has shown that it can be used (with certain attention) for simulation of filtration processes [6]. The quasi steady state Biot equation reads

$$\sigma_p = 2\mu e\left(\mathbf{u}_p\right) + \left(\hat{\lambda} \operatorname{div} \mathbf{u}_p - \alpha p_p\right) \mathrm{I}, \tag{3}$$

$$-\operatorname{div} \sigma_p = -\mu \triangle \mathbf{u}_p - \left(\hat{\lambda} + \hat{\mu}\right) \nabla \operatorname{div} \mathbf{u}_p + \alpha \nabla p_p = 0 \text{ in } \Omega_p, \tag{4}$$

$$\frac{\partial}{\partial t}\left(\beta p_p + \alpha \operatorname{div} \mathbf{u}_p\right) - \frac{K}{\mu} \triangle p_p = 0 \text{ in } \Omega_p. \tag{5}$$

We present here Biot equations in this form because often they appear in the literature in this way. In the steady state case considered here the time derivatives drop. It is important to note that this system describes the deformation in the (porous) solid body due to the flow through it as well as the effects of the movement of the solid structure to the fluid flow. With σ_p we denote the stress tensor, $\mathbf{u}_p$ is the vector field of effective displacements in the porous region and p_p is the effective pore pressure, $\hat{\lambda}$ and $\hat{\mu}$ are respectively the first and second Lamé constants, α is the effective stress coefficient and β is the inverse of the Biot's modulus.

Interface Conditions. In order to ensure conservation of mass and conservation of momentum we impose as interface conditions continuity of the normal components of the fluid velocity and of the stress tensor on the deformed interface $\partial \tilde{\Omega}_{pf}$. In the case of rigid porous media we are using these conditions since many years, see, e.g., [2,3] for details. As we consider dead end filtration in media with very small permeability, the velocity near the interface is perpendicular to the interface. Thus the fluid velocity on the interface is equal to its normal component. Using the inverse of the mapping T we can write the interface conditions in Lagrangian coordinates. We use the notations $\mathbf{v}_f = T^{-1}\left(\tilde{\mathbf{v}}_f\right)$, $p_f = T^{-1}\left(\tilde{p}_f\right)$ and $\sigma_f = T^{-1}\left(\tilde{\sigma}_f\right)$ to denote the fluid velocity, pressure and stress tensor transformed in Lagrangian coordinates.

$$-\frac{K}{\mu}\nabla p_p = \mathbf{v}_f \text{ on } \partial\Omega_{pf}, \tag{6}$$

$$\sigma_p \cdot \mathbf{n} = \sigma_f \cdot \mathbf{n} \text{ on } \partial\Omega_{pf}. \tag{7}$$

By $\mathbf{n}$ we denote the vector normal to $\partial\Omega_{pf}$ in the direction of the porous media. The fluid stress tensor in Lagrangian coordinates has the form

$$\sigma_f = p_f \mathrm{I} + \mu\left(\nabla \mathbf{v}_f F^{-1} + F^{-T} \nabla \mathbf{v}_f{}^T\right). \tag{8}$$

Full 3D Model. To summarize, the full 3D model of the flow consist of the Navier-Stokes equations (1) in the plain fluid region $\tilde{\Omega}_f$, the Biot system of equations (4), (5) in the porous media Ω_p, the interface conditions (6), (7) on $\partial\Omega_{pf}$ and respective boundary conditions on $\partial\Omega$.

3   Approximate Models

As we are interested in the equilibrium state of the system (i.e., the time derivative in (5) drops), we can analytically determine what type of function is the effective pressure p_p. To illustrate this let us assume that the porous media is initially flat and horizontal in an orthonormal coordinate system $Oxyz$. As we assume for the media to be thin, its z dimension is small. We denote this by $z = \varepsilon\bar{z}$. We can then write the Eq. 5 in the following way

$$\frac{\partial^2 p_p}{\partial x^2} + \frac{\partial^2 p_p}{\partial y^2} + \frac{1}{\varepsilon^2}\frac{\partial^2 p_p}{\partial\bar{z}^2} = 0 \text{ in } \Omega_p. \tag{9}$$

In the asymptotic analysis we neglect all of the terms of order ε^2 which leads to p_p being linear function in z-direction. In the general case p_p is linear in direction which is normal to the middle surface of the porous media. This can be shown by rewriting Eq. (5) in local coordinates. Having the p_p being linear in the direction normal to its middle surface it could be analytically calculated from the pressure p_p on the interface $\partial\Omega_{pf}$, if the latter would be known. In any case, this leads to decoupling of the flow and the elasticity parts in the Biot's equation. We use this decoupling to write an approximation of the full 3D model which could be solved numerically significantly more efficient.

Flow in Plain and in Rigid Porous Media: Navier-Stokes-Brinkman Model. As described above we use the incompressible Navier-Stokes system of equations to describe the flow in the plain region $\tilde{\Omega}_f$. In the case of non-deformable porous media and dead end filtration (which is considered here), for slow flow and media with very low permeability, within the porous media the flow is locally perpendicular to the midsurface of the porous media. In this case, similarly to the consideration for Biot equation, we can show that the pressure is a linear function along the normal to the midsurface. The easiest way to conclude this is to drop the Brinkman terms and consider Darcy law. For this reason instead of solving separately the Navier-Stokes system of equations in $\tilde{\Omega}_f$ and the pressure equation in Ω_p we can use the Navier-Stokes-Brinkman model to describe the flow problem in the full domain of interest $\Omega = \tilde{\Omega}_f \cup \tilde{\Omega}_p$. By doing this we fully separate the flow from the elasticity in the full 3D system, and the coupling remains on the interface only.

Poroelasticity: Shell Model. Solving the 3D Biot model requires significant computational efforts. In many of the industrial oil and air filters the filtration media is thin as on Fig. 1. Therefore the porous structure can be also modeled as poroelastic plate or shell, thus substantially decreasing the required computational time. Heuristic derivation of poroelastic plate model can be found in

[7,8]. Mathematically rigorous derivation of poroelastic plate model, based on asymptotic homogenization method, can be found in [9]. In previous studies we have validated the poroelastic plate model in comparison with an experiment (see, e.g., [6]) and have used it in simulation of filtration processes in the case of flat porous media. A rigorous derivation of poroelastic flexural three-dimensional shell model from the Biot's system through asymptotic analysis was done recently by Mikelić and Tambača in [10].

However as we are interested in the equilibrium state of the system and the elastic part from the Biot's system decouples form the flow part, we are free to choose a different type of shell model to represent the displacements in the porous media. Instead of the flexural shell we use a Naghdi type of shell described in details by Zang in [11]. This type of shell describes not just the normal and tangential displacements w and u but also the rotation θ of the line perpendicular to the middle surface. Without the Kirchoff hypothesis of small rotation θ this model could be used for a wider area of applications. In [11] one can find the weak formulation of the model. We use the strong form of the model given by the following system of equations

$$\epsilon^2 \left(-\theta'' - b'u' - bu'' + 2bb'w + b^2w'\right) + \frac{5}{6}\left(\theta + bu + w'\right) = \frac{\epsilon^2}{6\hat{\mu}}p_0^2 b', \tag{10}$$

$$\epsilon^2 \left(-b'\theta' - b\theta'' - 2bb'u' - b^2u'' + 3b^2b'w + b^3w'\right) + 3\left(-u'' + b'w + bw'\right)$$
$$+ \frac{5}{6}\left(b\theta + b^2u + bw'\right) - \frac{\epsilon^2}{6\hat{\mu}}\left(b'p_e^2 + 2bb'p_0^2\right), \tag{11}$$

$$\epsilon^2 \left(-b^2\theta' - b^3u' + b^4w\right) + 3\left(-bu' + b^2w\right) + \frac{5}{6}\left(-\theta' - b'u - bu' - w''\right) =$$
$$= \frac{1}{\hat{\mu}}\left(\frac{p_0^2 b}{2} + p_e^2 - \frac{b^2\epsilon^2 p_e^2}{6} - \frac{b^3\epsilon^2 p_0^2}{6}\right). \tag{12}$$

With b we denote the curvature of the middle surface of the porous media and with ϵ its thickness. To define the boundary conditions for the poroelastic displacements p_0^2 and p_e^2 we distinguish between the interfaces in the inlet and outlet region $\partial\Omega_{pf} = \partial\Omega_{pf+} \cup \partial\Omega_{pf-}$. We use the notations $\sigma_{f+} = \sigma_f$ on $\partial\Omega_{pf+}$, $\sigma_{f-} = \sigma_f$ on $\partial\Omega_{pf-}$, $\mathbf{n}_+ = \mathbf{n}$ on $\partial\Omega_{pf+}$ and $\mathbf{n}_- = \mathbf{n}$ on $\partial\Omega_{pf-}$. Let $\mathbf{t}_+$ and $\mathbf{t}_-$ be unit vectors perpendicular to $\mathbf{n}_+$ and $\mathbf{n}_-$ respectively. Also let the couples $\mathbf{n}_+, \mathbf{t}_+$ and $\mathbf{n}_-, \mathbf{t}_-$ be positively oriented such that they define a contravariant basis. Let us write the normal component of the stress tensor in local coordinates on the inlet and outlet interfaces.

$$\sigma_+ \cdot \mathbf{n}_+ = f_n^+ \frac{1}{1 - b\epsilon}\mathbf{n}_+ + f_t^+ \frac{1}{1 - b\epsilon}\mathbf{t}_+, \tag{13}$$

$$\sigma_- \cdot \mathbf{n}_- = f_n^- \frac{1}{1 + b\epsilon}\mathbf{n}_- + f_t^- \frac{1}{1 + b\epsilon}\mathbf{t}_-. \tag{14}$$

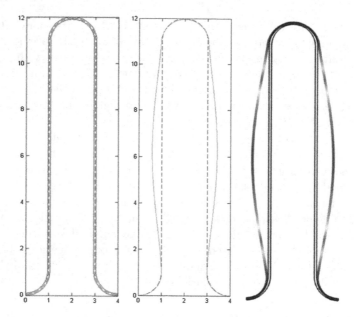

Fig. 2. Initial state of a single pleat, deflection given by the shell model and deflection given by the Biot's system

Using the local coordinates we define p_0^2 and p_e^2 in the following way

$$p_0^2 = \frac{f_n^+ - f_n^-}{2}, p_e^2 = \frac{f_n^+ + f_n^-}{2\epsilon}. \tag{15}$$

Numerical Validation. The use of plate and shell models is only applicable to geometries where one of the dimensions of the porous media is substantially smaller comparing to the others. In order to test the limitations of the poroelastic plate and shell models we have performed multiple numerical simulations testing the shell model against the 3D Biot's system for different values of the thickness of the porous media. The results of these simulations show that as long as the thickness of the porous media is under 4% of the other physical dimensions, the shell model gives solutions as accurate as the grid allows for. Using fine grid one can achieve relative difference between the reduced (shell) and the full (3D) models less then 10^{-6}. On Fig. 2 we show a comparison between the shell model and Biot's system for one case of thin media.

4   Numerical Algorithm

As described above instead of solving the full coupled 3D system of the Navier-Stoks system of equations, Biot's system, the interface and boundary conditions, we solve separately the Navier-Stokes-Brinkman system of equations and the poroelastic shell model and couple them via the interface conditions (e.g., via the shape of the pleats). The interface conditions are imposed by iterating between the two problems until convergence is achieved.

Solving the Coupled System. To describe more accurately the iterative algorithm we make use of a simulation of a single pleat from a pleated filter cartridge (Fig. 6). The blue and red colors represent respectively the plain fluid region Ω_f and the porous media region Ω_p, respectively. The left edge of the rectangular domain is an inlet and the right one is an outlet. On the top and bottom edges symmetry conditions are imposed for the flow and no displacement for the porous media. Starting with this initial shape we first freeze the porous media (i.e., consider it rigid for a while) and calculate the fluid pressure (Fig. 7) and velocity (Fig. 8) solving the Navier-Stokes-Brinkman system of equations (2). After this first step we calculate the stress σ_f. As we are interested in dead-end filtration with very low permeability of the porous media and relatively slow flow, the pressure is in orders of magnitude larger than the velocity derivatives in (8) and the stress tensor can be approximated by $\sigma_f \approx p_f I$. We set the normal component of σ_f as force acting on the elastic body. After calculating the displacements on the middle surface using the shell model (10), (11), (12), we reconstruct the porous region after the deflection and update the plain fluid region and the porous media region, $\hat{\Omega}_f$ and $\hat{\Omega}_p$ respectively (see Fig. 9). On the updated domain $\hat{\Omega}_f \cup \hat{\Omega}_p$ we deform respectively the grid and repeat the described procedure. As mentioned above, we are not interested in solving contact problems, and under our assumptions no topological changes in the computational domain occur. We continue with obtaining the fluid flow and the pressure in the updated domain, calculate the stress tensor, the elastic deformation and repeat this until convergence is achieved.

Discretization. For the Navier-Stokes-Brinkman system of equations we use the Finite Volume Method. To accurately capture the shape of the filtering media in its initial state, as well as after the deformation, we use unstructured quadrilateral grids which are fitted to the porous media. To ensure a robust discretization of the equations with large jump of the permeability on complex (rough) grids, we use multipoint flux-approximation method, MPFA, which was originally proposed for a scalar equation (see, e.g., [12]) and which we adopted for Stokes-Brinkman problem (see [13]). The discretized Navier-Stokes-Brinkman problem is solved by a variant of the Chorin projection method. The linear systems at each iteration are solved with a robust algebraic multigrid method available in a commercial software.

To discretize the system of equations (10), (11) and (12) describing the shell model, we use finite difference method. To solve the linear system of equations we use a direct solver.

Computational Grids. Two issues are important in selecting a meshing approach. The first one is the fact that in dead end filtration problems the main pressure jump occurs within the porous media, therefore resolving accurately the geometry of the filtering media is important. To tackle with this, as mentioned above, we use quadrilateral grid fitted to the plain-porous media interface. The second issue is that the solution of FSI problems in many cases requires remeshing the computational flow domain, what could take more computational time than the solution of the discretized system. For the physical systems of interest here, we have developed a very fast technique to update the computational grid with no need of remeshing. It relies on the fact that no change in the topology occurs for the regimes which we consider. Of course, the quality of the grid may become worse during deformation, therefore we have selected a robust discretization approach which works well on skewed and stretched grids.

5   Numerical Experiments

We have performed numerous simulations to ensure grid convergence for variety of physical parameters and different domains. We would like to focus on one particularly interesting result of these studies. As we focus on media with small permeability and slow flow, we are in Darcy regime. Therefore we can then compare the numerically obtained pressure drop with the analytically predicted Darcy pressure drop. We do this for a case when the pleats are not very narrow and the Darcy pressure drop dominates. For narrow pleats the pressure drop along the channel can be comparable to the Darcy one. The use of the multi-point flux approximation and interface capturing grids allows us to obtain accurate solutions using very few grid elements. On Figs. 3, 4 and 5 we show the numerically obtained solution for the pressure, starting with a grid having 200 grid elements along the straight part of a pleat, and using less and less grid elements. Even for the case of only three elements along the straight side of the pleat, the relative difference between the numerical simulations and the analytical solution is only 0.32 %. In this case we have used only 308 quadrilateral elements to discretize the full domain. Numerical tests performed with another software tool show that around 10 000 elements on uniform Cartesian grids are needed in order to achieve the same accuracy.

Fig. 3. Pressure drop though single pleat when using 200 grid elements along the straight part

For the case of initially flat porous media the numerical method was also tested against physical experiments in [6]. The numerical simulations gave very accurate predictions of the displacement of the porous media. This is another evidence for the accuracy of the modeling and numerical techniques used here.

Deflection Effects. To demonstrate how even small deflection can significantly change the flow within a filter element, we set a numerical simulation of the

Fig. 4. Pressure drop though single pleat when using 20 grid elements along the straight part

Fig. 5. Pressure drop though single pleat when using 3 grid elements along the straight part

deflection of a single pleat (Fig. 6). For the pleat geometry we chose the length of the straight part of the pleat to be 20 mm, the thickness of the pleat to be 0.5 mm and the distance between the two straight parts to be 0.8 mm. The left edge of the domain is an inlet with prescribed fluid velocity of 300 mm/s in horizontal direction for every point. The right edge of the domain is the outlet with prescribed pressure of 0 Pa. On the top and bottom edges of the region we set symmetry boundary conditions. For the dynamic viscosity of the fluid we set $\mu = 3.31e^{-6}$ kg.mm^{-1}.s^{-1}. For the permeability of the porous region $\tilde{\Omega}_p$ we use $K = 3.4e^{-5}$.

Fig. 6. Initial shape of a single pleat

As described in Sect. 4 we first calculate the pressure (Fig. 7) and velocities (Fig. 8) in the initial shape of the porous media. Next we use the computed pressure values to impose the forces acting on the porous region and to calculate the deflection of the elastic body.

After we have calculated the displacements on the middle surface of the porous media and the rotations θ, we can reconstruct the position and the shape of the filtering media after the deflection (Fig. 9). At this stage we also update the computational grid.

Having the updated computational grid we can now calculate the pressures (Fig. 10) and velocities (Fig. 11) in $\tilde{\Omega}_f$ and $\tilde{\Omega}_s$. On Figs. 8 and 11 we use the same color ranges to represent the magnitude of the velocity. Due to the deflection of the pleat, in some regions the fluid flow is now faster and in others - slower. This illustrates how even small deflection of the porous material can lead to significant changes in the fluid flow, and as a result, in the efficiency of a filter

Fig. 7. Pressure in the initial state

Fig. 8. Velocity magnitude in the initial state

Fig. 9. The shape of the single pleat after displacement

Fig. 10. Pressure after displacement

Fig. 11. Velocity magnitude after displacement

element. In this particular case, however, the pressure drop through the porous media does not change significantly after the deflection of the pleat. This leads to very fast convergence of the iterative scheme used to solve the system of fluid flow equations and elastic deformation equations.

6   Conclusion

The numerical results above show how the use of general quadrilateral grids, advanced discretization techniques like the MPFA and advanced solvers like the algebraic multigrid lead to robust and accurate solutions of the system of equations describing fluid interacting with a porous structure. The use of appropriate approximated models, like the shell ellipticity model, instead of more general models even further increases the efficiency of the simulation. By using the techniques presented one can simulate not only the behavior of a single pleat, as demonstrated in the numerical experiments, but also deflection of wide variety of thin filtering media during dead end filtration.

References

1. Richter, T.: Numerical Methods for Fluid-Structure Interaction Problems, Universität Heidelberg (2010)
2. Iliev, O., Kirsch, R., Lakdawala, Z., Rief, S., Steiner, K.: Modeling and simulation of filtration processes, In: Neunzert, H., Prätzel-Wolters. D. (eds.) Currents in Industrial Mathematics. Springer, Heidelberg (2015, to appear). ISBN 978-3-662-48257-5
3. Iliev, O., Laptev, V.: On numerical simulation of flow through oil filters. Comput. Visual. Sci. **6**, 139–146 (2004)
4. Biot, M.: Theory of elasticity and consolidation for a porous anisotropic solid. J. Appl. Phys. **26**, 182–185 (1955)
5. Biot, M.: Theory of stability and consolidation of a porous medium under initial stress. J. Math. Mech. **12**, 521–542 (1963)
6. Grosjean, N., Iliev, D., Iliev, O., Kirsch, R., Lakdawala, Z., Lance, M., Michard, M., Mikelić, A.: Experimental and numerical study of the interaction between fluid flow and filtering media on the macroscopic scale. Sep. Purif. Technol. (2015, in press)
7. Taber, L.: A theory for transverse deflection of poroelastic plates. J. Appl. Mech. **59**, 628–634 (1992)
8. Taber, L., Andrew, P.: Poroelastic plate and shell theories. In: Selvadurai, A.P.S. (ed.) Mechanics of Poroelastic Media, pp. 323–337. Springer, Dordrecht (1996)
9. Marciniak-Czochra, A., Mikelić, A.: A Rigorous Derivation of the Equations for the Clamped Biot-Kirchhoff-Love Poroelastic plate, arXiv:1211.6456 (2012)
10. Mikelić A, Tambača J (2015), Derivation of a poroelastic flexural shell model, arXiv:1504.06097v1
11. Zhang, S.: A linear shell theory based on variational principles. Ph.D. thesis, The Pennsylvania State University (2001)
12. Aavatsmark, I.: Multipoint flux approximation methods for quadrilateral grids. In: 9th International Forum on Reservoir Simulation (2007)
13. Iliev, O., Kirsch, R., Lakdawala, Z., Printsypar, G.: MPFA algorithm for solving Stokes-Brinkman equations on quadrilateral grids. In: Fuhrmann, J. (ed.) Finite Volumes for Complex Applications VII - Elliptic, Parabolic and Hyperbolic Problems. Springer Proceedings in Mathematics & Statistics, pp. 547–654. Springer, Cham (2014). ISBN 978-3-319-05590-9 (Print), ISBN 978-3-319-05591-6 (Online)

Spin-Based CMOS-Compatible Devices

Viktor Sverdlov[(✉)] and Siegfried Selberherr

Institute for Microelectronics, TU Wien, Vienna, Austria
{sverdlov,selberherr}@iue.tuwien.ac.at

Abstract. With CMOS feature size rapidly approaching scaling limits the electron spin attracts attention as an alternative degree of freedom for low-power non-volatile devices. Silicon is perfectly suited for spin-driven applications, because it is mostly composed of nuclei without spin and is characterized by weak spin-orbit interaction. Elliot-Yafet spin relaxation due to phonons' mediated scattering is the main mechanism in bulk silicon at room temperature. Uniaxial stress dramatically reduces the spin relaxation, particularly in thin silicon films. Lifting the valley degeneracy completely in a controllable way by means of standard stress techniques represents a major breakthrough for spin-based devices. Despite impressive progress regarding spin injection, the larger than predicted signal amplitude is still heavily debated. In addition, the absence of a viable concept of spin manipulation in the channel by electrical means makes a practical realization of a device working similar to a MOSFET difficult. An experimental demonstration of such a spin field-effect transistor (SpinFET) is pending for 25 years now, which at present is a strong motivation for researchers to look into the subject. Commercially available CMOS compatible spin-transfer torque magnetic random access memory (MRAM) built on magnetic tunnel junctions possesses all properties characteristic to universal memory: fast operation, high density, and non-volatility. The critical current for magnetization switching and the thermal stability are the main issues to be addressed. A substantial reduction of the critical current density and a considerable increase of the thermal stability are achieved in structures with a recording layer between two vertically sandwiched layers, where the recording layer is composed of two parts in the same plane next to each other. MRAM can be used to build logic-in-memory architectures with non-volatile storage elements on top of CMOS logic circuits. Non-volatility and reduced interconnect losses guarantee low-power consumption. A novel concept for non-volatile logic-in-memory circuits utilizing the same MRAM cells to store and process information simultaneously is proposed.

1 Introduction

The breathtaking increase in density, speed, and performance of modern integrated circuits has been supported by the continuous miniaturization of CMOS devices. Numerous outstanding technological challenges have been resolved on this exciting journey. However, even though the transistor size is scaled down, the load capacitance per unit area of a circuit stops decreasing. This suggests

© Springer International Publishing Switzerland 2015
I. Lirkov et al. (Eds.): LSSC 2015, LNCS 9374, pp. 42–49, 2015.
DOI: 10.1007/978-3-319-26520-9_4

that the on-current must stay constant in order to maintain appropriate high speed operation. Even more, in ultra-scaled MOSFETs with semi-ballistic transport in the channel the conductance determined by the number of transversal propagating modes ceases to depend on the channel length. This results in an approximately constant power dissipation of a single MOSFET regardless of its channel length, which would lead to a rapid increase of dissipated heat with the transistor density further increased. An obvious saturation of MOSFET miniaturization puts clear foreseeable limitations to the continuation of the increase in the performance of integrated circuits. Thus, research for finding alternative technologies and computational principles is paramount.

The principle of MOSFET operation is fundamentally based on the charge degree of freedom of an electron: the electron charge interacts with the gate induced electric field which can close the transistor by creating a potential barrier. Another intrinsic electron property, the electron spin, attracts at present much attention as a possible candidate for complementing or even replacing the charge degree of freedom in future electron devices. It is characterized by two possible projections on a given axis and can be potentially used in digital information processing. In addition, only a small amount of energy is needed to alter the spin orientation, which is necessary for low power applications. The electron spin as a vector may be pointed not only up or down but rather in any direction on a unit Bloch sphere. This opens the way to use the whole Bloch sphere of states to process and store information by initializing, manipulating, and detecting the spin orientation. A successful implementation of a quantum computer utilizing the spin states on the Bloch sphere requires the possibility of efficient spin initiation, coherent manipulation, and reliable read-out. Although encouraging results were achieved, the development of a robust two- and three-qubit gate is a pressing challenge for proceeding to a larger computational network.

Silicon predominantly (92 %) consists of ^{28}Si nuclei with zero magnetic spin. The spin-orbit interaction is also weak in the silicon conduction band. Because of these properties electron spin states of conduction electrons in silicon should show better stability, lower decoherence and longer spin lifetime, which makes silicon a perfect candidate for spin-driven device applications. Even though these features are promising and silicon processing technology is well established the demonstration of basic elements necessary for spin related applications, such as injection of spin-polarized currents into silicon, spin transport, and detection, were demonstrated only recently. Although it should be straightforward to inject spin-polarized carriers into silicon from a ferromagnetic contact, due to a fundamental conductivity mismatch between a ferromagnetic metal contact and the semiconductor, the problem was without solution for a long time. A special technique [1] based on the attenuation of hot electrons with spins anti-parallel to the magnetization of the ferromagnetic film allows creating an imbalance between the electrons with spin-up and spin-down in silicon thus injecting spin-polarized current. The spin-coherent transport through the device was studied by applying an external magnetic field causing precession of spins during their propagation from source to drain. The detection is performed with a similar hot electron spin filter. Although the drain current is fairly small due to the carriers' attenuation in the source and drain filters, as compared to the current of injected spins, the experimental set-up represents a first spin-driven device which can be envisaged working at room temperature. Contrary to the MOSFET, however, the described structure is a two-terminal device. Nevertheless, the first demonstration of coherent spin transport through an undoped $350 \mu m$ thick silicon wafer [2] has triggered a systematic study of spin transport properties in silicon [3].

2   Silicon SpinFET

The SpinFET is a future semiconductor spintronic device with a superior performance. A SpinFET is composed of ferromagnetic source and drain contacts, linked by a non-magnetic semiconductor channel region [4]. The effective spin-orbit interaction in the channel depends on the perpendicular electric field, so that the spin of an electron injected from the source starts precessing. Only the electrons with their spin aligned to the drain magnetization can leave the channel contributing to the current. The current modulation is achieved by tuning the strength of the spin-orbit interaction by applying the gate voltage. In order to realize the SpinFET, an efficient spin injection and detection, spin propagation, and spin manipulation by purely electrical means must be achieved [5]. Spin injection in silicon from a ferromagnetic metal electrode is compromised by an impedance mismatch problem [6]. A solution to this impedance mismatch problem is the introduction of a potential barrier between the ferromagnetic metal and the semiconductor [7]. An experimental demonstration of a signal which should correspond to spin injection in doped silicon at room temperature was first performed in 2009 [8] using an $Ni_{80}Fe_{20}/Al_2O_3$ tunnel contact. Electrical signals at temperatures as high as $500\,K$ have been reported in [9]. Regardless of the success in demonstrating the signal which should correspond to the spin injection at room temperature, the magnitude of the signal obtained within the three-terminal measurement scheme is a several orders of magnitude larger than the theoretical value [3]. The reasons for the discrepancies are heavily debated [3,10]. A plausible explanation suggested recently [11] interprets the large signal to be rather due to the trap assisted tunneling magnetoresistance.

When spin is injected, the possibility to transfer the excess spin injected from the source to the drain electrode is essential. The excess spin is not a conserved quantity. While diffusing, it gradually relaxes to its equilibrium value which is zero in a non-magnetic semiconductor. An estimation for the spin lifetime at room temperature obtained is ranging between 0.1 to 10ns [3], depending on doping. This corresponds to the spin diffusion length $l = 0.2$-$2 \mu m$. The spin lifetime is determined by the spin-flip processes [12,13]. In silicon the spin relaxation due to the hyperfine interaction of spins with the magnetic moments of the ^{29}Si nuclei (the natural abundance is 4.7 %) is only important at low temperature, while the spin relaxation by the Elliot-Yafet mechanism [12,13] due to electron-phonon scattering is dominant at room temperature. The Elliot-Yafet mechanism is mediated by the intrinsic interaction between the orbital motion of an electron and its spin and electron scattering. When the microscopic spin-orbit interaction is taken into account, the Bloch function with a fixed spin projection is not an eigenfunction of the total Hamiltonian. Because the eigenfunction contains a contribution with an opposite spin projection, even spin-independent scattering with phonons generates a small probability of spin flips [14].

In bulk silicon the main contribution to spin relaxation is due to optical phonon scattering between the valleys residing along different crystallographic axis, or f-phonons scattering [15,16]. A relatively large spin relaxation reported in electrically-gated lateral-channel silicon structures [17,18] indicates that the extrinsic interface induced spin relaxation mechanism is important. This may pose an obstacle in realizing spin-driven CMOS-compatible devices, and a deeper understanding of the fundamental spin relaxation mechanisms in silicon inversion layers, thin films, and fins is needed.

The theory of spin relaxation must account for the most relevant scattering mechanisms in thin silicon-in-insulator films: electron-phonon interaction and surface roughness scattering. In order to evaluate the corresponding spin relaxation matrix elements, the wave functions with opposite spin projections must be found. We employ the Hamiltonian which takes into account the only relevant valley pair along the [1]-axis [19]. The Hamiltonian must include confinement and, most importantly, an effective spin-orbit interaction. Shear strain lifts the degeneracy between the unprimed subbands [20]. The enhanced valley splitting rapidly reduces the main contribution to spin relaxation due to inter-valley scattering. This results in a giant spin lifetime enhancement shown in Fig. 1. Shear strain used to enhance the performance of modern MOSFETs is extremely efficient in enhancing the spin lifetime and the spin diffusion length in silicon thin films.

Silicon is characterized by weak spin-orbit interaction and is not considered as a candidate for a SpinFET channel material. In actual thin films the inversion symmetry is broken, and a relatively large value for the spin-orbit coupling [21] is predicted by atomistic calculation $\beta \approx 2\mu e \text{Vnm}$ [22], in agreement with the value reported experimentally [23]. The channel length needed to achieve a substantial tunneling magnetoresistance (TMR) modulation is close to a micron [24].

3   Spin-Transfer Torque Magnetic RAM

The basic element of magnetic random access memory (MRAM) is a magnetic tunnel junction (MTJ). The three-layer MTJ represents a sandwich of two magnetic layers separated by a thin spacer which forms a tunnel barrier. While the magnetization of the pinned layer is fixed, the magnetization orientation of the recording layer can be switched between the two stable states parallel and anti-parallel to the fixed magnetization direction. A memory cell based on MTJs is scalable, exhibits relatively low operating voltages, low power consumption, high operation speed, high endurance, and a simple structure. Switching between the two states is induced by spin-polarized current flowing through the MTJ [25,26]. The recording layer magnetization switching, by means of the

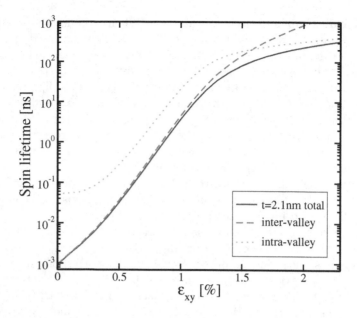

Fig. 1. Dependence of spin lifetime on shear strain for T = 300 K in a film of 2.1 nm thickness. The inter- and intra-valley contributions are also shown...

spin-transfer torque (STT), makes STT-MRAM a promising candidate for future universal memory.

Because the spin-polarized current is only a fraction of the total charge current passing through the cell, the reduction of the current density required for switching and the increase of the switching speed are the most important challenges in STT-MRAM developments. We demonstrated by micromagnetic simulations that, if the recording layer is composed of two parts, a nearly three time faster switching is achieved [27]. Thanks to the removal of the central part in the recording layer the magnetization switching occurs in-plane. This use of the composite structure of the recording layer allows to decrease the switching energy while preserving the thermal stability [28,29].

4   STT-MRAM Based Logic-in-Memory

The introduction of non-volatile logic could help to reduce significantly the heat generation, especially at stand-by, booting, and resuming stages. It is extremely attractive to use the same elements as memory and latches to reduce the time delay and energy waste while transferring data between CPU and memory blocks. The MRAM technology is promising for building logic-in-memory configurations which combine non-volatile memory cells and logic circuits [30,31].

It is even more attractive to use the memory arrays to carry logic operations. In [32,33] MTJ-based reprogrammable logic gates realize the basic Boolean logic

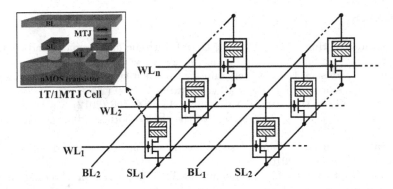

Fig. 2. The common STT-MRAM architecture based on the one-transistor/one-MTJ (1T/1MTJ) structure.

operations AND, OR, NAND, NOR, and the Majority operation. A material implication (IMP) logic gate [34] can be implemented by using any two MRAM memory cells from an MRAM array (Fig. 2) to perform the Boolean IMP operation. A sequence of logical IMP supplemented with FALSE allows to perform any given Boolean operation.

Compared to the TiO_2 memristive switches [35], MRAM provides a higher endurance. Furthermore, the bistable resistance state of the MRAM cell eliminates the need for refreshing circuits. The logic implementation using MRAM cells relies on a conditional switching of MTJs caused by the state-dependent current modulation on the output (target) MTJ. The resistance modulation between the high and low resistance states in the MTJ is proportional to the TMR ratio. The error probability of MTJ-based operations decreases with increasing TMR ratio which is thus the most important device parameter for the reliability [36].

5   Summary and Conclusion

Recent ground-breaking experimental and theoretical findings regarding spin injection and transport in silicon make spin an attractive option to supplement or to replace the charge degree of freedom for computations. Uniaxial stress employed to enhance the electron mobility can also be used to boost the spin lifetime significantly. CMOS-compatible STT-MRAM cells built on magnetic tunnel junctions with a composite recording layer demonstrate a three-fold improvement of the switching time as compared to similar cells with a monolithic layer. The realization of an intrinsic non-volatile logic-in-memory architecture by using MRAM arrays is demonstrated.

Acknowledgements. This work is supported by the European Research Council through the grant #247056 MOSILSPIN.

References

1. Appelbaum, I., Huang, B., Monsma, D.J.: Electronic measurement and control of spin transport in Silicon. Nature **447**, 295–298 (2007)
2. Huang, B., Monsma, D.J., Appelbaum, I.: Coherent spin transport through a 350 micron thick silicon wafer. Phys. Rev. Lett. **99**, 177209 (2007)
3. Jansen, R.: Silicon spintronics. Nat. Mater. **11**, 400–408 (2012)
4. Datta, S., Das, B.: Electronic analog of the electro-optic modulator. Appl. Phys. Lett. **56**, 665–667 (1990)
5. Sugahara, S., Nitta, J.: Spin-transistor electronics: an overview and outlook. Proc. IEEE **98**, 2124–2154 (2010)
6. Schmidt, G., Ferrand, D., Molenkamp, L.W., Filip, A.T., van Wees, B.J.: Fundamental obstacle for electrical spin injection from a ferromagnetic metal into a diffusive semiconductor. Phys. Rev. B **62**, R4790–R4793 (2000)
7. Rashba, E.I.: Theory of electrical spin injection: tunnel contacts as a solution of the conductivity mismatch problem. Phys. Rev. B **62**, R16267–R16270 (2000)
8. Dash, S.P., Sharma, S., Patel, R.S., de Jong, M.P., Jansen, R.: Electrical creation of spin polarization in silicon at room temperature. Nature **462**, 491–494 (2009)
9. Li, C., van't Erve, O., Jonker, B.: Electrical injection and detection of spin accumulation in silicon at 500K with magnetic metal/silicon dioxide contacts. Nat. Commun. **2**, 245 (2011)
10. Jansen, R., Deac, A.M., Saito, H., Yuasa, S.: Injection and detection of spin in a semiconductor by tunneling via interface states. Phys. Rev. B **85**, 134420 (2012)
11. Song, Y., Dery, H.: Magnetic-field-modulated resonant tunneling in ferromagnetic-insulator-nonmagnetic junctions. Phys. Rev. Lett. **113**, 047205 (2014)
12. Zutic, I., Fabian, J., Das Sarma, S.: Spintronics: fundamentals and applications. Rev. Mod. Phys. **76**, 323–410 (2004)
13. Fabian, J., Matos-Abiaguea, A., Ertler, C., Stano, P., Zutic, I.: Semiconductor spintronics. Acta Phys. Slovaca **57**, 565–907 (2007)
14. Cheng, J.L., Wu, M.W., Fabian, J.: Theory of the spin relaxation of conduction electrons in silicon. Phys. Rev. Lett. **104**, 016601 (2010)
15. Li, P., Dery, H.: Spin-orbit symmetries of conduction electrons in silicon. Phys. Rev. Lett. **107**, 107203 (2011)
16. Song, Y., Dery, H.: Analysis of phonon-induced spin relaxation processes in silicon. Phys. Rev. B **86**, 085201 (2012)
17. Li, J., Appelbaum, I.: Modeling spin transport in electrostatically-gated lateral-channel silicon devices: role of interfacial spin relaxation. Phys. Rev. B **84**, 165318 (2011)
18. Li, J., Appelbaum, I.: Lateral spin transport through bulk silicon. Appl. Phys. Lett. **100**, 162408 (2012)
19. Osintsev, D., Baumgartner, O., Stanojevic, Z., Sverdlov, V., Selberherr, S.: Subband splitting and surface roughness induced spin relaxation in (001) silicon SOI MOSFETs. Solid-State Electron. **90**, 34–38 (2013)
20. Sverdlov, V.: Strain-Induced Effects in Advanced MOSFETs. Springer, Wien - New York (2011)
21. Jancu, J.M., Girard, J.C., Nestoklon, M.O., Lemaître, A., Glas, F., Wang, Z.Z., Voisin, P.: STM images of subsurface Mn Atoms in GaAs: evidence of hybridization of surface and impurity states. Phys. Rev. Lett. **101**, 196801 (2008)
22. Prada, M., Klimeck, G., Joynt, R.: Spin-orbit splittings in Si/SiGe quantum wells: from ideal Si membranes to realistic heterostructures. New J. Phys. **13**, 013009 (2011)

23. Wilamowski, Z., Jantsch, W.: Suppression of spin relaxation of conduction electrons by cyclotron motion. Phys. Rev. B **69**, 035328 (2004)
24. Osintsev, D., Sverdlov, V., Stanojeviè, Z., Makarov, A., Selberherr, S.: Temperature dependence of the transport properties of spin field-effect transistors built with InAs and Si channels. Solid-State Electron. **71**, 25–29 (2012)
25. Slonczewski, J.: Current-driven excitation of magnetic multilayers. J. Magn. Magn. Mater. **159**, L1–L7 (1996)
26. Berger, L.: Emission of spin waves by a magnetic multilayer traversed by a current. Phys. Rev. B **54**, 9353–9358 (1996)
27. Makarov, A., Sverdlov, V., Osintsev, D., Selberherr, S.: Reduction of switching time in pentalayer magnetic tunnel junctions with a composite-free layer. Phys. Status Solidi - Rapid Res. Lett. **5**, 420–422 (2011)
28. Makarov, A., Sverdlov, V., Selberherr, S.: Magnetic tunnel junctions with a composite free layer: a new concept for future universal memory. In: Luryi, S., Xu, J., Zaslavsky, A. (eds.) Future Trends in Microelectronics, pp. 93–101. Wiley, New York (2013)
29. Makarov, A.: Modeling of emerging resistive switching based memory cells. Dissertation, Institute for Microelectronics, TU Wien (2014)
30. Endoh, T.: STT-MRAM technology and its NV-logic applications for ultimate power management. In: 2014 CMOS Emerging Technologies Research (CMOSETR), p. 14 (2014)
31. Natsui, M., Suzuki, D., Sakimura, N., Nebashi, R., Tsuji, Y., Morioka, A., Sugibayashi, T., Miura, S., Honjo, H., Kinoshita, K., Ikeda, S., Endoh, T., Ohno, H., Hanyu, T.: Nonvolatile logic-in-memory array processor in 90 nm MTJ/MOS achieving 75% leakage reduction using cycle-based power gating. In: 2013 IEEE International Solid-State Circuits Conference Digest of Technical Papers (ISSCC), pp. 194–195 (2013)
32. Lyle, A., Harms, J., Patil, S., Yao, X., Lilja, D.J., Wang, J.P.: Direct communication between magnetic tunnel junctions for nonvolatile logic fan-out architecture. Appl. Phys. Lett. **97**, 152504 (2010)
33. Lyle, A., Patil, S., Harms, J., Glass, B., Yao, X., Lilja, D., Wang, J.: Magnetic tunnel junction logic architecture for realization of simultaneous computation and communication. IEEE Trans. Magn. **47**, 2970–2973 (2011)
34. Mahmoudi, H., Windbacher, T., Sverdlov, V., Selberherr, S.: Implication logic gates using spin-transfer-torque-operated magnetic tunnel junctions for intrinsic logic-in-memory. Solid-State Electron. **84**, 191–197 (2013)
35. Borghetti, J., Snider, G., Kuekes, P., Yang, J., Stewart, D., Williams, R.: Memristive switches enable stateful logic operations via material implication. Nature **464**, 873–876 (2010)
36. Mahmoudi, H., Windbacher, T., Sverdlov, V., Selberherr, S.: Reliability analysis and comparison of implication and reprogrammable logic gates in magnetic tunnel junction logic circuits. IEEE Trans. Magn. **49**, 5620–5628 (2013)

Multilevel Methods on Graphs

Shortest-Path Queries in Planar Graphs on GPU-Accelerated Architectures

Guillaume Chapuis$^{(\boxtimes)}$ and Hristo Djidjev

Los Alamos National Laboratory, Los Alamos, NM 87545, USA
{gchapuis,djidjev}@lanl.gov

Abstract. We develop an efficient parallel algorithm for answering shortest-path queries in planar graphs and implement it on a multi-node CPU-GPU clusters. The algorithm uses a divide-and-conquer approach for decomposing the input graph into small and roughly equal subgraphs and constructs a distributed data structure containing shortest distances within each of those subgraphs and between their boundary vertices. For a planar graph with n vertices, that data structure needs $O(n)$ storage per processor and allows queries to be answered in $O(n^{1/4})$ time.

Keywords: Shortest path problems · Graph algorithms · Distributed computing · GPU computing · Graph partitioning

1 Introduction

Finding shortest paths (SPs) in graphs has applications in transportation, social network analysis, network routing, and robotics, among others. The problem asks for a path of shortest length between one or more pairs of vertices. There are many algorithms for solving SP problems sequentially. Dijkstra's algorithm [2] finds the distances between a source vertex v and all other vertices of the graph in $O(m \log n)$ time, where n and m are the numbers of the vertices and edges of the graph, respectively. It can also be used to find efficiently the distance between a pair of vertices. This algorithm is nearly optimal (within a logarithmic factor), but has irregular structure, which makes it hard to implement efficiently in parallel. Floyd-Warshall's algorithm, on the other hand, finds the distances between all pairs of vertices of the graph in $O(n^3)$ time, which is efficient for dense $(m = \Theta(n^2))$ graphs, has a regular structure good for parallel implementation, but is inefficient for sparse $(m = O(n))$ graphs such as planar graphs.

In this paper we are considering the query version of the problem. It asks to construct a data structure that will allow to answer any subsequent distance query fast. A distance query asks, given an arbitrary pair of vertices v, w, to compute $\text{dist}(v, w)$. This problem has applications in web mapping services such as MapQuest and Google Maps. There is a tradeoff between the size of the data structure and the time for answering a query. For instance, Dijkstra's algorithm gives a trivial solution of the query version of the SP problem with (small) $O(n + m)$ space (for storing the input graph), but large $O(m \log n)$ query time

© Springer International Publishing Switzerland 2015
I. Lirkov et al. (Eds.): LSSC 2015, LNCS 9374, pp. 53–60, 2015.
DOI: 10.1007/978-3-319-26520-9_5

(for running Dijkstra's algorithm with a source the first query vertex). On the other end of the spectrum, Floyd-Warshall's algorithm can be used to construct a (large) $O(n^2)$ data structure (the distance matrix) allowing (short) $O(1)$ query time (retrieving the distance from the data base). However, for very large graphs, the $O(n^2)$ space requirement is impractical. We are interested in an algorithm that needs significantly less than $O(n^2)$ space, but will answer queries faster than Disjkstra's algorithm. Our algorithm will use the structure of planar graphs for increased efficiency, as most road networks are planar or near-planar, and will also be highly parallelizable, making use of the features available in modern high-performance clusters and specialized processors such as the GPUs.

The query version for shortest path queries in planar graphs was proposed in [3] and after that different aspects of the problem were studied by multiple authors, e.g., [1,9]. Here we present the first distributed implementation for solving the problem that is designed to make use of the potential for parallelism offered by GPUs. Our solution makes use of the fast parallel algorithm for computing shortest paths in planar graphs from [5], resulting in asymptotically faster and also shown to be efficient in practice.

2 Preliminaries

Given a graph G with a weight $\text{wt}(e)$ on each edge e, the length of a path p is the sum of the weights of the edges of the path. The *single-pair shortest path problem* (*SPSP*) is, given a pair v, w of vertices of G, to find a path between v and w, called *shortest path* (SP), with minimum length. The length of that path is called *distance* between v and w and is denoted as $\text{dist}(v, w)$. For any subgraph H of G, the distance between v and w in H is denoted as $\text{dist}_H(v, w)$. The *single-source shortest path problem* (*SSSP*) is to find SPs from a fixed vertex v to all other vertices of G. Finally, the *all-pairs shortest path problem* (*APSP*) is to find SPs between all pairs of vertices. There are distance versions of SPSP, SSSP, and APSP, which are more commonly studied, where the objective is to compute the corresponding distances instead of SPs. Most distance algorithms allow the corresponding SPs to be retrieved in additional time proportional to the number of the edges of the path. In this paper, by SPSP, SSSP, and APSP we mean the distance versions of these problems.

A k-partition $\mathcal{P}$ of G is a set $V_1, \ldots, V_k$ of subsets of $V(G)$, the set of the vertices of G, such that $V_i \cap V_j = \emptyset$ if $i \neq j$ and $\bigcup_{i=1}^{k} V_i = V(G)$. We call the subgraphs of G induced by V_i *components* of $\mathcal{P}$. The *boundary* of the partition consists of all vertices of G that have at least one neighbor in a different component. We denote by $BG(G)$ or simply by BG the subgraph of G whose vertices are the boundary vertices of G and there is an edge between two vertices v and w of BG iff there is an edge between them in G or if they belong to the same component of $\mathcal{P}$. In the next section we will assign appropriate weights on the edges of BG and solve the APSP problem on it. For any $C \in \mathcal{P}$, we denote by $B(C)$ the set of all boundary vertices that are from C. For any planar graph of n vertices and bounded ($O(1)$ as a function of n) vertex degree, one can find in $O(n)$ time a k-partition $\mathcal{P}$ with $|B(C)| = O(\sqrt{n/k})$ for each component $C \in \mathcal{P}$ [6].

3   Algorithm Overview and Analysis

Our algorithm works in two modes: preprocessing mode, during which a data structure is computed that allows efficient SP queries, and the query mode that uses that data structure to compute the distance between a query pair of vertices. We assume that the input is a planar graph G of n vertices and bounded vertex degree and the cluster has p nodes.

3.1   Preprocessing Mode

The preprocessing algorithm (Algorithm 1) has three phases. During the first phase (line 1), the graph is partitioned and each component is assigned to a distinct cluster node. During the second phase (lines 2–5), the APSP problem is solved for each component C independently and in parallel and the computed distance matrix APSP(C) is stored at the same node. Finally, in the third phase (lines 6–10), the boundary graph BG is constructed and the APSP is solved for BG. That computation is done distributedly such that the distances from vertex $v \in BG$ to all other vertices of BG are computed at the node containing v, by using Dijkstra's algorithm [2]. The computed distance matrix is stored at the node that has done the computations. Hence, at the end of the algorithm, the node $N(C)$ contains two matrices: one containing the SP distances in C and the other containing all SP distances in BG with source a vertex in $BG \cap C$.

One can think of BG as a compressed version of G where the non-boundary vertices are removed, but are implicitly represented in BG by the information encoded in its edge weights. Note however that the distances APSP(C) (and the corresponding edge weights of BG) are not distances in G; the reason is that a shortest path between two vertices v and w from C might pass through vertices not in C. The next lemma is based on the observation that a shortest path between two vertices of BG has structure $p_1, e_1, p_2, e_2, \ldots, e_{s-1}, p_s$, where p_i denotes a shortest path of vertices from the same component C (and hence is a distance stored in APSP(C)) and e_i denotes an edge joining vertices from different components.

Lemma 1. *[5] For any two vertices $v, w \in BG$ the distance between v and w in BG is equal to the distance between v and w in G.*

We will next estimate the time and space (memory) required to run the algorithm. As G is planar and of bounded vertex degree (as a function of n), it can be divided in $O(n)$ time into k parts so that each part has no more than (n/k) vertices and $O(\sqrt{n/k})$ boundary vertices [7]. We will estimate the requirements of each phase. Since the maximum amount of coarse-grained parallelism of Algorithm 1 is $\min\{p, k\}$, we assume without loss of generalization that $p \leq k$.

Phase 1 requires $O(n)$ running time and $O(n)$ space [7].

The complexity of Phase 2 is dominated by the time for computing distances in line 3. We assume that we are using the algorithm from [5] that can be implemented efficiently on a GPU-accelerated architecture and has complexity $O(N^{9/4})$. Then Phase 2 requires $O((k/p)(n/k)^{9/4}) = O(n^{9/4}/(pk^{5/4}))$ time and $kO((n/k)\sqrt{n/k}) = O(n^{3/2}/k^{1/2})$ total space. The space per processor is $kO((n/k)^2) = O(n^2/k)$.

For Phase 3, the number of the vertices of BG is $kO(\sqrt{n/k}) = O(\sqrt{nk})$ and the number of the edges is $kO((\sqrt{n/k})^2) = O(n)$. One execution of line 8 (for one component C) takes $(k/p)|BG \cap C||E(BG)|\log(|BG|) = (k/p)O(\sqrt{n/k})O(n\log n)$ time and $O(n)$ space. The space needed for one iteration of Step 9 is $|BG \cap C||BG| = O(\sqrt{n/k}\sqrt{nk}) = O(n)$. Hence Phase 3 requires $O((k/p)n^{3/2}/k^{1/2}\log n) = O(n^{3/2}k^{1/2}\log n/p)$ time and $O(nk/p)$ space per processor.

Summing up the requirements for Phases 1, 2, and 3, we get $O(n^{9/4}/(pk^{5/4}) + n^{3/2}k^{1/2}\log n/p))$ time and $O(n + n^2/(pk) + nk/p)$ space per processor needed for Algorithm 1. Assuming space is more important in this case than time (since nodes have limited memory), we find that $k = n^{1/2}$ minimizes the function $n^2/k + nk$. Hence we have the following result.

Lemma 2. *With $k = \lceil n^{1/2} \rceil$ and $p \leq k$, Algorithm 1 runs in $O(n^{7/4}\log n/p)$ time and uses $O(n^{3/2}/p)$ space per processor. With $p = k$, the time and space are $O(n^{5/4})$ and $O(n)$, respectively.*

The time bound of Lemma 2 is conservative as it doesn't take into account our use of fine-grain parallelism due to multi-threading, e.g., by the GPUs.

Algorithm 1. Preprocessing algorithm

Input: A planar graph G

Output: A data structure for efficient shortest path queries in G

/* Partitioning */

1: Construct a k-partition $\mathcal{P}$ of G and assign each component C to a distinct node $N(C)$

/* Solve the APSP problem for each component */

2: **for all** components $C \in \mathcal{P}$ **do in parallel**

3:     Solve APSP for C and save the distances in a table $\text{APSP}(C)$

4:     For each pair of boundary vertices $v, w \in C$ define edge (v, w), if not already in G, and assign a weight $\text{wt}(v, w) = \text{dist}_C(v, w)$

5: **end for**

/* Solve the APSP problem for the boundary graph */

6: Define a boundary graph BG with vertices all boundary vertices of G and edges as defined in the previous step and store it at each node

7: **for all** components $C \in \mathcal{P}$ **do in parallel**

8:     Solve SSSP in BG for each vertex of $C \cap BG$

9:     Store the distances from all vertices of $C \cap BG$ to all vertices of BG in a table $\text{APSP}_{BG}(C)$

10: **end for**

3.2   Query Mode

The query algorithm (Algorithm 2) is based on the fact that if $C_1 \neq C_2$, then any path between v_1 and v_2 should cross both $B(C_1)$ and $B(C_2)$. Let π be a shortest path between v_1 and v_2. Then π can be divided into three parts: from v_1 to a vertex b_1 from $B(C_1)$, from b_1 to a vertex b_2 on p from $B(C_2)$, and from b_2 to v_2. Vertices b_1 and b_2 minimizing the length of p are found as follows: in the loop on lines 2–7, for each b_2 an optimal b_1 and $\mathrm{dist}(v_1, b_2)$ are found; in lines 10–12 an optimal b_2 is found.

Algorithm 2. Query algorithm

Input: Vertices v_1, v_2 of G, a k-partition $\mathcal{P}$ of G, tables $\mathrm{APSP}(C)$ and $\mathrm{APSP}_{BG}(C)$
    for all $C \in \mathcal{P}$
Output: $\mathrm{dist}(v_1, v_2)$
1: Determine components C_1 and C_2 such that $v_1 \subset C_1$, $v_2 \subset C_2$
2: **for all** vertices $b_2 \in B(C_2)$ **do** in parallel
            /* Compute $\mathrm{dist}(v_1, b_2)$ */
3:    $\mathrm{dist}(v_1, b_2) = \infty$
4:    **for all** vertices $b_1 \in B(C_1)$ **do**
5:       $\mathrm{dist}(v_1, b_2) = \min\{\mathrm{dist}(v_1, b_2), \mathrm{dist}_{C_1}(v_1, b_1) + \mathrm{dist}_{BG}(b_1, b_2)\}$
6:    **end for**
7: **end for**
8: **If** $N(C_1) \neq N(C_2)$ **then** transfer the column of $\mathrm{SP}(C_2)$ corresponding to v_2 from
    $N(C_2)$ to $N(C_1)$.
            /* Now we can compute $\mathrm{dist}(v_1, v_2)$ */
9: $\mathrm{dist}(v_1, v_2) = \infty$
10: **for all** vertices $b_2 \in B(C_2)$ **do**
11:    $\mathrm{dist}(v_1, v_2) = \min\{\mathrm{dist}(v_1, v_2), \mathrm{dist}(v_1, b_2) + \mathrm{dist}_{C_2}(b_2, v_2)\}$
12: **end for**
13: **If** $C_1 = C_2$ **then** $\mathrm{dist}(v_1, v_2) = \min\{\mathrm{dist}(v_1, v_2), \mathrm{dist}_{C_1}(v_1, v_2)\}$, where the distance
    $\mathrm{dist}_{C_1}(v_1, v_2)$ is taken from $\mathrm{APSP}(C_1)$.

Lemma 3. *Algorithm 2 correctly computes* $\mathrm{dist}(v_1, v_2)$ *and its running time is* $O(n^{1/4})$ *with* $k = \lceil n^{1/2} \rceil$ *and* $p \geq \lceil n^{1/4} \rceil$.

Proof. Given in the extended version [4].

4 Implementation Details

In this section, we describe how the preprocessing and query modes are implemented on a hybrid CPU-GPU cluster. We use a distance matrix to represented both the input graph G and the output. Such a 2-dimensional matrix contains in cell (i, j) the value of the distance from vertex i to vertex j. Initially, cell (i, j) contains $\mathrm{wt}(i, j)$ if an edge (i, j) is present in G, or infinity otherwise. These values are updated as the algorithm progresses. At the end of the algorithm, cell (i, j) contains $\mathrm{dist}(i, j)$.

In phase 1 of the preprocessing mode, we construct a k-partition of G using the METIS library [8]. Based on that partition, we reorder the vertices of G so

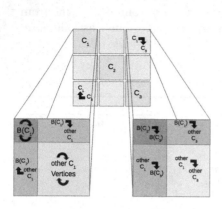

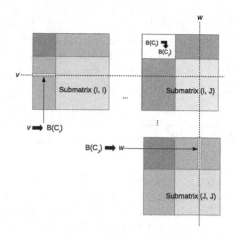

Fig. 1. Vertices from the same component are stored contiguously starting with boundary vertices. In the two bottom submatrices, dark-grey regions (red in the color version) are part of the boundary matrix. The light-grey region of the bottom right submatrix is not computed during preprocessing (Color figure online).

Fig. 2. The distances required to compute dist(v, w), shown in white, are scattered in three submatrices: two diagonal ones, for component I and for component J, and a non-diagonal submatrix (I, J) (Color figure online).

that vertices from the same component have consecutive indices and boundary vertices of each components have the lowest indices – see Fig. 1.

In phase 2, we compute the shortest distances within each of the components. For k components, this phase gives a total k independent tasks that can be executed in parallel. Computations at this phase are already balanced across nodes as components contain roughly the same number of vertices and the APSP algorithm from [5] ensures the same $O(N^{9/4})$ complexity with respect to the number of nodes.

Finally, phase 3 consists in computing the shortest distances within the boundary graph using Dijkstra's algorithm. Computations at this phase may be imbalanced between nodes for two reasons. First, the number of boundary vertices in two components may differ and, second, the complexity of Dijkstra's algorithm does not solely depend on the number of vertices in the graph, but also on the number of edges, which may vary even more than the number of vertices between two components' boundary graphs.

In the query mode, we are interested in finding dist(v, w), where v and w are from components I and J, respectively. The required values for that computation are scattered in three submatrices, as illustrated in Fig. 2. For such a query, assuming $k = p$, node i, holding the required values from diagonal submatrix I and non-diagonal submatrix (I, J), will be in charge of the computations. Required values from diagonal submatrix J are held by node j and need to be transfered to node i.

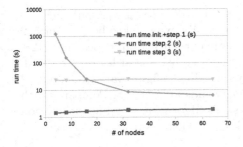 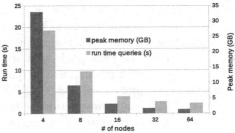

Fig. 3. Preprocessing run times for a fixed graph size of 256 k vertices and increasing number of nodes.

Fig. 4. Peak memories and run times for 10k queries for a fixed graph size of 256 k vertices and increasing number of parts/processors.

5   Experimental Evaluation

In this section we describe experiments designed to test our algorithm and its implementation. Specifically, we are going to test the strong scaling properties by running our code on a fixed graph size and a varying number p of cluster nodes and number k of components. All computations are run on a 300 node cluster. Each cluster node is comprised of 2 × Eight-Core Intel Xeon model E5-2670 @ 2.6 GHz and two GPGPU Nvidia Tesla M2090 cards connected to PCIe-2.0 × 16 slots. In order to make full use of the available GPUs, each node is assigned at least two graph components so that the two associated diagonal submatrices can be computed simultaneously on the two GPUs.

For the strong-scaling experiment, the graph size is fixed to 256k vertices. Preprocessing and queries are run with increasing numbers of nodes ranging from 4 to 64. Each node handles 2 components (one per available GPU); therefore the number of components k ranges from 8 to 128.

Figure 3 shows the run times for the preprocessing mode. For low numbers of nodes and thus low values of k, preprocessing time is dominated by step 2 - the computation of the shortest distances within each component - since lower k values means larger components. For higher numbers of nodes and thus higher values of k, preprocessing time becomes dominated by step 3 - the computation of the boundary graph - as more components mean higher numbers of incident edges and thus larger boundary graphs. Note that while the figure seems to show supralinear speedup, that is not the case (and similarly for the memory usage). The reason is that, with increasing the number of processors p, the number k of parts is increased too (as it is tied to p in this implementation) and hence the complexity of the algorithm is also reduced.

Figure 4 shows the query times and peak memory usage per node. The run times are given for 10, 000 queries from random sources to random targets. Note that in the query mode only fine-grain (node-level) parallelism is used, while multiple nodes are still needed for distributed storage and, optionally, to handle multiple queries in parallel (not implemented in the current version). For the memory usage, the optimal value for k, theoretically expected to be $\sqrt{n}$ – or 512 for this instance – is not reached in this experiment since k only goes up to 128. We can however see that peak memory usage per node is still dropping with increasing values of k up to 128. The query times in the figure vary from about 2 milliseconds per query for $k = 8$ to 0.25 milliseconds for $k = 128$. Compared with the Boost library implementation of Dijkstra's algorithm, our implementation answers queries on the largest instances about 1000 times faster.

6   Conclusion

We developed and implemented a distributed algorithm for shortest path queries in planar graphs with good scalability. It allows answering SP queries in $O(n^{1/4})$ time by using $O(\sqrt{n})$ processors with $O(n)$ space per processor and $O(n^{5/4})$ preprocessing time. Our implementation on 300 node CPU-GPU cluster has preprocessing time of less than 10 seconds using 32 or more nodes and 0.25 milliseconds per query using two nodes. Interesting tasks for future research is implementing a version allowing parallel queries and reducing the query time of the implementation to $O(\log n)$ by using properties of graph planarity.

References

1. Chen, D.Z., Xu, J.: Shortest path queries in planar graphs. In: Proceedings of the Thirty-Second Annual ACM Symposium on Theory of Computing, STOC 2000, pp. 469–478. ACM, New York (2000)
2. Dijkstra, E.W.: A note on two problems in connexion with graphs. Numer. Math. **1**(1), 269–271 (1959)
3. Djidjev, H.: Efficient algorithms for shortest path queries in planar digraphs. In: d'Amore, F., Franciosa, P.G., Marchetti-Spaccamela, A. (eds.) Graph-Theoretic Concepts in Computer Science. Lecture Notes in Computer Science, vol. 1197, pp. 151–165. Springer, Heidelberg (1997)
4. Djidjev, H., Chapuis, G.: Shortest-path queries in planar graphs on GPU-accelerated architectures (arXiv). CoRR (2015). http://arxiv.org
5. Djidjev, H., Thulasidasan, S., Chapuis, G., Andonov, R., Lavenier, D.: Efficient multi-GPU computation of all-pairs shortest paths. In: IPDPS, pp. 360–369 (2014)
6. Frederickson, G.N.: Fast algorithms for shortest paths in planar graphs, with applications. SIAM J. Comput. **16**(6), 1004–1022 (1987)
7. Frederickson, G.N.: Planar graph decomposition and all pairs shortest paths. J. ACM **38**(1), 162–204 (1991)
8. Karypis, G., Kumar, V.: Multilevel k-way partitioning scheme for irregular graphs. J. Parallel Distrib. Comput. **48**(1), 96–129 (1998)
9. Mozes, S., Sommer, C.: Exact distance oracles for planar graphs. In: Proceedings of the 23rd Annual ACM-SIAM Symposium on Discrete Algorithms, pp. 209–222 (2012)

Mathematical Modeling and Analysis of PDEs Describing Physical Problems

A Numerical Approach to Price Path Dependent Asian Options

Tatiana Chernogorova[1]($\boxtimes$) and Lubin Vulkov[2]

[1] Faculty of Mathematics and Informatics, University of Sofia, Sofia, Bulgaria
chernogorova@fmi.uni-sofia.bg
[2] Faculty of Natural Sciences and Education, University of Rousse, Rousse, Bulgaria
lvalkov@uni-ruse.bg

Abstract. In this paper we develop a parabolic-hyperbolic splitting method for resolving the degeneracy of order γ, $0 < \gamma \leq 2$ in the ultra-parabolic equation of path dependent Asian options. For the space discretization of the parabolic subproblem we have used two approximations. The first one is the finite volume difference scheme of S. Wang [11], while the second one is the monotone difference scheme of A.A. Samarskii [9]. Some computation results and a comparison between the two methods are presented.

Keywords: Asian options · Ultra-parabolic equation · Degeneracy · Finite volume method

1 Introduction

An Asian option is a derivative product the payoff of which depends on the average of an underlying asset price over some time period. The average is less exposed to sudden crashes or rallies in stock price and over time is harder to manipulate than a single stock price. Thus the Asian options are less expensive than comparable plain vanilla options [3,12].

A number of techniques to price Asian options have been proposed: Monte-Carlo method, analytical approximations, Laplace/Fourier transform approach, modified binomial tree approach, PDE approach, etc., see e. g. [1,2,6–8,10,12,13] and references there in. In this paper we discuss numerical methods for general PDE models of path dependent options.

Let $\bar{S}$ represent the underlying stock price. Then a path dependent option in a constant elasticity variance environment can be modeled by the stochastic differential equation given by

$$d\bar{S} = \mu_\gamma \bar{S} dt + \sigma_1 \bar{S}^\gamma dz, \; 0 < \gamma \leq 2,$$

where μ_γ is the average option, $\mu_\gamma(\bar{S}) = \bar{S}$ (arithmetic average options) or $\mu_\gamma(\bar{S}) = \log \bar{S}$ (geometric average options) and σ_1 is the volatility. Using standard

© Springer International Publishing Switzerland 2015
I. Lirkov et al. (Eds.): LSSC 2015, LNCS 9374, pp. 63–71, 2015.
DOI: 10.1007/978-3-319-26520-9_6

arguments [3, 12], the value V of an option depending on $\bar{x} = \bar{x}(t) = \int\limits_0^t \bar{S}(v)dv$ is given by

$$\frac{\partial V}{\partial \tau} = \frac{1}{2}\sigma_1^2 \bar{S}^\gamma \frac{\partial^2 V}{\partial \bar{S}^2} + r\bar{S}\frac{\partial V}{\partial \bar{S}} - \mu_\gamma \frac{\partial V}{\partial \bar{x}} - rV, \quad \tau = T - t, \tag{1}$$

where $(\bar{S}, \bar{x}, \tau) \in (0, \infty) \times (0, \infty) \times [0, T]$, r is the interest rate, t is the time and τ is the time to maturity T. The difficulties that arise at the numerical treatment of Asian options governed by ultra-parabolic Eq. (1) are: the degeneracy on the boundary $\bar{S} = 0$; unbounded domain; non-smooth (even discontinuous) payoff (initial) function; small volatility σ_1 causes boundary layers, etc. In the case of Black-Scholes operator ($\gamma = 2$) the exponential change $\bar{S} = e^{\bar{x}}$ removes the degeneracy and is often used for construction of numerical methods [1, 6–8, 12].

Following [4, 5] we take the initial and boundary conditions for the localized problem:

$$V(\bar{S}, \bar{x}, 0) = \max\{X(\bar{x}),\ 0\} \equiv V_0(\bar{S}, \bar{x}), \tag{2}$$

$$V(0, \bar{x}, \tau) = e^{-r\tau}\max\{X(\bar{x}),\ 0\} \equiv V_1(\bar{x}, \tau), \tag{3}$$

$$V(S_{\max}, \bar{x}, \tau) = \max\left\{e^{-r\tau}X(\bar{x}) + S_{\max}\left(1 - e^{-r\tau}\right)/rT,\ 0\right\} \equiv V_2(\bar{x}, \tau), \tag{4}$$

$$V(\bar{S}, 0, \tau) = \bar{S}\left(1 - e^{-r\tau}\right)/rT \equiv V_3(\bar{S}, \tau), \tag{5}$$

where $(\bar{S}, \bar{x}, \tau) \in [0, S_{\max}] \times [0, x_{\max}] \times [0, T]$, $X(\bar{x}) = (x_{\max} - \bar{x})/T - K$ and K is the exercise strike price of the option.

Using the transformations $S = \bar{S}/x_{\max}$, $x = \bar{x}/x_{\max}$ and the notations $S_0 = S_{\max}/x_{\max}$, $\sigma = \sigma_1(x_{\max})^{\frac{\gamma-2}{2}}$, the problem (1)–(5) can be written as

$$\frac{\partial V}{\partial \tau} = \frac{1}{2}\sigma^2 S^\gamma \frac{\partial^2 V}{\partial S^2} + rS\frac{\partial V}{\partial S} - \mu_\gamma \frac{\partial V}{\partial x} - rV, \ (S, x, \tau) \in (0, S_0) \times (0, 1) \times (0, T], \tag{6}$$

$$V(S, x, 0) = V_0(S, x), \tag{7}$$

$$V(0, x, \tau) = V_1(x, \tau), \quad V(S_0, x, \tau) = V_2(x, \tau), \quad V(S, 0, \tau) = V_3(S, \tau). \tag{8}$$

The rest of the paper is organized as follows. In Sect. 2 we describe the splitting method while in the next one we derive two difference schemes for the parabolic subproblem. In Sect. 4 we construct the difference scheme for the hyperbolic subproblem and present some theoretical results. Some of the computational results obtained by the two numerical methods are presented and compared in Sect. 5. Finally, Sect. 6 summarizes our conclusions.

2   The Splitting Method

We will describe a splitting method for the problem (6)–(8) into two subproblems: the first with respect to (S, τ) and the second one - with respect to (x, τ). Let us introduce the non-uniform mesh in time: $0 = \tau_1 < \tau_2 < \cdots <$

$\tau_{P+1} = T$, $\triangle \tau_n = \tau_{n+1} - \tau_n$. Starting with the initial condition (7), we solve consequently two problems on each of the subintervals $(\tau_n, \tau_{n+1}]$, $n = 1, 2, \ldots, P$:

Parabolic Subproblem: For given $V(S, x, \tau_n)$, find the solution

$$u(S, x, \tau), \quad (S, x, \tau) \in (0, S_0) \times (0, 1) \times (\tau_n, \tau_{n+1/2}], \quad x\text{-fixed}$$

of the problem

$$\frac{1}{2}\frac{\partial u}{\partial \tau} = \frac{1}{2}\sigma^2 S^\gamma \frac{\partial^2 u}{\partial S^2} + rS\frac{\partial u}{\partial S} - ru, \tag{9}$$

$$u(S, x, \tau_n) = V(S, x, \tau_n), \tag{10}$$

$$u(0, x, \tau_{n+1/2}) = V_1(x, \tau_{n+1/2}), \tag{11}$$

$$u(S_0, x, \tau_{n+1/2}) = V_2(x, \tau_{n+1/2}); \tag{12}$$

Hyperbolic Subproblem: For given $u(S, x, \tau_{n+1/2})$ find the solution

$$V(S, x, \tau), \quad (S, x, \tau) \in (0, S_0) \times (0, 1) \times (\tau_{n+1/2}, \tau_{n+1}], \quad S\text{-fixed}$$

of the problem

$$\frac{1}{2}\frac{\partial V}{\partial \tau} + \mu_\gamma \frac{\partial V}{\partial x} = 0, \tag{13}$$

$$V(S, 0, \tau_{n+1}) = V_3(S, \tau_{n+1}), \tag{14}$$

$$V(S, x, \tau_{n+1/2}) = u(S, x, \tau_{n+1/2}). \tag{15}$$

Further on, we take advantage of the computational cost reduction yielded by the use of the parabolic (9)–(12) and hyperbolic (13)–(15) problems splitting and the robust difference schemes of S. Wang [11] and A.A. Samarskii [9] for the degenerate parabolic Eq. (9).

3 Difference Approximations of the Parabolic Subproblem

When a standard finite-difference method is applied to (9)–(12), one has to deal with the degeneracy at $S = 0$, a possible small volatility σ, etc. Therefore, one has to use an appropriate approximation to overcome these difficulties.

3.1 First Difference Approximation

First we will implement the S. Wang difference scheme [11]. Let us consider the problem (9)–(12). We rewrite Eq. (9) in the divergent form

$$\frac{1}{2}\frac{\partial u}{\partial \tau} = \frac{\partial}{\partial S}\left(aS\frac{\partial u}{\partial S} + bu\right) - cu, \tag{16}$$

where $a(S) = \frac{1}{2}\sigma^2 S^{\gamma-1}$, $b(S) = rS - \gamma a(S)$, $c(S) = 2r - \frac{1}{2}\gamma(\gamma-1)S^{\gamma-2}\sigma^2$. We divide interval $(0, S_0)$ on N subintervals $I_i = [S_i, S_{i+1}]$, $i = 1, 2, \ldots, N$ by nodes $0 = S_1 < S_2 < \ldots < S_{N+1} = S_0$. Let $h_i = S_{i+1} - S_i$, $i = 1, 2, \ldots, N$. Let us introduce the secondary mesh $S_{i-1/2} = 0.5(S_{i-1} + S_i)$, $i = 2, 3, \ldots, N+1$, $S_{i+1/2} = 0.5(S_i + S_{i+1})$, $i = 2, 3, \ldots, N$.

We integrate Eq. (16) over the interval $\left[S_{i-1/2}, S_{i+1/2}\right]$, $i = 2, 3, \ldots, N$ and apply the central rectangular formula to the integrals to get

$$\frac{1}{2}\frac{\partial u}{\partial \tau}\bigg|_{S_i} \hbar_i = \rho(u)|_{S_{i+1/2}} - \rho(u)|_{S_{i-1/2}} - c_i u_i \hbar_i, \tag{17}$$

where $\hbar_i = S_{i+1/2} - S_{i-1/2}$, $u_i = u(S_i, x, \tau)$, $c_i = c(S_i)$, $\rho(u) = aS\frac{\partial u}{\partial S} + bu$. In order to obtain an approximation of the flux $\rho(u)$ at $S_{i+1/2}$, $i = 2, 3, \ldots, N$ for fixed x and τ, we consider the boundary value problem

$$\left(a_{i+1/2}Sw' + b_{i+1/2}w\right)' = 0, \quad a_{i+1/2} = a(S_{i+1/2}), \quad b_{i+1/2} = b(S_{i+1/2}), \tag{18}$$

$$w(S_i) = u_i, \quad w(S_{i+1}) = u_{i+1}, \quad S \in I_i. \tag{19}$$

We solve the problem (18)–(19) for w and then find

$$\rho_i(u) = b_{i+1/2}\frac{S_{i+1}^{\alpha_i}u_{i+1} - S_i^{\alpha_i}u_i}{S_{i+1}^{\alpha_i} - S_i^{\alpha_i}}, \quad \alpha_i = \frac{b_{i+1/2}}{a_{i+1/2}}, \tag{20}$$

that is an approximation of the flux at $S_{i+1/2}$, $i = 2, 3, \ldots, N$. For the flux at $S_{i-1/2}$ for $i = 3, 4, \ldots, N$ we have an analogical expression.

This analysis is not applicable on the interval $I_1 = [S_1, S_2] \equiv [0, S_2]$, because (18) degenerates. Instead of the problem (18)–(19) we consider the problem

$$\left(a_{3/2}Sw' + b_{3/2}w\right)' = C_1, \quad S \in I_1, \quad w(0) = u_1, \quad w(S_2) = u_2.$$

We solve this problem and find for $\rho_1(u)$

$$\rho_1(u) = \frac{1}{2}\left[\left(a_{3/2} + b_{3/2}\right)u_2 - \left(a_{3/2} - b_{3/2}\right)u_1\right]. \tag{21}$$

Now we insert the expressions of the flux (20), (21) into (17) to obtain at a fixed x the expressions:

$$\frac{1}{2}\frac{\partial u}{\partial \tau}\bigg|_{S=S_2} \hbar_2 = b_{5/2}\frac{S_3^{\alpha_2}u_3 - S_2^{\alpha_2}u_2}{S_3^{\alpha_2} - S_2^{\alpha_2}} - \frac{1}{2}\left[\left(a_{3/2} + b_{3/2}\right)u_2 - \left(a_{3/2} - b_{3/2}\right)u_1\right]$$

$$-\hbar_2 c_2 u_2,$$

$$\frac{1}{2}\frac{\partial u}{\partial \tau}\bigg|_{S=S_i} \hbar_i = b_{i+1/2}\frac{S_{i+1}^{\alpha_i}u_{i+1} - S_i^{\alpha_i}u_i}{S_{i+1}^{\alpha_i} - S_i^{\alpha_i}} - b_{i-1/2}\frac{S_i^{\alpha_{i-1}}u_i - S_{i-1}^{\alpha_{i-1}}u_{i-1}}{S_i^{\alpha_{i-1}} - S_{i-1}^{\alpha_{i-1}}}$$

$$-\hbar_i c_i u_i, \quad i = 3, 4, \ldots, N.$$

Let us introduce the non-uniform mesh $0 = x_1 < x_2 < \ldots < x_j < x_{j+1} < \ldots < x_{M+1} = 1$, $h_j^x = x_{j+1} - x_j$. With respect to time we construct the fully implicit scheme. Then we have:

$$\frac{\bar{u}_{2,j} - u_{2,j}}{\Delta \tau_n} \hbar_2 = b_{5/2} \frac{S_3^{\alpha_2} \bar{u}_{3,j} - S_2^{\alpha_2} \bar{u}_{2,j}}{S_3^{\alpha_2} - S_2^{\alpha_2}}$$

$$-\frac{1}{2} \cdot \left[(a_{3/2} + b_{3/2}) \bar{u}_{2,j} - (a_{3/2} - b_{3/2}) \bar{u}_{1,j} \right] - \hbar_2 c_2 \bar{u}_{2,j}, \quad j = 1, 2, \ldots, M, \quad (22)$$

$$\frac{\bar{u}_{i,j} - u_{i,j}}{\Delta \tau_n} \hbar_i = b_{i+1/2} \frac{S_{i+1}^{\alpha_i} \bar{u}_{i+1} - S_i^{\alpha_i} \bar{u}_i}{S_{i+1}^{\alpha_i} - S_i^{\alpha_i}} - b_{i-1/2} \frac{S_i^{\alpha_{i-1}} \bar{u}_i - S_{i-1}^{\alpha_{i-1}} \bar{u}_{i-1}}{S_i^{\alpha_{i-1}} - S_{i-1}^{\alpha_{i-1}}}$$

$$-\hbar_i c_i \bar{u}_{i,j}, \quad i = 3, 4, \ldots, N, \quad j = 2, 3, \ldots, M \quad (23)$$

$$u_{i,j} = V(S_i, x_j, \tau_n), \quad i = 1, 2, \ldots, N+1; \quad j = 1, 2, \ldots, M+1, \quad n = 1, 2, \ldots P, \quad (24)$$

$$V(S_i, x_j, 0) = V_0(S_i, x_j), \quad i = 1, 2, \ldots, N+1; \quad j = 1, 2, \ldots, M+1, \quad (25)$$

$$\bar{u}_{1,j} = V_1(S_1, x_j), \quad j = 1, 2, \ldots, M, \quad (26)$$

$$\bar{u}_{N+1,j} = V_2(S_{N+1}, x_j), \quad j = 1, 2, \ldots, M, \quad (27)$$

where $\bar{u}$ is the approximate solution on the $n + 1/2$-th time level and u is the approximate solution on the n-th time level. The truncation error of the scheme (22)–(27) is of order $O(\Delta \tau + h)$, where $h = \max\limits_{1 \leq j \leq M} h_j$, $\Delta \tau = \max\limits_{1 \leq n \leq P} \Delta \tau_n$. This system has a three-diagonal matrix and its solution can be found by Thomas procedure. Applying the discrete maximum principle [9] to (22)–(27) one can prove the following assertion:

Lemma 1. *Suppose that* $u_{i,j} > 0$, $i = 1, 2, \ldots, N+1$, $j = 1, 2, \ldots, M+1$. *Then for sufficiently small* $\Delta \tau_n$ *we have* $\bar{u}_{i,j} \geq 0$, $i = 1, 2, \ldots, N+1$, $j = 1, 2, \ldots, M+1$.

3.2 Second Difference Approximation

Now we will construct the classical monotone scheme of A. A. Samarskii [9].

We rewrite Eq. (9) in the form

$$\frac{1}{2} \frac{\partial u}{\partial \tau} = \frac{\partial}{\partial S} \left(k(S) \frac{\partial u}{\partial S} \right) + p(S) \frac{\partial u}{\partial S} - ru, \quad (28)$$

where

$$k(S) = 0.5 \sigma^2 S^\gamma, \quad p(S) = rS - 0.5 \gamma S^{\gamma - 1} \sigma^2. \quad (29)$$

The initial and boundary conditions are (10)–(12). Let us introduce the uniform mesh

$$\bar{\omega}_h = \{ S_i = (i-1)h, \quad i = 1, 2, \ldots, N+1, \quad h = S_0/N \}.$$

For the problem (10)–(12), (28), (29) we derive the fully implicit monotone difference scheme with local approximation error $O(\triangle\tau + h^2)$:

$$\frac{\bar{u}_{i,j} - u_{i,j}}{\triangle\tau_n} = \bar{\rho}_i \frac{1}{h} \left[a_{i+1} \frac{\bar{u}_{i+1,j} - \bar{u}_{i,j}}{h} - a_i \frac{\bar{u}_{i,j} - \bar{u}_{i-1,j}}{h} \right] + b_i^+ a_{i+1} \frac{\bar{u}_{i+1,j} - \bar{u}_{i,j}}{h}$$

$$+ b_i^- a_i \frac{\bar{u}_{i,j} - \bar{u}_{i-1,j}}{h} - r\bar{u}_{i,j}, \quad i = 2, 3, \dots, N, \quad j = 1, 2, \dots, M,$$

where $\bar{\rho}_i = \frac{1}{1 + \frac{1}{2}h\frac{|p(S_i)|}{k(S_i)}}$, $a_i = k(S_i - \frac{h}{2})$, $b_i^+ = \frac{p^+(S_i)}{k(S_i)}$, $b_i^- = \frac{p^-(S_i)}{k(S_i)}$, $p^- = \frac{p-|p|}{2}$,

$p^+ = \frac{p+|p|}{2}$. The approximations for the initial and boundary conditions are the same as in the first method. This discrete problem has a three-diagonal matrix too and also has the non-negativity property of Lemma 1.

4   Full Discretization

We approximate the hyperbolic subproblem by implicit difference scheme. For the boundary condition (14) we have

$$\hat{V}_{i,1} = V_3(S_i, x_1), \quad i = 2, 3, \dots, N. \tag{30}$$

The approximation for the initial condition is

$$V(S_i, x_j, \tau_{n+1/2}) = u(S_i, x_j, \tau_{n+1/2}). \tag{31}$$

For Eq. (13) we construct an implicit backward difference scheme

$$\frac{\hat{V}_{i,j} - \bar{u}_{i,j}}{\triangle\tau_n} + \mu_{\gamma,i} \frac{\hat{V}_{i,j} - \hat{V}_{i,j-1}}{h_{j-1}^x} = 0, \quad i = 2, 3, \dots, N, \quad j = 2, 3, \dots, M+1. \tag{32}$$

The truncation error of the scheme (30)–(32) is of order $O(\tau + h)$ and is unconditionally stable.

In view of Lemma 1 we have the following

Theorem 1. *For sufficiently small τ the numerical solutions, obtained by the two methods, are non-negative.*

Next, let us discuss the convergence of the numerical methods. From the discretization in Sect. 3.1, one can see that the consistency of the first difference scheme lies on the consistency of the flux $\rho(u)$ approximation. In a similar way as it is been shown in [11] for the case $\gamma = 2$ the discretization (22)–(27) admits a finite element formulation with special trial space S_h. Then in our case the following analog of the estimate in Lemma 4.2 in [11] holds:

$$\|\rho(w) - \rho_h(w)\|_{\infty, I_i} \leq C \|\rho'(w)\|_{\infty, I}, \quad i = 1, 2, \dots, N,$$

where w is a sufficiently smooth function and w_I is the S_h interpolant. Now, using a technique similar to those in Sect. 4 of [11], one can obtain the following convergence result:

Theorem 2. *Let V be the exact solution of (6)–(8) and $\{V^n\}$ be the numerical solution of the difference scheme (22)–(27). Then, there exists a positive constant C, independent of N, M and τ, such that the global error satisfies*

$$\|V(t_n)|_{\Omega_h} - V^n\|_\infty \leq C(\tau + h),$$

where $V(t_n)|_{\overline{\Omega}_h}$ is the restriction of the exact solution on the product of the meshes with respect to S and x.

5 Numerical Experiments

In order to observe the behaviour of the accuracy and the rate of convergence for the two methods we use the analytical solution

$$V_a(S, x, \tau) = (2 - x)\,(S/S_0)^2\,e^{-r\tau}.$$

We choose this function because its character is similar to the character of the exact solution of the problem under consideration when $\gamma = 2$ [2]. For all examples, presented in this paper, we use the following fixed values of the parameters: $S_0 = 2$, $T = 1$, $K = 1$, $r = 0.05$, $\sigma = 0.4$. Numerical experiments were performed for the different values of $\gamma \in (0, 2]$ and for $\mu_\gamma(S) = S$. For every one of the experiments the time-step decreases until establishment of the first four significant digits of the relative C-norm of the error (RCN) at the last level $\tau = T$ is reached. The rate of convergence (RC) is calculated using the double mesh principle.

The results from the numerical investigations of the discretizations are the following. The scheme, constructed in Sect. 3.1, works properly for $\gamma \in [0.8, 2]$ and in this interval, in general, it is more accurate and has higher rate of convergence than the discretization from Sect. 3.2 (see Tables 1, 2 and 3). For the values $\gamma \in (0, 0.8)$ the dominator in (20) is equal to zero in the computer and the discretization from Sect. 3.1 is not applicable. The discretization from Sect. 3.2 is applicable for all values $\gamma \in (0, 2]$ (see Tables 1, 2 and 3). For the two discretizations the rate of convergence decreases, when γ decreases (Tables 1, 2, 3 and 4).

Table 1. $\gamma = 1.5$

Space steps		0.1	0.05	0.025	0.0125	0.00625
Discretization 3.1	RCN	1.440 E-4	3.836 E-5	9.986 E-6	2.563 E-6	6.489 E-7
	RC	-	1.91	1.94	1.96	1.98
Discretization 3.2	RCN	4.338 E-4	1.263 E-4	3.462 E-5	9.144 E-6	2.369 E-6
	RC	-	1.78	1.87	1.92	1.95

Table 2. $\gamma = 1$

Space steps		0.1	0.05	0.025	0.0125	0.00625
Discretization 3.1	RCN	1.406 E-3	5.388 E-4	1.655 E-4	4.434 E-5	1.152 E-5
	RC	-	1.38	1.70	1.90	1.94
Discretization 3.2	RCN	1.472 E-3	6.354 E-4	2.468 E-4	8.485 E-5	2.467 E-5
	RC	-	1.21	1.36	1.55	1.78

Table 3. $\gamma = 0.8$

Space steps		0.1	0.05	0.025	0.0125	0.00625
Discretization 3.1	RCN	1.675 E-3	7.956 E-4	3.452 E-4	1.248 E-4	3.698 E-5
	RC	-	1.08	1.20	1.47	1.75
Discretization 3.2	RCN	1.496 E-3	7.190 E-4	3.231 E-4	1.338 E-4	4.967 E-5
	RC	-	1.06	1.16	1.27	1.43

Table 4. $\gamma = 0.1$, Discretization 3.2

Space steps	0.1	0.05	0.025	0.0125	0.00625
RCN	9.692 E-4	4.962 E-4	2.508 E-4	1.256 E-4	6.240 E-5
RC	-	0.96	0.99	1.00	1.01

6 Conclusions

We derive and implement two splitting techniques to price path dependent Asian options. At the first splitting we use a fitted finite volume method for the spatial discretization of the one-dimensional parabolic subproblem, while the second splitting is a Samarskii's scheme. The numerical experiments confirm the positivity preserving of the numerical solutions and their efficiency in dependence on the order of degeneracy γ.

Acknowledgement. This work was partially supported by the European Union under Grant Agreement number 304 617 (FP7 Marie Curie Action Project Multi-ITN Strike - Novel Methods in Computational Finance) and the Bulgarian Fund of Sciences under Grants No. FNI I 02/9-2014 and No. FNI I 02/20-2014.

References

1. Cen, Z., Le, A., Xu, A.: Finite difference scheme with a moving mesh for pricing Asian options. Appl. Math. Comp. **219**(16), 8667–8675 (2013)
2. Chernogorova, T.P., Vulkov, L.G.: Two splitting methods for a fixed strike Asian option. In: Dimov, I., Faragó, I., Vulkov, L. (eds.) NAA 2012. LNCS, vol. 8236, pp. 214–221. Springer, Heidelberg (2013)

3. Cox, J., Rubinstein, M.: Option Markets. Prentice-Hall, Englewood Cliffs (1985)
4. Hugger, J.: A fixed strike Asian option and comments on its numerical solution. ANZIAM J. **45**(E), C215–C231 (2004)
5. Hugger, J.: Wellposedness of the boundary value formulation of a fixed strike Asian option. J. Comp. Appl. Math. **105**, 460–481 (2006)
6. Marcozzi, M.D.: An addaptive extrapolation discontinuous Galerkin method for the valuation of Asian options. J. Comp. Math. **235**, 3632–3645 (2011)
7. Meyer, G.H.: On pricing American and Asian options with PDE methods. Acta Math. Univ. Comenianne LXX **1**, 153–165 (2000)
8. Oosterlee, C.W., Frish, J.C., Gaspar, F.J.: TVD, WENO and blended BDF discretizations for Asian options. Comp. Visual. Sci. **6**, 131–138 (2004)
9. Samarskii, A.A.: The Theory of Difference Schemes. Marcel Dekker, New York (2001)
10. Sengypta, I.: Pricing Asian options in financial markets using Melling transformations. EJDE **2014**(234), 1–9 (2014)
11. Wang, S.: A novel fitted finite volume method for Black-Scholes equation governing option pricing. IMA J. Numer. Anal. **24**, 699–720 (2004)
12. Wilmott, P., Dewyne, J., Howison, S.: Option Pricing: Mathematical Models and Computation. Oxford Financial Press, Oxford (1993)
13. Zvan, R., Forsyth, P.A., Vetzal, K.: Robust numerical methods for PDE models of Asian options. J. Comp. Finance **1**(2), 39–78 (1998)

Operator-Difference Scheme
with a Factorized Operator

Petr N. Vabishchevich[1,2]([⊠])

[1] Nuclear Safety Institute, 52, B. Tulskaya, 115191 Moscow, Russia
[2] North-Eastern Federal University, 58, Belinskogo, 677000 Yakutsk, Russia
vabishchevich@gmail.com

Abstract. In the study of difference schemes for time-dependent problems of mathematical physics, the general theory of stability (well-posedness) for operator-difference schemes is in common use. At the present time, the exact (matching necessary and sufficient) conditions for stability are obtained for a wide class of two- and three-level difference schemes considered in finite-dimensional Hilbert spaces.

The main results of the theory of stability for operator-difference schemes are obtained for problems with self-adjoint operators. In this work, we consider difference schemes for numerical solution of the Cauchy problem for first order evolution equation, where non-self-adjoint operator is represented as a product of two non-commuting self-adjoint operators. We construct unconditionally stable regularized schemes based on the solution of a grid problem with a single operator multiplier on the new time level.

1 Introduction

Samarskii's theory of stability (well-posedness) for operator-difference schemes [1,2] is the theoretical basis of the study of numerical methods for solving time-dependent problems. In general, stability conditions for difference schemes with self-adjoint operators can be formulated in a more simple way, compared to the non self-adjoint case. Some important classes of two- and three-level difference schemes having non-self-adjoint operators are considered. In particular, special attention should be paid to difference schemes with a subordinate skew-symmetric part.

Difference schemes with weights have a special meaning for computational practice. In some cases, weighting factors can be varying both in time and space. These schemes have non-self-adjoint original operators, but they can be transformed into schemes with symmetric operators, particularly, by a suitable choice of special norms. This symmetrization possibility allows us to study these schemes on the basis of the general results of the theory of stability (well-posedness) for operator-difference schemes [3].

The earlier considered schemes with factorized factors refer to problems, where factorized structure is formed by a special approximation in space and/or time. In some cases, it is of interest to highlight factorized structure of the

© Springer International Publishing Switzerland 2015
I. Lirkov et al. (Eds.): LSSC 2015, LNCS 9374, pp. 72–79, 2015.
DOI: 10.1007/978-3-319-26520-9_7

main operator of problem. Here we investigate a new class of operator-difference schemes with a factorized operator. We consider the Cauchy problem

$$\frac{du}{dt} + A_1 A_2 u = f(t), \quad 0 < t \le T, \tag{1}$$

$$u(0) = u_0, \tag{2}$$

with self-adjoint positive definite operators A_α, $\alpha = 1, 2$. In the operator $A = A_1 A_2$ we separate the two multipliers A_1 and A_2, and on the basis of one of these we easily construct implicit approximation, while for the second one we focus on explicit approximation.

Unconditionally stable regularized schemes are constructed here. Their numerical implementation involves only the operator A_1. Such schemes, for example, can be used to solve problems with anomalous diffusion, where A_1 corresponds to the standard diffusion operator, whereas A_2 is a fractional operator $(A_2 = (-\triangle)^{-\beta}, 0 < \beta < 1)$.

2   Cauchy Problem

Let H be the finite-dimensional Hilbert space, where the scalar product and norm are $(\cdot, \cdot)$ and $\| \cdot \|$, respectively. We seek the solution $u(t)$ of the first order evolution Eq. (1). In (1) $f(t) \subset L_2(0, T; H)$ is given, and A_1, A_2 are linear and time-independent operators from H to H ($A_\alpha : H \to H$, $\alpha = 1, 2$). Eq. (1) is supplemented with the initial condition (2).

We consider the Cauchy problem (1), (2) under the conditions that the operators A_1 and A_2 are self-adjoint and positive definite in H:

$$A_\alpha = A_\alpha^* \ge \delta_\alpha E, \quad \delta_\alpha > 0, \quad \alpha = 1, 2, \tag{3}$$

where E is the identity operator in H. The peculiarity of the problem (1)–(3) is that the operator $A = A_1 A_2$ is non-self-adjoint since the operators A_1 and A_2 are non-commuting. However, such problems can be easily symmetrized, for example, by multiplying Eq. (1) by the operator A_2.

First, we obtain the simplest a priori estimates for the solution of the Cauchy problem (1)–(3), which will serve as a basis in the study of the operator-difference schemes. For $D = D^* > 0$ by H_D we denote the space H with the scalar product $(y, w)_D = (Dy, w)$ and norm $\|y\|_D = (Dy, y)^{1/2}$.

We multiply, for example, Eq. (1) in H by $A_2 \dfrac{du}{dt}$ and get

$$\left(A_2 \frac{du}{dt}, \frac{du}{dt} \right) + \frac{1}{2} \frac{d}{dt} (A_1 A_2 u, A_2 u) = \left(f, A_2 \frac{du}{dt} \right).$$

Taking into account (3) and

$$\left(f, A_2 \frac{du}{dt} \right) \le \left(A_2 \frac{du}{dt}, \frac{du}{dt} \right) + \frac{1}{4} (A_2 f, f),$$

we obtain the inequality

$$\frac{d}{dt}\|A_2u\|_{A_1}^2 \leq \frac{1}{2}\|f\|_{A_2}^2.$$

For it we get the desired a priori estimate

$$\|A_2u(t)\|_{A_1}^2 \leq \|A_2u_0\|_{A_1}^2 + \frac{1}{2}\int_0^t \|f(\theta)\|_{A_2}^2 d\theta, \qquad (4)$$

which ensures the stability of the solution of the problem (1)–(3) with respect to the initial data and right-hand side.

Along with (4) we also present an a priori estimate in a more simple norm. Multiplying Eq. (1) in H by A_2u, we get

$$\frac{1}{2}\frac{d}{dt}(A_2u, u) + (A_1A_2u, A_2u) = (f, A_2u).$$

Taking into account

$$(f, A_2u) < (A_1A_2u, A_2u) + \frac{1}{4}(A_1^{-1}f, f),$$

we obtain

$$\frac{d}{dt}\|u\|_{A_2}^2 \leq \frac{1}{2}\|f\|_{A_1^{-1}}^2.$$

This inequality implies the a priori estimate

$$\|u(t)\|_{A_2}^2 \leq \|u_0\|_{A_2}^2 + \frac{1}{2}\int_0^t \|f(\theta)\|_{A_1^{-1}}^2 d\theta \qquad (5)$$

for the solution of the problem (1)–(3).

The symmetrization of Eq. (1) can be performed not only by multiplying by the operator A_2. The second approach is associated with multiplying by A_1^{-1}. In this case from (1) we obtain the equation

$$A_1^{-1}\frac{du}{dt} + A_2u = A_1^{-1}f(t), \quad 0 < t \leq T,$$

for which we can obtain a priori estimates slightly different from (4), (5).

3   Scheme with Weights

For the numerical solution of the differential-operator problem (1), (2) we use the usual scheme with weights [1]. We define a uniform in time grid

$$\overline{\omega}_\tau = \omega_\tau \cup \{T\} = \{t^n = n\tau, \quad n = 0, 1, ..., N, \quad \tau N = T\}$$

with time step $\tau > 0$ and denote $y^n = y(t^n)$, $t^n = n\tau$. When using a two-level scheme, Eq. (1) is approximated by the following difference equation

$$\frac{y^{n+1} - y^n}{\tau} + A_1A_2(\sigma y^{n+1} + (1-\sigma)y^n) = \varphi^n, \qquad (6)$$

where σ is a numerical parameter (weight), which usually satisfies $0 \le \sigma \le 1$. Concerning φ^n, one possibility is to set $\varphi^n = f(\sigma t^{n+1} + (1 - \sigma)t^n)$. Taking into account (2) we supplement (6) with the initial condition

$$y^0 = u_0. \tag{7}$$

A detailed study of the two- and three-level schemes with weights (e.g., necessary and sufficient conditions for stability, the choice of the norm) for problems with self-adjoint operators was held in [2,3]. Taking into account the fact that, as explained above, the Eq. (1) can be symmetrized, these results can be used also in the analysis of difference schemes for the problem (1)–(3). Here, we restrict ourselves to the simplest estimates of the stability of the operator-difference scheme (6), (7). The estimates (4), (5) for the solution of the differential problem (1)–(3) will be our starting point.

Theorem 1. *For $\sigma \ge 1/2$, the operator-difference scheme (6), (7) is unconditionally stable in H_D, $D = A_2 A_1 A_2, Q$, where*

$$Q = A_2 + \left(\sigma - \frac{1}{2}\right)\tau A_2 A_1 A_2,$$

and the solution satisfies the a priori estimates

$$\|A_2 y^{n+1}\|_{A_1}^2 \le \|A_2 u_0\|_{A_1}^2 + \frac{1}{2}\sum_{k=0}^{n}\tau\|A_2\varphi^k\|_{Q^{-1}}^2, \tag{8}$$

$$\|y^{n+1}\|_Q^2 \le \|u_0\|_Q^2 + \frac{1}{2}\sum_{k=0}^{n}\tau\|\varphi^k\|_{A_1^{-1}}^2. \tag{9}$$

After symmetrization, using the introduced notation, we can write the scheme (6) as

$$Q\frac{y^{n+1} - y^n}{\tau} + A_2 A_1 A_2 \frac{y^{n+1} + y^n}{2} = A_2\varphi^n. \tag{10}$$

We multiply this equation in H by $2(y^{n+1} - y^n)$ and obtain the equality

$$2\tau\left(Q\frac{y^{n+1} - y^n}{\tau}, \frac{y^{n+1} - y^n}{\tau}\right)$$

$$+(A_1 A_2 y^{n+1}, A_2 y^{n+1}) - (A_1 A_2 y^n, A_2 y^n) = 2\tau\left(A_2\varphi^n, \frac{y^{n+1} - y^n}{\tau}\right).$$

Using the following inequality

$$\left(A_2\varphi^n, \frac{y^{n+1} - y^n}{\tau}\right) \le \left(Q\frac{y^{n+1} - y^n}{\tau}, \frac{y^{n+1} - y^n}{\tau}\right)$$

$$+\frac{1}{4}\left(Q^{-1}A_2\varphi^n, A_2\varphi^n\right),$$

we get the estimate

$$\|A_2 y^{n+1}\|_{A_1}^2 \le \|A_2 y^n\|_{A_1}^2 + \frac{\tau}{2}\|A_2 \varphi^n\|_{Q^{-1}}^2.$$

This inequality leads to the desired a priori estimate (8).

The estimate (9) is proved in a similar way. We multiply Eq. (10) in H by $\tau(y^{n+1} + y^n)$ and obtain

$$(Qy^{n+1}, y^{n+1}) - (Qy^n, y^n) + \frac{\tau}{2}(A_1 A_2(y^{n+1} + y^n), A_2(y^{n+1} + y^n))$$

$$= \tau(\varphi^n, A_2(y^{n+1} + y^n)).$$

For the right-hand side we have

$$(\varphi^n, A_2(y^{n+1} + y^n)) \le \frac{1}{2}(A_1 A_2(y^{n+1} + y^n), A_2(y^{n+1} + y^n)) + \frac{1}{2}(A_1^{-1}\varphi^n, \varphi^n).$$

Thus, at $t = t^{n+1}$ we obtain the estimate

$$\|y^{n+1}\|_Q^2 \le \|y^n\|_Q^2 + \frac{\tau}{2}\|\varphi^n\|_{A_1^{-1}}^2,$$

which implies (9).

The estimates (8), (9) are grid analogues of estimates (4), (5) and ensure unconditional stability of difference scheme with weights (6), (7) under natural restriction $\sigma \ge 1/2$. Considering the corresponding problem for the error, we make sure that the solution of the operator-difference problem (6), (7) converges to the solution of the differential-difference problem (1), (2) in H_D for $\sigma \ge 1/2$ with $\mathcal{O}((2\sigma - 1)\tau + \tau^2)$. For $\sigma = 1/2$ we get second order of convergence with respect to τ, for the other values of σ we get first order convergence.

In the scheme with weights (6), (7) the transition to a new time level is provided by solving the problem

$$(E + \sigma \tau A_1 A_2) y^{n+1} = \psi^n, \tag{11}$$

i.e. by inverting the operator $E + \sigma \tau A_1 A_2$. This discrete problem, in general, is much more complicated to solve, compare to

$$(E + \sigma \tau A_\alpha) y^{n+1} = \psi_\alpha^n, \tag{12}$$

where $\alpha = 1$ or $\alpha = 2$. The problem (11) is more difficult to solve, in particular, because we have to invert a non-self-adjoint operator. In this situation a reasonable task is to try to construct regularized schemes, based on solving the more simple problem, (12).

4   Regularized Scheme

We are interested in a scheme, for which the implicit approximation is associated with the operator A_1 ($\alpha = 1$ in (12)). In this case the operator A_2 has to be bounded.

$$A_2 \le \Delta_2 E. \tag{13}$$

In this particular case we suppose that the operator A_1 in (1), (2) is the main operator (and more easy to invert), while the operator A_2 is more difficult to invert.

For the numerical solution of the problem (1)–(3), (13) we use the scheme

$$(E + \sigma\tau A_1)\frac{y^{n+1} - y^n}{\tau} + A_1 A_2 y^n = \varphi^n. \tag{14}$$

In this case we can only rely on the first order of convergence with respect to time for any value of the weight σ.

The scheme (14) can be considered as a regularized scheme

$$\frac{y^{n+1} - y^n}{\tau} + (E + \sigma\tau A_1)^{-1} A_1 A_2 y^n = \varphi^n. \tag{15}$$

In accordance with the principle of regularization of operator-difference schemes [1] the stability is ensured by a multiplicative perturbation of the operator A_1. The main result of stability of this scheme is formulated in the following theorem.

Theorem 2. *For* $\sigma \geq \Delta_2/2$ *the operator-difference scheme (7), (13), (15) is unconditionally stable in* H_D, $D = \tilde{A}, Q$, *where*

$$\tilde{A} = A_2(E + \sigma\tau A_1)^{-1} A_1 A_2, \quad Q = A_2 - \frac{1}{2}\tau\tilde{A} > 0.$$

At the same time the solution satisfies the a priori estimates

$$\|y^{n+1}\|_{\tilde{A}}^2 \leq \|u_0\|_{\tilde{A}}^2 + \frac{1}{2}\sum_{k=0}^{n}\tau\|A_2\varphi^k\|_{Q^{-1}}^2, \tag{16}$$

$$\|y^{n+1}\|_Q^2 \leq \|u_0\|_Q^2 + \frac{1}{2}\sum_{k=0}^{n}\tau\|A_2\varphi^k\|_{\tilde{A}^{-1}}^2. \tag{17}$$

The proof is similar to that of Theorem 1. After symmetrization from (15) we get

$$A_2\frac{y^{n+1} - y^n}{\tau} + A_2(E + \sigma\tau A_1)^{-1} A_1 A_2 y^n = A_2\varphi^n.$$

Taking into account the introduced notation, we write this scheme as

$$Q\frac{y^{n+1} - y^n}{\tau} + \tilde{A}\frac{y^{n+1} + y^n}{2} = A_2\varphi^n. \tag{18}$$

The most important element of our consideration is associated with the proof of positive definiteness of the operator Q. We have

$$Q = A_2 - \frac{1}{2}\tau\tilde{A} = A_2 - \frac{1}{2}\tau A_2(E + \sigma\tau A_1)^{-1} A_1 A_2.$$

Taking into account that for the positive definite operator A_1

$$(E + \sigma\tau A_1)^{-1} A_1 < \frac{1}{\sigma\tau}E,$$

and (13) we obtain

$$Q > A_2 - \frac{1}{2\sigma} A_2^2 \geq \left(1 - \frac{\Delta_2}{2\sigma}\right) A_2.$$

Thereby, for $\sigma \geq \Delta_2/2$ we have $Q > 0$.

Multiplying Eq. (18) scalarly in H by $2(y^{n+1} - y^n)$, we get

$$\|y^{n+1}\|_{\tilde{A}}^2 - \|y^n\|_{\tilde{A}}^2 + 2\tau \left\|\frac{y^{n+1} - y^n}{\tau}\right\|_Q^2 = 2\tau \left(A_2\varphi^n, \frac{y^{n+1} - y^n}{\tau}\right).$$

In the standard way we obtain the inequality

$$\|y^{n+1}\|_{\tilde{A}}^2 \leq \|y^n\|_{\tilde{A}}^2 + \frac{\tau}{2}\|A_2\varphi^n\|_{Q^{-1}}^2,$$

which implies the estimate (16).

Similarly, multiplying (18) scalarly in H by $\tau(y^{n+1} + y^n)$, we get

$$\|y^{n+1}\|_Q^2 - \|y^n\|_Q^2 + \frac{\tau}{2}\|y^{n+1} + y^n\|_{\tilde{A}}^2 = \tau(A_2\varphi^n, y^{n+1} + y^n).$$

Estimating the right-hand side in the standard way, we obtain the inequality

$$\|y^{n+1}\|_Q^2 \leq \|y^n\|_Q^2 + \frac{\tau}{2}\|A_2\varphi^n\|_{\tilde{A}^{-1}}^2,$$

which leads to the estimate (17).

Special attention should be paid to the case when the main operator is A_2, and for the operator A_1 we have (see (13)) the inequality

$$A_1 \leq \Delta_1 E. \tag{19}$$

In this case, we focus on the regularization of the operator A_2. Similar to (15), we use regularized scheme:

$$\frac{y^{n+1} - y^n}{\tau} + A_1(E + \sigma\tau A_2)^{-1}A_2 y^n = \varphi^n. \tag{20}$$

The computational implementation of this scheme is based on the solution of the problem (12) for $\alpha = 2$. The following assertion holds.

Theorem 3. *For $\sigma \geq \Delta_1/2$ the operator-difference scheme (7), (19), (20) is unconditionally stable in H_D, $D = \tilde{A}, Q$, where*

$$\tilde{A} = (E + \sigma\tau A_2)^{-1}A_2, \quad Q = A_1^{-1} - \frac{1}{2}\tau\tilde{A} > 0.$$

and the solution satisfies the a priori estimates

$$\|y^{n+1}\|_{\tilde{A}}^2 \leq \|u_0\|_{\tilde{A}}^2 + \frac{1}{2}\sum_{k=0}^{n} \tau\|A_1^{-1}\varphi^k\|_{Q^{-1}}^2, \tag{21}$$

$$\|y^{n+1}\|_Q^2 \leq \|u_0\|_Q^2 + \frac{1}{2}\sum_{k=0}^{n} \tau\|A_1^{-1}\varphi^k\|_{\tilde{A}^{-1}}^2. \tag{22}$$

The proof is based on a slightly different symmetrization of scheme (20). We multiply it by A_1^{-1} and pass from (20) to the scheme

$$A_1^{-1}\frac{y^{n+1} - y^n}{\tau} + (E + \sigma\tau A_2)^{-1}A_2 y^n = A_1^{-1}\varphi^n.$$

Using the new notation this scheme takes the form

$$Q\frac{y^{n+1} - y^n}{\tau} + \widetilde{A}\frac{y^{n+1} + y^n}{2} = A_1^{-1}\varphi^n. \tag{23}$$

For $\sigma \geq \Delta_1/2$ we have $Q > 0$. For the scheme (23) the following inequality holds

$$\|y^{n+1}\|_{\widetilde{A}}^2 \leq \|y^n\|_{\widetilde{A}}^2 + \frac{\tau}{2}\|A_1^{-1}\varphi^n\|_{Q^{-1}}^2,$$

$$\|y^{n+1}\|_Q^2 \leq \|y^n\|_Q^2 + \frac{\tau}{2}\|A_1^{-1}\varphi^n\|_{\widetilde{A}^{-1}}^2.$$

This leads us to the estimates (21), (22) for the difference scheme (7), (19), (20).

Acknowledgements. This work was supported by RFBR (project 14-01-00785)

References

1. Samarskii, A.A.: The Theory of Difference Schemes. Marcel Dekker, New York (2001)
2. Samarskii, A.A., Gulin, A.V.: Stability of Difference Schemes. URSS, Moscow (2004). In Russian
3. Samarskii, A.A., Matus, P.P., Vabishchevich, P.N.: Difference Schemes with Operator Factors. Springer, Dordrecht (2002)

Computational Identification of the Right Hand Side of the Parabolic Equations in Problems of Filtration

V.I. Vasil'ev[✉], M.V. Vasil'eva, A.M. Kardashevsky, and D.Ya. Nikiforov

NEFU Named After M.K. Ammosov, 58 Belinski, 677000, Yakutsk, Russia
vasvasil@mail.ru, kardam123@gmail.com

Abstract. In this paper, we will consider the right-hand side of a parabolic equation in a multidimensional domain, which depends only on time. For the numerical solution of the initial boundary value problem, a homogeneous implicit differential scheme is used. The problem at a particular time level is solved on the basis of a special decomposition into two standard elliptic boundary value problems. We discuss the results of numerical experiments for a model problem of filtration theory.

Keywords: Inverse problem · Identification of the coefficient · Parabolic partial differential equation · Difference scheme

1   Introduction

In this paper, we consider the inverse problem of determination of dependence of the parabolic equation right-hand side on time with known values of the solution at an interior point in the entire range of time. The problem is solved by finite difference method using an unconditionally stable homogeneous purely implicit difference scheme.

The non-classical problem on the new time layer is solved using the solution decomposition, which leads to the solution of two conventional elliptic problems. The facilities of the proposed computational algorithm are illustrated by numerical experiments carried out on a model non-stationary problem of filtration with a single well.

2   Problem Statement

We consider the two-dimensional problem for a parabolic equation defined in a rectangular domain

$$\Omega = \{x \mid x = (x_1, x_2), \quad 0 < x_\alpha < l_\alpha, \quad \alpha = 1, 2\}.$$

I. Lirkov et al. (Eds.): LSSC 2015, LNCS 9374, pp. 80–87, 2015.
DOI: 10.1007/978-3-319-26520-9_8

Let us formulate the direct problem as follows. We search function $u(x,t)$, $x \in \Omega$, $0 \leqslant t \leqslant T$, $T > 0$, which satisfies the parabolic equation

$$\frac{\partial u}{\partial t} = \text{div}(k(x)\text{grad}(u)) + \sum_{i=1}^{M} q_i(t)f_i(x), \quad x \in \Omega, \quad 0 < t \leqslant T, \qquad (1)$$

and the initial and boundary conditions:

$$k(x)\frac{\partial u}{\partial n} = 0, \quad x \in \partial\Omega, \quad 0 < t \leqslant T, \qquad (2)$$

$$u(x,0) = u_0(x), \quad x \in \Omega, \qquad (3)$$

where n is the outward normal to $\partial\Omega$. Here, $\partial\Omega$ is the boundary of Ω. The formulation (1)–(3) corresponds to a direct problem, in which all input parameters are known, i.e., the functions $k(x)$, $u_0(x)$, $q_i(t)$, $f_i(x)$, $i = 1, ..., M$ are given. Now, we consider the inverse problem in which the coefficients $q_i(t)$, $i = 1, ..., M$ in (1) must be determined. To find them we define additional conditions in the form of

$$u(x_i^*, t) = p_i(t), \quad x_i^* \in \Omega, \quad i = 1, ..., M, \quad 0 < t \leqslant T. \qquad (4)$$

Thus, we have formulated the inverse problem of finding the function $u(x,t)$, $q(t)_i$, $i = 1, ..., M$, satisfying Eqs. (1)–(3) with additional conditions (4) and there are reasons to consider this problem as well-posed.

The initial boundary value problem (1)–(4) is a mathematical model of the flow of a weakly compressible liquid in an elastically deformable porous medium with M wells, which are point sources/sinks with coordinates x_i^*, and given functions

$$f_i(x) = \delta(x - x_i^*), \quad i = 1, ..., M,$$

where $\delta(x - x_i^*)$ is the Dirac delta function. In this case, the values of bottom hole pressures $p_i(t)$ for each well are given, and the well debits must be determined $q_i(t)$, $i = 1, ..., M$. The undoubted advantage of this formulation of the problem is the possibility of constructing homogeneous differential schemes.

In this work, we consider questions of construction of computational algorithm for solution of the inverse problem and we formulate the conditions of algorithm applicability. In addition, under more strict conditions on the parameters of the problem we obtain an estimate of stability of the approximate solution, based on which, taking into account the linearity of the inverse problem, we show the convergence of the approximate solution to the exact solution of the differential problem.

3   Algorithm for Solving the Direct Problem

Numerical solution of the parabolic problem (1)–(3) is carried out using the finite-difference method. For this, in Ω we introduce the uniform rectangular grid ω:

$$\omega_\alpha = \left\{ x_\alpha = i_\alpha h_\alpha, \ i_\alpha = 1, 2, ..., N_\alpha - 1, \ N_\alpha h_\alpha = l_\alpha \right\}, \quad \alpha = 1, 2, \quad \omega = \omega_1 \times \omega_2.$$

We introduce the Hilbert space of the grid functions $y, v \in H = L_2(\omega)$, in which the inner product and the norm are defined as follows:

$$[y, w] \equiv \sum_{x \in \bar{\omega}} y(x) w(x) \hbar_1 \hbar_2, \quad \|y\| \equiv \sqrt{[y, y]},$$

where $\hbar_{\alpha i}$, $\alpha = 1, 2$ are the grid steps in the corresponding directions:

$$\hbar_{\alpha i} = \begin{cases} h_\alpha, & i = 1, 2, ..., N_\alpha - 1, \\ h_\alpha/2 & i = 0, N_\alpha. \end{cases}$$

For a grid analogue of the two-dimensional elliptic operator $\mathcal{A}$ we use the additive representation:

$$\mathcal{A} = \sum_{\alpha=1}^{2} \mathcal{A}_\alpha, \quad \alpha = 1, 2, \quad x \in \omega, \tag{5}$$

where $\mathcal{A}_\alpha$ is a discrete analogue of the differential operator of the original problem (1)–(2) for α direction, where $\alpha = 1, 2$.

Under the assumption of sufficient smoothness of the coefficient $k(x)$ the grid operator $\mathcal{A}_1$ in accordance with homogeneous Neumann boundary condition (2) can be written as follows:

$$\mathcal{A}_1 y = \begin{cases} \dfrac{2}{h_1} k(0.5h_1, x_2) \dfrac{y(h_1, x_2) - y(0, x_2)}{h_1}, & x_2 \in \omega_2, \\ \dfrac{1}{h_1} \left(k(x_1 + 0.5h_1, x_2) \dfrac{y(x_1 + h_1, x_2) - y(x_1, x_2)}{h_1} \right. & \\ \left. + k(x_1 - 0.5h_1, x_2) \dfrac{y(x_1, x_2) - y(x_1 - h_1, x_2)}{h_1} \right), & x \in \omega, \\ \dfrac{2}{h_1} k(l_1 - 0.5h_1, x_2) \dfrac{y(l_1, x_2) - y(l_1 - h_1, x_2)}{h_1}, & x_2 \in \omega_2. \end{cases} \tag{6}$$

Similarly, we construct the grid operator $\mathcal{A}_2$. The direct solutions show self-adjointness and positive definition of operators $\mathcal{A}_\alpha$:

$$\mathcal{A}_\alpha = \mathcal{A}_\alpha^* \geqslant 0, \quad \alpha = 1, 2.$$

The grid operator $\mathcal{A} = \mathcal{A}_1 + \mathcal{A}_2$ approximates the corresponding differential operator with an accuracy of $O(|h|^2)$, as in the differential case, are self-adjoint and positive definite in H:

$$\mathcal{A} = \mathcal{A}^* \geqslant 0. \tag{7}$$

Taking into account (7), we obtain the corresponding a priori estimate for the solution of the boundary value problem (1)–(3).

After the approximation in space of the problem (1)–(3) we come to the Cauchy problem:

$$\frac{dy}{dt} + \mathcal{A}y = \sum_{i=1}^{M} q_i(t) f_i(x), \quad 0 < t \leqslant T, \tag{8}$$

$$y(0) = u_0. \tag{9}$$

For the solution of the Cauchy problem (8), (9) we define the a priori estimate that shows the stability of solutions according to the initial data and the right-hand side

$$\|y(t)\| \leqslant \|u_0\| + \sum_{i=1}^{M} \|f_i\| \int_0^t |p(\theta)| d\theta. \tag{10}$$

It should be noted that the estimate (10) holds also in the Banach space of grid functions $L_\infty(\omega)$, in which

$$\| \cdot \| = \| \cdot \|_\infty, \quad \|y\|_\infty \equiv \max_{x \in \omega} |y|.$$

This fact can be established on the basis of the maximum principle for grid functions and relevant comparison theorems, taking into account the non-strict diagonal dominance of the matrix (operator) $\mathcal{A}$. We will use a uniform grid in time: $t^n = n\tau$, $n = 0, 1, ..., N$, $\tau N = T$ and denote $y^n = y(t^n)$, $t^n = n\tau$. Let's start the time discretization for the direct problem (8), (9). For the approximate solution of the initial-boundary value problem for the parabolic Eq. (1) we use an unconditionally stable purely implicit scheme

$$\frac{y^{n+1} - y^n}{\tau} + \mathcal{A}y^{n+1} - \sum_{i=1}^{M} q_i^{n+1} f_i, \quad n = 0, 1, ..., N - 1. \tag{11}$$

The initial condition (9) gives

$$y^0 = u_0. \tag{12}$$

The grid solution of the problem (11), (12) satisfies the following evaluation in a Banach space $L_\infty(\omega)$:

$$\|y^{n+1}\| \leqslant \|y^n\| + \tau \sum_{i=1}^{M} \|f_i\| |q_i^{n+1}|, \quad n = 0, 1, ..., N - 1. \tag{13}$$

The estimate (13) acts as a grid analogue of (11) for the solution of problem (8)–(10). To prove (13) we can apply the maximum principle for the grid functions. The second possibility of obtaining a priori estimates of stability (13) is based on introducing the concept of logarithmic norm.

4   Algorithm for Solving the Inverse Problem

After a full discretization (in space and time) of the direct problem (1)–(3) we can proceed to the identification of coefficients $q_i(t)$, $i = 1, ..., M$ of the right-hand side. Due to the additional conditions (4) in the internal nodes $x_i^* \in \omega$ we have

$$y^{n+1}(x_i^*) = p_i^{n+1}, \quad i = 1, ..., M, \quad n = 0, 1, ..., N - 1. \tag{14}$$

For approximate solution of the problem (11), (12), (14) at the new time level we will use the decomposition proposed in [11]:

$$y^{n+1}(x) = v^{n+1}(x) + \sum_{i=1}^{M} \dot{q}_i^{n+1} w_i(x), \quad x \in \overline{\omega}. \tag{15}$$

Another variant of the decomposition, which is applicable to a narrower class of inverse problems, was considered earlier in works [8,9]. To determine $v^{n+1}(x)$ the following homogeneous grid equation is solved

$$\frac{v^{n+1} - y^n}{\tau} + \mathcal{A}v^{n+1} = 0, \quad n = 0, 1, ..., N-1. \tag{16}$$

The grid functions $w_i^{n+1}(x)$ are determined by the operator equations (solving boundary value problems for elliptic equations with finite right-hand sides)

$$\frac{1}{\tau}w_i + \mathcal{A}w_i = f_i, \quad i = 1, ..., M. \tag{17}$$

When using the decomposition (15)–(17) the Eq. (11) is solved at any p_i^{n+1}, $i = 1, ..., M$. To determine them we involve additional conditions (14). Thus, the substitution of (15) into (14) gives, in general, a system of linear algebraic equations

$$\sum_{i=1}^{M} q_i^{n+1} w_i(x_i^*) = p^{n+1} - v^{n+1}(x_i^*), \quad i = 1, ..., M. \tag{18}$$

It should be noted that the solution of the operator Eq. (17), which is the second boundary value problem for an inhomogeneous elliptic equation in the case of problems of filtration theory, is a finite function in a small neighborhood of wells. Therefore, neglecting the mutual influence of wells, we have explicit formulas for determining well debits

$$q_i^{n+1} = \frac{p^{n+1} - v^{n+1}(x_i^*)}{w_i(x_i^*)}, \quad i = 1, ..., M. \tag{19}$$

The fundamental point of applicability of this algorithm is associated with the condition $w_i^{n+1}(x_i^*) \neq 0$. The grid function $w_i^{n+1}(x_i^*)$ is defined as solution of the grid elliptic Eq. (17). The property of preservation sign $w_i^{n+1}(x_i^*)$ is satisfied, in particular, when the following condition is fulfilled

$$f_i(x) \geqslant 0, \quad x \in \Omega, \quad i = 0, 1, ..., M,$$

and we obtain

$$\|f_i\| > 0, \quad i = 0, 1, ..., M.$$

In the case of filtration theory, these conditions are always satisfied, moreover

$$\|f_i\| = 1, \quad i = 0, 1, ..., M.$$

Therefore, no doubts about applicability of the proposed computational algorithm for filtration problems.

5   Computational Experiments

To demonstrate the efficiency of the proposed computational algorithm for identification of the right-hand side of a parabolic equation, we consider a model non-stationary problem of filtration of weakly compressible liquid in an elastically deformable porous medium with a single well drilled in the center of the domain. Below we present the results of solving the problem in a square $(l_1 = l_2 = 10)$. Assume, that

$$k(x) = 1, \quad f(x) = \delta(x - x^*), \quad u_0(x) \equiv 1, \quad x \in \overline{\Omega}, \quad x^* = (l_1/2, l_2/2).$$

For the synthetic experiment the coefficient $q(t)$ is taken in the form of

$$q(t) = \frac{0.5 \sin(\pi t)}{1. + e^{g(t-0.6T)}}, \quad t \in (0, T]. \tag{20}$$

For large values of the parameter g the function $q(t)$ is close to a discontinuous function with discontinuity at $t = 0.6\,T$. In the examples below we set $T = 1$. To investigate the accuracy of the identification of well debit using a given bottom hole pressure at each time steps, we first solve the direct problem, where we determine the dependence of the bottom hole pressure $p(t)$ from time with a given debit $q(t)$, $t \in (0, T]$.

The direct problem is solved using different spatial and time grids, the number of nodes varies as $n = N_1 = N_2 = 50, 100, 200, 400$; $N = 50, 100, 200$. Figure 1 shows the dynamics of bottom hole pressure $p(t)$ for various spatial steps $(h_1 = h_2 = 0.2, 0.1, 0.05, 0.025)$ and time steps $(\tau = 0.01, N = 100)$ with $g = 20$ (left) and $g = 1000$ (right).

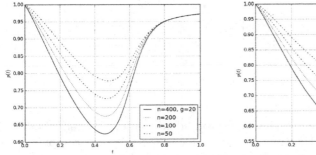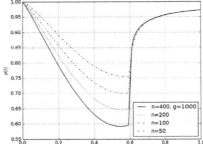

Fig. 1. The bottom hole pressure $p(t)$ for different n

Since the direct problem is sufficiently difficult it is not possible to determine the exact bottom hole pressure on such coarse spatial grids. It makes sense to examine the accuracy of the identification of less smooth coefficients on the right-hand side of parabolic equations. Within the framework of synthetic experiment we investigate the inverse problem with $g = 1000$ (20).

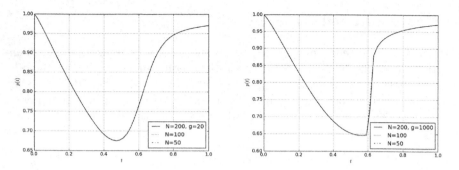

Fig. 2. The bottom hole pressure $p(t)$ for different N

Thus, we simulate the situation of the approximate solution of the inverse problem with a discontinuous coefficient. Figure 2 shows similar graphs of bottom hole pressure for different time grids $\tau = 0.02, 0.01, 0.005$, $N = 50, 100, 200$ and the same spatial grid $n = N_1 = N_2 = 200$. From these graphs it is clear that there is no significant effect of the time grid on the accuracy of the solution of direct problems, even when the well debit $q(t)$ is close to a discontinuous function with a discontinuity at $t = 0.6\ T$. The obtained values of bottom hole pressure are used to verify the accuracy of the proposed computational algorithm.

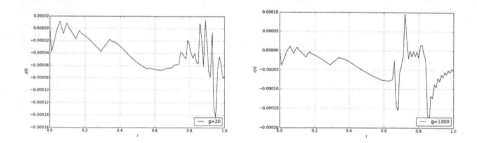

Fig. 3. The solution error of inverse problem

Figure 3 shows results of the numerical identification of well debit $q(t)$ throughout the range of the time $t \in (0, T]$, using the proposed computational algorithm for different values of the constants $g = 20, 1000$ with $N_1 = 100$, $N_2 = 100$, $N = 100$. It shows the error of determining the well debit at a given time interval. This error is estimated as $z(t) = q_h(t) - q(t)$, where $q_h(t)$ is the approximate solution, and $q(t)$ is the exact value given by analytical formula (20). The graphs show a very good identification accuracy of the proposed algorithm. It should be noted that the decrease in half of the time step increases the accuracy by one order.

The proposed computational algorithm (15)–(19) is well defined what is justified by the fact that the right-hand side of Eq. (19) is nonsingular (the denominator is always different from zero by construction). The auxiliary function $w(\mathbf{x})$ for different values of the time step $\tau = 0.02$ ($N = 50$) $\tau = 0.0025$ ($N = 400$) on the spatial grid with $N_1 = N_2 = 200$ at the observation point $w(\mathbf{x}_*)$ is significantly different from zero, so the bottom hole pressure $q(t)$ is bounded above.

Acknowledgments. The authors express their sincere gratitude to Professor P.N.Vabischevich for ideas of problem formulation and fruitful discussions. This work is financially supported by RFBR (projects 13-01-00719, 15-31-20856).

References

1. Alifanov, O.M.: Inverse Heat Transfer Problems. Springer, Heidelberg (2011)
2. Aster, R.C.: Parameter Estimation and Inverse Problems. Elsevier Science, Burlington (2011)
3. Kabanikhin, S.I.: Inverse and Ill-Posed Problems, Theory and Applications, p. 459. De Gruyter, Germany (2011)
4. Lavrent'ev, M.M., Romanov, V.G., Shishatskii, S.P.: Ill-posed problems of mathematical physics and analysis. Am. Math. Soc. **19**(1), 332–337 (1986)
5. Isakov, V.: Inverse Problems for Partial Differential Equations. Springer, New York (1998)
6. Prilepko, A.I., Orlovsky, D.G., Vasin, I.A.: Methods for Solving Inverse Problems in Mathematical Physics. Marcel Dekker Inc., New York (2000)
7. Vogel, C R.: Computational methods for inverse problems. Society for Industrial and Applied Mathematics, Philadelphia (2002)
8. Samarskii, A.A., Vabishchevich, P.N.: Numerical Methods for Solving Inverse Problems of Mathematical Physics. De Gruyter, Berlin (2007)
9. Borukhov, V.T., Vabishchevich, P.N.: Numerical solution of the inverse problem of reconstructing a distributed right-hand side of a parabolic equation. Comput. Phys. Commun. **126**(1), 32–36 (2000)
10. Vabishchevich, P.N., Vasil'ev, V.I.: Computational determination of the lowest order coefficient in a parabolic equation. Doklady Math. **89**(2), 179 181 (2014)
11. Vabishchevich, P.N., Vasilyeva, M.V., Vasilyev, V.I.: Computational algorithm for identification of the right-hand side of the parabolic equation. In: Dimov, I., Faragó, I., Vulkov, L. (eds.) FDM 2014. LNCS, vol. 9045, pp. 385–392. Springer, Heidelberg (2015)

Numerical Methods for Multiphysics Problems

Algebraic Multigrid Based Preconditioners for Fluid-Structure Interaction and Its Related Sub-problems

Ulrich Langer and Huidong Yang[✉]

Johann Radon Institute for Computational and Applied Mathematics,
Austrian Academy of Sciences, Linz, Austria
{ulrich.langer,huidong.yang}@ricam.oeaw.ac.at

Abstract. This work is devoted to the development and testing of algebraic multigrid based preconditioners for the linearized coupled fluid-structure interaction problem using low order finite element basis functions, and the compressible and nearly incompressible elasticity sub-problems in mixed displacement-pressure form using higher-order finite element basis functions. The preconditioners prove to be robust with respect to the mesh size, time step size, and other material parameters.

Keywords: Preconditioner · FSI · AMG · Taylor-Hood element

1 Introduction

The aim of this work is to construct and test efficient preconditioners for the coupled fluid-structure interaction (FSI) problems, which turn out to be robust with respect to discretization and material parameters. We use an approach based on the *LDU* block factorization of the coupled system matrix, that follows the blocks of multiphysics. Such a FSI preconditioner demands efficient solution methods for sub-problems. For finite element equations using low (equal) order basis functions, a class of algebraic multigrid (AMG) methods [3,10] used for the discrete elliptic and saddle point sub-problems in FSI have been reported in [6,11,13]. Using higher-order (and mixed-order) basis functions in FSI simulation has advantages with respect to the inf-sup stable discretization [2]. Meanwhile, it poses a number of challenges for the solution method to the discrete sub-problems, that is reported in this work. The remainder of the paper is organized as follows. In Sect. 2, we present the FSI system arising from the discretization and linearization of the model FSI problem, and the saddle point system from the discrete elasticity sub-problem for compressible and nearly incompressible materials using the Taylor-Hood element. The FSI preconditioners and the AMG preconditioner for the sub-problem are described in Sect. 3. In Sect. 4, we show the performance of the preconditioners for both the coupled FSI and saddle point systems. Finally, some conclusions are drawn in Sect. 5.

© Springer International Publishing Switzerland 2015
I. Lirkov et al. (Eds.): LSSC 2015, LNCS 9374, pp. 91–98, 2015.
DOI: 10.1007/978-3-319-26520-9_9

2   Preliminaries

2.1   The Coupled FSI System

We consider the coupled FSI system in strong form on the reference domain $\Omega^0 \subset \mathbb{R}^3$ composed of the fluid and structure sub-domains $\bar{\Omega}^0 = \bar{\Omega}_f^0 \cup \bar{\Omega}_s^0$, $\Omega_f^0 \cap \Omega_s^0 = \emptyset$: Find the fluid domain displacement $d_f : \bar{\Omega}_f^0 \mapsto \mathbb{R}^3$, fluid velocity $u : \bar{\Omega}_f^0 \mapsto \mathbb{R}^3$, fluid pressure $p : \bar{\Omega}_f^0 \mapsto \mathbb{R}$ and structure displacement $d_s : \bar{\Omega}_s^0 \mapsto \mathbb{R}^3$ for $t \in (0, T]$ such that

$$-\Delta d_f = 0 \quad \text{in} \quad \Omega_f^0, \tag{1a}$$

$$d_f = d_s \quad \text{on} \quad \Gamma^0, \tag{1b}$$

$$\rho_f J_f \partial_t u + \rho_f J_f ((u - w_f) \cdot F_f^{-1} \nabla) u - \nabla \cdot (J_f \sigma_f(u, p) F_f^{-T}) = 0 \quad \text{in} \quad \Omega_f^0, \tag{1c}$$

$$\nabla \cdot (\rho_f J_f F_f^{-1} u) = 0 \quad \text{in} \quad \Omega_f^0, \tag{1d}$$

$$\rho_s \partial_{tt} d_s - \nabla \cdot (F_s S) = 0 \quad \text{in} \quad \Omega_s^0, \tag{1e}$$

$$u = \partial_t d_s, \quad J_f \sigma_f(u, p) F_f^{-T} n_f + F_s S n_s = 0 \quad \text{on} \quad \Gamma^0, \tag{1f}$$

where $\Gamma^0 = \partial \Omega_f^0 \cap \partial \Omega_s^0$ denotes the interface, $F_f = I + \nabla d_f$ and $F_s = I + \nabla d_s$ the fluid and structure deformation gradient tensor, respectively, $J_f = \det F_f$ and $J_s = \det F_s$ their corresponding determinants, ρ_f and ρ_s the fluid and structure density, respectively, n_f and n_s the fluid and structure outerward unit normal vector, respectively, $\sigma_f(u, p) := \mu(\nabla u + \nabla^T u) - pI$ the Cauchy stress tensor with the dynamic viscosity term μ, $S = \lambda_s \text{tr}(E_s) I + 2\mu_s E_s$ the second Piola-Kirchhoff stress tensor for the St. Venant - Kirchhoff material, with $E_s = 0.5(F_s^T F_s - I)$ representing the Green-Lagrange strain tensor with the Lamé constant λ_s and the shear modulus μ_s. To complete the system, proper boundary and initial conditions are needed; see more details in [5,7].

As in [5], for the linearization, we use Newton's method. For the time discretization, we use the first order implicit Euler scheme and a first order Newmark-β scheme for the fluid and structure sub-problem, respectively. For the space discretization, we use the stabilized $P_1 - P_1$ finite element discretization with standard hat basis functions for both the fluid velocity and pressure interpolations, and the P_1 finite element discretization with the standard hat function for both the fluid and structure displacement interpolations. After linearization and discretization, we obtain the following linearized coupled FSI system of finite element equations at each Newton iteration:

$$Kx = b, \tag{2}$$

where

$$K = \begin{bmatrix} A_m & A_{ms} & 0 \\ 0 & A_s & A_{sf} \\ A_{fm} & A_{fs} & A_f \end{bmatrix}, \quad x = \begin{bmatrix} x_m \\ x_s \\ x_f \end{bmatrix}, \quad b = \begin{bmatrix} b_m \\ b_s \\ b_f \end{bmatrix}, \tag{3}$$

with the stiffness matrices A_m, A_s and A_f for the mesh movement, structure and fluid sub-problems on the diagonal and the coupling on the off-diagonal, x_m, x_s and x_f being the corrections of the fluid domain displacement, structure displacement, and the fluid velocity and pressure, respectively; see [5] for details.

2.2   The Linear Elasticity Sub-problem in Mixed Form

To study the robustness and efficiency of the AMG preconditioner for the discrete sub-problems using higher-order basis functions, we consider the (stationary) linear elasticity sub-problem in the classical mixed displacement-pressure form: Find the displacement $d_s : \bar{\Omega}_s^0 \mapsto \mathbb{R}^3$ and pressure $p_s : \bar{\Omega}_s^0 \mapsto \mathbb{R}$ such that

$$-\nabla \cdot (2\mu\varepsilon(d_s)) + \nabla p_s = 0 \quad \text{in } \Omega_s^0, \tag{4a}$$
$$-\nabla \cdot d_s - (1/\lambda)p_s = 0 \quad \text{in } \Omega_s^0, \tag{4b}$$

with the boundary conditions $d_s = g_D$ on Γ_D and $(2\mu\varepsilon(d_s) - p_s I)n = g_N$ on Γ_N, and $\varepsilon(d_s) = (\nabla d_s + \nabla^T d_s)$. In this mixed displacement-pressure form, the displacement and pressure are associated by the relation $p_s = -\lambda\nabla \cdot u_s$; see, e.g., [1]. In the limit case $\lambda \to +\infty$, the system (4) goes to the Stokes case, that somehow covers part of the fluid sub-problem. The application of such a model to FSI has been reported in [11] for equal order mixed finite element discretization. Here we consider robust and efficient AMG preconditioners for the finite element equations using the classical Taylor-Hood element, that fulfills the inf − sup stability condition; see, e.g., [2].

After discretization, the following saddle point system arises:

$$\underbrace{\begin{bmatrix} A & B^T \\ B & -C \end{bmatrix}}_{=:K} \begin{bmatrix} u \\ p \end{bmatrix} = \begin{bmatrix} K_{ll} & K_{ql}^T & B_{ll}^T \\ K_{ql} & K_{qq} & B_{lq}^T \\ \hline B_{ll} & B_{lq} & -C_{ll} \end{bmatrix} \begin{bmatrix} u_l \\ u_q \\ \hline p_l \end{bmatrix} = \begin{bmatrix} f_l \\ f_q \\ \hline g_l \end{bmatrix} = \begin{bmatrix} f \\ g \end{bmatrix}. \tag{5}$$

For convenience of presentation, as suggested in [12], we have reordered the system according to the linear and quadratic degrees of freedom (DOF), where the subscripts l and q denote the DOF associated to the linear and quadratic basis functions, respectively.

3   Preconditioners

3.1   A Preconditioner for the FSI Problem

Based on the complete LDU factorization of the system matrix K, we propose the following FSI preconditioner $\hat{K}$ for (2):

$$
\begin{aligned}
\hat{K} = \hat{L}\hat{D}\hat{U} &:= \begin{bmatrix} I & 0 & 0 \\ 0 & I & 0 \\ A_{fm}\hat{A}_m^{-1} & \hat{A}_{fs}\hat{A}_s^{-1} & I \end{bmatrix} \begin{bmatrix} \hat{A}_m & 0 & 0 \\ 0 & \hat{A}_s & 0 \\ 0 & 0 & \hat{S} \end{bmatrix} \begin{bmatrix} I & \hat{A}_m^{-1}A_{ms} & 0 \\ 0 & I & \hat{A}_s^{-1}A_{sf} \\ 0 & 0 & I \end{bmatrix} \\
&= \begin{bmatrix} \hat{A}_m & 0 & 0 \\ 0 & \hat{A}_s & 0 \\ A_{fm} & \hat{A}_{fs} & \hat{S} \end{bmatrix} \begin{bmatrix} I & \hat{A}_m^{-1}A_{ms} & 0 \\ 0 & I & \hat{A}_s^{-1}A_{sf} \\ 0 & 0 & I \end{bmatrix},
\end{aligned}
\tag{6}
$$

where $\hat{A}_m$, $\hat{A}_s$ and $\hat{S}$ are algebraic multigrid (AMG) preconditioners [4] for the sub-problems of mesh movement A_m, structure A_s and modified fluid

$S = A_f - \hat{A}_{fs}\hat{A}_s^{-1}A_{sf}$, respectively. As in [7], the Schur complement S is constructed by choosing $\hat{A}_s = \text{diag}[A_s]$ and $\hat{A}_{fs} = A_{fs} - A_{fm}\hat{A}_m^{-1}A_{ms}$ with $\hat{A}_m = \text{diag}[A_m]$, where "diag" denotes the block diagonal of the corresponding matrix. Since the system matrix K is unsymmetric and indefinite, we use the preconditioned GMRES method [9] as outer iteration to solve (2). In each preconditioning step, we have to evaluate $r = \hat{K}^{-1}b$ for a given vector b, that requires the action of the inverses of $\hat{A}_m$ and $\hat{A}_s$ and $\hat{S}$ to some vector that is nothing but an AMG W-cycle with a zero initial guess. In particular, we use the AMG [3] for the mesh movement and structure sub-problems, and the AMG [10] for the fluid sub-problem, respectively.

3.2   An AMG Preconditioner for the Saddle Point Sub-problems

We construct an AMG preconditioner for the saddle point systems using the classical Taylor-Hood élement. In this AMG method, we propose a new strategy for coarsening the linear and quadratic DOF which avoids mixing different orders of DOF on coarse levels. The idea was inspired by [10], where a separation coarsening strategy has been considered for the saddle point problem, in order to avoid mixing of the velocity and pressure DOF on coarse levels. In our new strategy, we further separate the linear and quadratic DOF of displacement. The interpolation matrix is easy to construct: We first construct the graph connectivity of the matrix, and then use the coarsening strategy [3] to construct the interpolation matrices for the linear and quadratic DOF of the displacement, and the linear DOF of the pressure. More precisely, we construct the prolongation matrix P_{l+1}^l from the coarse level $l+1$ to the next finer level l in form of

$$P_{l+1}^l = \begin{bmatrix} I_{l+1}^l & & \\ & J_{l+1}^l & \\ & & H_{l+1}^l \end{bmatrix}, \tag{7}$$

where the prolongation matrices, $I_{l+1}^l : (\mathbb{R}^3)^{n_{l+1}} \to (\mathbb{R}^3)^{n_l}$ and $J_{l+1}^l : (\mathbb{R}^3)^{m_{l+1}} \to (\mathbb{R}^3)^{m_l}$ are defined for the linear and quadratic displacement DOF, respectively, with n_l and m_l being the number of linear and quadratic displacement DOF on level l, respectively, $H_{l+1}^l : \mathbb{R}^{k_{l+1}} \to \mathbb{R}^{k_l}$ for the pressure DOF, with k_l being the number of pressure DOF on level l. The restriction matrix from the finer level l to the next coarser level $l+1$ is constructed as $(P_{l+1}^l)^T$. The system matrix K_{l+1} on the level $l+1$ is constructed by the Galerkin projection method, i.e., $K_{l+1} = (P_{l+1}^l)^T K_l P_{l+1}^l$. For the smoothing procedure, we have used the Braess-Sarazin-type smoother [14]. To test the robustness of the AMG preconditioner, we use the V-cycle preconditioned GMRES method.

4   Numerical Results

4.1   Numerical Results for a FSI Model Problem

The computational FSI domain is a channel with an obstacle inside, as illustrated in Fig. 1. The x-, y- and z-coordinates represent the lateral, anterior-posterior,

Fig. 1. The configuration of the geometry (left), the fluid mesh (middle) and the structure mesh (right).

and the vertical directions. respectively. The channel has the size $[0, 12]$ cm, $[0, 2]$ cm and $[0, 2]$ cm, in the x-, y- and z-direction, respectively. The obstacle is composed of two quarter cylinders with radius 0.8 cm. When the flow goes from the left to right in the lateral direction, the FSI interaction occurs on the obstacle surface. We uses Netgen [8] to generate the finite element meshes with the conforming grids on the interface; see a mesh example in Fig. 1. The information of the number of nodes (#Nod), tetrahedral elements (#Tet), and degrees of freedom (#Dof) on the coarse mesh (C), intermediate mesh (I), fine mesh (F) and very fine mesh (V) is summarized in Table 1. In all the numerical tests, we set the relative residual error $1.0e - 09$ in corresponding norms as stopping criteria. We use the nonlinear isotropic and homogeneous hyperelastic model of the St. Venant - Kirchhoff material, where the elastic constants are $\lambda_s = 1.73e{+}06$ dyne/cm^2 and $\mu_s = 1.15e{+}06$ dyne/cm^2. The density of the structure is $\rho_s = 1$ g/cm^3. The fluid kinematic viscosity is $\nu = 0.035$ cm^2/s. We set the fluid density $\rho_f \in \{1.146, 0.1146, 0.01146, 0.001146\}$ g/cm^3, that covers a large range of flows, e.g., the water, blood and air flow. The structure is fixed on the boundaries. The inflow boundary condition is $u = 30$ cm/s on Γ_{in} (the left hand side of the channel). For the outflow on the right hand side of the channel, we use the zero Neumann boundary condition. On the rest, we use homogeneous Dirichlet boundary condition. We set the time step size $\Delta t = 0.125$ ms and 0.0625 ms. From the iteration numbers of the preconditioned GMRES method (#It (Pre_GMRES)) in Tables 2 and 3, we observe the robustness of the preconditioners with respect to the mesh size, time step size and the material parameters, i.e., the iteration numbers stay in a similar range. In addition, the computational CPU time (measured in seconds (s)) scales very well with respect to the number of DOFs. For more detailed numerical study, we refer to [7].

Table 1. Four levels of finite element meshes.

	#Nod	#Tet	#Dof
Coarse mesh (C)	1307	4880	5781
Intermediate mesh (I)	8347	39040	37295
Fine mesh (F)	59155	312320	265811
Very fine mesh (V)	444323	2498560	1667778

Table 2. The iteration numbers of the preconditioned GMRES (#It (Pre_GMRES)) and the CPU time in seconds (s) using $\Delta t = 0.125\,\mathrm{ms}$.

ρ_f	Level			
	C	I	F	V
1.146	3 (4.7 s)	3 (38.9 s)	3 (328.5 s)	3 (2644.4 s)
0.1146	3 (4.7 s)	2 (30.9 s)	3 (327.5 s)	3 (2640.2 s)
0.01146	2 (3.7 s)	2 (31.5 s)	2 (261.8 s)	2 (2109.2 s)
0.001146	2 (3.9 s)	2 (31.2 s)	2 (263.5 s)	2 (2115.2 s)

Table 3. The iteration numbers of the preconditioned GMRES (#It (Pre_GMRES)) and the CPU time in seconds (s) using $\Delta t = 0.0625\,\mathrm{ms}$.

ρ_f	Level			
	C	I	F	V
1.146	3 (4.7 s)	3 (39.2 s)	3 (332.5 s)	3 (2645.2 s)
0.1146	3 (4.7 s)	2 (31.3 s)	3 (332.0 s)	3 (2636.5 s)
0.01146	2 (3.8 s)	2 (31.4 s)	2 (263.7 s)	3 (2664.7 s)
0.001146	2 (3.7 s)	2 (31.0 s)	2 (262.6 s)	2 (2141.8 s)

4.2 Numerical Results for Compressible and Nearly Incompressible Elasticity Model Problems

In order to test the robustness of our proposed AMG preconditioner for the saddle point sub-problems in Sect. 3.2, we consider a unit cube $(0, 1)^3$ as the computational domain for the linear elasticity problem. The domain is subdivided into tetrahedra with four levels of mesh refinement $L_1 - L_4$. The number of tetrahedron (#Tet) and nodes (#Nodes), and the total number of DOF of the saddle

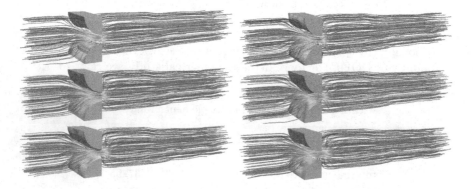

Fig. 2. The visualization of the fluid velocity streamlines and the structure deformation at different time level $t = 3k$ ms, $k = 1, 2, ..., 6$ from up left to down right.

Fig. 3. Numerical results of the displacement (left) and pressure (right) for the linear elasticity problem in mixed form.

Table 4. Number of tetrahedron (#Tet) and nodes (#Nodes), and total number of DOF for saddle point (#DOF) systems on four levels $L_1 - L_4$.

Level	L_1	L_2	L_3	L_4
#Tet	64	512	4096	32768
#Nodes	125	729	4913	35937
#DOF	2312	15468	112724	859812

Table 5. Performance of the V-cycle preconditioned GMRES solver for the linear elasticity problem in mixed form using one pre/post Braess-Sarazin smoother.

Level	L_1	L_2	L_3	L_4
#It ($\nu = 0.3003$, $E = 2990800$)	33 (11.7s)	33 (10.4s)	33 (51.4s)	32 (461.8s)
#It ($\nu = 0.49999$, $E = 2990800$)	42 (15.1s)	39 (12.5s)	38 (58.5s)	37 (531.5s)
#It ($\nu = 0.499999999$, $E = 2990800$)	42 (14.4s)	39 (12.1s)	38 (56.8s)	37 (532.4s)

point (#DOF) systems are shown in Table 4. We fix the bottom of the domain, i.e., $u = [0, 0, 0]^T$ at $z = 0$, prescribe a Dirichlet data on the top, i.e., $u = [0, 0, 1]^T$ at $z = 1$, and use zero Neumann condition on the rest of the boundaries. Using color, we show the value $\|d_s\|_{\mathbb{R}^3}$ and p_s on the left and right plots in Fig. 3 as an illustration, respectively. The elasticity deformation is scaled by a factor of 0.5. We set $\mu = 1.15e + 06$, $\lambda = 1.73e + 06$ (corresponding to the Poisson ratio $\nu = 0.3003$, Young's modulus $E = 2990800$) for the compressible material, and $\mu = 9.9694e + 05$, $\lambda = 4.9846e + 10$ and $\mu = 9.9693e + 05$, $\lambda = 4.9847e + 14$ (corresponding to $\nu = 0.49999$, $E = 2990800$ and $\nu = 0.499999999$, $E = 2990800$, respectively) for the nearly incompressible material. We observe the robustness with respect to the mesh size of the V-cycle preconditioned GMRES solver using the Braess-Sarazin smoother; see iteration numbers of one V-cycle preconditioned GMRES method and the computational CPU time in seconds (s) in Table 5 for both the compressible (#It ($\nu = 0.3003$, $E = 2990800$)) and nearly incompressible material (#It ($\nu = 0.49999$, $E = 2990800$) and #It ($\nu = 0.499999999$, $E = 2990800$)). From the iteration numbers of the V-cycle preconditioned GMRES solver in Table 5, we also observe the robustness of the AMG preconditioner with respect to the near incompressibility.

5 Conclusions

We considered robust and efficient preconditioners for the linearized coupled FSI system arising from lower-order finite element discretization. From the numerical study, the preconditioned GMRES method has shown the robustness and efficiency with respect to the mesh size, time step size and varying fluid density. Furthermore, the AMG V-cycle preconditioned GMRES method for the saddle point system of the elasticity sub-problem discretized with Taylor-Hood element has demonstrated the robustness and efficiency with respect to the mesh size and near incompressibility.

References

1. Braess, D.: Finite Elements - Theory, Fast Solvers, and Applications in Solid Mechanics. Cambridge University Press, Cambridge (2007)
2. Brezzi, F., Fortin, M.: Mixed and Hybrid Finite Element Methods. Springer, New York (1991)
3. Kickinger, F.: Algebraic multigrid for discrete elliptic second-order problems. In: Hackbusch, W., Wittum, G. (eds.) Multigrid Methods V. Lecture Notes in Computational Sciences and Engineering, pp. 157–172. Springer, Heidelberg (1998)
4. Haase, G., Langer, U.: Multigrid methods: from geometrical to algebraic versions. In: Bourlioux, A., Gander, M.J., Sabidussi, G. (eds.) Modern Methods in Scientific Computing and Applications. Series II. Mathematics, Physics and Chemistry, vol. 75, pp. 103–154. Kluwer Academic Press, Dordrecht (2002)
5. Langer U., Yang H.: Numerical simulation of fluid-structure interaction problems with hyperelastic models: a monolithic approach (2014). arXiv:1408.3737
6. Langer, U., Yang, H.: Partitioned solution algorithms for fluid-structure interaction problems with hyperelastic models. J. Comput. Appl. Math. **276**, 47–61 (2015)
7. Langer, U., Yang, H.: A note on robust preconditioners for monolithic fluid-structure interaction systems of finite element equations (2015). arXiv:1412.6845
8. Schöberl, J.: NETGEN - an advancing front 2D/3D-mesh generator based on abstract rules. Comput. Visual. Sci. **1**, 41–52 (1997)
9. Saad, Y., Schultz, M.H.: GMRES: a generalized minimal residual algorithm for solving nonsymmetric linear systems. SIAM J. Sci. Stat. Comput. **7**, 856–869 (1986)
10. Wabro, M.: Coupled algebraic multigrid methods for the Oseen problem. Comput. Visual. Sci. **7**, 141–151 (2004)
11. Yang, H.: Partitioned solvers for the fluid-structure interaction problems with a nearly incompressible elasticity model. Comput. Visual. Sci. **14**(5), 227–247 (2011)
12. Yang, H.: An algebraic multigrid method for quadratic finite element equations of elliptic and saddle point systems in 3D (2015). arXiv:1503.01287
13. Yang, H., Zulehner, W.: Numerical simulation of fluid-structure interaction problems on hybrid meshes with algebraic multigrid methods. J. Comput. Appl. Math. **235**, 5367–5379 (2011)
14. Zulehner, W.: A class of smoothers for saddle point problems. Computing **65**(3), 227–246 (2000)

Control and Uncertain Systems

Functional Differential Model of an Anaerobic Biodegradation Process

Milen K. Borisov[1], Neli S.Dimitrova[1](✉), and Mikhail I. Krastanov[1,2]

[1] Institute of Mathematics and Informatics, Bulgarian Academy of Sciences, Acad. G. Bonchev Str. Bl. 8, 1113 Sofia, Bulgaria
{milen_kb,nelid}@math.bas.bg
[2] Faculty of Mathematics and Informatics, Sofia University, 5 James Bourchier Blvd., 1164 Sofia, Bulgaria
krastanov@fmi.uni-sofia.bg

Abstract. In this paper we study a nonlinear functional differential model of a biological digestion process, involving two microbial populations and two substrates. We establish the global asymptotic stability of the model solutions towards a previously chosen equilibrium point and in the presence of two different discrete delays. Numerical simulation results are also included.

1 Introduction

We consider a well-known anaerobic digestion model for biological treatment of wastewater in a continuously stirred tank bioreactor (cf. for example [2,3]). Here we include discrete time delays in the equations to model the delay in the conversion of nutrient consumed by the viable biomass. For more detailed motivation see [13,14] and the references therein. The model is described by the following nonlinear differential equations:

$$
\begin{aligned}
\frac{d}{dt}s_1(t) &= u(s_1^i - s_1(t)) - k_1\mu_1(s_1(t))x_1(t) \\
\frac{d}{dt}x_1(t) &= e^{-\alpha u\tau_1}\mu_1(s_1(t-\tau_1))x_1(t-\tau_1) - \alpha u x_1(t) \\
\frac{d}{dt}s_2(t) &= u(s_2^i - s_2(t)) + k_2\mu_1(s_1(t))x_1(t) - k_3\mu_2(s_2(t))x_2(t) \\
\frac{d}{dt}x_2(t) &= e^{-\alpha u\tau_2}\mu_2(s_2(t-\tau_2))x_2(t-\tau_2) - \alpha u x_2(t).
\end{aligned}
\tag{1}
$$

The state variables s_1, s_2 and x_1, x_2 denote substrate and biomass concentrations, respectively: s_1 is the organic substrate, characterized by its chemical oxygen demand (COD), s_2 denotes the volatile fatty acids (VFA), x_1 and x_2 are the acidogenic and methanogenic bacteria respectively; s_1^i and s_2^i are the input substrate concentrations. The constants $\tau_j \geq 0$, $j = 1, 2$, stand for the time delay

This research has been partially supported by the Sofia University "St Kl. Ohridski" under contract No. 08/26.03.2015.

I. Lirkov et al. (Eds.): LSSC 2015, LNCS 9374, pp. 101–108, 2015.
DOI: 10.1007/978-3-319-26520-9_10

in conversion of the corresponding substrate to viable biomass for the jth bacterial population. Here $e^{-\alpha u \tau_j} x_j(t - \tau_j)$, $j = 1, 2$, represents the biomass of those microorganisms that consume nutrient τ_j units of time prior to time t and that survive in the chemostat the τ_j units of time necessary to complete the process of converting the nutrient to viable biomass at time t. The parameter $\alpha \in (0, 1)$ represents the proportion of bacteria that are affected by the dilution rate u. The constants k_1, k_2 and k_3 are yield coefficients related to COD degradation, VFA production and VFA consumption respectively. For biological evidence, s_1^i and s_2^i as well as all parameters in (1) are assumed to be positive.

The functions $\mu_1(s_1)$ and $\mu_2(s_2)$ model the specific growth rates of the bacteria. Following [9] we impose the following assumption on μ_1 and μ_2:

Assumption A1. For each $j = 1, 2$ the function $\mu_j(s_j)$ is defined for $s_j \in [0, +\infty)$, $\mu_j(0) = 0$, and $\mu_j(s_j) > 0$ for each $s_j > 0$; the function $\mu_j(s_j)$ is bounded and Lipschitz continuous for all $s_j \in [0, +\infty)$.

The Eq. (1) with $\tau_1 = \tau_2 = 0$ have been already investigated by the authors; thereby, global stabilizability via feedback control is proposed in [4], whereas [5] considers the case of global stabilization of the solutions using constant dilution rate u. This second approach is now extended to model (1) involving discrete delays $\tau_j > 0$, $j = 1, 2$. More precisely, in this paper we define a suitable positive constant u_b and prove that for any (admissible) value of the dilution rate $u \in (0, u_b)$ there exists an equilibrium point which is globally asymptotically stable for system (1). To our knowledge, such investigations have not been carried out for this model.

2   Global Asymptotic Stabilizability of the Model

We set $u_b = \max\left\{u: u\alpha e^{\alpha u \tau_1} \leq \mu_1(s_1^i), u\alpha e^{\alpha u \tau_2} \leq \mu_2(s_2^i)\right\}$ and make the following

Assumption A2. For each point $\bar{u} \in (0, u_b)$ there exist points $s_1(\bar{u}) = \bar{s}_1 \in (0, s_1^i)$ and $s_2(\bar{u}) = \bar{s}_2 \in (0, s_2^i)$, such that the following equalities hold true

$$\bar{u} = \frac{e^{-\alpha \bar{u} \tau_1}}{\alpha} \mu_1(\bar{s}_1) = \frac{e^{-\alpha \bar{u} \tau_2}}{\alpha} \mu_2(\bar{s}_2).$$

A similar assumption is called in [7] regulability of the system.

Let $\bar{s}_1$ and $\bar{s}_2$ be determined according to Assumption A2. Compute further

$$x_1(\bar{u}) = \bar{x}_1 = \frac{s_1^i - \bar{s}_1}{\alpha k_1 e^{\alpha \bar{u} \tau_1}}, \quad x_2(\bar{u}) = \bar{x}_2 = \frac{s_2^i - \bar{s}_2 + \alpha k_2 \bar{x}_1}{\alpha k_3 e^{\alpha \bar{u} \tau_2}}. \tag{2}$$

Then the point $p(\bar{u}) = \bar{p} = (\bar{s}_1, \bar{x}_1, \bar{s}_2, \bar{x}_2)$ is a nontrivial (positive) equilibrium point for system (1).

Assumption A3. There exist positive numbers ν_1 and ν_2 such that the following inequalities hold true

$$\mu_1(s_1^-) < \mu_1(\bar{s}_1) < \mu_1(s_1^+), \quad \mu_2(s_2^-) < \mu_2(\bar{s}_2) < \mu_2(s_2^+)$$

for each

$$s_1^- \in (0, \bar{s}_1), s_1^+ \in (\bar{s}_1, s_1^i + \nu_1], s_2^- \in (0, \bar{s}_2) \text{ and } s_2^+ \in (\bar{s}_2, s_2^i + \nu_2].$$

Assumption A3 is always fulfilled when the functions $\mu_j(\cdot)$, $j = 1, 2$, are monotone increasing (like the Monod specific growth rate). If at least one function $\mu_j(\cdot)$ is not monotone increasing (like the Haldane law) then the points $\bar{s}_j$ have to be chosen sufficiently small in order to satisfy Assumption A3.

Denote by R^+ the set of all positive real numbers and by C_τ^+ – the nonnegative cone of continuous functions $\varphi : [-\tau, 0] \to R^+$, where $\tau = \max\{\tau_1, \tau_2\}$, and set $C_\tau^4 := \{\varphi = (\varphi_{s_1}, \varphi_{x_1}, \varphi_{s_2}, \varphi_{x_2}) \in C_\tau^+ \times C_\tau^+ \times C_\tau^+ \times C_\tau^+\}$.

Let $\bar{u} \in (0, u_b)$ be chosen in such a way that Assumptions A2 and A3 are satisfied. Denote by Σ the system obtained from (1) by substituting the parameter u by $\bar{u}$. Using the Schauder fixed-point theorem it is easy to prove that for each $\varphi \in C_\tau^4$ there exists $\varrho > 0$ and a unique solution $\Phi(t, \varphi) = (s_1(t, \varphi), x_1(t, \varphi), s_2(t, \varphi), x_2(t, \varphi))$ of (1) defined on $[-\tau, \varrho)$ such that $\Phi(t, \varphi) = \varphi(t)$ for each $t \in [-\tau, 0]$ (cf. Theorem 2.1 in [8]).

We shall prove below that the equilibrium point $\bar{p}$ is globally asymptotically stable for system Σ.

Theorem 1. *Let the Assumptions A1, A2 and A3 be fulfilled and let φ_0 be an arbitrary element of C_τ^4. Then the corresponding solution $\Phi(t, \varphi_0)$ is well defined on $[-\tau, +\infty)$ and converges asymptotically towards $\bar{p}$.*

Proof. We fix an arbitrary $\varphi_0 \in C_\tau^4$. Then there exists $\varrho > 0$ such that the corresponding solution $\Phi(t, \varphi_0)$ of Σ (denoted by $\Phi(t) := (s_1(t), x_1(t), s_2(t), x_2(t))$ for simplicity) is defined on $[-\tau, \varrho)$. The proof uses some ideas from [13, 14]. For the reader's convenience we subdivide the proof in five claims.

Claim 1. *The components of $\Phi(t)$ take positive values for each $t \in [-\tau, \varrho)$.*

Proof of Claim 1. If $s_1(t) = 0$ for some $t \in [0, \varrho)$, then $\dot{s}_1(t) > 0$. This implies that $s_1(t) > 0$ for each $t \in [-\tau, \varrho)$. Analogously one can obtain that $s_2(t) > 0$ for each $t \in [-\tau, \varrho)$. Since

$$x_j(t) = \varphi_{x_j}(0)e^{-\alpha \bar{u}t} + \int_0^t e^{-\alpha \bar{u}(t-\sigma)} \mu_j(s_j(\sigma - \tau_j))x_j(\sigma - \tau_j)d\sigma, \; j = 1, 2,$$

then $x_j(t) > 0$ for each $t \in [-\tau, \varrho)$. This completes the proof of Claim 1.   ◇

Claim 2. *The solution $\Phi(t)$ of Σ is defined for each $t \in [-\tau, +\infty)$ and is bounded.*

Proof of Claim 2. Denote

$$s(t) := k_2 e^{-\alpha \bar{u}\tau_1} s_1(t) + k_1 e^{-\alpha \bar{u}\tau_1} s_2(t) \quad \text{and} \quad s^i = k_2 e^{-\alpha \bar{u}\tau_1} s_1^i + k_1 e^{-\alpha \bar{u}\tau_1} s_2^i.$$

Then $s(t)$ satisfies the differential equation

$$\dot{s}(t) = \bar{u}(s^i - s(t)) - k_1 k_3 e^{-\alpha \bar{u}\tau_1} \mu_2(s_2(t))x_2(t).$$

We set $q_1(t) := s(t) + k_1 k_3 e^{-\alpha \bar{u}(\tau_1 - \tau_2)} x_2(t + \tau_2) - s^i/\alpha$ and $q_2(t) := s(t) + k_1 k_3 x_2(t + \tau_2) - s^i$. Then

$$\dot{q}_1(t) = \bar{u}\left[s^i - s(t) - \alpha k_1 k_3 e^{-\alpha \bar{u}(\tau_1 - \tau_2)} x_2(t + \tau_2)\right]$$
$$\leq \bar{u}\left[s^i - \alpha\left(s(t) + k_1 k_3 e^{-\alpha \bar{u}(\tau_1 - \tau_2)} x_2(t + \tau_2)\right)\right] = -\alpha \bar{u} q_1(t),$$

and hence

$$q_1(t) \leq q_1(0) \cdot e^{-\alpha \bar{u} t}. \tag{3}$$

The latter inequality shows that $q_1(t)$ is bounded. Using the fact that the values of $s_1(t)$, $s_2(t)$ and $x_2(t)$ are positive, it follows that $s_1(t)$, $s_2(t)$ and $x_2(t)$ are bounded as well. Analogously one can obtain that

$$q_2(t) \geq q_2(0) \cdot e^{-\bar{u} t}. \tag{4}$$

The estimates (3), (4) and the definition of $s(\cdot)$ imply that for each $\varepsilon > 0$ there exists $T_\varepsilon > 0$ such that for each $t \geq T_\varepsilon$ the following inequalities hold true

$$s^i - \varepsilon < k_2 s_1(t) + k_1 s_2(t) + k_1 k_3 e^{-\alpha \bar{u}(\tau_1 - \tau_2)} x_2(t + \tau_2) < \frac{s^i}{\alpha} + \varepsilon. \tag{5}$$

It is easy to see (in the same way as the estimates (5)) that for each $\varepsilon > 0$ there exists a finite time $T_\varepsilon > 0$ such that for all $t \geq T_\varepsilon$ the following inequalities hold

$$s_1^i - \varepsilon < s_1(t) + k_1 e^{\alpha \bar{u} \tau_1} x_1(t + \tau_1) < \frac{s_1^i}{\alpha} + \varepsilon. \tag{6}$$

The inequalities (6) imply that $x_1(t)$ is also bounded. Thus the trajectory $\Phi(t)$ of Σ is well defined and bounded for all $t \geq -\tau$ (cf. also Theorem 3.1 of [8]). This completes the proof of Claim 2. $\diamond$

Claim 3. There exists $T_0 > 0$ such that $s_1(t) < s_1^i$ and $s_2(t) < s_2^i + k_2 s_1^i/k_1$ for each $t \geq T_0$.

Proof of Claim 3. First let us assume that there exists $\bar{t} > 0$ such that $s_1(t) \geq s_1^i$ for all $t \geq \bar{t}$. Then we have

$$\dot{s}_1(t) = \bar{u}(s_1^i - s_1(t)) - k_1 \mu_1(s_1(t)) x_1(t) < 0.$$

Since $s_1(\cdot)$ and $x_1(\cdot)$ are bounded differentiable functions defined on $[-\tau, +\infty)$, then $\dot{s}_1(\cdot)$ is an uniformly continuous function. Barbălat's Lemma (cf. [6]) leads to

$$0 = \lim_{t \to \infty} \dot{s}_1(t) = \lim_{t \to \infty} [\bar{u}(s_1^i - s_1(t)) - k_1 \mu_1(s_1(t)) x_1(t)].$$

Because $s_1^i - s_1(t) \leq 0$ and $x_1(t) > 0$, the above equalities imply that $s_1(t) \downarrow s_1^i$ and $x_1(t) \downarrow 0$ as $t \uparrow \infty$. On the other hand, if we set (cf. Lemma 2.2 of [14])

$$z_1(t) := x_1(t) + \int_{t - \tau_1}^{t} e^{-\alpha \bar{u} \tau_1} \mu_1(s_1(\sigma)) x_1(\sigma) d\sigma,$$

we obtain according to Assumption 3 that

$$\dot{z}_1(t) = x_1(t)(e^{-\alpha\bar{u}\tau_1}\mu_1(s_1(t)) - \alpha\bar{u}) > 0 \quad \text{for all } t \geq \bar{t},$$

and so $z_1(t) \uparrow z_1^* > 0$ as $t \uparrow \infty$. But this is impossible according to the definition of $z_1(\cdot)$ and because we have already shown that $x_1(t) \downarrow 0$ as $t \uparrow \infty$.

Hence, there exists a sufficiently large $T_0 > 0$ with $s_1(T_0) \leq s_1^i$. Moreover, if the equality $s_1(\bar{t}) = s_1^i$ holds true for some $\bar{t} \geq T_0$, then we have

$$\dot{s}_1(\bar{t}) = \bar{u}(s_1^i - s_1(\bar{t})) - k_1\mu_1(s_1(\bar{t}))x_1(\bar{t}) = -k_1\mu_1(s_1(\bar{t}))x_1(\bar{t}) < 0.$$

The last inequality shows that $s_1(t) < s_1^i$ for each $t > T_0$.

Further with $s(t) = k_2e^{-\alpha\bar{u}\tau_1}s_1(t) + k_1e^{-\alpha\bar{u}\tau_1}s_2(t)$ and $s^i = k_2e^{-\alpha\bar{u}\tau_1}s_1^i + k_1e^{-\alpha\bar{u}\tau_1}s_2^i$ we obtain

$$\dot{s}(t) = \bar{u}(s^i - s(t)) - k_1k_3e^{-\alpha\bar{u}\tau_1}\mu_2(s_2(t))x_2(t).$$

One can show in the same way as above that $s(t) < s^i$ for each $t \geq T_0$ (if necessary T_0 can be enlarged), i. e. $k_2e^{-\alpha\bar{u}\tau_1}s_1(t) + k_1e^{-\alpha\bar{u}\tau_1}s_2(t) \leq k_2e^{-\alpha\bar{u}\tau_1}s_1^i + k_1e^{-\alpha\bar{u}\tau_1}s_2^i$. Since $0 < s_1(t) < s_1^i$, it follows that $s_2(t) \leq s_2^i + k_2s_1^i/k_1$. This establishes Claim 3. $\diamond$

Claim 4. Denote

$$\gamma_j := \limsup_{t\uparrow\infty} x_j(t), \quad \delta_j := \liminf_{t\uparrow\infty} x_j(t), \quad j = 1, 2$$

$$v_1(t) := s_1(t) + k_1x_1(t + \tau_1), \quad v_2(t) := k_2s_1(t) + k_1s_2(t) + k_1k_3x_2(t + \tau_2),$$

$$\alpha_j := \limsup_{t\uparrow\infty} v_j(t), \quad \beta_j := \liminf_{t\uparrow\infty} v_j(t), \quad j = 1, 2.$$

Then the following relations hold true: $\delta_1 > 0$, $\alpha_1 = \beta_1$ and $\gamma_1 = \delta_1$, $\alpha_2 = \beta_2$ and $\gamma_2 = \delta_2$.

Proof of Claim 4. Let us assume that $\delta_1 = 0$. Choose an arbitrary $\varepsilon \in (0, (s_1^i - \bar{s}_1)/(1 + e^{\alpha\bar{u}\tau_1}k_1))$. According to Claim 2 (see (6)) there exists $T_\varepsilon > 0$ such that for all $t \geq T_\varepsilon$ the following inequalities hold true

$$s_1^i - \varepsilon < s_1(t - \tau_1) + k_1e^{\alpha\bar{u}\tau_1}x_1(t) < \frac{s_1^i}{\alpha} + \varepsilon. \tag{7}$$

Since $\delta_1 = 0$ there exists $t_0 > \max(T_\varepsilon, T_0)$ such that $x_1(t_0) < \varepsilon$. We set (cf. Lemma 3.5 of [14])

$$\sigma := \min\{x_1(t) : t \in [t_0 - \tau_1, t_0]\}$$

$$\bar{t} := \sup\{t \geq t_0 - \tau_1 : x_1(\tau) \geq \sigma \text{ for all } \tau \in [t_0 - \tau_1, t]\}.$$

Clearly $\sigma \in (0, \varepsilon]$, $\bar{t} \in [t_0 - \tau_1, +\infty)$, $x_1(t) \geq \sigma$ for all $t \in [t_0 - \tau_1, \bar{t}]$ and

$$x_1(\bar{t}) = \sigma, \quad \dot{x}_1(\bar{t}) \leq 0. \tag{8}$$

Taking into account (7) and the choice of ε, we obtain consecutively

$$s_1^i > s_1(\bar{t} - \tau_1) \geq s_1^i - k_1 e^{\alpha \bar{u} \tau_1} x_1(\bar{t}) - \varepsilon \geq$$
$$\geq s_1^i - (1 + e^{\alpha \bar{u} \tau_1} k_1)\varepsilon > \bar{s}_1,$$

$$\dot{x}_1(\bar{t}) = e^{-\alpha \bar{u} \tau_1} \mu_1(s_1(\bar{t} - \tau_1))x_1(\bar{t} - \tau_1) - \alpha \bar{u} x_1(\bar{t}) > \alpha \bar{u} \sigma - \alpha \bar{u} \sigma = 0.$$

The last inequality contradicts (8), which means that $\delta_1 > 0$.

The proof of the equalities $\alpha_j = \beta_j$ and $\gamma_j = \delta_j$, $j = 1, 2$, is based on similar ideas used in the proofs of Lemma 4.3 of [14] and Theorem 3.1 of [13], so we omit it here due to the limited paper length. $\Diamond$

Claim 5. The equilibrium point $\bar{p}$ is locally asymptotically stable for all values of the delays $\tau_1 \geq 0$ and $\tau_2 \geq 0$.

Proof of Claim 5. Denote for simplicity $a = k_1 \mu_1'(\bar{s}_1)\bar{x}_1$ and $b = k_3 \mu_2'(\bar{s}_2)\bar{x}_2$. It follows from Assumption A3 that $a > 0$ and $b > 0$ hold true. It is straightforward to see that the characteristic equation of Σ corresponding to the equilibrium point $\bar{p}$ has the form

$$0 = P(\lambda; \tau_1, \tau_2) = P_1(\lambda; \tau_1) \times P_2(\lambda; \tau_2),$$

where λ is a complex number and

$$P_1(\lambda; \tau_1) = \lambda^2 + (\bar{u} + a + \alpha \bar{u})\lambda + \alpha \bar{u}(\bar{u} + a) - \alpha \bar{u}(\bar{u} + \lambda)e^{-\lambda \tau_1},$$
$$P_2(\lambda; \tau_2) = \lambda^2 + (\bar{u} + b + \alpha \bar{u})\lambda + \alpha \bar{u}(\bar{u} + b) - \alpha \bar{u}(\bar{u} + \lambda)e^{-\lambda \tau_2}.$$

First it is straightforward to see that if $\tau_1 = \tau_2 = 0$ then there exist no roots λ of $P(\lambda; \tau_1, \tau_2) = 0$ with $Re(\lambda) \geq 0$. Let $\tau_1 > 0$ and $\tau_2 > 0$. We are looking for purely imaginary roots $\lambda = i\omega$ of $P_j(\lambda; \tau_j) = 0$ with $\omega > 0$, $j = 1, 2$. For $P_1(i\omega; \tau_1) = 0$ we obtain

$$-\omega^2 + (\bar{u} + a + \alpha \bar{u})i\omega + \alpha \bar{u}(\bar{u} + a) - \alpha \bar{u}(\bar{u} + i\omega)e^{-i\omega \tau_1} = 0,$$
$$-\omega^2 + (\bar{u} + a + \alpha \bar{u})i\omega + \alpha \bar{u}(\bar{u} + a) - \alpha \bar{u}(\bar{u} + i\omega)(\cos(\tau_1 \omega) - i\sin(\tau_1 \omega)) = 0.$$

Separating the real and the imaginary parts of the last equation implies

$$-\omega^2 + \alpha \bar{u}(\bar{u} + a) = \alpha \bar{u}^2 \cos(\tau_1 \omega) + \alpha \bar{u} \omega \sin(\omega \tau_1),$$
$$(\bar{u} + a + \alpha \bar{u})\omega = -\alpha \bar{u}^2 \sin(\tau_1 \omega) + \alpha \bar{u} \omega \cos(\omega \tau_1). \tag{9}$$

Squaring both sides of the Eq. (9) and adding leads to

$$\omega^4 + (\bar{u} + a)^2 \omega^2 + \alpha^2 \bar{u}^2 a(2\bar{u} + a) = 0.$$

Obviously, the latter equation does not possess positive real roots since $a > 0$. The same conclusion holds true for $P_2(i\omega; \tau_2) = 0$. Therefore, $P(\lambda; \tau_1, \tau_2) = 0$ does not have purely imaginary roots for any $\tau_1 > 0$ and $\tau_2 > 0$. Applying Lemma 2 from [10] (see also [11,12] for similar results) to the exponential polynomial $P(\lambda; \tau_1, \tau_2)$ we obtain that the characteristic equation does not have roots with

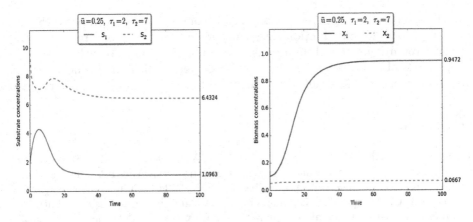

Fig. 1. Time evolution of $s_1(t)$, $s_2(t)$ (left) and $x_1(t)$, $x_2(t)$ (right)

nonnegative real parts. This means that for any $\tau_1 \geq 0$ and $\tau_2 \geq 0$ the equilibrium $\bar{p}$ is locally asymptotically stable.   ◇

The local asymptotic stability of the equilibrium $\bar{p}$ together with the convergence of the solution $\Phi(t)$ and the attractivity of $\bar{p}$, proved above throughout Claims 1 to 4, imply that $\bar{p}$ is globally asymptotically stable.

The proof of Theorem 1 is completed.   ◆

3   Computer Simulation

Consider the following specific growth rate functions in the model (1), taken from [1–3]:

$$\mu_1(s_1) = \frac{m_1 s_1}{k_{s_1} + s_1} \text{ (Monod law)}, \quad \mu_2(s_2) = \frac{m_2 s_2}{k_{s_2} + s_2 + (s_2/k_I)^2} \text{ (Haldane law)}.$$

In the simulation process we shall use the following numerical values for the model coefficients, which are obtained by real experiments and given in [1]:

$k_1 = 10.53$	$k_2 = 28.6$	$k_3 = 1074$	$s_1^i = 7.5$	$s_2^i = 75$	$\alpha = 0.5$
$m_1 = 1.2$	$k_{s_1} = 7.1$	$m_2 = 0.74$	$k_{s_2} = 9.28$	$k_I = 16$	

As an example let us take $\tau_1 = 2$ and $\tau_2 = 7$. Within the above coefficient values we compute the admissible upper bound $u_b = 0.646$ for u, thus $u \in (0, 0.646)$.

Consider $\bar{u} = 0.25$. Then the corresponding internal equilibrium is $\bar{p} = (1.096, 0.9472, 6.432, 0.06674)$. Using the initial conditions $\varphi_{s_1}(t) = 2$, $\varphi_{x_1}(t) = 0.1$ for $t \in [-\tau_1, 0]$, and $\varphi_{s_2}(t) = 10$, $\varphi_{x_2}(t) = 0.05$ for $t \in [-\tau_2, 0]$, the numerical outputs are visualized in Fig. 1.

4   Conclusion

In this paper we investigate a bioreactor model for wastewater treatment by anaerobic digestion. The model Eq. (1) involve discrete delays, describing the time delay in nutrient conversion to viable biomass. Using a properly chosen admissible value for the dilution rate $\bar{u}$ we prove the global convergence of the solutions towards an equilibrium point, corresponding to $\bar{u}$. To authors' knowledge, such kind of investigations have not been yet fulfilled for this delay bioreactor model. Numerical simulation is included to confirm the theoretical results.

References

1. Alcaraz-González, V., Harmand, J., Rapaport, A., Steyer, J.-P., González-Alvarez, V., Pelayo-Ortiz, C.: Software sensors for highly uncertain WWTPs: a new apprach based on interval observers. Water Res. **36**, 2515–2524 (2002)
2. Bernard, O., Hadj-Sadok, Z., Dochain, D.: Advanced monitoring and control of anaerobic wastewater treatment plants: dynamic model develop- ment and identification. In: Proceedings of Fifth IWA International Sympposium WATERMATEX, Gent, Belgium, pp. 3.57-3.64 (2000)
3. Bernard, O., Hadj-Sadok, Z., Dochain, D., Genovesi, A., Steyer, J.-P.: Dynamical model development and parameter identification for an anaerobic wastewater treatment process. Biotechnol. Bioeng. **75**, 424–438 (2001)
4. Dimitrova, N.S., Krastanov, M.I.: On the asymptotic stabilization of an uncertain bioprocess model. In: Lirkov, I., Margenov, S., Waśniewski, J. (eds.) LSSC 2011. LNCS, vol. 7116, pp. 115–122. Springer, Heidelberg (2012)
5. Dimitrova, N.S., Krastanov, M.I.: Model-based optimization of biogas production in an anaerobic biodegradation process. Comput. Math. Appl. **68**, 986–993 (2014)
6. Gopalsamy, K.: Stability and Oscillations in Delay Differential Equations of Population Dynamics. Kluwer Academic Publishers, Dordrect (1992)
7. Grognard, F., Bernard, O.: Stability analysis of a wastewater treatment plant with saturated control. Water Sci. Technol. **53**, 149–157 (2006)
8. Hale, J.K.: Theory of Functional Differential Equations. Applied Mathematical Sciences, vol. 3. Springer, New York (1977)
9. Maillert, L., Bernard, O., Steyer, J.-P.: Robust regulation of anaerobic digestion processes. Water Sci. Technol. **48**(6), 87–94 (2003)
10. Ruan, S.: On nonlinear dynamics of predator-prey models with discrete delay. Math. Model. Nat. Phenom. **4**(2), 140–188 (2009)
11. Ruan, S., Wei, J.: On the zeroes of transcendental functions with applications to stability of delay differential equations. Dynam. Contin. Impuls. Syst. **10**, 863–874 (2003)
12. Smith, H.: An Introduction to Delay Differential Equations with Applications to the Life Sciences. exts in Applied Mathematics, vol. 57. Springer, New York (2011)
13. Wang, L., Wolkowicz, G.: A delayed chemostat model with general nonmonotone response functions and differential removal rates. J. Math. Anal. Appl. **321**, 452–468 (2006)
14. Wolkowicz, G., Xia, H.: Global asymptotic behavior of a chemostat model with discrete delays. SIAM J. Appl. Math. **57**(4), 1019–1043 (1997)

Time-Optimal Control Problem in the Space of Probability Measures

Giulia Cavagnari[1]([✉]) and Antonio Marigonda[2]

[1] Department of Mathematics, University of Trento,
Via Sommarive 14, 38123 Povo, TN, Italy
giulia.cavagnari@unitn.it

[2] Department of Computer Sciences, University of Verona,
Strada Le Grazie 15, 37134 Verona, Italy
antonio.marigonda@univr.it

Abstract. We are going to define a time optimal control problem in the space of probability measures. Our aim is to model situations in which the initial position of a particle is not exactly known, even if the evolution is assumed to be deterministic. We will study some natural generalization of objects commonly used in control theory, proving some interesting properties. In particular we will focus on a comparison result between the classical minimum time function and its natural generalization to the probability measures setting.

Keywords: Optimal transport · Differential inclusions · Time optimal control

1 Introduction

Usual finite-dimensional time optimal control problem can be stated as follows: given a set-valued map $F : \mathbb{R}^d \rightrightarrows \mathbb{R}^d$ satisfying some structural assumptions, and a nonempty closed subset $S \subseteq \mathbb{R}^d$, we consider the solutions of differential inclusion starting from a given point $x_0 \in \mathbb{R}^d$, namely

$$
\begin{cases}
\dot{x}(t) \in F(x(t)), & t > 0, \\
\\
x(0) = x_0 \in \mathbb{R}^d.
\end{cases}
\tag{1}
$$

Then we can define the *minimum time function* $T : \mathbb{R}^d \to [0, +\infty]$ by setting for every $x_0 \in \mathbb{R}^d$

$$
T(x_0) := \inf\{T > 0 : \exists x(\cdot) \text{ solving (1) such that } x(T) \in S\}.
\tag{2}
$$

The study of the minimum time function and of its properties is a central topic in control theory, and the related literature is huge.

The present work is motivated by a natural consideration: in many real-world applications the *starting position* x_0 of the moving particle is known only up to

© Springer International Publishing Switzerland 2015
I. Lirkov et al. (Eds.): LSSC 2015, LNCS 9374, pp. 109–116, 2015.
DOI: 10.1007/978-3-319-26520-9_11

some uncertainties. For example it can be obtained only by an averaging of many measurement processes. It is worth noticing that this situation can happen even if we assume a pure deterministic evolution of the system.

A natural choice to model our knowledge about the particle's starting position is to consider it as a probability measure $\mu_0 \in \mathscr{P}(\mathbb{R}^d)$. The case in which μ_0 is a Dirac delta function concentrated at a point x_0 corresponds of course to the classical case in which perfect knowledge of the starting position is assumed.

This fact leads us to formulate directly our problem as regarding time-dependent measures, i.e. curves in $\mathscr{P}(\mathbb{R}^d)$. In this sense, the evolution of the starting *measure* μ_0 gives us a *macroscopic* point of view on the system, while the single (classical) trajectory corresponds to a *microscopic* point of view.

A natural requirement for the evolving measure $t \mapsto \mu_t$ is that at every time $t \in [0, T]$ we must have $\int_{\mathbb{R}^d} d\mu_t = 1$, since the probability to find *somewhere* the classical particle must be always equal to 1. This leads us to consider the evolution of the measure ruled by the following *continuity equation*, to be understood in the distributional sense

$$\begin{cases} \partial_t \mu_t + \mathrm{div}(v_t \mu_t) = 0, & t > 0, \\ \\ \mu_{|t=0} = \mu_0, \end{cases} \tag{3}$$

where $v_t(\cdot)$ is a time-depending Borel vector field belonging to $L^1_{\mu_t}(\mathbb{R}^d; \mathbb{R}^d)$ for a.e. $t \in [0, T]$.

It is well known that if $v_t(\cdot)$ is Lipschitz continuous, we can consider the *characteristics system*

$$\begin{cases} \dfrac{d}{dt} T_t(x) = v_t(T_t(x)), \\ \\ T_0(x) = x, \end{cases} \tag{4}$$

and the *unique* solution of (3) can be expressed by the *push-forward* of the initial measure μ_0 by the time-depending vector field $T_t(\cdot)$ solving (4), i.e., $\mu_t = T_t \sharp \mu_0$, where the push-forward $X \sharp \mu$ of a measure μ by a Borel vector field X is defined as

$$\int_{\mathbb{R}^d} \varphi(x) \, d(X \sharp \mu) = \int_{\mathbb{R}^d} \varphi \circ X(x) \, d\mu, \quad \forall \varphi : \mathbb{R}^d \to \mathbb{R} \text{ bounded Borel function.}$$

However (3) has been proven to be well-posed even in situations in which the regularity of the vector field v_t is not sufficient to guarantee uniqueness of the solutions of (4). Heuristically, this is due to the fact that the evolution of the measure is not affected by singularities in a μ_t-negligible set. Following [2], we recall that the integrability assumption $\|v_t\|_{L^p_\mu(\mathbb{R}^d)} \in L^1([0, T])$ yields the existence of a solution of (3) in the sense of a continuous curve $t \mapsto \mu_t$ in the space of probability measures endowed with the weak* topology induced by the duality with continuous and bounded functions $\varphi \in C^0_b(\mathbb{R}^d)$ (i.e., a *narrowly continuous curve* in the space of probability measures).

In many cases, the solutions of (3) can be constructed as *superpositions* of characteristics in the following sense: every probability measure η on the product space $\mathbb{R}^d \times \Gamma_T$, where Γ_T is the space of continuous curves in $\mathbb{R}^d$ defined on $[0,T]$, concentrated on the integral solutions of (4) (without assuming any uniqueness of the latter), can be used to define a solution of (3). Conversely, also every solution of (3) admits such a representation. We refer to [1,2] for such kind of results (see also Theorem 5.8 in [3] and Theorem 8.2.1 in [2]).

In a control-theoretic framework, in order to find a proper generalization of (1), it seems a natural choice to couple the dynamics (3) with the nonholonomic constraint $v_t(x) \in F(x)$ for a.e. $t \in [0,T]$ and μ_t-a.e. $x \in \mathbb{R}^d$, i.e., to ask that the driving vector field for the time-dependent measure μ_t is a suitable Borel selection of the set-valued map F. This is motivated also by the fact that in this case for *smooth* vector field v_t, the solutions of the characteristics system (4) turns out to be admissible trajectories of (1).

The link between the solutions of (3) and (4) has been extensively studied in the last years, we refer to [1] for a detailed presentation of the related issues. In particular, sufficient conditions are provided in order to grant existence and uniqueness in special classes of measures of the solutions of (3) also in cases where the corresponding (4) fails to provide uniqueness of the solutions. Moreover, also strict relationships between (3) and optimal transport theory have been already studied by many authors, and we refer to [2,3,5] for further details.

If we focus our attention on the set $\mathscr{P}_p(\mathbb{R}^d)$ of Borel probability measures with finite p-*moment*, i.e. measures μ satisfying $|\cdot| \in L^p_\mu(\mathbb{R}^d)$, we can consider also the metric structure induced by the p-*Wasserstein distance* $W_p(\cdot,\cdot)$ between measures. We refer the reader to [2] for all the details on Wasserstein distance.

In order to state our time-optimal control problem, we need also a convenient generalization of the target set S of the classical case. To introduce it, we consider the following heuristic argument (closely related to some interpretation of quantum mechanics), which follows the probabilistic motivation which led us to consider the controlled continuity equation as a good replacement for the differential inclusion.

Suppose to have an observer who makes some measurements on the system. The only quantity which we can consider is an average of the results of the measurements. From a mathematical point of view, we can model a measurement as a continuous map $\phi \in C^0(\mathbb{R}^d)$, thus the average result of the measurement of a system whose state is described by $\mu \in \mathscr{P}(\mathbb{R}^d)$ is given by the expected value of ϕ, namely $\int_{\mathbb{R}^d} \phi(x)\,d\mu(x)$.

A natural choice for the target set is to fix a threshold for each measurement and try to steer the system into states where the results of such measurements is below that threshold. Without loss of generality, we can fix the threshold to be 0 for all the measurements in which we are interested, thus the generalized target can be defined as follows: fix a subset $\Phi \subseteq C^0(\mathbb{R}^d;\mathbb{R})$ (which corresponds to the measurements in which we are interested) and define the generalized target to be

$$\tilde{S}^\Phi := \left\{ \mu \in \mathscr{P}(\mathbb{R}^d) : \int_{\mathbb{R}^d} \phi(x)\,d\mu(x) \le 0 \text{ for all } \phi \in \Phi \right\}.$$

In general some additional requirements on Φ are needed in order to have a good definition of the integral. We will deal mainly with the case in which for all $\phi \in \Phi$ there exist constants $A, B > 0$ and $p \geq 1$ such that $\phi(x) \geq A|x|^p - B$. An important example of this situation is given by fixing $S \subseteq \mathbb{R}^d$ and considering $\Phi = \{d_S(\cdot)\}$, i.e., we are going to measure the average distance from S. With this definition we have

$$\tilde{S}^{\{d_S\}} := \left\{ \mu \in \mathscr{P}(\mathbb{R}^d) : \int_{\mathbb{R}^d} d_S(x)\, d\mu(x) \leq 0 \right\} = \{\mu \in \mathscr{P}(\mathbb{R}^d) : \operatorname{supp}\mu \subseteq S\}.$$

Another interpretation of our framework in this case can be given in terms of *pedestrian dynamics*: suppose to have initially a crowd of people represented by a (normalized) probability measure μ_0 and to be able to identify a *safety zone* $S \subseteq \mathbb{R}^d$, while $F(\cdot)$ represents some (possible) nonholonomic constraints to the motion. Then if our aim in case of danger is to steer all the crowd to the safety zone in the minimum amount of time, we can choose $\Phi = \{d_S(\cdot)\}$. In a more realistic situation, it may not be possible to steer *all* the crowd to S. If we fix $\alpha \in [0,1]$ and choose $\Phi = \{d_S(\cdot) - \alpha\}$, we are still satisfied for example if the ratio between the number of people in the safe zone and all the people is above $1 - \alpha$, or if we can take the people sufficiently near to the safe zone.

Having defined the set of admissible trajectories and the target set in the space of probability measures, the definition of generalized minimum time function at a probability measure μ_0 is the straigthforwardly generalization of the classical one, i.e., the infimum of all the times T for which there exists an admissible trajectory defined on $[0, T]$ and satisfying $\mu_T \in \tilde{S}^\Phi$.

The paper is structured as follows: in Sect. 2 we introduce precise definitions of the generalized objects we are going to study, together with some of their properties. In Sect. 3 we prove the main results of the paper, finally in Sect. 4 we give some insight into the current work.

2   Generalized Objects and Their Properties

Definition 1 (Standing Assumption). *We will say that a set-valued function* $F : \mathbb{R}^d \rightrightarrows \mathbb{R}^d$ *satisfies the assumption* (F_j), $j = 0, 1$ *if the following hold true*

(F_0) $F(x) \neq \emptyset$ *is compact and convex for every* $x \in \mathbb{R}^d$, *moreover* $F(\cdot)$ *is continuous with respect to the Hausdorff metric, i.e. given* $x \in X$, *for every* $\varepsilon > 0$ *there exists* $\delta > 0$ *such that* $|y - x| \leq \delta$ *implies* $F(y) \subseteq F(x) + B(0, \varepsilon)$ *and* $F(x) \subseteq F(y) + B(0, \varepsilon)$.

(F_1) $F(\cdot)$ *has linear growth, i.e. there exist nonnegative constants* L_1 *and* L_2 *such that* $F(x) \subseteq \overline{B(0, L_1|x| + L_2)}$ *for every* $x \in \mathbb{R}^d$.

Definition 2 (Generalized targets). *Let* $p \geq 1$, $\Phi \subseteq C^0(\mathbb{R}^d, \mathbb{R})$ *such that the following property holds*

(T_E) *there exists* $x_0 \in \mathbb{R}^d$ *with* $\phi(x_0) \leq 0$ *for all* $\phi \in \Phi$.

We define the generalized targets $\tilde{S}^{\Phi}$ and $\tilde{S}_p^{\Phi}$ as follows

$$\tilde{S}^{\Phi} := \left\{ \mu \in \mathscr{P}(\mathbb{R}^d) : \int_{\mathbb{R}^d} \phi(x) \, d\mu(x) \leq 0 \text{ for all } \phi \in \Phi \right\},$$

$$\tilde{S}_p^{\Phi} := \tilde{S}^{\Phi} \cap \mathscr{P}_p(\mathbb{R}^d).$$

We define also the generalized distance from $\tilde{S}_p^{\Phi}$ as $\tilde{d}_{\tilde{S}_p^{\Phi}}(\cdot) := \inf_{\mu \in \tilde{S}_p^{\Phi}} W_p(\cdot, \mu)$.

For further use, we will say that Φ satisfies property (T_p) with $p \geq 0$ if the following holds true

(T_p) for all $\phi \in \Phi$ there exist $A_\phi, C_\phi > 0$ such that $\phi(x) \geq A_\phi |x|^p - C_\phi$.

The following proposition states some straightforward properties of the generalized targets. Its proof is immediate from the definition of generalized target.

Proposition 1 (Properties of the generalized targets). *Let $p \geq 0$ and $\Phi \subseteq C^0(\mathbb{R}^d, \mathbb{R})$ be such that (T_E) and (T_0) hold. Then $\tilde{S}^{\Phi}$ and $\tilde{S}_p^{\Phi}$ are convex, moreover $\tilde{S}^{\Phi}$ is w^*-closed in $\mathscr{P}(\mathbb{R}^d)$, while $\tilde{S}_p^{\Phi}$ is closed in $\mathscr{P}_p(\mathbb{R}^d)$ endowed with the p-Wasserstein metric $W_p(\cdot, \cdot)$. If moreover (T_p) holds for some $p \geq 1$, then $\tilde{S}^{\Phi} = \tilde{S}_p^{\Phi}$ is compact in the w^*-topology and in the W_p-topology.*

Definition 3 (Admissible curves). *Let $\Gamma : \mathbb{R}^d \rightrightarrows \mathbb{R}^d$ be a set-valued function, $I = [a, b]$ a compact interval of $\mathbb{R}$, $\alpha, \beta \in \mathscr{P}(\mathbb{R}^d)$. We say that a Borel family of probability measures $\boldsymbol{\mu} = \{\mu_t\}_{t \in I}$ is an admissible trajectory (curve) defined in I for the system joining α and β, if there exists a family of Borel vector fields $v = \{v_t(\cdot)\}_{t \in I}$ such that*

1. *$\boldsymbol{\mu}$ is a narrowly continuous solution in the distributional sense of the continuity equation $\partial_t \mu_t + \mathrm{div}(v_t \mu_t) = 0$, with $\mu_{|t=a} = \alpha$ and $\mu_{|t=b} = \beta$.*
2. *$J_F(\boldsymbol{\mu}, v) < +\infty$, where $J_F(\cdot)$ is defined as*

$$J_F(\boldsymbol{\mu}, v) := \begin{cases} \displaystyle\int_I \int_{\mathbb{R}^d} \left(1 + I_{F(x)}\left(v_t(x)\right)\right) d\mu_t(x) \, dt, & \text{if } \|v_t\|_{L^1_{\mu_t}} \in L^1([0,T]), \\ +\infty, & \text{otherwise,} \end{cases} \tag{5}$$

where $I_{F(x)}$ is the indicator function of the set $F(x)$, i.e., $I_{F(x)}(\xi) = 0$ for all $\xi \in F(x)$ and $I_{F(x)}(\xi) = +\infty$ for all $\xi \notin F(x)$.

In this case, we will also shortly say that $\boldsymbol{\mu}$ is driven by v.

When $J_F(\cdot)$ is finite, this value expresses the time needed by the system to steer α to β along the trajectory $\boldsymbol{\mu}$ with family of velocity vector fields v.

Definition 4 (Generalized minimum time). *Let $\Phi \in C^0(\mathbb{R}^d; \mathbb{R})$ and $\tilde{S}^{\Phi}$, $\tilde{S}_p^{\Phi}$ ($p \geq 1$) be the corresponding generalized targets defined in Definition 2. In* *analogy with the classical case, we define the* generalized minimum time function $\tilde{T}^\Phi : \mathscr{P}(\mathbb{R}^d) \to [0, +\infty]$ *by setting*

$$\tilde{T}^\Phi(\mu_0) := \inf \left\{ J_F(\boldsymbol{\mu}, v) : \boldsymbol{\mu} \text{ is an admissible curve in } [0, T], \right. \tag{6}$$

$$\left. \text{driven by } v, \text{ with } \mu_{|t=0} = \mu_0, \ \mu_{|t=T} \in \tilde{S}^\Phi \right\},$$

where, by convention, $\inf \emptyset = +\infty$.

Given $\mu_0 \in \mathscr{P}(\mathbb{R}^d)$, *an admissible curve* $\boldsymbol{\mu} = \{\mu_t\}_{t \in [0, \tilde{T}^\Phi(\mu_0)]} \subseteq \mathscr{P}(\mathbb{R}^d)$, *driven by a time depending Borel vector-field* $v = \{v_t\}_{t \in [0, \tilde{T}^\Phi(\mu_0)]}$ *and satisfying* $\mu_{|t=0} = \mu_0$ *and* $\mu_{|t=\tilde{T}^\Phi(\mu_0)} \in \tilde{S}^\Phi$ *is* optimal *for* μ_0 *if*

$$\tilde{T}^\Phi(\mu_0) = J_F(\boldsymbol{\mu}, v).$$

Given $p \geq 1$, *we define also a generalized minimum time function* $\tilde{T}^\Phi_p : \mathscr{P}_p(\mathbb{R}^d) \to [0, +\infty]$ *by replacing in the above definitions* $\tilde{S}^\Phi$ *by* $\tilde{S}^\Phi_p$ *and* $\mathscr{P}(\mathbb{R}^d)$ *by* $\mathscr{P}_p(\mathbb{R}^d)$. *Since* $\tilde{S}_p \subseteq \tilde{S}$, *it is clear that* $\tilde{T}^\Phi(\mu_0) \leq \tilde{T}^\Phi_p(\mu_0)$.

3   Main Results

Theorem 1 (First comparison between $\tilde{T}^\Phi$ and T). *Consider the generalized minimum time problem as in Definition 4 assuming* (F_0), (F_1), *and suppose that there exists* $S \subseteq \mathbb{R}^d$ *such that* $\tilde{S}^\Phi = \tilde{S}^{\{d_S\}}$. *Then for all* $\mu_0 \in \mathscr{P}(\mathbb{R}^d)$ *we have*

$$\tilde{T}^\Phi(\mu_0) \geq \|T\|_{L^\infty_{\mu_0}},$$

where $T : \mathbb{R}^d \to [0, +\infty]$ *is the classical minimum time function for the system* $\dot{x}(t) \in F(x(t))$ *with target* S.

Proof. For sake of clarity, in this proof we will simply write $\tilde{T}$ and $\tilde{S}$, thus omitting Φ.

If $\tilde{T}(\mu_0) = +\infty$ there is nothing to prove, so assume $\tilde{T}(\mu_0) < +\infty$. Fix $\varepsilon > 0$ and let $\boldsymbol{\mu} = \{\mu_t\}_{t \in [0,T]} \subseteq \mathscr{P}(\mathbb{R}^d)$ be an admissible curve starting from μ_0, driven by $v = \{v_t\}_{t \in [0,T]}$ such that $T = J_F(\boldsymbol{\mu}, v) < \tilde{T}(\mu_0) + \varepsilon$ and $\mu_{|t=T} \in \tilde{S}$. In particular, we have that $v_t(x) \in F(x)$ for μ_t-a.e. $x \in \mathbb{R}^d$ and a.e. $t \in [0,T]$, hence $|v_t(x)| \leq (L_1 + L_2)(1 + |x|)$ for μ_t-a.e $x \in \mathbb{R}^d$. Accordingly,

$$\int_0^T \int_{\mathbb{R}^d} \frac{|v_t(x)|}{1 + |x|} \, d\mu_t \, dt \leq T(L_1 + L_2) < +\infty.$$

By the superposition principle (Theorem 5.8 in [3] and Theorem 8.2.1 in [2]), we have that there exists a probability measure $\boldsymbol{\eta} \in \mathscr{P}(\mathbb{R}^d \times \Gamma_T)$ satisfying

1. $\boldsymbol{\eta}$ is concentrated on the pairs $(x, \gamma) \in \mathbb{R}^d \times \Gamma_T$ such that γ is absolutely continuous and

$$\gamma(t) = x + \int_0^t v_t(\gamma(s)) \, ds$$

2. for all $t \in [0, T]$ and all $\varphi \in C_b^0(\mathbb{R}^d)$

$$\int_{\mathbb{R}^d} \varphi(x) d\mu_t(x) = \iint_{\mathbb{R}^d \times \Gamma_T} \varphi(\gamma(t)) \, d\boldsymbol{\eta}(x, \gamma).$$

Evaluating the above formula at $t = 0$, we have that if $x \notin \operatorname{supp} \mu_0$ or $\gamma(0) \neq x$, then $(x, \gamma) \notin \operatorname{supp} \boldsymbol{\eta}$.

Let $\{\psi_n\}_{n \in \mathbb{N}} \in C_c^\infty(\mathbb{R}^d; [0, 1])$ with $\psi_n(x) = 0$ if $x \notin B(0, n+1)$ and $\psi_n(x) = 1$ if $x \in B(0, n)$. By Monotone Convergence Theorem, since $\{\psi_n(\cdot) d_S(\cdot)\}_{n \in \mathbb{N}} \subseteq C_b^0(\mathbb{R}^d)$ is an increasing sequence of nonnegative functions pointwise convergent to $d_S(\cdot)$, we have for every $t \in [0, T]$

$$\iint_{\mathbb{R}^d \times \Gamma_T} d_S(\gamma(t)) \, d\boldsymbol{\eta}(x, \gamma) = \lim_{n \to \infty} \iint_{\mathbb{R}^d \times \Gamma_T} \psi_n(\gamma(t)) d_S(\gamma(t)) \, d\boldsymbol{\eta}(x, \gamma)$$

$$= \lim_{n \to \infty} \int_{\mathbb{R}^d} \psi_n(x) d_S(x) \, d\mu_t(x)$$

By taking $t = T$, we have that the last term vanishes because $\mu_{|t=T} \in \tilde{S}$ and so $\operatorname{supp} \mu_{|t=T} \subseteq S$, therefore

$$\iint_{\mathbb{R}^d \times \Gamma_T} d_S(\gamma(T)) \, d\boldsymbol{\eta}(x, \gamma) = 0.$$

In particular, we necessarily have that $\gamma(T) \in S$ and $\gamma(0) = x$ for $\boldsymbol{\eta}$-a.e. $(x, \gamma) \in \mathscr{P}(\mathbb{R}^d \times \Gamma_T)$, whence $T \geq T(x)$ for μ_0-a.e. $x \in \mathbb{R}^d$, since $T(x)$ is the infimum of the times needed to steer x to S along trajectories of the system. Thus, $\tilde{T}(\mu_0) + \varepsilon \geq T(x)$ for μ_0-a.e. $x \in \mathbb{R}^d$ and, by letting $\varepsilon \to 0$, we conclude that $\tilde{T}(\mu_0) \geq \|T\|_{L_{\mu_0}^\infty}$.   □

It can be shown that the inequality appearing in Theorem 1 may be strict without further assumptions, however the following result states a relevant case in which equality holds, justifying also the name of *generalized minimum time problem* we gave.

Lemma 1 (Second comparison result). *Assume the same hypotheses and notation as in Theorem 1. Then, for every $x_0 \in \mathbb{R}^d$ we have $\tilde{T}^\Phi(\delta_{x_0}) = \tilde{T}_p^\Phi(\delta_{x_0}) = T(x_0)$ for all $p \geq 1$.*

Proof. Let us use the same notation as before, thus omitting Φ.

By Theorem 1 we have $\tilde{T}(\delta_{x_0}) \geq \|T\|_{L_{\delta_{x_0}}^\infty} = T(x_0)$. Conversely, let $\gamma_\varepsilon(\cdot)$ be a solution of $\dot{x}(t) \in F(x(t))$ such that $\gamma_\varepsilon(0) = x_0$ and $\gamma_\varepsilon(T(x_0) + \varepsilon) \in S$. Set $\mu_t^\varepsilon = \gamma_\varepsilon(t) \sharp \delta_{x_0}$ and $\boldsymbol{\mu}^\varepsilon = \{\mu_t^\varepsilon\}_{t \in [0, T(x_0)+\varepsilon]}$. By Theorem 8.3.1 in [2], we have that there exists a Borel vector field $v_t^\varepsilon : [0, T] \times \mathbb{R}^d \to \mathbb{R}^d$ such that $\partial_t \mu_t^\varepsilon + \operatorname{div}(v_t^\varepsilon \mu_t^\varepsilon) = 0$. Moreover, by construction we have that $\dot{\gamma}_\varepsilon(t) = v_t^\varepsilon(\gamma_\varepsilon(t)) \in F(\gamma_\varepsilon(t))$, thus $v_t^\varepsilon(x) \in F(x)$, for μ_t^ε-a.e. $x \in \mathbb{R}^d$ and a.e. $t \in [0, T]$. We conclude that μ_t^ε is an admissible curve steering δ_{x_0} to $\tilde{S}$ in time $T(x_0) + \varepsilon$, hence $\tilde{T}(\delta_{x_0}) \leq T(x_0) + \varepsilon$. By letting $\varepsilon \to 0^+$, we obtain the desired equality.   □

4   Conclusion

The study of generalized minimum time function in the space of probability measures is still largely in progress. In the forthcoming paper [4], more general cases will be treated, together with a dynamic programming principle and a result of existence of optimal trajectories in the space of probability measures.

We plan also to extend the definition of minimum time by possibly adding some terms in the functional penalizing the concentration of the mass, in order to treat more realistic problems coming from pedestrian dynamics.

Finally, the characterization of the generalized minimum time as solution of a suitable infinite-dimensional Hamilton-Jacobi-Bellmann equation seems to be quite hard, as well as to state a result comparable to Pontryagin Maximum Principle for this kind of problems. In [6] a similar problem was addressed, trying to characterize the infimum in the space of curves on $\mathscr{P}_2(\mathbb{R}^d)$ of an action-like functional (without control) starting from a given measure. In the same paper, they obtained that this value is a viscosity subsolution of a suitable HJB equation, while the supersolution part was proved only in dimension 1 using a special representation of the optimal transport map on $\mathbb{R}$.

Acknowledgements. The authors have been supported by INdAM - GNAMPA Project 2015: *Set-valued Analysis and Optimal Transportation Theory Methods in Deterministic and Stochastics Models of Financial Markets with Transaction Costs.*

References

1. Ambrosio, L.: The flow associated to weakly differentiable vector fields: recent results and open problems. In: Bressan, A., Chen, G.-Q.G., Lewicka, M., Wang, D. (eds.) Nonlinear Conservation Laws and Applications. The IMA Volumes in Mathematics and its Applications, vol. 153, pp. 181–193. Springer, New York (2011)
2. Ambrosio, L., Gigli, N., Savaré, G.: Gradient Flows in Metric Spaces and in the Space of Probability Measures. Lectures in Mathematics ETH Zürich, 2nd edn. Birkhäuser Verlag, Basel (2008)
3. Bernard, P.: Young measures, superpositions and transport. Indiana Univ. Math. J. **57**(1), 247–276 (2008)
4. Cavagnari, G., Marigonda, A., Nguyen, K.T., Priuli, F.S.: Generalized control systems in the space of probability measures (Submitted)
5. Dolbeault, J., Nazaret, B., Savaré, G.: A new class of transport distances between measures. Calc. Var. Partial. Differ. Equ. **34**(2), 193–231 (2009)
6. Gangbo, W., Nguyen, T., Tudorascu, A.: Hamilton-Jacobi equations in the Wasserstein space. Methods Appl. Anal. **15**(2), 155–184 (2008)

Sufficient Conditions for Small Time Local Attainability for a Class of Control Systems

Antonio Marigonda[1]($\boxtimes$) and Thuy Thi Le[2]

[1] Department of Computer Science, University of Verona,
Strada Le Grazie, 15, 37134 Verona, Italy
antonio.marigonda@univr.it
[2] Department of Pure and Applied Mathematics, University of Padova,
Via Trieste, 63, 35121 Padova, Italy

Abstract. We deal with the problem of *small time local attainability* (STLA) for nonlinear finite-dimensional time-continuous control systems. More precisely, given a nonlinear system $\dot{x}(t) = f(t, x(t), u(t))$, $u(t) \in U$, possibly subjected to state constraints $x(t) \in \Omega$ and a closed set S, our aim is to provide sufficient conditions to steer to S every point of a suitable neighborhood of S along admissible trajectories of the system, respecting the constraints, and giving also an upper estimate of the minimum time needed for each point near S to reach S.

Keywords: Geometric control theory · Small-time local attainability · State constraints

1 Introduction

We consider a finite-dimensional control system

$$
\begin{cases}
\dot{y}(t) = f(y(t), u(t)), & \text{for a.e. } t > 0, \\
y(0) = x, \\
u(t) \in U, & \text{for a.e. } t > 0.
\end{cases}
\tag{1}
$$

where U is a given compact subset of $\mathbb{R}^m$, $x \in \mathbb{R}^d$, $u(\cdot) \in \mathscr{U} := \{v : [0, +\infty[\to U \text{ such that } v \text{ is measurable}\}$, and $f : \mathbb{R}^d \times U \to \mathbb{R}^d$ is continuous on $\mathbb{R}^d \backslash S$ and such that for every compact $K \subseteq \mathbb{R}^d \backslash S$ there exists $L = L_K > 0$ with

$$
\|f(x, u) - f(y, u)\| \leq L_K \|x - y\|, \text{ for all } x, y \in K, \ u \in U.
$$

Given a closed subset $S \subseteq \mathbb{R}^d$, called *the target set*, the *minimum time function* $T : \mathbb{R}^d \to [0, +\infty]$ is defined as follows:

$$
T(x) := \inf\{T > 0 : \exists y(\cdot) \text{ solution of (1) satisfying } y(0) = x, \ y(T) \in S\}, \tag{2}
$$

where we set $\inf \emptyset = +\infty$ by convention.

© Springer International Publishing Switzerland 2015
I. Lirkov et al. (Eds.): LSSC 2015, LNCS 9374, pp. 117–125, 2015.
DOI: 10.1007/978-3-319-26520-9_12

We are interested in the following property, called *small-time local attainability* (STLA): given $T > 0$ there exists an open set $U \subseteq \mathbb{R}^d$ such that $U \supseteq S$ and $T(x) \leq T$ for all $x \in U$. This amounts to say that for every fixed time $T > 0$ there is a neighborhood of the target whose points can be steered to the target itself along admissible trajectories of the system in a time less than T. STLA may be formulated also in this way: for every $\bar{x} \in \partial S$ there exists $\delta_{\bar{x}} > 0$ and a continuous function $\omega_{\bar{x}} : [0, +\infty[\to [0, +\infty[$ such that $\omega(r) \to 0$ as $r \to 0$ and $T(x) \leq \omega_{\bar{x}}(d_S(x))$ for all $x \in B(\bar{x}, \delta_{\bar{x}})$, where $d_S(\cdot)$ denotes the Euclidean distance function from S.

STLA has been studied by several authors, and it turned out that estimates of this type have consequences also in *regularity* *property* of the minimum time function. One of the most important results on this line was found in [6], where it was proved that a controllability condition known as *Petrov's condition* yields an estimate $T(x) \leq \omega_{\bar{x}}(d_S(x))$ with $\omega_{\bar{x}}(r) = C_{\bar{x}}r$, for $C > 0$, and this is equivalent to local Lipschitz continuity of $T(\cdot)$ in $U \backslash S$ for a suitable neighborhood U of S.

For a compact target S, Petrov's condition can be formulated as follows: there exist $\delta, \mu > 0$, such that for every $x \in \mathbb{R}^d \backslash S$ whose distance $d_S(x)$ from S is less than δ there exist $u \in U$ and a point $\bar{x} \in S$ with $\|x - \bar{x}\| = d_S(x)$ and

$$\langle x - \bar{x}, f(x, u) \rangle \leq -\mu d_S(x). \tag{3}$$

From a geometric point of view, the underlying idea is the following: for every point near to S, there is an admissible velocity *pointing* *toward* S sufficiently fast, i.e., whose component in direction of S is sufficiently large. Since Petrov's condition involves only admissible velocities (i.e. *first order term* in the expansion of the trajectories) we refer to it as a *first-order condition* for STLA.

If we assume that the distance is smooth around S, we can give also another version of Petrov's condition: for every x near to S we require the existence of an admissible C^1-trajectory $\gamma_x(\cdot)$ of (1) satisfying $\gamma_x(0) = x$ and $\dfrac{d}{dt}(d_S \circ \gamma_x)$ $(0) < -\mu$. Accordingly, due to the smoothness of γ_x and d_S, we have also that for $t > 0$ sufficiently small we have $\dfrac{d}{dt}(d_S \circ \gamma_x)(t) < -\mu$. This formulation enhances the infinitesimal decreasing properties of the distance along at least one admissible trajectories contained in (3).

Natural steps toward the generalization of this condition are the following:

1. consider instead of the distance its square, since it is well known that the square of the distance enjoys more regularity properties then the distance itself.

2. take an integral version of the infinitesimal decreasing property, thus obtaining

$$d_S^2(\gamma_x(t)) - d_S^2(x) < -\frac{\mu}{2}t d_S(x) + o(t).$$

3. notice that instead of $\gamma_x(t)$ we can consider any point $y_t \in \mathscr{R}_x(t)$, where

$$\mathscr{R}_x(t) := \{z \in \mathbb{R}^d : \text{there exists a trajectory } \gamma \text{ of (1) with}$$
$$\gamma(0) = x, \ \gamma(t) = y\}$$

The crucial fact is that the map $t \mapsto y_t$ is *no longer required to be necessarily an admissible trajectory*, even if $y_t \in \mathscr{R}_x(t)$ for all t. Such kinds of curves will be called $\mathscr{A}$-trajectories starting from x.

In this way the problem is reduced to estimate the rate of decreasing of the distance along $\mathscr{A}$-trajectories. The first paper in which this point of view was introduced is [1], where all the above generalization were performed. More precisely, it is assumed that there exists $\mu > 0$ such that for every x near to S and t sufficiently small we can find an $\mathscr{A}$-trajectory (in the original paper is called R-trajectory) y_t such that

$$y_t = x + a(t; x) + t^\alpha A(x) + o(t^\alpha; x),$$

where

1. $a(\cdot)$, $A(\cdot)$ are smooth functions,
2. the reminder satisfies a uniform estimate $\|o(t^\alpha; x)\| \leq K t^{\alpha+\beta}$ with K, β suitable positive constants independent of x,
3. $\|a(\cdot)\|$ is bounded from above by $M t^s d_S(x)$ where M is a suitable constant,
4. there exists a point $\bar{x} \subset S$ with $\|x - \bar{x}\| = d_S(x)$ and $\langle x - \bar{x}, A(x) \rangle \leq -\mu d_S(x)$.

Roughly speaking, we require the infinitesimal decreasing property of Petrov's condition for the *essential leading term* of at least an $\mathscr{A}$-trajectory which now is a term of order $\alpha \geq 1$. The name "essential leading term", introduced by [1], is motivated by the fact that as long as x is taken near to S, we have that $\|a(\cdot)\|$ vanishes. By the equivalency between Petrov's condition and local Lipschitz continuity of $T(\cdot)$ we can not expect any more an estimate like $T(x) \leq C d_S(x)$ in the case $\alpha > 1$, however it turns out that a similar estimate holds true, yielding $T(x) \leq C d_S^{1/\alpha}(x)$. We refer to this conditions as *higher order Petrov-like conditions* for STLA.

In [4] was treated the case in which the constant μ appearing in Petrov's condition is a function $\mu = \mu(d_S(x))$ allowed to slowly vanish as $d_S(x) \to 0$. This was not covered by [1], since there was assumed μ to be always constant. From a geometric point of view, this means that we are allowed to arrive tangentially to the target. There was obtained an estimate $T(x) \leq C d_S^\beta(x)$ also involving the dependency of $\mu(\cdot)$ on $d_S(\cdot)$, but under additional geometrical assumptions on the target, which were removed in a later paper [2] by Krastanov, where the results of [1,4] are subsumed in a unique formulation, but still under strong smoothness hypothesis on the terms appearing in the expression of y_t and taking into account a decay of $r \mapsto \mu(r)$ only as suitable powers of r.

The recent paper [5] weakened some smoothness assumptions required in [1,2] on the terms appearing in the expression of the $\mathscr{A}$-trajectory $t \mapsto y_t$, but instead of them, the authors assumed more regularity on the target set than in [2]. With even more regularity, in [5] is also defined a *generalized curvature* by means of suitable generalized gradients of higher order of the distance function. This allows to consider not only first-order expansion of the distance along an $\mathscr{A}$-trajectory, but also second-order effects, improving further STLA sufficient conditions. This was in the spirit of [1], in which was pointed out that STLA cannot be reduced to attainability of the *single points* of the target, but needs to take into account also the *geometrical properties* of the target.

We present here a STLA result removing the smoothness assumptions on the terms appearing in the expression of the $\mathscr{A}$-trajectory $t \mapsto y_t$, as in [5], but without any additional regularity hypothesis on the target set used in [5], thus fully generalizing the results of [2] also in presence of the additional state constraint $y(t) \in \overline{\Omega}$, where Ω is an open subset of $\mathbb{R}^d$ with $\Omega \backslash S \neq \emptyset$.

In general a complete description of the set of $\mathscr{A}$-trajectories, on which higher order conditions must be checked, turns to be very difficult. For control affine systems of the form

$$
\begin{cases}
\dot{x}(t) = f_0(x(t)) + \sum_{i=1}^{M} u_i(t) f_i(x(t)), \\
x(0) = x,
\end{cases}
\tag{4}
$$

where $f_i(\cdot)$ are smooth vector fields, and $u_i : [0, +\infty[\to [-1, 1]$ are measurable, additional information on $\mathscr{A}$-trajectories can be obtained by the study of the Lie algebra generated by $\{f_i\}_{i=1,\dots,M}$, as performed in various degree of generality in the papers [1,2,4,5]. In the forthcoming paper [3] the analysis of such kind of systems is performed also in presence of state constraints, in order to provide explicit higher order conditions for STLA.

The paper is structured as follows: in Sect. 2 we formulate and prove the main result on STLA, and in Sect. 3 we compare this result with some other similar results from [2,5].

2   A General Result on STLA

Throughout the paper, given a set $Z \subseteq \mathbb{R}^d$ and a positive number δ, we set $Z_\delta = \{y \in \mathbb{R}^d : d_Z(y) \leq \delta\}$, moreover we denote by $\partial^P d_S(x)$ the proximal superdifferential of d_S at x. Given an open set $\Omega \subseteq \mathbb{R}^d$, we consider the $\overline{\Omega}$-*state constrained problem*, i.e., we add to system (1) the condition $x(t) \in \overline{\Omega}$. Consequently, we can define the *state constrained reachable set from* $x_0 \in \overline{\Omega}$ at time $\tau \geq 0$:

$$
\mathscr{R}_{x_0}^{\Omega}(\tau) := \left\{ y(\tau) : y(\cdot) \text{ is a solution of (1) defined on } [0, \tau] \text{ with } y([0, \tau]) \subseteq \overline{\Omega} \right\}.
$$

The *state constrained minimum time function* from $x_0 \in \overline{\Omega}$ is

$$
T_\Omega(x_0) := \begin{cases}
+\infty, & \text{if } \mathscr{R}_{x_0}^{\Omega}(\tau) \cap S = \emptyset \text{ for all } \tau \geq 0, \\
\inf\{\tau \geq 0 : \mathscr{R}_{x_0}^{\Omega}(\tau) \cap S \neq \emptyset\}, & \text{otherwise.}
\end{cases}
$$

Lemma 1. *Let $\delta > 0$ be a constant, $\lambda : \mathbb{R}^2 \to \mathbb{R}$, $\theta : \mathbb{R} \to \mathbb{R}$ be continuous functions such that*

1. $r \mapsto \dfrac{\theta(r)r}{\lambda(\theta(r),r)}$ is bounded from above by a nonincreasing function $\beta(\cdot) \in L^1(]0,\delta[)$;
2. $\lambda(\theta(r),r) > 0$ for $0 < r < \delta$, and $\lambda(0,r) = 0$ for $r > 0$.

Consider any sequence $\{r_i\}_{i \in \mathbb{N}}$ in $[0,\delta]$ satisfying for all $i \in \mathbb{N}$:

(S_1) $r_{i+1}^2 - r_i^2 \leq -\lambda(\theta(r_i),r_i)$, (S_2) $\theta(r_i) \neq 0$ implies $r_i \neq 0$.

Then we have: a) $r_i \to 0$; b) $\displaystyle\sum_{i=0}^{\infty} \theta(r_i) \leq 2 \int_0^{r_0} \beta(r)\,dr$.

Proof. According to (S_1), the sequence $\{r_i\}_{i \in \mathbb{N}}$ is monotone and bounded from below, thus it admits a limit r_∞ satisfying $0 \leq r_\infty < \delta$. Assume by contradiction that $r_\infty > 0$. By passing to the limit for $i \to +\infty$ in (S_1), since $\lambda(\cdot,\cdot) \in C^0$ and $\lambda(\theta(r_i),r_i) \geq 0$, we obtain that $0 = \lambda(\theta(r_\infty),r_\infty)$ contradicting the assumptions on λ, thus $r_\infty = 0$. Since if $\theta(r_i) \neq 0$ we have $r_i \neq 0$ and $\dfrac{r_i^2 - r_{i+1}^2}{\lambda(\theta(r_i),r_i)} \geq 1$, we obtain

$$\sum_{i=0}^{\infty} \theta(r_i) = \sum_{\substack{i=0 \\ \theta(r_i)\neq 0}}^{\infty} \theta(r_i) \leq \sum_{\substack{i=0 \\ \theta(r_i)\neq 0}}^{\infty} \frac{\theta(r_i)}{\lambda(\theta(r_i),r_i)}(r_i^2 - r_{i+1}^2)$$

$$\leq \sum_{\substack{i=0 \\ \theta(r_i)\neq 0}}^{\infty} \frac{\theta(r_i)}{\lambda(\theta(r_i),r_i)}(r_i + r_{i+1})(r_i - r_{i+1}) \leq 2 \sum_{\substack{i=0 \\ \theta(r_i)\neq 0}}^{\infty} \frac{\theta(r_i)r_i}{\lambda(\theta(r_i),r_i)}(r_i - r_{i+1})$$

$$\leq 2 \sum_{\substack{i=0 \\ \theta(r_i)\neq 0}}^{\infty} \beta(r_i)(r_i - r_{i+1}) \leq 2 \int_0^{r_0} \beta(r)\,dr,$$

recalling the monotonicity property of $r \mapsto \beta(r)$.

Theorem 1 (General attainability). Consider the system (1). Let $\delta_0 > 0$ be a positive constant, $\sigma, \mu : [0,+\infty[\times[0,+\infty[\to [0,+\infty[$, and $\tau, \theta : [0,+\infty[\to [0,+\infty[$ be continuous functions. Let $Q : [0,+\infty[\times\mathbb{R}^d \to [0,+\infty[$ be a function such that $t \mapsto Q(t,x)$ is continuous for every $x \in S_{\delta_0}\backslash S$.
    We assume that:

(1) $\tau(r) = 0$ iff $r = 0$, $0 < \theta(r) \leq \tau(r)$ for every $0 < r < \delta_0$;
(2) for any $x \in (S_{\delta_0} \cap \overline{\Omega})\backslash S$ and $0 < t \leq \tau(d_S(x))$ the following holds
    (2.a) $\mathscr{R}_x^{\Omega}(t) \cap S_{2\delta_0} \neq \{x\}$,
    (2.b) if $\mathscr{R}_x^{\Omega}(t) \cap S = \emptyset$, there exists $y_t \in \mathscr{R}_x^{\Omega}(t) \cap B(x,\chi(t,d_S(x)))$ with

$$\min_{\zeta \in \partial^P d_S(x)} \langle d_S(x)\zeta, y_t - x\rangle + \|y_t - x\|^2 \leq -\mu(t,d_S(x)) + \sigma(t,d_S(x));$$

    (2.c) if S is not compact, then $\left(\mathscr{R}_x^{\Omega}(t) \cap S_{2\delta_0}\right)\backslash S \subseteq B(0,Q(t,x))$
(3) the continuous function $\lambda : [0,+\infty[\times[0,+\infty[\to \mathbb{R}$, defined as $\lambda(t,r) := 2\mu(t,r) - 2\sigma(t,r)$, satisfies the following properties:
    (3.a) $0 < 2\lambda(\theta(r),r) < r^2$, $\lambda(0,r) = 0$ for all $0 < r < \delta_0$;

(3.b) $r \mapsto \dfrac{\theta(r)r}{\lambda(\theta(r), r)}$ is bounded from above by a nonincreasing function $\beta(\cdot) \in L^1(]0, \delta_0[)$.

Then, if we set $\omega(r_0) := 2\displaystyle\int_0^{r_0} \beta(r)\, dr$, we have that $T_\Omega(x) \leq \omega(d_S(x))$ for any $x \in S_{\delta_0} \cap \overline{\Omega}$.

Before proving the result, we make some remarks on the assumptions. Assumption (2.a) requires that from every x in the feasible set and sufficiently near to S we can move remaining inside the feasible set and not too far from S. Moreover, given a time $t < T(x)$ (thus $\mathscr{R}_x(t) \cap S = \emptyset$), in (2.b) we assume the existence of a y_t in the reachable set, not too far from x (2.c), such that the square of the distance from the target is decreased of at least $\lambda(t, d_S(x))/2 = \mu(t, d_S(x)) - \sigma(t, d_S(x))$. Assumption (3) requires λ to satisfy the requests of Lemma 1, thus concluding the proof.

Proof (of Theorem 1). We define a sequence of points and times $\{(x_i, t_i, r_i)\}_{i \in \mathbb{N}}$ by induction as follows. We choose $x_0 \in (S_{\delta_0} \cap \overline{\Omega}) \backslash S$, and set $r_0 = d_S(x_0)$, $t_0 = \min\{T_\Omega(x_0), \theta(r_0)\}$. Suppose to have defined x_i, t_i, r_i. We distinguish the following cases:

1. if $x_i \in S$, we define $x_{i+1} = x_i$, $t_{i+1} = 0$, $r_{i+1} = 0$.
2. if $x_i \notin S$ and $t_i \geq T_\Omega(x_i)$, in particular we have $T_\Omega(x_i) < +\infty$, thus we can choose $x_{i+1} \in \mathscr{R}_{x_i}^\Omega(T_\Omega(x_i)) \cap S$ and define $r_{i+1} = 0$, $t_{i+1} = 0$.
3. if $x_i \notin S$ and $t_i < T_\Omega(x_i)$, we choose $x_{i+1} \in \mathscr{R}_{x_i}^\Omega(t_i)$ such that

$$\min_{\zeta_i \in \partial^P d_S(x_i)} \langle r_i \zeta_i, x_{i+1} - x_i \rangle + \|x_{i+1} - x_i\|^2 \leq -\mu(t_i, r_i) + \sigma(t_i, r_i),$$

and define $r_{i+1} = d_S(x_{i+1})$, $t_{i+1} = \min\{T_\Omega(x_{i+1}), \theta(r_{i+1})\}$. According to the semiconcavity of $d_S^2(\cdot)$ (with semiconcavity constant 2) and recalling that $\zeta_x \in \partial^P d_S(x)$ iff $2\zeta_x d_S(x) \in \partial^P d_S(\cdot)$, we have that there exists $\zeta_x \in \partial d_S(x)$ such that

$$r_{i+1}^2 - r_i^2 \leq \langle 2\zeta_i r_i, x_{i+1} - x_i \rangle + 2\|x_{i+1} - x_i\|^2 \leq -\lambda(t_i, r_i). \qquad (5)$$

We notice that in this case $x_{i+1} \notin S$ since $x_{i+1} \in \mathscr{R}_{x_i}^\Omega(t_i)$ and $t_i = \theta(r_i) < T_\Omega(x_i)$, thus $t_{i+1} > 0$ and $r_{i+1} > 0$.

The assumptions of Lemma 1) are satisfied:

1. $r_{i+1}^2 - r_i^2 \leq -\lambda(\theta(r_i), r_i)$,
2. it is obvious that $\theta(r_i) \neq 0$ implies $r_i \neq 0$. Indeed, assume that $r_i = 0$. Since $0 \leq \theta(r) \leq \tau(r)$, and $\tau(r) = 0$ iff $r = 0$, we have $\theta(0) = 0$.
3. by assumption, there exists $\beta \in L^1(]0, \delta_0[)$ such that $\dfrac{\theta(s)s}{\lambda(\theta(s), s)} \leq \beta(r)$.

Applying Lemma 1, we have that a) $r_i \to 0$, b) $\sum_{i=0}^{\infty} \theta(r_i) \leq 2 \int_0^{r_0} \beta(r)\, dr$. Since

$\sum_{i=0}^{\infty} t_i \leq \sum_{i=0}^{\infty} \theta(s_i)$, we have $\sum_{i=0}^{\infty} t_i \leq 2 \int_0^{r_0} \beta(r)\, dr$. If S is compact, since $d_S(x_i) \to 0$, we have that $\{x_i\}_{i \in \mathbb{N}}$ is bounded. If S is not compact, for every $j \in \mathbb{N}$, we notice that $x_j \in \left(S_{2\delta_0} \cap \mathscr{R}_{x_0}^{\Omega}\left(\sum_{i=0}^{j} t_i \right) \right) \setminus S \subset B\left(0, Q\left(\sum_{i=0}^{j} t_i, x \right) \right)$. Since $\sum_{i=0}^{j} t_i$ converges, we have that there exists $R > 0$ such that $Q\left(\sum_{i=0}^{j} t_i, x \right) \leq R$ for all $j \in \mathbb{N}$, thus also in this case $\{x_i\}_{i \in \mathbb{N}}$ is bounded. Up to subsequence, still denoted by $\{x_i\}_{i \in \mathbb{N}}$, we have that there exists $\bar{x} \subset \mathbb{R}^d$ such that $x_i \to \bar{x}$. Since $d_S(x_i) \to 0$, we have $\bar{x} \in S$ and so $T_{\Omega}(x_0) \leq \sum_{i=1}^{\infty} t_i \leq \omega(d_S(x_0))$, which concludes the proof.

At this level the state constraints play no role, since their presence is hidden in Assumption (2) of Theorem 1, which requires the knowledge of at least an approximation of the reachable set in time t. In the control-affine case (4) this can be obtained by studying the Lie algebra generated by the vector fields appearing in the dynamics, since, as well known, noncommutativity of the flows of such vector fields will generate further direction along which the system can move, and so more $\mathscr{A}$-trajectories. Indeed, up to an higher order error, such $\mathscr{A}$-trajectories can be described by mean of their generating Lie brackets at the initial point, and so it is possible to impose the decreasing condition of Assumption (2) of Theorem 1 directly on such Lie brackets. This gives a tool to check it in many interesting cases. State constraints may reduce the number of feasible $\mathscr{A}$-trajectory generated by Lie bracket operations, since in order to construct each of them we have to concatenate several flows, thus possibly exiting from the feasible region after a certain time. This problem can be faced for instance by imposing a sort of *inward pointing conditions* (see e.g. [3]) in order to prevent such a situation, forcing all the flows involved the construction of the bracket to remain inside the feasible region. Finally, we notice that the distance to the boundary of the feasible region may be estimate by a semiconcavity inequality similar to the one used to estimate the decreasing of the distance from the target, thus at each step of the construction in the proof of Theorem 1 it is possible to estimate also the distance from the boundary of the feasible region.

3   Comparison with Other Results

Example 1. The ground space is $\mathbb{R}$, and set $S = \{z_k : k \in \mathbb{N}\} \cup \{0\}$. Since S does not satisfy the internal sphere condition, the results of [5] cannot be applied. Take $U = [-1, 1]$ and define $f(x, u) = \frac{u}{\log |x|}$ for $0 < |x| < 1/2$. We have that $f \in C_{loc}^{1,1}(S_{1/2} \setminus S) \times [-1, 1])$ and w.l.o.g. we can extend it to a function $C_{loc}^{1,1}((\mathbb{R} \setminus S) \times [-1, 1])$, still denoted by f. Clearly, for any $0 < \bar{x} < 1/2$ the optimal control corresponds to $u(t) \equiv 1$, and for $-1/2 < \bar{x} < 0$ the optimal control is -1. We restrict our attention only to $x > 0$ due to the symmetry of the system. Consider now any $\mathscr{A}$-trajectory $\sigma_{\bar{x}}(\cdot)$ starting from $\bar{x}$ of the form $\sigma_{\bar{x}}(t) = \bar{x} + a(t, \bar{x}) + t^\alpha A(\bar{x}) + o(t^\alpha, \bar{x})$, where $A(\cdot)$ is a Lipschitz continuous map, $\|a(t,x)\| \leq t^s c(x)$ for $s > 0$ and a Lipschitz map $c(\cdot)$ satisfying $c(x) \to 0$ when $d_S(x) \to 0$, i.e. of the same structure as in [1,2]. If $\sigma_{\bar{x}}(t) > \bar{x}$ for all $t > \tau_0$ the $\mathscr{A}$-trajectory do not approach the target. Excluding this case, and up to a time shift, we can restrict to $\mathscr{A}$-trajectories satisfying $0 \leq \sigma_{\bar{x}}(t) \leq \bar{x}$ for $t > 0$. In particular, we have that $|\sigma_{\bar{x}}(t) - \bar{x}| \leq \frac{2t}{\lceil \log \bar{x} \rceil}$ since all trajectories contained in $[0, \bar{x}]$ have modulus of speed which cannot exceed $\frac{1}{\lceil \log \bar{x} \rceil}$. By letting $\bar{x} \to 0^+$, we obtain for all $t > 0$ that $\|t^\alpha A(0) + o(t^\alpha, 0)\| \leq 0$, thus, dividing by t^α and letting $t \to 0^+$ we obtain $A(0) = 0$ and thus by Lipschitz continuity of $A(\cdot)$ we have $|A(x)| \leq C|x|$. In particular, the results of [1] cannot be applied because the essential leading term vanishes as we approach the target. Theorem 3.1, which is the main result of [2], requires the existence of $0 \leq \lambda < \dfrac{2\alpha}{2\alpha - 1}$ for the $\mathscr{A}$-trajectory $\sigma_{\bar{x}}(\cdot)$ such that $\langle x - \pi_S(x), A(x) \rangle \leq -\delta d_S^\lambda(x)$, where $\pi_S(x)$ is the projection of x on S. For $x > 0$, we obtain $d_S(x) = x$ and $\pi_S(x) = 0$, and together with $|A(x)| \leq C|x|$, this implies $\lambda \geq 2$, but $\dfrac{2\alpha}{2\alpha - 1} \leq 2$ since it is assumed that $\alpha \geq 1$, thus also this result cannot be applied. We consider the optimal solution $\gamma_{\bar{x}}(t) = \bar{x} + \frac{t}{\log \bar{x}} + o(t)$ corresponding to the control $u = 1$. Take $y_{\bar{x}}(t) = \bar{x} + \frac{t}{2 \log \bar{x}}$. It can be easily proved that $\bar{x} > y_{\bar{x}}(t) > \gamma_{\bar{x}}(t)$, and from this that $y_{\bar{x}}(\cdot)$ is an $\mathscr{A}$-trajectory. Assumption of Theorem 1 are satisfied with $\delta_0 = 1/2$, $\theta(r) = \tau(r) = r|\log r|$, $\mu(t, r) = \frac{rt}{2|\log r|}$, $\sigma = \frac{t^2}{4 \log^2 x}$, $\beta(r) = |\log r|$, providing the estimate $T(x) \leq 4(x - x \log x)$. Indeed, we can compute exactly the minimum time function in this case, which turns out to be $T(x) = x - x \log x$ for $0 < x < 1/2$ (and in general $T(x) = |x| - |x| \log |x|$ for $|x| < 1/2$).

Acknowledgements. A. Marigonda—The first author has been supported by INdAM - GNAMPA Project 2015: *Set-valued Analysis and Optimal Transportation Theory Methods in Deterministic and Stochastics Models of Financial Markets with Transaction Costs.*

References

1. Krastanov, M.I., Quincampoix, M.: Local small time controllability and attainability of a set for nonlinear control system. ESAIM Control Optim. Calc. Var. **6**, 499–516 (2001)
2. Krastanov, M.I.: High-order variations and small-time local attainability. Control Cybern. **38**(4B), 1411–1427 (2009)
3. Le, T.T., Marigonda, A., Small-time local attainability for a class of control systems with state constraints (submitted)
4. Marigonda, A.: Second order conditions for the controllability of nonlinear systems with drift. Commun. Pure Appl. Analy. **5**(4), 861–885 (2006)

5. Marigonda, A., Rigo, S.: Controllability of some nonlinear systems with drift via generalized curvature properties. SIAM J. Control Optim. **53**(1), 434–474 (2015)
6. Veliov, V.M.: Lipschitz continuity of the value function in optimal control. J. Optim. Theor. Appl. **94**(2), 335–363 (1997)

Financing the Reduction of Emissions from Deforestation: A Differential Game Approach

Bernadette Riesner and Gernot Tragler[✉]

Institute of Statistics and Mathematical Methods in Economics,
Vienna University of Technology, 1040 Vienna, Austria
e0927319@student.tuwien.ac.at, gernot.tragler@tuwien.ac.at

Abstract. This paper analyzes and compares two versions of a mechanism that aims at mitigating climate change through REDD (Reduced Emissions from Deforestation and Forest Degradation). In this mechanism industrialised countries compensate countries with rainforests if they reduce their deforestation, because it is more cost efficient than restricting carbon emissions from domestic production. The initial question is, which funding possibility yields the best environmental results and is most beneficial for the involved parties. For this purpose, differential games are developed, in which industrialized countries and countries with rainforests denote the two players. Solutions are obtained by applying Pontryagin's Maximum Principle and the concept of Nash and Stackelberg Equilibria. Due to the model assumptions, analytical solutions can be found. It turns out that both versions of the mechanism can be a valuable contribution in the battle against climate change. Moreover, most advantages and disadvantages of the two variants turn out to be robust w.r.t. parameter changes and small modifications of the model.

Keywords: REDD · Climate policy · Differential game · Optimal control

1 Introduction

In search of promising strategies to combat climate change, REDD is one of the most debated proposals. The basic idea is to pay money to forest owners so that they do not cut down their forest and hence avoid greenhouse gas emissions [4].[1] Rainforests are important CO_2-sinks and deforestation causes approximately 20 % of global CO_2-emissions [10]. Therefore, the preservation of rainforests within the scope of a REDD-mechanism can play a vital role in the battle against climate change. Already in the Bali Roadmap 2007 the fundamental decision for the implementation of a REDD mechanism was written down,

---

[1] The currently discussed REDD+ mechanism operates at a national level and does not directly compensate individual forest owners.

© Springer International Publishing Switzerland 2015
I. Lirkov et al. (Eds.): LSSC 2015, LNCS 9374, pp. 126–133, 2015.
DOI: 10.1007/978-3-319-26520-9_13

but up to now no agreement on the financing could be achieved. The discussion there focuses on the choice between a market-based and a fund-based approach. In both scenarios, the forest owners receive money if their deforestation is below a certain reference rate. The reference rate should describe how much they would have deforested in absence of a REDD mechanism. In a market-based scenario, forest owners can generate certificates if their deforestation is below the reference rate. That means that reduced deforestation is converted into reduced carbon emissions and these certificates can be sold on an international certificate market. Buyers who have a reduction obligation, imposed by the Kyoto Protocol or a succeeding agreement, can use these certificates towards their emissions reductions compliance targets. That means, if the price of the certificate is lower than the domestic carbon avoidance costs, the buyer can comply with her reduction target in a cheaper way. A fund-based solution implies that a fund is implemented into which everybody, in practice mainly industrial countries though, can pay money. The thus arising sum will be distributed among rainforest countries, according to their reduced deforestation. Donors with reduction obligations are not allowed to count the emission reduction they financed as their own reduction. The strongest advocate of a market-based solution is the Coalition for Rainforest Nations. They argue that only in a market-solution with certificate trading, industrial countries have a monetary incentive to invest into the preservation of the rainforest and thus only in this approach sufficient money can be raised. On the other hand, the largest rainforest country Brasil and the insular state Tuvalu belong to the most vehement opponents of a market-based approach. Their main point is that a market-solution only helps industrial countries to cheaply comply with their reduction obligations but that it does not lead to additional emission reductions [4]. However, both reasonings fall short, as REDD cannot be analysed independently from the negotiations for new reduction obligations in the scope of a successive treaty of the Kyoto Protocol. It can be assumed that industrial countries will be willing to accept more stringent reduction targets if they are able to fulfill them relatively cheaply with the help of certificate trading as part of REDD. Which effect prevails, the increased willingness to transfer money to the South if emission certificates are thereby generated and to accept low emission caps in a market-based approach, or the additional reductions beyond the obligations in a fund-based solution is the starting point for this paper. To the best of our knowledge, this is the first mathematical paper that focuses on this specific topic.

2 The Model

2.1 The Baseline Scenario

To analyse this issue, two agents will be considered that interact in a finite period $[0, T]$. The first agent will be called *north* and represents industrial countries that do not own forest. The second one, *south*, represents developing countries with rainforests.

The modelling of the south is similar to [7]. In the absence of an international agreement, the south faces the following optimisation problem:

$$\max_{D(t)} \int_0^T e^{-r_s t} \left\{ \left[\overline{P} - \theta D(t) \right] D(t) + Y \left[F_0 - F(t) \right] \right\} dt + e^{-r_s T} \phi F(T) \tag{1}$$
$$\text{s.t.} \quad \dot{F}(t) = -D(t), \quad F(0) = F_0.$$

It is here assumed that deforestation in the south $D(t)$ yields two kinds of income: Firstly, timber can be sold at a price that is linearly decreasing in the amount sold. Then revenues through timber sale at time t amount to $\left[\overline{P} - \theta D(t) \right] D(t)$, where $\overline{P}$ is the maximal market price obtained when $D(t)$ tends towards zero and θ is a positive parameter that determines the steepness of the demand curve. Secondly, the deforested areas can be used for agricultural production. Here, like in [7], it is assumed that the yield per cultivated land is constant. Thereby agricultural income can be modelled as $Y \left[F_0 - F(t) \right]$, where F_0 denotes the initial size of the rainforest and $F(t)$ is the size of the rainforest at time t.

Let $\overline{D}(t)$ be the function that optimizes problem (1) and serves as a reference rate for the definition of *reduced deforestation*.

The economic utility of the north is modelled in a more schematic way. In absence of an international agreement, the north faces the following optimisation problem:

$$\max_{E_n(t)} \int_0^T e^{-r_n t} \left\{ a E_n(t) - b E_n(t)^2 - c_n \left[S(t) - \underline{S} \right]^2 \right\} dt - e^{-r_n T} \psi S(T) \tag{2}$$
$$\text{s.t.} \quad \dot{S}(t) = E_n(t) + \gamma \overline{D}(t), \quad S(0) = S_0.$$

It is important here to reflect the two-edged role of greenhouse gas emissions for the north: On the one hand, emissions are closely linked to production and thereby economic welfare. On the other hand, excessive emissions of greenhouse gases lead to climate change and all the negative impacts related to it. Analogously to [1,3] it is assumed that production and greenhouse gas emissions of the north $E_n(t)$ grow proportionally and that the utility derived from production is concave. Pinning down the economic utility of the north as $a E_n(t) - b E_n(t)^2$ fulfills both requirements. The last term in curly brackets in (2) reflects the damage caused by the accumulated stock of greenhouse gases $S(t)$. As in [1–3] it is assumed that the damage caused by a certain concentration of greenhouse gases is convex in the stock. As there has always been CO_2 in the atmosphere it is not the existence but the concentration above a threshold $\underline{S}$ that causes damage. The damage function used here reflects those two observations, and c_n weights the damages, in comparison to economic utility.

The dynamic constraint in (2) describes the assumption that the accumulated stock of greenhouse gases in the atmosphere increases linearly in the emissions of the north, $E_n(t)$, and those of the south, $E_s(t)$. In this model, the emissions of the south solely stem from deforestation activities. Therefore, deforestation has to be converted into corresponding greenhouse gas emissions. The most natural way to do so is to assume that a certain area of rainforest on average stores a certain amount of CO_2, which will be released after logging. This leads to $E_s(t) = \gamma D(t)$, the emissions of the south $E_s(t)$ are thus proportional to deforestation.

2.2 A Market-Based Approach

As in [5], a two-stage game is considered. In the first stage, the north agrees on emission caps. In the second stage, emission certificates are traded, or more specifically, the north will buy emission certificates from the south in order to fulfill its emission caps with less restrictions for domestic production.

As this problem will be solved using backwards induction, it is more intuitive to start the detailed description at the second stage. Let $O_n(t)$ denote the emission cap of the north at time t that results from the first stage. The north can emit more than the cap $O_n(t)$, but the transgression has to be compensated through the purchase of emission certificates $Z_n(t)$, thus $E_n(t) = O_n(t) + Z_n(t)$. Increasing emissions according to (2) result in increasing domestic production, but the corresponding certificates have to be bought at market price $p_z(t)$. To comply with the caps in the cheapest possible way the north faces the following optimisation problem:

$$\max_{Z_n(t)} \int_0^T e^{-r_n t}\Big\{ a\big[O_n(t) + Z_n(t)\big] - b\big[O_n(t) + Z_n(t)\big]^2 - p_z(t)Z_n(t)\Big\}dt.$$

Here, no terms for the damages caused by excessive pollution appear, as the trading of emission certificates only redistributes emissions between traders but does not change the overall sum of them. The total amount of pollution will thus be discerned at the first stage, and the associated damage will be considered by the north then.

If the south emits less than in the baseline scenario, it can sell certificates $Z_s(t) = \gamma(\overline{D}(t) - D(t))$. The south now has to balance the utility derived from deforestation (1) and the income from selling certificates:

$$\max_{D(t)} \int_0^T e^{-r_s t}\Big\{ \big[\overline{P} - \theta D(t)\big]D(t) + Y\big[F_0 - F(t)\big] + p_z(t)\gamma\big[\overline{D}(t) - D(t)\big]\Big\}dt$$

$$+ e^{-r_s T}\phi F(T)$$

$$\text{s.t.}\quad \dot{F}(t) = -D(t), \quad F(0) = F_0.$$

The equilibrium market price $p_z(t)$ is the price that leads to $Z_n(t) = Z_s(t)$ $\forall t \in [0,T]$. Let $Z_n^*\big[O_n(t),t\big]$, $D^*\big[O_n(t),t\big]$ be the (Nash-Equilibrium-)solutions of this game.

Now the first stage can be considered. At the first stage, the trade-off between economic utility and damage through emissions is optimized. The north faces:

$$\max_{O_n(t)} \int_0^T e^{-r_n t} \left\{ a\left\{ O_n(t) + Z_n^*[O_n(t), t]\right\} - b\left\{ O_n(t) + Z_n^*[O_n(t), t]\right\}^2 \right.$$

$$\left. - p_z(t) Z_n^*[O_n(t), t] - c_n[S(t) - \underline{S}]^2 \right\} dt - e^{-r_n T} \psi S(T)$$

$$\text{s.t.} \quad \dot{S}(t) = O_n(t) + \gamma \overline{D}(t), \quad S(0) = S_0.$$

The constraint results from the fact that trade only redistributes the emissions between regions but does not directly change overall emissions, thus $E_n(t) + E_s(t) = O_n(t) + Z_n(t) + \gamma \overline{D}(t) - Z_s(t) = O_n(t) + \gamma \overline{D}(t)$.

2.3   A Fund-Based Approach

For the modelling of a fund-based approach again two stages will be required. In the first stage, the north agrees on emission caps and assigns a price to a certain area of saved rainforest. As the north (Stackelberg leader) is able to foresee the reaction of the south (follower) it thus also determines how much it is willing to pay into the fund for the preservation of the rainforest. In the second stage, the south optimizes deforestation according to the money in the fund and the assigned price.

Just like in the market-based approach, it is more intuitive to start the detailed description in the second stage, in which the emission caps $O_n(t)$ and the prices offered by the north $p_f(t)$ are already set. The south therefore considers the following problem:

$$\max_{D(t)} \int_0^T e^{-r_s t} \left\{ [\overline{P} - \theta D(t)] D(t) + Y[F_0 - F(t)] + p_f(t) \gamma [\overline{D}(t) - D(t)] \right\} dt$$

$$+ e^{-r_s T} \phi F(T)$$

$$\text{s.t.} \quad \dot{F}(t) = -D(t), \quad F(0) = F_0.$$

Let $D^*[O_n(t), p_f(t), t]$ be the optimal path of deforestation.

Now, at the first stage, the north optimally chooses the emission caps and the price offers. As in the market-based scenario, economic interests and the avoidance of damages have to be balanced:

$$\max_{O_n(t), p_f(t)} \int_0^T e^{-r_n t} \left\{ aO_n(t) - bO_n(t)^2 - p_f(t) \gamma \left\{ \overline{D}(t) - D^*[O_n(t), p_f(t), t]\right\} \right.$$

$$\left. - c_n[S(t) - \underline{S}]^2 \right\} dt - e^{-r_n T} \psi S(T)$$

$$\text{s.t.} \quad \dot{S}(t) = O_n(t) + \gamma D^*[O_n(t), p_f(t), t], \quad S(0) = S_0.$$

As there is no carbon trading, reduced emissions from reduced deforestation directly lead to decreased overall emissions. Thus it holds that $E_n(t) + E_s(t) = O_n(t) + \gamma D(t)$, which leads to the dynamic constraint above.

Table 1. Parameter values used for the analysis

T	$\overline{P}$	θ	F_0	Y	r_n	r_s	γ	ϕ	ψ	$\underline{S}$	S_0	a	b	c_n
10	120	0.1	$3.9 \cdot e9$	11	0.05	0.05	6.66	10	10	$2.2 \cdot e6$	$3 \cdot e6$	300	0.05	$5 \cdot e{-}7$

3   Results

For the presentation of results parameters were chosen as specified in Table 1. Parameters either represent a real price or quantity or are chosen to fulfill the following criterion. In the baseline scenario, north and south should emit 27,600 and 6,900 million tonnes of CO_2, respectively. This criterion results from [8,10]. The remaining freedom in the choice of parameters was used to ensure interpretable behaviour of both players in the two REDD scenarios.

All optimisation problems from Sect. 2 can be solved analytically using the Matlab toolbox *Symbolic*. However, the display of each of the closed-form formulae would require more than 25,000 characters and is therefore omitted, while Fig. 1 reveals the most important information.

The upper left plot in Fig. 1 shows that the introduction of a market-based as well as a fund-based REDD mechanism leads to a significant decline in deforestation. In the market-scenario, deforestation decreases on average by 30 %. A fund leads to approximately 25 % less deforestation. It is in line with theory that a market-based REDD mechanism leads to a sharper decline in deforestation, because the north has more incentive to buy certificates than to donate money.

In the upper left plot in Fig. 1, it can be seen that actual emissions of the north do not decrease in any of the REDD scenarios relative to the baseline. In the market-based approach, the north agrees on lower emission caps than in the fund or the baseline scenario. However, the north purchases large quantities of certificates from the south, with the result that it actually emits more in the market-based approach than in the fund or baseline scenario.

The remaining question, which of these contrary effects prevails, is answered in the lower plot in Fig. 1. It shows that both REDD mechanisms can lead to a decline in global total emissions. In detail, the fund-based approach is able to reduce emissions more effectively than the market-based approach. The additional reductions beyond the obligations in the fund-based solution is thus more substantial than the fact that more emissions through deforestation are avoided in the market based approach.

These findings seem to be very stable with respect to the choice of parameters. Sensitivity analyses have been carried out for all parameters. Regardless of the considered combination of parameters, market-based REDD leads to less deforestation while fund-based REDD results in less total global emissions. In a longer version of this paper [9] it is shown that these key findings are also robust w.r.t. small modifications of the model.

However, the chances that an agreement becomes implemented rather depend on the benefit that the involved parties derive from it than on the benefit for the environment. It can be shown that both regions benefit from the introduction

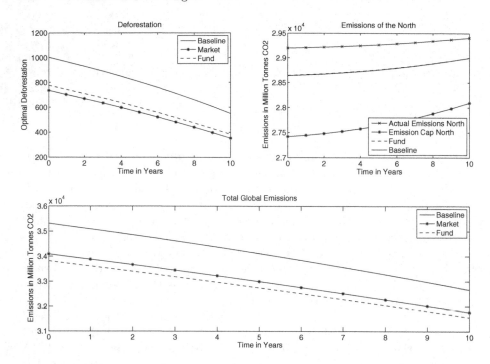

Fig. 1. Deforestation (u.l.), emissions of the north (u.r.), total global emissions

of any REDD mechanism. For the south, the welfare gains are larger in the case of market-based REDD. This reproduces the fact that most rainforest countries favour financing of REDD through carbon trading [4]. For the north, fund-based REDD results in a higher welfare gain. The EU officially does not prefer any of the two mechanisms. However, Norway and Germany already started paying into a REDD-like fund [6]. The reason for this result might be that in a fund-based approach the north, as the donor, can set the price of reduced deforestation whereas in a market scenario the price results from supply and demand.

This finding, however, is not independent of the choice of parameters. That means, a fund only results in more profit gain for the north as long as environmental awareness c_n is below a threshold. For $c_n = 10^{-6}$ and above, market-based REDD becomes more profitable. For the south it is the other way round. If the north's environmental awareness is relatively high ($c_n = 5 \cdot 10^{-6}$), the south changes its preference towards a fund.

4   Conclusion

The starting point of this paper is to analyse the main distinguishing features of market-based and fund-based REDD. For that purpose, a model that reproduces the main ideas of both REDD mechanisms is developed and the results are analyzed. It is shown that the introduction of any of the two mechanisms leads to less deforestation and less global emissions, and can thereby be a valuable contribution in the battle against climate change. As widely believed in the real discussion, also the model shows that market-based REDD can reduce deforestation more effectively and industrialized countries are willing to pay higher compensation payments. However, also as assumed in public discussion, the therewith avoided emissions are partly compensated by increasing emissions of the industrialized countries. In fund-based REDD, industrialized countries hardly increase their emissions, because it is not possible to use the avoided emissions from avoided deforestation towards their own emissions-reduction compliance targets. This effect is strong enough to offset the upside of the market-based approach, and total global emissions are lower in the fund-based approach than in the market-scenario. Therefore, given our current understanding, we would advocate a fund-based mechanism. However, much more research can be done to put suggestions of this sort on a sound scientific basis. A first future extension of the model could be to add the damage caused by climate change to the utility function of the south. This significantly raises the complexity of the model but might yield further interesting results.

References

1. Andrés-Domenech, P., Martín-Herrán, G., Zaccour, G.: An empirical differential game for sustainable forest management (2008)
2. Benchekroun, H., Long, N.V.: On the multiplicity of efficiency-inducing tax rules. Econ. Lett. **76**, 331–336 (2002)
3. Breton, M., Zaccour, G., Zahaf, M.: A differential game of joint implementation of environmental projects. Automatica **41**, 1737–1749 (2005)
4. Dooley, K., Leal, I., Ozinga, S.: An overview of selected REDD proposals (2008). http://www.fern.org/sites/fern.org/files/media/documents/document_4314_4315.pdf
5. Helm, C.: International emissions trading with endogenous allowance choices. J. Public Econ. **87**, 2737–2747 (2003)
6. Internet Platform of the Amazon Fund. http://www.amazonfund.gov.br
7. Fredj, K., Martín-Herrán, G., Zaccour, G.: Slowing deforestation pace through subsidies: a differential game. Automatica **40**, 301–309 (2008)
8. Olivier, J. G. J., Janssens-Maenhout, G., Muntean, M., Peters. J. A.H.W.: Trends in global CO2 emissions: 2013 Report. PBL Netherlands Environmental Assessment Agency (2013)
9. Riesner, B.: Financing the reduction of emissions from deforestation: a differential game approach. Master thesis, TU Wien (2015)
10. Stern, N.: The Economics of Climate Change: The Stern Review. Cambridge University Press, Cambridge (2007)

Relaxation of Euler-Type Discrete-Time Control System

Vladimir M. Veliov[✉]

Institute of Statistics and Mathematical Methods in Economics,
Vienna University of Technology, Vienna, Austria
veliov@tuwien.ac.at

Abstract. This paper investigates what is the Hausdorff distance between the set of Euler curves of a Lipschitz continuous differential inclusion and the set of Euler curves for the corresponding convexified differential inclusion. It is known that this distance can be estimated by $O(\sqrt{h})$, where h is the Euler discretization step. It has been conjectured that, in fact, an estimation $O(h)$ holds. The paper presents results in favor of the conjecture, which cover most of the practically relevant cases. However, the conjecture remains unproven, in general.

1 Introduction

In this paper we address the problem of convexification of finite-difference inclusions resulting from Euler discretization of the differential inclusion

$$\dot{x}(t) \in F(x(t)), \quad x(0) = x_0, \quad t \in [0,1], \tag{1}$$

where $x \in \mathbb{R}^n$, $x_0 \in \mathbb{R}^n$ is given, and $F : \mathbb{R}^n \Rightarrow \mathbb{R}^n$ is a set-valued mapping. Standing assumptions will be that F is compact-valued, bounded (by a constant denoted further by $|F|$) and Lipschitz continuous with a Lipschitz constant L with respect to the Hausdorff metric.[1]

Denote by S the set of all solutions of (1), and by $R := \{x(1) : x(\cdot) \in S\}$ the reachable set at $t = 1$. In parallel, we consider the convexified differential inclusion

$$\dot{y}(t) \in \operatorname{co} F(y(t)), \quad y(0) = x_0, \quad t \in [0,1], \tag{2}$$

and denote by S^{co} and R^{co} the corresponding solution set and reachable set. Now we consider the Euler discretizations of (1) and (2):

$$x_{k+1} \in x_k + hF(x_k), \quad k = 0, \dots, N-1, \tag{3}$$

This research is supported by the Austrian Science Foundation (FWF) under grant P 26640-N25.

[1] In fact, the global boundedness and Lipschitz continuity can be replaced with local ones if all solutions of (1) are contained in a bounded set. Then the formulations of some of the claims in the paper should be somewhat modified. The standing assumptions above are made simpler for more transparency.

I. Lirkov et al. (Eds.): LSSC 2015, LNCS 9374, pp. 134–141, 2015.
DOI: 10.1007/978-3-319-26520-9_14

and

$$y_{k+1} \in y_k + h \operatorname{co} F(y_k), \quad k = 0, \ldots, N-1, \quad y_0 = x_0, \tag{4}$$

where N is a natural number and $h = 1/N$ is the mesh size. Denote by S_h and S_h^{co} the sets of (discrete) solutions of these inclusions, respectively, and by R_h and R_h^{co} the corresponding reachable sets.

It is well known that $S^{co} = \operatorname{cl} S$. This paper investigates what is the Hausdorff distance between S_h and S_h^{co}, and also between R_h and R_h^{co}. The former is defined as

$$H(S_h, S_h^{co}) = \sup_{(y_0, \ldots y_N) \in \mathcal{S}_h^{co}} \inf_{(x_0, \ldots x_N) \in \mathcal{S}_h} \max_{i=0,\ldots,N} |y_i - x_i| = \sup_{y \in \mathcal{S}_h^{co}} \inf_{x \in \mathcal{S}_h} \|x - y\|_{l_\infty}.$$

Results by Tz. Donchev [1] and G. Grammel [4] imply that $H(S_h, S_h^{co}) = O(\sqrt{h})$. The unpublished author's report [9] contains the following

Conjecture: There exists a constant c such that for every natural number N

$$H(S_h, S_h^{co}) \le ch. \tag{5}$$

This conjecture has been proved in a number of special cases (see Sect. 3), but not in general. It is important to clarify what the constant c depends on. A stronger form of the conjecture is that c depends only on $|F|$, L, and the dimension of the space, n. However, in some of the results presented below the constant c will depend also on some geometric properties of $F(x)$. Therefore we speak about the weak and the strong form of the conjecture. We mention that there is an even stronger form of the conjecture, where Lipschitz continuity is required for $\operatorname{co} F$ instead of F. This case will be only partly discussed in Part 2 of Sect. 3.

Clearly, (5) implies the same estimation for $H(R_h, R_h^{co})$, but the inverse implication does not need to be true. (Here, and at some places below we use the symbol H also for the Hausdorff distance between compact subsets of $\mathbb{R}^n$, which will be clear from the context.)

The problems mentioned above are relevant for many engineering applications, where switched systems [5] or mixed-integer control problems (see [6–8] and the references therein) arise. The mixed-integer control problems can be formulated as

$$\min_{u(\cdot)} \left\{ p(x(1)) + \int_0^1 q(x(t)) \, dt \right\} \tag{6}$$

$$\dot{x}(t) = \varphi(x(t), u(t)), \quad x(0) = x_0, \quad u(t) \in U, \quad t \in [0,1], \tag{7}$$

where some of the components of the control u are restricted in a convex set, the remaining components take values in a discrete set. Thus the set U is nonconvex. The problem becomes combinatorial and due to the high dimension of its discretized counterpart (obtained, say, by the Euler method with mesh size h) is hard to be solved numerically. For this reason, in the above mentioned papers the authors propose to solve the convexified version of the problem and then from the numerically obtained optimal control to construct another, piecewise constant one, that takes values in U only, and such that the loss of performance is small (relative to the discretization step h). It is easy to see that the loss of performance (compared with the optimal performance of the convexified problem) can be estimated by $H(S_h, S_h^{co}) + O(h)$, and in the case $q = 0$ by $H(R_h, R_h^{co}) + O(h)$, provided that p and q are Lipschitz continuous. This gives one motivation for the question formulated above.

In the next section we prove a result related to the problem posed above (but not implying validity of the conjecture), while in Sect. 3 we present cases in which the conjecture is proved under some additional conditions.

2   A Related Result

The next result deviates from the conjecture formulated in the introduction, but has practical relevance in view of the control problem (6), (7).

Theorem 1. *There exists a constant C such that for every natural number N and for every $y = (y_0, y_1, \ldots, y_N) \in S_h^{co}$ there exist positive numbers $h_1, \ldots h_N$ with $\sum_{k=1}^N h_k = 1$ and a solution $x = (x_0, \ldots x_N)$ of*

$$x_{k+1} \in x_k + h_k F(x_k), \quad k = 0, \ldots, N - 1, \tag{8}$$

such that

$$\|x - y\|_{l_\infty} \le (4n + 1)|F|e^L \, h.$$

Proof. Obviously co F is Lipschitz and bounded with the same constants as F.

Let $y = (y_0, y_1, \ldots, y_N) \in S_h^{co}$. Then there exist $\xi_i \in \mathrm{co}\, F(y_i)$ such that

$$y_{i+1} = y_i + h\xi_i, \quad i = 0, \ldots, N - 1. \tag{9}$$

We split the points $y_0, \ldots, y_N$ into groups of $n + 1$ successive elements, the last one containing possibly a smaller number of elements. Let m be the number of groups, not counting the last one if it contains less than $n + 1$ elements. Thus m is the largest integer for which $m(n + 1) \le N$.

We shall define a trajectory $(x_0, x_1, \ldots, x_N)$ of (8) successively for each group of indexes. Namely, since x_0 is given and $y_0 = x_0$, we set $\Delta_0 = |x_0 - y_0| = 0$, then we assume that $x_{i(n+1)}$ is already defined, together with the corresponding steps h_j, $j = 0, \ldots, i(n + 1)$. Denote $\Delta_i = |x_{i(n+1)} - y_{i(n+1)}|$.

Due to (9) we have that for $j = 0, \ldots, n$

$$\xi_{i(n+1)+j} \in \mathrm{co}\, F(y_{i(n+1)+j}) = \mathrm{co}\, F\left(y_{i(n+1)} + h\sum_{s=0}^{j-1} \xi_{i(n+1)+s}\right)$$

$$\subset \mathrm{co}\, F(y_{i(n+1)}) + hjL|F|\mathbf{B},$$

where $\mathbf{B}$ is the unit ball in $\mathbb{R}^n$. Then there exist $\tilde{\xi}_{i(n+1)+j} \in \mathrm{co}\, F(y_{i(n+1)})$ such that

$$|\tilde{\xi}_{i(n+1)+j} - \xi_{i(n+1)+j}| \le hjL|F|, \quad j = 0, \ldots, n, \tag{10}$$

where we have set $\tilde{\xi}_{i(n+1)} = \xi_{i(n+1)}$. Since $\tilde{\xi}_{i(n+1)+j} \in \operatorname{co} F(y_{i(n+1)})$, we have also that

$$\frac{1}{n+1} \sum_{j=0}^{n} \tilde{\xi}_{i(n+1)+j} \in \operatorname{co} F(y_{i(n+1)}).$$

According to the Carathéodory theorem, there exist $\tilde{\eta}_{i(n+1)+j} \in F(y_{i(n+1)})$ and $\alpha_j \geq 0$, $\sum_{j=0}^{n} \alpha_j = 1$, such that

$$\sum_{j=0}^{n} \alpha_j \tilde{\eta}_{i(n+1)+j} = \frac{1}{n+1} \sum_{j=0}^{n} \tilde{\xi}_{i(n+1)+j}. \tag{11}$$

Let us define $h_{i(n+1)+j} = h_j := (n+1)h\alpha_j$. Due to the Lipschitz continuity of F, there exists $\eta_{i(n+1)} \in F(x_{i(n+1)})$ such that

$$|\eta_{i(n+1)} - \tilde{\eta}_{i(n+1)}| \leq L\Delta_i.$$

To extend the trajectory $x_0, \ldots, x_{i(n+1)}$ we set

$$x_{i(n+1)+1} = x_{i(n+1)} + \bar{h}_0 \eta_{i(n+1)}.$$

Since

$$H(F(x_{i(n+1)+1}), F(y_{i(n+1)})) \leq$$
$$H(F(x_{i(n+1)+1}), F(x_{i(n+1)})) + H(F(x_{i(n+1)}), F(y_{i(n+1)})) \leq \bar{h}_0 L|F| + L\Delta_i,$$

there exists $\eta_{i(n+1)+1} \in F(x_{i(n+1)+1})$ such that

$$|\eta_{i(n+1)+1} - \tilde{\eta}_{i(n+1)+1}| \leq \bar{h}_0 L|F| + L\Delta_i.$$

Then we define

$$x_{i(n+1)+2} = x_{i(n+1)+1} + \bar{h}_1 \eta_{i(n+1)+1}.$$

Continuing in the same way we define for every $j = 0, \ldots, n$ the vectors $\eta_{i(n+1)+j}$ and $x_{i(n+1)+j+1}$ such that

$$\eta_{i(n+1)+j} \in F(x_{i(n+1)+j}),$$
$$x_{i(n+1)+j+1} = x_{i(n+1)+j} + \bar{h}_j \eta_{i(n+1)+j},$$
$$|\eta_{i(n+1)+j} - \tilde{\eta}_{i(n+1)+j}| \leq L|F| \sum_{k=0}^{j-1} \bar{h}_k + L\Delta_i. \tag{12}$$

In this way the trajectory of (3) is extended to the discrete time $(i+1)(n+1)$. The next estimations follow from (10), (11), (12):

$$\Delta_{i+1} = |x_{(i+1)(n+1)} - y_{(i+1)(n+1)}|$$

$$\leq |x_{i(n+1)} - y_{i(n+1)}| + \left| \sum_{j=0}^{n} \bar{h}_j \eta_{i(n+1)+j} - h \sum_{j=0}^{n} \xi_{i(n+1)+j} \right|$$

$$\leq \Delta_i + \sum_{j=0}^{n} \bar{h}_j \left| \eta_{i(n+1)+j} - \tilde{\eta}_{i(n+1)+j} \right| + \left| \sum_{j=0}^{n} \bar{h}_j \tilde{\eta}_{i(n+1)+j} - h \sum_{j=0}^{n} \tilde{\xi}_{i(n+1)+j} \right|$$

$$+ h \sum_{j=1}^{n} \left| \tilde{\xi}_{i(n+1)+j} - \xi_{i(n+1)+j} \right|$$

$$\leq \Delta_i + \sum_{j=0}^{n} \bar{h}_j L \Delta_i + L|F| \sum_{j=1}^{n} \bar{h}_j \sum_{k=0}^{j-1} \bar{h}_k$$

$$+ \left| (n+1)h \sum_{j=1}^{n} \alpha_j \tilde{\eta}_{i(n+1)+j} - h \sum_{j=1}^{n} \tilde{\xi}_{i(n+1)+j} \right| + h \sum_{j=0}^{n} hj L|F|$$

$$\leq (1 + (n+1)Lh)\Delta_i + (n+1)^2 L|F|h^2 + \frac{n(n+1)}{2} L|F|h^2$$

$$\leq (1 + (n+1)Lh)\Delta_i + (n+1)(2n+1)L|F|h^2.$$

Since this holds for any $i < m$ it implies in a standard way the inequality

$$\Delta_i \leq (2n+1)|F|e^{i(n+1)Lh}h \leq (2n+1)|F|e^{m(n+1)Lh}h \leq (2n+1)|F|e^L h.$$

Then taking into account the errors that can be made within n intermediate steps, or in the last $N - m(n+1) \leq n$ steps we obtain for the above defined solution of (8)

$$|x_k - y_k| \leq (2n+1)|F|e^L h + 2n|F|h \qquad \forall k = 0, \ldots, N.$$

The proof is complete.   □

Obviously the above theorem does not give an answer to the main question in this paper, since the time-steps in (8) need not be uniform. Although the total number of jumps is N, there could be much smaller distance between the jumps, which may be trouble for practical implementations. Moreover, as it is clear from the proof, in the terms of the control problem (6)–(7), the choice of $u_k \in U$ at step k depends on n future values of the optimal control of the convexified problem (that is, it is anticipative). However, this is in line with the model predictive control methodology used in practice. Moreover, the construction in the proof of the theorem can be viewed as an alternative of the "adaptive control grid" proposed in [7] where $2N$ jump points of the control are used (instead of N).

3   Cases in Which the Conjecture is Proved

The proofs of the results in this section are available in the Research Report 2015-10 at http://orcos.tuwien.ac.at/research/research_reports/.

Part 1. First we consider the case of a constant mapping F, that is, the inclusion

$$\dot{x}(t) \in V, \quad x(0) = x_0, \ t \in [0,1], \tag{13}$$

where $V \subset I\!\!R^n$ is compact. This case will be embodied later in more general considerations.

We mention that conjecture (5) has not been proved even in this "simple" case. However, for constant mappings $F(x) = V$ it holds that

$$H(R_h, R_h^{co}) \leq ch,$$

where the constant c depends only on $|V|$ and n. This can be proved (and has been proved by several mathematicians in private communications with the author: Z. Artstein, M. Brokate, E. Farkhi, T. Donchev) in different ways, the simplest of which uses the Shapley-Folkmann theorem (see e.g. [3, Appendix 1]). Now, we consider the case of a set V consisting of finite number of points:

$$V = \{v_1, \dots, v_s\}, \quad v_i \in I\!\!R^n. \tag{14}$$

The proof is given in the research report [9] and is somewhat modified below.

Proposition 1. *For differential inclusion (13) with the constant mapping V specified in (14) the estimation*

$$H(S_h, S_h^{co}) \leq 2s|V|h,$$

holds for every $h = 1/N$, $N \in I\!\!N$.

The proof of the above proposition is constructive and the construction similar to what is called in [7,8] *Sum Up Rounding Strategy.*

We mention that the constant $c = 2s|V|$ in Proposition 1 depends on the number of elements of V, that is, only the weaker form of Conjecture (5) is proved (the constant c depends on the geometric properties of V). In particular, it does not help to deal with sets V for which the boundary of $co\,V$ contains curved pieces. The next result is capable to capture some such cases.

Part 2. In this part we consider the general inclusion (1), weakening a bit the standing assumptions. Namely, instead of assuming Lipschitz continuity of F we assume that $co\,F$ is Lipschitz continuous.

Notice that all sequences in S_h and S_h^{co} are contained in the compact set $X := \{x \in I\!\!R^n : |x - x_0| \leq M\}$. Let there exist functions $l_i : X \to I\!\!R^n$, $i = 1, \dots, n$ such that:

(i) l_i are Lipschitz continuous;
(ii) the vectors $l_i(x)$, $i = 1, \dots, n$, are linearly independent and $|l_i(x)| = 1$ for every $x \in X$;
(iii) for every $x \in X$, every $\bar{v} \in co\,F(x)$ and every $\sigma_1, \dots, \sigma_n \in \{-1, 1\}$ there exists $v \in F(x)$ such that

$$\sigma_i \alpha_i(x; v - \bar{v}) \leq 0, \ i = 1, \dots, n,$$

where $\alpha_i(x;z)$ is the i-th coordinate of $z \in \mathbb{R}^n$ in the basis $\{l_i(x)\}$. Clearly, the numbers $\alpha_i(x;z)$ are uniquely defined from

$$z = \sum_{i=1}^{n} \alpha_i(x;z) \, l_i(x). \tag{15}$$

Proposition 2. *Under the suppositions made in Part 2 there exists a constant C such that*

$$H(S_h, S_h^{co}) \leq Ch$$

for every $h = 1/N$, $N \in \mathbb{N}$.

The next is a simple consequence of the above proposition.

Corollary 1. *Under the conditions of Proposition 2, let F satisfy*

$$F(x) = \partial(co\,F(x)) \quad \forall x \in \mathbb{R}^n,$$

where ∂Y denotes the boundary of Y. Then the conclusion of Proposition 2 holds true.

Indeed, we may take an arbitrary fixed orthonormed basis $\{l_i(x) = l_i\}$. Let us take an arbitrary $\bar{v} \in co\,F(x)$ and $\sigma_i \in \{-1, 1\}$. If $\bar{v} \notin \partial F(x)$, then moving from $\bar{v}$ along the vector $-(\sigma_1 l_1 + \ldots + \sigma_n l_n)$ we shall reach a point $v \in \partial F(x)$ for which (iii) is obviously satisfied.

One example (that was considered as non-trivial) is the inclusion (13) with V being the semi-circle in $\mathbb{R}^2$ (a semi-sphere in $\mathbb{R}^n$ can be treated in the same way):

$$V = \{(v_1, v_2) : (v_1)^2 + (v_2)^2 = 1,\ v_2 \geq 0\}.$$

The claim of the conjecture (5) for this example follows from Proposition 2. Indeed, one may take $l_i = e_i$ – the standard basis in $\mathbb{R}^2$. For any $\bar{v} \in co\,V$ and $\sigma_1, \sigma_2 \in \{-1, 1\}$ define $v_2 = \bar{v}_2$, $v_1 = -\sigma_1 \sqrt{1 - \bar{v}_2^2}$. Then $\sigma_1 \alpha_1(v) = -\sqrt{1 - \bar{v}_2^2} \leq 0$ and $\sigma_2 \alpha_2(v) = 0$. Assumption (iii) of Proposition 2 is fulfilled.

Part 3. Now, we consider a differential inclusion of the form

$$\dot{x}(t) \in G(x)V, \quad x(0) = x_0, \tag{16}$$

where $G(x)$ is an $(n \times m)$-matrix and $V \subset \mathbb{R}^m$.

Proposition 3. *Let V be compact and $G(\cdot)$ be Lipschitz continuous with constant $L > 0$, and bounded by a constant M, both with respect to the operator norm of G. Let (5) holds for the differential inclusion (13) with some constant c. Then for the differential inclusion (16) the estimation*

$$H(S_h, S_h^{co}) \leq cM(1 + L)e^{L|V|}h,$$

holds for every $h = 1/N$, $N \in \mathbb{N}$.

This proposition is an extension of [8, Theorem 2] where it is assumed that G is differentiable and V is a box. The proof below is a discrete-time adoption of that in [8].

Part 4. Following [8], one can use the above proposition to obtain an estimation as in (5) for non-affine inclusions of the form

$$\dot{x} \in f(x, U), \tag{17}$$

where $U \in I\!\!R^m$ consists of finite number of points; $U = \{u_1, \ldots, u_s\}$, and $f(\cdot, u_i) : I\!\!R^n \to I\!\!R^n$.

Proposition 4. *Let the functions $f(\cdot, u_i)$ be Lipschitz continuous with constant $L > 0$, and bounded by a constant M. Then for the differential inclusion (17) the estimation*

$$H(S_h, S_h^{\mathrm{co}}) \leq 2s^{3/2} M (1 + \sqrt{s}L) e^{\sqrt{s}L} h, \tag{18}$$

holds for every $h = 1/N$, $N \in I\!\!N$.

This propositions extends [8] in that $f(\cdot, u)$ is not assumed differentiable. The constant in (18) depends on the number of elements of U, which means that only the weak form of Conjecture (5) is proved in the considered special case. On the other hand, the proposition covers most of the practically interesting cases.

4 Conclusion

To the author's knowledge, the conjecture that $H(S_h, S_h^{\mathrm{co}}) = O(h)$ is still open (both in its stronger and weaker form). We stress that the conjecture has not been proved even in the case of a constant mapping $F(x) = V \subset I\!\!R^n$. However, the partial results in this paper cover most of the practically important cases.

References

1. Donchev, T.: Approximation of lower semicontinuous differential inclusions. Numer. Funct. Anal. Optim. **22**(1–2), 55–67 (2001)
2. Dontchev, A., Farkhi, E.: Error estimates for discretized differential inclusion. Computing **41**(4), 349–358 (1989)
3. Ekeland, I., Temam, R.: Convex Analysis and Variational Problems. Elsevier, Amsterdam (1976)
4. Grammel, G.: Towards fully discretized differential inclusions. Set-Valued Anal. **11**(1), 1–8 (2003)
5. Liberzon, D.: Switchings in Systems and Control. Birkhäuser, Boston (2003)
6. Sager, S.: Numerical Methods for Mixed-Integer Optimal Control Problems. Der andere Verlag, Tönning, Lübeck, Marburg (2005)
7. Sager, S., Bock, H.G., Reinelt, G.: Direct methods with maximal lower bound for mixed-integer optimal control problems. em Math. Program. Ser. A **118**, 109–149 (2009)
8. Sager, S., Bock, H.G., Diehl, M.: The integer approximation error in mixed-integer optimal control. em Math. Program. **133**, 1–23 (2012)
9. Veliov, V.M.: Relaxation of Euler-type discrete-time control systems. ORCOS Research Report 273, Vienna (2003)

Enabling Exascale Computation

Uncertainty Quantification for Porous Media Flow Using Multilevel Monte Carlo

Jan Mohring[1]([✉]), René Milk[2], Adrian Ngo[3], Ole Klein[3], Oleg Iliev[1], Mario Ohlberger[2], and Peter Bastian[3]

[1] Fraunhofer ITWM, Fraunhofer-Platz 1, 67663 Kaiserslautern, Germany
jan.mohring@itwm.fraunhofer.de
[2] Institute for Computational and Applied Mathematics, University of Münster, Orleans-Ring 10, 48149 Münster, Germany
[3] Interdisciplinary Center for Scientific Computing, University of Heidelberg, Im Neuenheimer Feld 368, 69120 Heidelberg, Germany

Abstract. Uncertainty quantification (UQ) for porous media flow is of great importance for many societal, environmental and industrial problems. An obstacle for progress in this area is the extreme computational effort needed for solving realistic problems. It is expected that exa-scale computers will open the door for a significant progress in this area. We demonstrate how new features of the Distributed and Unified Numerics Environment DUNE [1] address these challenges. In the frame of the DFG funded project EXA-DUNE the software has been extended by multiscale finite element methods (MsFEM) and by a parallel framework for the multilevel Monte Carlo (MLMC) approach. This is a general concept for computing expected values of simulation results depending on random fields, e.g. the permeability of porous media. It belongs to the class of variance reduction methods and overcomes the slow convergence of classical Monte Carlo by combining cheap/inexact and expensive/accurate solutions in an optimal ratio.

Keywords: Uncertainty quantification · Multilevel Monte Carlo · Multiscale finite elements · Porous media · Random permeability · Exascale · DUNE

1 Introduction

We demonstrate how newly developed DUNE modules for multilevel Monte Carlo, multiscale finite elements, and for generating random permeability fields can be combined to solve hard problems of uncertainty quantification in porous media flow using large scale computer clusters.

2 Model Problem

A simple model problem which is still able to illustrate the combination of parallel MLMC, MsFEM, and efficient generation of random permeability fields is

© Springer International Publishing Switzerland 2015
I. Lirkov et al. (Eds.): LSSC 2015, LNCS 9374, pp. 145–152, 2015.
DOI: 10.1007/978-3-319-26520-9_15

the stationary single phase flow through a cubic cell in non-dimensional form:

$$-\nabla \cdot [k(x,\omega)\,\nabla p(x,\omega)] = 0 \quad \text{for } x \in D = (0,1)^d, \quad \omega \in \Omega$$

$$p\big|_{x_1=0} = 1, \quad p\big|_{x_1=1} = 0, \quad \frac{\partial p}{\partial n} = 0 \text{ on other boundaries,} \qquad (1)$$

with dimension $d \in \{2,3\}$, real pressure p, real scalar permeability k, and random vector ω. Let $K(x,\omega) = \log k(x,\omega)$ the logarithm of permeability. We assume that its expected value and spacial covariance are shift invariant and satisfy

$$\mathrm{E}\,[K(x,\cdot)] = 0, \quad \mathrm{E}\,[K(x,\cdot)\,K(y,\cdot)] = C(x-y) = C(y-x) \text{ for } x,y \in D. \quad (2)$$

The quantity of interest is the total flux through the unit cube

$$Q(\omega) := \int\limits_{x_1=0} k(x,\omega)\,\frac{\partial}{\partial n}p(x,\omega)\,dS(x). \qquad (3)$$

A dimensional problem on cube $[0,L]^d$ with input pressure p_0 and reference permeability $k_0 = \exp(\mathrm{E}\,[\log k'(0,\cdot)])$ may be converted into form (1)–(3) by setting $x = x'/L$, $p = p'/p_0$, $k = k'/k_0$, and $Q = Q'/(p_0 k_0 L^{d-2})$, where $'$ marks dimensional variables. In particular, the effective constant permeability, which leads to the same flux as the random field, reads $k'_{\text{eff}}(\omega) = k_0\,Q(\omega)$. Here, we consider a practically relevant covariance of type (2) proposed in [2]:

$$C(h) = \sigma^2 \exp\left(-\,\|h\|_2 /\lambda\right), \quad h \in [-1,1]^d \qquad (4)$$

with variance $1 \leq \sigma^2 \leq 4$ and correlation length $0.05 \leq \lambda \leq 0.3$. The bounds are not strict. However, for higher σ^2 computing mean values of Q becomes irrelevant. In case of larger λ measurements may allow a deterministic reconstruction of k and for smaller λ homogenization techniques should be applied.

3   Generation of Random Permeability Fields

Literature provides several ways of generating random permeability fields, most of which cannot be applied to our setting. Factoring covariance functions are assumed in [3], which is not true for (4). Methods based on truncated Karhunen-Loève expansions [4] are inexact as eigenvalues do not decay rapidly. Therefore, we have implemented a parallel version of the circulant embedding algorithm introduced in [5], which is exact down to grid size, still fast, and works for a wide class of covariance functions. The idea is as follows. Let $N \in \mathbb{N}$, $J = \{0,\ldots,2N-1\}^d$, $\{\psi_k : J \ni n \mapsto \exp\left(i\pi\frac{k\cdot n}{N}\right) \mid k \in J\}$ the Fourier basis on J, $\tilde{C}$ the 2-periodic continuation of the covariance function C to $\mathbb{R}^d$, $x_n = \frac{n}{N}$, $n \in \mathbb{Z}^d$ discrete grid points, and $c_k = \sum_{n \in J} \tilde{C}(x_n)\,\overline{\psi}_k(n)$ the Fourier coefficients of C. As C is symmetric the c_k are real. Under some additional conditions, which are stated in [5] and which are satisfied by covariance (4) for sufficiently small λ, the c_k are also non-negative. Then we can construct

$$F(x_n,\omega) = \sum_{k \in J} \mu_k\,\omega_k\,\psi_k(n) \text{ with } n \in J, \quad \mu_k = \sqrt{(2N)^{-d}\,c_k} \qquad (5)$$

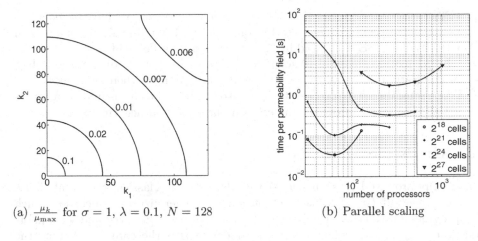

(a) $\frac{\mu_k}{\mu_{\max}}$ for $\sigma = 1$, $\lambda = 0.1$, $N = 128$ (b) Parallel scaling

Fig. 1. Creating random permeability fields by circulant embedding

and random numbers $\omega \in \mathbb{C}^J$ satisfying $w_k = \alpha_k + i\,\beta_k$ with independent standard normally distributed α_k and β_k. Finally, a simple calculation shows that the (discrete) permeability fields $k_1 = \log \Re(F)$ and $k_2 = \log \Im(F)$ satisfy (2). The following aspects of our implementation [6,7] are crucial:

1. The Fourier transform of C and the subsequent computation of the μ_k is done only once during initialization. Creating a new pair of permeability fields requires only one inverse Fourier transform, cf. Eq. (5).
2. The parallel **FFTW3** package is used for Fourier transforms. As illustrated in Fig. 1(b), this package does not scale properly. Rather, there are optimal numbers of processors for a certain problem size, where local data can be kept in fast memory, but communication is not dominant, yet.

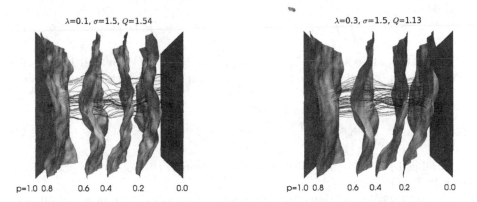

Fig. 2. Pressure p and flux Q for different correlation lengths, same random ω.

3. FFTW3 uses a one-dimensional domain decomposition and we have to redistribute data between processors to be consistent with the domain decomposition used by the PDE solver (additional effort $\sim$ number of grid cells).
4. The μ_k plotted in Fig. 1(a) for $d = 2$ are about the roots of the eigenvalues of the Karhunen-Loève expansion, which demonstrates their slow decay.
5. The permeability field generator can also be applied to measured covariances satisfying Eq. (2), provided the c_k turn out to be non-negative.

4   Multiscale Methods

Originally proposed by Hou and Wu in 1997 [8], the classical multiscale finite element method (MsFEM) saw extension and generalization since, for which [9] provides an overview. Calculating local correctors that carry fine-scale information of $k(x, \omega)$ to enhance a coarse nodal basis is the core idea behind the MsFEM. These updated basis functions are predicted to show superior approximation qualities. We will concisely describe this process in three steps below. There we will confine ourselves to consider only triangular meshes, but the generalization to hexahedral/quadrilateral meshes is of course straightforward. Let us now consider a regular triangulation $\mathcal{T}_H$ of D and define $V_H := P_1(\mathcal{T}_H) \cap H_0^1(D)$. With this in place we can summarize the classical MsFEM as follows:

1. A suitable choice for V_H, the coarse finite element space is required, i.e. the spaces' dimensionality still allows for a cost-effective discretization of Eq. (1).
2. Then, for each coarse cell $K \in \mathcal{T}_H$, we incorporate fine scale information into all coarse nodal basis functions that have support on K. Let $\Phi_i \in V_H$ be one of those basis functions. We then define the so called 'corrected basis function' by the following two properties:

$$\Phi_i^{\mathrm{ms}}|_{\partial K} = \Phi_i|_{\partial K} \tag{6}$$

and

$$\int_K k(x, \omega) \nabla \Phi_i^{\mathrm{ms}}|_K \cdot \nabla \phi = 0 \qquad \forall \phi \in H_0^1(K). \tag{7}$$

3. Lastly, we define as $V^{\mathrm{MsFEM}} := \mathrm{span}\{\Phi_i^{\mathrm{ms}}| \, 1 \leq i \leq N\}$ our multiscale finite element space and in it look for the Galerkin approximation p as presented by: find $p^{\mathrm{MsFEM}} \in V^{\mathrm{MsFEM}}$ such that

$$\int_\Omega k(x, \omega) \nabla p^{\mathrm{MsFEM}} \cdot \nabla v = \int_\Omega fv \qquad \forall v \in V^{\mathrm{MsFEM}}. \tag{8}$$

We shall call p^{MsFEM} the MsFEM approximation of the solution p of the PDE given in Eq. (1).

The boundary conditions (6) for the local problems (7) are prone to produce rapid oscillations close to ∂K, which may greatly diminish the overall quality of the MsFEM approximation. To rectify this we employ oversampling [10,11], meaning we solve local problems on a slightly expanded domain $U(K) \supset K$, but restrict the obtained solutions back to K and use only the information available there to correct the coarse basis functions. The troublesome boundary layer is thereby pushed to $U(K) \setminus K$ and ignored, improving the MsFEM approximation at the cost of the discretization becoming non-conforming. For a detailed, rigorous study of different oversampling strategies we refer the reader to [12].

Based on DUNE [1] in general and the DUNE Generic Discretization Toolbox and DUNE Stuff modules [13,14] in particular, our implementation in the DUNE Multiscale [15] module is available as open source[1]. First results on the parallel scalability of the MsFEM implementation have been demonstrated in [16].

5   Multilevel Monte Carlo

In many problems of porous media flow the permeability field is unknown and we can only guess its probability distribution from a few measurements. Usually, we are interested in some macroscopic aggregate quantity, e.g. the total flux through a given sector of the ground. As we do not know the exact permeability field we can only deduce the distribution of this quantity from the distribution of permeabilities. This kind of UQ is usually performed by MC methods due to the high number of random parameters. However, standard MC methods converge slowly and MLMC [4,17,18] can be used to improve performance by variance reduction. More precisely, the same kind of problems are solved by both, accurate but slow PDE solvers and inexact but fast ones. The variance of the aggregate quantity is split into the variance of the coarse solution, which can be reduced by many quick evaluations, and the variance of the difference of fine and coarse solutions, which is typically small and requires only few realizations to be reduced as needed.

In order to deal with a large variety of problems in UQ we have implemented quite a general MLMC module using DUNE [7]. It expects classes providing the aggregate quantity computed by a coarse solver and classes providing the difference of results by two methods of increasing accuracy for the same random field. The distribution of sample problems to different processor groups and choosing the optimal number of samples per level are done automatically. Before getting into details we repeat the principle of MLMC.

Let ω be a random field characterizing the scalar permeability $k(\cdot,\omega)$. Assume we have $L+1$ different numerical methods available to approximate the flux $Q(\omega)$ by values $Q_l(\omega)$, $l = 0,\ldots,L$. Let the methods be ordered by increasing accuracy and cost. Then we can rewrite $Q(\omega)$ as telescoping sum:

$$Q(\omega) = \underbrace{Q_0(\omega)}_{Y_0(\omega)} + \underbrace{Q_1(\omega) - Q_0(\omega)}_{Y_1(\omega)} + \cdots + \underbrace{Q_L(\omega) - Q_{L-1}(\omega)}_{Y_L(\omega)} + \underbrace{Q(\omega) - Q_L(\omega)}_{Z_L(\omega)}.$$

Let $\boldsymbol{\omega}$ be an n-dimensional vector of random fields ω_i distributed like ω. Then

$$Y_{ln}(\boldsymbol{\omega}) := \frac{1}{n}\sum_{i=1}^{n} Y_l(\omega_i) \text{ satisfies } \mathrm{E}\left[Y_{ln}\right] = \mathrm{E}\left[Y_l\right], \ \mathrm{Var}\left[Y_{ln}\right] = \frac{1}{n}\mathrm{Var}\left[Y_l\right].$$

---

[1] https://github.com/wwu-numerik/DUNE-Multiscale/, BSD-2 licensed.

Given n_l realizations on level l we can construct the following estimator of $\mathrm{E}\,[Q]$:

$$\hat{Q}\left(\omega^0\ldots\omega^L\right) = \sum_{l=0}^{L} Y_{l n_l}\left(\omega^l\right) \text{ with } \mathrm{E}\left[Q - \hat{Q}\right] = \mathrm{E}\left[Z_L\right],\ \mathrm{Var}\left[\hat{Q}\right] = \sum_{l=0}^{L}\frac{1}{n_l}\mathrm{Var}\left[Y_l\right].$$

Let method L be chosen so accurate that $|\mathrm{E}\left[Z_L\right]| \leq \varepsilon$. Our goal is to have

$$\mathrm{E}\left[\left(\hat{Q}-\mathrm{E}\left[Q\right]\right)^2\right] = \mathrm{Var}\left[\hat{Q}\right] + \left(\mathrm{E}\left[\hat{Q}-Q\right]\right)^2 \leq 2\,\varepsilon^2 \text{ following from } \sum_{l=0}^{L}\frac{1}{n_l}\mathrm{Var}\left[Y_l\right] = \varepsilon^2.$$
$$(9)$$

This condition may be achieved by different combinations of numbers n_l and we choose the one with minimal CPU time. Let $v_l = \mathrm{Var}\left[Y_l\right]$, t_l the mean time computing difference Y_l once, and $T = \sum_{l=0}^{L} n_l t_l$ the total time computing $\hat{Q}$. Minimizing T under constraint (9) and turning to integers gives

$$n_l = \mathrm{ceil}\left[\alpha\sqrt{v_l/t_l}\right] \text{ with Lagrangian multiplier } \alpha = \frac{1}{\varepsilon^2}\sum_{l=0}^{L}\sqrt{v_l t_l}.\quad (10)$$

In practice, the n_l are computed based on estimates of t_l and v_l. Obviously, MLMC is well-suited for parallel computing as samples can be computed independently. Turning to a parallel implementation we have to answer two main questions: How many processors p_l are used per sample on a given level? This is mainly determined by the parallel performance of the permeability generator, cf. Fig. 1(b). And further: How can we minimize communication estimating mean times and variances? This is done by an outer loop i over a few breaks, e.g. $n_\mathrm{b} = 3$, and an inner loop over the levels. Here, Y_l are computed many times in parallel by groups of p_l processors until time T_l^i when statistical moments are exchanged between groups. At the beginning of a new outer loop we compute optimal n_l as in Eq. (10) and new stopping times as $T_l^i = (n_l - n_l')\,t_l\frac{p_l}{p}\frac{i}{n_b}$, where n_l' denotes the number of samples on level l created so far and p is the total number of processors.

6   Numerical Results

In order to demonstrate the flexibility of the parallel MLMC framework we have used it to implement a two-level scheme where MsFEM is used on level 0 and a standard continuous Galerkin method on level 1. The two approximations of a typical pressure field are compared in Fig. 3(a). The speedup of MLMC by solving samples in parallel is illustrated for $p_l = 1$, $\varepsilon = 0.03$ in Fig. 3(b).

Showing performance, we have set up a three-level scheme solving realizations of the model problem in 3D (Yasp-grid, Q1 elements, AMG solver, cf. Fig. 2) on the ITWM cluster for mesh sizes 2^{-k}, $k = 5, 6, 7$ on 1, 4, and 32 processors, respectively. MLMC is used on 1024 processors to compute the expected effective permeability (same as Q) for pairs (σ, λ) as illustrated in Fig. 4(a). The accuracy is $\varepsilon = 0.005$ except for $\varepsilon = 0.008$ at $(2, 0.25)$ and $(2, 0.3)$. In the latter case, 220160, 31744, and 54944 samples are required on levels 0, 1, and 2,

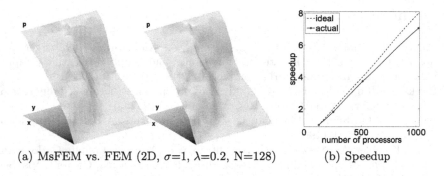

(a) MsFEM vs. FEM (2D, σ=1, λ=0.2, N=128)   (b) Speedup

Fig. 3. Parallel two-level Monte Carlo with MsFEM as fast solver

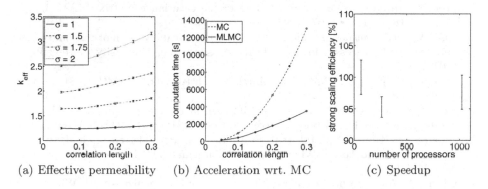

(a) Effective permeability   (b) Acceleration wrt. MC   (c) Speedup

Fig. 4. Computing effective permeabilities by three-level MC.

respectively. Computing one Y_2 takes about 7 s including 0.6 s for generating k. Creating Fig. 4(a) took 27.7 h, which corresponds to 3.2 years on a single processor. Figure 4(b) illustrates superiority of MLMC over standard MC especially for higher λ ($\sigma = 1.5$) and Fig. 4(c) the almost ideal speedup.

7   Conclusion

Combining new DUNE modules on MLMC, MsFEM, and random permeability fields we have computed effective permeabilities for a practically relevant covariance function over a wide range of correlation lengths and variances, which may serve as a future benchmark. Our MLMC framework allows combining different kinds of algorithms per level, in particular MsFEM as coarse solver. We put emphasis on the fact that the permeability generator determines a moderate optimal number of processors to be used per realization, which has influenced our strategy for parallelizing MLMC.

Acknowledgements. This research was funded by the DFG SPP 1648 Software for Exascale Computing.

References

1. Bastian, P., Blatt, M., Dedner, A., Engwer, C., Klöfkorn, R., Ohlberger, M., Sander, O.: A generic grid interface for parallel and adaptive scientific computing. Part I: abstract framework. Computing **82**(2–3), 103–119 (2008)
2. Hoeksema, R.J., Kitanidis, P.K.: Analysis of the spatial structure of properties of selected aquifers. Water Resour. Res. **21**(4), 563–572 (1985)
3. Ghanem, R.G., Spanos, P.D.: Stochastic Finite Elements: A Spectral Approach. Springer, New York (1991)
4. Cliffe, K., Giles, M., Scheichl, R., Teckentrup, A.L.: Multilevel Monte Carlo methods and applications to elliptic PDEs with random coefficients. Comput. Vis. Sci. **14**(1), 3–15 (2011)
5. Dietrich, C., Newsam, G.N.: Fast and exact simulation of stationary gaussian processes through circulant embedding of the covariance matrix. SIAM J. Sci. Comput. **18**(4), 1088–1107 (1997)
6. Ngo, A.Q.: Discontinuous Galerkin based geostatistical inversion of stationary flow and transport processes in groundwater. Dissertation, University of Heidelberg (2015)
7. Milk, R., Mohring, J.: DUNE mlmc, March 2015. http://dx.doi.org/10.5281/zenodo.16562
8. Hou, T.Y., Wu, X.H.: A multiscale finite element method for elliptic problems in composite materials and porous media. J. Comput. Phys. **134**(1), 169–189 (1997)
9. Efendiev, Y., Hou, T.Y.: Multiscale Finite Element Methods: Theory and Applications. Surveys and Tutorials in the Applied Mathematical Sciences, vol. 4. Springer, New York (2009)
10. Gloria, A.: An analytical framework for numerical homogenization. II. Windowing and oversampling. Multiscale Model. Simul. **7**(1), 274–293 (2008)
11. Hou, T.Y., Wu, X.H., Zhang, Y.: Removing the cell resonance error in the multiscale finite element method via a Petrov-Galerkin formulation. Commun. Math. Sci. **2**(2), 185–205 (2004)
12. Henning, P., Peterseim, D.: Oversampling for the multiscale finite element method. Multiscale Model. Simul. **11**(4), 1149–1175 (2013)
13. Schindler, F., Milk, R.: DUNE Generic Discretization Toolbox, March 2015. http://dx.doi.org/10.5281/zenodo.16563
14. Milk, R., Schindler, F.: DUNE Stuff, March 2015. http://dx.doi.org/10.5281/zenodo.16564
15. Milk, R., Kaulmann, S.: DUNE Multiscale, March 2015. http://dx.doi.org/10.5281/zenodo.16560
16. Bastian, P., et al.: EXA-DUNE: flexible PDE solvers, numerical methods and applications. In: Lopes, L., et al. (eds.) Euro-Par 2014, Part II. LNCS, vol. 8806, pp. 530–541. Springer, Heidelberg (2014)
17. Giles, M.B.: Multilevel Monte Carlo path simulation. Oper. Res. **56**(3), 607–617 (2008)
18. Efendiev, Y., Iliev, O., Kronsbein, C.: Multilevel Monte Carlo methods using ensemble level mixed MsFEM for two-phase flow and transport simulations. Comput. Geosci. **17**(5), 833–850 (2013)

Task-Based Parallel Sparse Matrix-Vector Multiplication (SpMVM) with GPI-2

Dimitar Stoyanov$^{(\boxtimes)}$, Rui Machado, and Franz-Josef Pfreundt

Fraunhofer ITWM, Kaiserslautern, Germany
{stoyanov,machado,pfreundt}@itwm.fraunhofer.de
http://www.itwm.fraunhofer.de

Abstract. We present a task-based implementation of SpMVM with the PGAS communication library GPI-2. This computational kernel is essential for the overall performance of the Krylov subspace solvers but its proper hybrid parallel design is nowadays still a challenge on hierarchical architectures consisting of multi- and many-core sockets and nodes. The GPI-2 library allows, by default and in a natural way, a task-based parallelization. Thus, our implementation is fully asynchronous and it considerably differs from the standard hybrid approaches combining MPI and threads/OpenMP. Here we briefly describe the GPI-2 library, our implementation of the SpMVM routine, and then we compare the performance of our Jacobi preconditioned Richardson solver against the PETSc-Richardson using Poisson BVP in a unit cube as a benchmark test. The comparison employs two types of domain decomposition and demonstrates the preemptive performance and better scalability of our task-based implementation.

Keywords: GASPI · GPI-2 · PGAS · Task-based hybrid parallelization · Sparse matrix-vector multiplication · Krylov subspace solvers · Performance

1 Introduction and Motivation

The so called pure- or flat-MPI programming (one MPI-process per core) is nowadays no longer the most appropriate approach on systems with multi-core and multi-socket nodes. A hybrid parallelization is considered a natural choice instead: it combines a coarser, inter-node distributed memory parallelization with the more fine-grained, intra-node shared memory parallelization. Particularly, a task-based parallelization, where inter-nodal exchange can be independently performed by each thread from within thread-parallel regions, seems to be the proper alternative to reveal and fully exploit the hierarchical parallelism of such architectures.

The classical and most often used variant of hybrid parallelization is to combine MPI and threads/OpenMP. Particularly with regards to SpMVM this approach is followed for instance in [5,6]. But such a combination imposes

© Springer International Publishing Switzerland 2015
I. Lirkov et al. (Eds.): LSSC 2015, LNCS 9374, pp. 153–160, 2015.
DOI: 10.1007/978-3-319-26520-9_16

certain restrictions and performance issues with respect to thread-safety [2–4]. For instance, the MPI 2.0 standard prescribes four interface levels of threading support, one of them (MPI_THREAD_MULTIPLE) allowing a more task-based parallelization. MPI 3.0 improves some aspects related to threading - e.g., using MPI_Probe when several threads share a rank, etc. But in general, the hybrid parallelization based on MPI is still a challenge, it contains certain open issues (see e.g. [9], where also the new hybrid MPI+MPI approach is discussed), and it is often the case that MPI implementations do not provide a high performance support for task-based multi-threading. Consequently, applications aren't usually developed for such support and hence there is a non-optimal usage of resources - say, of the growing capabilities of high-performance interconnects. Particularly, many numerical libraries still use flat-MPI, e.g. in PETSc [11] threading has only recently appeared in the developers version.

Another point is that currently the trend in computer systems architecture is to see an increasing number of cores per node, with Non-Uniform Memory Access (NUMA) and with heterogenous resources. This not only puts pressure on multi-threaded support but it creates a need for more dynamic and asynchronous execution, to hide the latency of inter-node communication as well as that of intra-node memory and synchronization operations.

The GASPI interface [8] was specified with the previous aspects in mind and GPI-2 [7] was implemented to cope with them. The focus on asynchronous, one-sided communication with multi-threaded support and weak synchronization semantic creates an opportunity for new, more scalable implementations of performance critical building blocks such as the SpMVM, which is crucial for the case of Krylov solvers.

In this work, we present a task-based parallel implementation of SpMVM that takes advantage of our communication library GPI-2. We demonstrate the potential of our approach on the solution of a Poisson Boundary Value Problem (BVP) in a unit cube and we compare the performance against PETSc using two different types of Domain Decomposition (DD). The results show a significant performance advantage and better scalability when using the appropriate DD based on graph partitioning methods (METIS).

The rest of the paper is organized as follows: first we briefly describe the features of GPI-2 and our task-based SpMVM implementation, which differs in many aspects from the classical hybrid approach; then we formulate the model problem and explain the DD used in the comparisons. Further, we present and comment the performance results, and finally some conclusions are drawn.

2   GASPI/GPI-2 and Task-Based Parallelization

GPI-2 is the implementation of the GASPI standard, a relatively recent interface specification which aims at providing a compact API for parallel computations. It consists of one-sided communication routines, notifications-based synchronization, passive communication, global atomics and collective operations. It also defines groups (which are similar to MPI communicators and are used in collective operations) and the concept of segments. Segments are contiguous blocks of memory and can be made accessible (to read and write) to all threads on all ranks of a GASPI program.

GPI-2 is thus a communication library for C/C++ and Fortran based on one-sided communication. It adopts a PGAS-like model where each rank owns one or more memory segments which are globally accessible. Moreover, in GPI-2 all communication routines are thread-safe, allowing a more asynchronous and fine-grained multi-threaded execution as opposed to a bulk-synchronous communication with a single (master) thread, responsible for communication.

From an implementation point of view, GPI-2 aims at introducing a minimal overhead by providing a very thin layer, close to and exploiting hardware capabilities such as RDMA. One focus aspect is to provide truly asynchronous communication, that progresses in parallel as soon as it is triggered. This allows a better overlap of communication and computation, hiding the latency of communication.

Our GPI-2 based SpMVM is implemented in a task-based fashion, where a GPI-2 process (with the corresponding rank) is started per available NUMA socket. Within each rank a pool of POSIX threads is then used. Each thread dynamically polls for tasks to perform: this can be transferring data or computing a locally available part. This ensures that all threads are busy and that communication is overlapped and hidden behind the computation.

Note that such a task-based implementation is applicable to other kinds of large-scale scientific computations. Although it often requires a re-formulation of the algorithm, the attained benefits are considerable (as it will be demonstrated here). Below we provide more details about how is this achieved for the SpMVM kernel.

3   SpMVM with GPI-2

SpMVM is a memory–bounded routine; the SpMVM-kernels perform poorly, achieving $\sim 10\%$ from the theoretical peak performance [1], being far from reaching the theoretical speedup even on SMP-architectures. The principal problems related to the SpMVM performance are known (see [1] and the references therein): (i) restricted temporal locality as there is little data reuse, e.g. the matrix elements are used once only; (ii) irregular access to the input vector; (iii) large number of matrix rows of a very short row-length to multiply; (iv) indirect memory access imposed by the sparse matrix storage formats; etc.

The numerical treatment of (systems of) PDEs on hybrid architectures usually uses hierarchical decomposition: the coarse grained parallelism is attained by domain decomposition (DD), while the fine-grained parallelism on the node is achieved by thread parallelization. Each subdomain (SD) is mapped to a computational node, in our case this is a GPI-2 rank associated with a NUMA socket. The DD defines the distribution of the vector of unknowns (and of the rows of the sparse matrix for row-wise distribution) over the SDs, also the disposition of the discretization nodes at the subdomain interface which gives the topology of the inter-nodal exchange in SpMVM within the Krylov solvers. The resulting communication pattern depends on this topology (i.e., on the sparsity structure of the matrix) and is entirely irregular and problem-dependent. Note, that it is neither reasonable nor possible on each SD to keep a local replica of the full SpMVM input vector. One should copy locally only the remote items of the input vector needed on this SD, i.e. requested by the non-zero matrix elements, distributed on this SD. Our solution of this issue is to gather this topological information at the stages of mesh partitioning and discretization and to create, for each SD, a set of buffers to be written (lists of indices of the mesh nodes at the SD-interface). Then during execution, when the SpMVM routine is invoked, these buffers are used to perform the transfer of the remote input vector items.

Assuming a row-wise matrix and vector distribution, we designate the locally distributed matrix rows as $\mathbf{A}$, the full input vector as $\mathbf{X}$, and the local part of the output vector as $\mathbf{Y}_{lcl}$. Thus, the SpMVM should calculate the expression $\mathbf{Y}_{lcl} = \mathbf{A} * \mathbf{X}$ on each SD. A standard way to overlap communication and computation in SpMVM (see e.g. [5]) is to decompose $\mathbf{A}$ into: (i) a local part $\mathbf{A}_{lcl}$, which multiplies the local part $\mathbf{X}_{lcl}$ of the input vector $\mathbf{X}$, and (ii) its complementary matrix-chunk $\mathbf{A}_{rmt}$, containing elements which multiply the "remote" part $\mathbf{X}_{rmt}$ of the input vector. The elements of $\mathbf{X}_{rmt}$ correspond to the mesh nodes at the interface of the neighbour SDs and should be locally transferred. Formally $\mathbf{X} = \mathbf{X}_{lcl} + \mathbf{X}_{rmt}$ holds and according to this decomposition the SpMVM operation can be written as:

$$\mathbf{Y}_{lcl} = \mathbf{A}_{lcl} * \mathbf{X}_{lcl} + \mathbf{A}_{rmt} * \mathbf{X}_{rmt} \tag{1}$$

The "standard" hybrid implementation of SpMVM usually uses a single "communication thread" per socket or node which runs an MPI-process and performs the inter-nodal exchange; the other threads are eventually "mapped" to it to access MPI, otherwise performing local computations to overlap the communication [5,6]. Our GPI-based SpMVM kernel uses the same idea but is differently organized; a brief sketch of it follows. Taking advantage of GPI-2, it uses task-based parallel, one-sided RDMA transfer of $\mathbf{X}_{rmt}$ overlapped by computation:

(1) Some number of threads - say, as many as the number of neighbour SDs are - start independently transferring $\mathbf{X}_{rmt}$, each thread communicating with one neigbour SD;

(2) All other threads start polling jobs to perform the local part $\mathbf{A}_{lcl} * \mathbf{X}_{lcl}$ in Eq. (1), where "job" means a subset of matrix rows to be multiplied. Note that the jobs are independent from each other;

(3) When the transfer of $\mathbf{X}_{rmt}$ is over all threads start polling jobs from both the local and remote parts of the multiplication;

(4) Locally synchronize all threads and then perform the addition in Eq. (1).

Distinguishing features of our approach are: (i) the transfer of $\mathbf{X}_{rmt}$ is task-based thread parallel; (ii) the multiplication in both local and remote parts of Eq. (1) is asynchronously parallel; (iii) independently on the matrix-sparsity pattern, the job-polling mechanism provides presumably a quasi-optimal dynamic load balancing, with no idle threads (but this feature should be further tested on different matrices); (iv) the threads are spawned in the beginning of the iterative solver routine and are joined at its very end - i.e., we do not have the usual thread fork/join overhead as in the MPI/OpenMP implementations. To shortly summarize: our task-based parallelization allows for effective communication/computation overlapping leading to a better performance.

4 Model Problem and Domain Decomposition

We solve a Boundary Value Problem (BVP) for the Poisson equation in a unit cube which allows an (easily constructed) exact solution. The discretization is on a regular rectangular mesh with second order finite differences. Then the $O(h^2)$-convergence of the numerical solution would indicate a correct implementation. If we discretize in the internal mesh-nodes only, the assembled matrix is symmetric and positive definite (SPD), and the linear system can be solved with the Conjugate Gradients (CG) method.

We apply two variants of Domain Decomposition (DD):

(i) Cutting planes approach (Z-slices): the cube is split via planes parallel to the (x,y)-coordinate plane, i.e. the cube is cut into subdomains (SDs) or slices perpendicular to the z-axis.
(ii) Graph partitioning using the METIS [10] library.

While METIS provides partitions of a quite high quality, the Z-slices approach is far from being optimal, because when the number of SDs increases (strong scaling) the thickness of each slice decreases and the communication/computation ratio gets higher, limiting scalability. On the other side, this DD approach is illustrative and appropriate for benchmarking and comparing different solvers.

5 Performance Results and Comments

The underlying architecture consist of computational nodes connected via FDR Infiniband, each node being composed of two Intel Xeon E5-2680v2 (IvyBridge) sockets, with 10 cores per socket and 64 GB RAM.

We compare our GPI-2 implementation vs. PETSc-3.4.4. linked against the Intel MPI and MKL libraries. The domain partitioning is identical in PETSc and in the GPI-2 cases: two SDs (with successive indices) are assigned to each physical node, both in the case of the Z-slices and the METIS-partitioning. Furthermore, in our case, when a SD is mapped to a GPI-2 rank, the discretization nodes belonging to it, are uniformly distributed over the computing threads. The distribution of the matrix rows over the GPI-2 ranks and then over the computing threads matches exactly this nodal distribution. In the case of PETSc, when two SDs have been assigned to a physical node, again all locally distributed mesh nodes are uniformly split over the MPI-processes running on this computing node.

Table 1. Problem Size 257^3, $||exact - appr||_C^{4000\ itrs.} = 5.129525e - 1$

Physical nodes		1	2	4	8	16	32	60
GPI-nodes		2	4	8	16	32	64	120
Total cores/MPI-procs.		20	40	80	160	320	640	1200
DD-type: Z-sclices	PETSc, exec. time [s]	359	184	95	52	30	20	15
	GPI-2, exec. time [s]	214	109	55	27	16	12	10
DD-type: METIS	PETSc, exec. time [s]	358	181	91	47	26	16	13
	GPI-2, exec. time [s]	216	111	55	27	15	8	5

We use the CRS-formatted matrix storage. Our library contains several iterative solvers (`CG`, `BiCGstab`, etc.), but we have chosen the Richardson method as a benchmark: it allows for a fair comparison because the calculations performed in the GPI-Richardson and PETSc-Richardson routines are identical - this can be shown by monitoring the residual at each iteration. For the resulting linear system of our model problem we have measured the execution time to perform 4000 Jacobi-preconditioned Richardson iterations. The initial approximation of the solution is in both cases zero and after 4000 iterations in both solvers we obtain identical values for the current residual $L2$-norm $||b - A * x||_{L2}$ and for the C-norm of the error $||exact - appr||_C$ (i.e., the C-norm of the difference between the exact and the numerical solutions).

We compare the execution times and the measured real speedup of GPI-2 based Richardson vs. PETSc-Richardson for two different problem sizes. The timings for the size 257^3 are presented in Table 1, while Fig. 1 depicts the obtained speedup (along with the ideal one) for the two DD-techniques we use and taking the execution on a single node as base.

Similarly, Table 2 contains the measurements for the size 351^3, with the obtained speedup presented in Fig. 2. On the finer mesh the convergence of

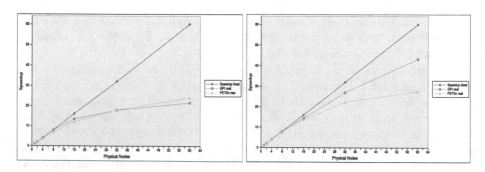

Fig. 1. Speedup GPI-2 vs. PETSc: Jacobi Preconditioned Richardson, 4000 itrs, size 257^3, partitioning using Z-slices (left) and METIS (right)

Table 2. Problem Size 351^3, $||exact - appr||_C{}^{4000\ itrs.} = 6.033188e - 1$

Physical nodes		1	2	4	8	16	32	60
GPI-2 ranks		2	4	8	16	32	64	120
Total cores/MPI-procs.		20	40	80	160	320	640	1200
DD-type: Z-sclices	PETSc, exec. time [s]	922	467	241	128	79	49	36
	GPI-2, exec. time [s]	566	282	148	76	37	24	21
DD-type: METIS	PETSc, exec. time [s]	918	460	233	117	61	33	22
	GPI-2, exec. time [s]	564	289	153	81	36	19	11

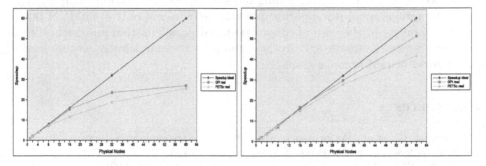

Fig. 2. Speedup GPI-2 vs. PETSc: Jacobi Preconditioned Richardson, 4000 itrs, size 351^3, partitioning using Z-slices (left) and METIS (right)

the Richardson method is slower and - after performing the same number of iterations - the difference with the exact solution is bigger.

In both cases the comparison has been done separately for our two types of DD. One easily sees that - independently of the type of partitioning - GPI-Richardson clearly outperforms PETSc, it is about twice faster, despite the fact that we use no hardware optimization (e.g. vectorization). Furthermore, although the inefficient Z-slices partitioning produces almost the same speedup for the two solvers, our GPI-2 version has shorter execution times.

About the partitioning one may say that compared to the Z-slices the METIS-DD is certainly more appropriate: it produces faster execution times starting from 8 (case 257^3) or 16 (case 351^3) physical nodes on. Using METIS-DD GPI-Richardson is not only faster but also scales better than PETSc-Richardson.

6   Conclusion

From an application point of view, a distinguishing property when working with GPI-2 is that it provides full freedom and flexibility to follow a task based parallelization. In this sense, the GPI-2 model meets the requirements and the challenges of the nowadays hierarchical architectures, proposing an alternative to both pure-MPI programming and the standard hybrid approaches with MPI and threads/OpenMP.

We have briefly sketched our GPI-2 implementation of the SpMVM kernel, which uses asynchronous communication and allows for fine-grained and better communication/computation overlap. We have used this kernel in a small library of Krylov subspace solvers. Using as a benchmark the Jacobi Preconditioned Richardson method to iterate the linear system arising after the discretization of a Poisson BVP in a unit cube, we have shown that our Richardson solver outperforms the Richardson solver of PETSc. We have confirmed this behaviour for two different types of domain decomposition: Z-slices-partitioning and graph partitioning with the METIS library. In the latter case, our version is not only faster than PETSc-Richardson but it also scales better.

As we noted, from a programming model point of view, conceptually similar implementations could bring performance advantages not only in SpMVM but - more generally - in the case of other DD-based parallelization approaches, e.g. additive Schwartz, where a truly asynchronous communication scheme could enable evident performance gains.

References

1. Gormas, G., et al.: Performance evaluation of the sparse matrix-vector multiplication on modern architectures. J. Supercomput. **50**, 36–77 (2009)
2. Gropp, W.D., Thakur, R.: Issues in developing a thread-safe MPI implementation. In: Mohr, B., Träff, J.L., Worringen, J., Dongarra, J. (eds.) PVM/MPI 2006. LNCS, vol. 4192, pp. 12–21. Springer, Heidelberg (2006)
3. Balaji, P., Buntinas, D., Goodell, D., Gropp, W.D., Thakur, R.: Toward efficient support for multithreaded MPI communication. In: Lastovetsky, A., Kechadi, T., Dongarra, J. (eds.) EuroPVM/MPI 2008. LNCS, vol. 5205, pp. 120–129. Springer, Heidelberg (2008)
4. Hagger, G., Wellein, G.: Introduction to High Performance Computing for Scientists and Engineers. CRC Press, Boca Raton (2010)
5. Lange, M., Gorman, G., Weiland, M., Mitchell, L., Southern, J.: Achieving efficient strong scaling with PETSc using hybrid MPI/OpenMP optimisation. In: Kunkel, J.M., Ludwig, T., Meuer, H.W. (eds.) ISC 2013. LNCS, vol. 7905, pp. 97–108. Springer, Heidelberg (2013)
6. Schubert, G., Fehske, H., Hager, G., Wellein, G.G.: Hybrid-parallel sparse matrix-vector multiplication with explicit communication overlap on current multicore-based systems. Parallel Process. Lett. **21**(3), 339–358 (2011)
7. http://www.gpi-site.com/gpi2/
8. http://www.gaspi.de/
9. http://openmp.org/wp/sc13-tutorial-hybrid-mpi-and-openmp-parallel-programming/
10. http://www-users.cs.umn.edu/karypis/metis/
11. http://www.mcs.anl.gov/petsc/

Efficient Algorithms for Hybrid HPC Systems

On the Preconditioned Quasi-Monte Carlo Algorithm for Matrix Computations

V. Alexandrov[1], O. Esquivel-Flores[2], S. Ivanovska[3], and A. Karaivanova[3(✉)]

[1] ICREA-BSC, C/Jordi Girona 29, 08034 Barcelona, Spain
[2] UNAM, Mexico and Barcelona Supercomputing Centre, Barcelona, Spain
[3] IICT-BAS, acad. G. Bonchev Street, Bl. 25A, 1113 Sofia, Bulgaria
anet@parallel.bas.bg

Abstract. In this paper we present a quasi-Monte Carlo Sparse Approximate Inverse (SPAI) preconditioner. In contrast to the standard deterministic SPAI preconditioners that use the Frobenius norm, Monte Carlo and quasi-Monte Carlo preconditioners rely on stochastic and hybrid algorithms to compute a rough matrix inverse (MI). The behaviour of the proposed algorithm is studied. Its performance is measured and compared with the standard deterministic SPAI and MSPAI (parallel SPAI) approaches and with the Monte Carlo approach. An analysis of the results is also provided.

1 Introduction

Recently Monte Carlo parallel numerical methods have been widely used for matrix computations due to their specific properties such as inherent parallelism, allowing minimal communication (e.g. being communication avoiding by design), high level of scalability as well as fault-tolerance and resilience in the parallel case. The other reason for the recent interest in MCMs is that the methods have evolved significantly since the early days. Much of the effort in the development of Monte Carlo methods has been in the construction of variance reduction techniques which speed up the computation by reducing the rate of convergence of crude MCM, which is $O(N^{-1/2})$. An alternative approach to acceleration is to change the type of random sequence, and hence improve the behavior by N. Quasi-Monte Carlo methods (QMCMs) use quasirandom (also known as low-discrepancy) sequences instead of pseudorandom sequences, with the resulting convergence rate for numerical integration being as good as $O((\log N)^k)N^{-1})$. Some results of using QMCMs for linear algebra problems can be found in [1].

Solving systems of linear equations is a well-known problem in engineering and sciences. Using iterative or direct methods to solve these systems may be a costly approach in both time and computational effort for certain classes of problems. One option of reducing the effort of solving these systems is to apply preconditioners before using an iterative method. Depending on the method used to compute the preconditioner, the savings and end-results vary. A very sparse preconditioner is computed quickly, but it is unlikely to improve the quality

© Springer International Publishing Switzerland 2015
I. Lirkov et al. (Eds.): LSSC 2015, LNCS 9374, pp. 163–171, 2015.
DOI: 10.1007/978-3-319-26520-9_17

of the solution. On the other hand, computing a rather dense preconditioner is computationally expensive and might be time or cost prohibitive. Therefore, finding a good preconditioner that is computationally efficient, while still providing substantial improvement to the iterative solution process, is a worthwhile research topic.

The next section gives and overview of related work. Monte Carlo and quasi-Monte Carlo methods, and the specific matrix inversion algorithm that is discussed as a SPAI preconditioner, are presented in Sect. 3. Section 4 provides information on results and findings from experiments with matrices of varying sizes and sparsity. The last section concludes and gives an outlook on the future work.

2   Related Work

Research efforts in the past have been directed towards optimizing the approach of sparse approximate inverse preconditioners. Improvements to the Frobenius norm have been proposed for example by concentrating on sparse pattern selection strategies [10], or building a symmetric preconditioner by averaging off-diagonal entries [11]. Further, it has been shown that the sparse approximate inverse preconditioning approach is also a viable course of action on large-scale dense linear systems [2].

In the past there have been differing approaches and advances towards a parallelisation of the SPAI preconditioner. The method that is used to compute the preconditioner provides the opportunity to be implemented in a parallel fashion. In recent years the class of Frobenius norm minimizations that has been used in the original SPAI implementation [5] was modified and is provided in a parallel SPAI software package. One implementation of it, by the original authors of SPAI, is the Modified SParse Approximate Inverse (MSPAI [15]).

This version provides a class of modified preconditioners such as MILU (modified ILU), interface probing techniques and probing constraints to the original SPAI, apart from a more efficient, parallel Frobenius norm minimization. Further, this package also provides two novel optimization techniques. One option is using a dictionary in order to avoid redundant calculations, and to serve as a lookup table. The second possibility is using an option in the program to switch to a less computational intensive, sparse QR decomposition whenever possible. This optimized code runs in parallel, together with a dynamic load balancing.

Further discussion of additional advances, which are building upon the SPAI software suite, will be presented in the next section.

2.1   SParse Approximate Inverse Preconditioner (SPAI)

The SPAI algorithm [13] is used to compute a sparse approximate inverse matrix M for a given sparse input matrix B. This is done by minimizing $||BM - I||$ in the Frobenius norm. The algorithm explicitly computes the approximate inverse, which is intended to be applied as a preconditioner of an iterative method.

The SPAI application provides the option to fix the sparsity pattern of the approximate inverse a priori or capture it automatically.

Since the introduction of the original SPAI in 1996, several advances, building upon the initial implementation, have been made. Two newer implementations are provided by the original authors, the before mentioned MSPAI, and the highly scalable Factorized SParse Approximate Inverse (FSPAI [14]). The intended use of both differs depending on the problem at hand.

Whereas MSPAI is used as a preconditioner for large sparse and ill-conditioned systems of linear equations, FSPAI is applicable only to symmetric positive definite systems of this kind. FSPAI is based around an inherently parallel implementation, generating the approximate inverse of the Cholesky factorization for the input matrix. MSPAI on the other hand is using an extension of the well-known Frobenius norm minimization that has been introduced in the original SPAI.

2.2   Stochastic SParse Approximate Inverse Preconditioner

In [7] stochastic SPAI preconditioner has been presented and extensively studied. It uses a Monte Carlo algorithm for approximate matrix inverse. In the general case we proceed in the following way: Assume the general case where $\|B\| > 1$ and consider the splitting

$$B = \hat{B} - C, \tag{1}$$

where the off-diagonal elements of $\hat{B}$ are the same as those of B, and the diagonal elements of $\hat{B}$ are defined as $\hat{b}_{ii} = b_{ii} + \alpha_i\|B\|$, choosing in most cases $\alpha_i > 1$ for $i = 1, 2, ..., n$. For the simplicity of the algorithm it is often easier to fix α rather than altering it over the rows of the matrix [6,8,12].

From (1) compute $A = B_1^{-1}B_2$, which satisfies $\|A\| < 1$. Further, by careful choice, of $\hat{B}$, it is possible to make $\|A\| < \frac{1}{2}$, which gives faster convergence of the MC. Then generate the inverse of $\hat{B}$ by

$$m_{rr'}^{(-1)} \approx \frac{1}{N} \sum_{s=1}^{N} \left[\sum_{(j|s_j=r')} W_j \right], \tag{2}$$

where $(j|s_j = r')$ means that only

$$W_j = \frac{a_{rs_1} a_{s_1 s_2} \cdots a_{s_{j-1} s_j}}{p_{rs_1} p_{s_1 s_2} \cdots p_{s_{j-1} s_j}},$$

for which $s_j = r'$ are included in the sum (2). Calculating $\|B\|$ can be an expensive operation and, so, any a priori information allowing for a reasonable estimate here is useful. From this it is then necessary to work back and recover B^{-1} from $\hat{B}^{-1}$. To do this recursive process ($k = n - 1, n - 2, \ldots, 0$) is used on $\hat{B}^{-1}$:

$$B_k^{-1} = B_{k+1}^{-1} + \frac{B_{k+1}^{-1} S_{k+1} B_{k+1}^{-1}}{1 - trace\left(B_{k+1}^{-1} S_{k+1}\right)}, \tag{3}$$

where $B_n^{-1} = \hat{B}^{-1}$ and S_i is all zero except for the $\{ii\}^{th}$ component, which is from the matrix $S = \hat{B} - B$. Then $B_0^{-1} = B^{-1}$.

The make up of matrix S means that while (3) looks complicated it is, in fact, reasonably simple. This means that it is not as computationally complex and when transferred to code there are obvious simplifications possible to make sure that many multiplications by zero are not performed. This method of splitting and recovery leads to the algorithm presented in [18], which details a MC algorithm for inverting general matrices.

3   Quasi-Monte Carlo Approach

We recall some basic concepts of QMCMs, [9]. First, for a sequence of N points $\{x_n\}$ in the d-dimensional half-open unit cube I^d define

$$R_N(J) = \frac{1}{N}\#\{x_n \in J\} - m(J)$$

where J is a rectangular set and $m(J)$ is its volume. Then define star discrepancy

$$D_N = sup_{J \in E}|R_N(J)|,$$

where E is the set of all rectangular subsets in I^d and $E^\star$ is the set of all rectangular subsets in I^d with one vertex at the origin.

The basis for analyzing QMC quadrature error is the Koksma-Hlawka inequality:

Theorem (Koksma-Hlawka, [9]**):** For any sequence $\{x_n\}$ and any function f of bounded variation (in the Hardy-Krause sense), the integration error is bounded as follows

$$\left|\frac{1}{N}\sum_{n=1}^{N} f(x_n) - \int_{I^d} f(x)\,dx\right| \leq V(f)D_N^\star. \tag{4}$$

The star discrepancy of a point set of N truly random numbers in one dimension is $O(N^{-1/2}(\log\log N)^{1/2})$, while the discrepancy of N quasirandom numbers in s dimensions can be as low as $O(N^{-1}(\log N)^{s-1})$. Most notably there are the constructions of Halton, Soboĺ, Faure, and Niederreiter, and their modifications for producing quasirandom numbers. Description of these can be found for example in Niederreiter's monograph [17]. Different kinds of quasi-random sequences exist. The theoretical properties of these point sets look promising, but are only valid asymptotically. Therefore, only when an "almost infnite" number of points is used, can one rely on the theory to compare quasi-random point sets and sequences. For smaller and more practical ranges of the number of points, there can be side-effects and the results with quasi-random points may not always be what the theory at infinity predicts.

Now recall that Monte Carlo methods for linear algebra problems are based on computing matrix-vector products $h^T A^i f$ (see [16]). But computing $h^T A^i f$

is equivalent to computing an $(i+1)$-dimensional integral. Thus we may analyze using QRNs with bounds from numerical integration. We do not know A^i explicitly, but we do know A and we perform random walks on the elements of the matrix to compute approximately $h^T A^i f$.

Let A be a general sparse matrix with d_i nonzero elements per row. The following mapping procedure corresponds to importance sampling approach:

$$G = [0, 1)$$

$$G_i = [\frac{\sum_{k'=1}^{i-1} |a_{ik'}|}{\sum_{k'=1}^{n} |a_{ik'}|}, \frac{\sum_{k'=1}^{i} |a_{ik'}|}{\sum_{k'=1}^{n} |a_{ik'}|}), \quad i = 1, \ldots, n$$

and summation on k' means summation only on nonzero elements:

$$a(x, y) = a_{ij}, x \in G_i, \ y \in G_j, \ i = 1, \ldots, n, \ j = 1, \ldots, d.$$

Often, the vectors f and h are chosen to be $(1, 1, \ldots, 1)$, so $h(x) = 1, x \in G$, $f(x) = 1, x \in G$.

In this case after similar calculation we prove that the bound on the error (for non-normalized matrix) is given by:

$$|h^T A f - \frac{1}{N} \sum_{s=1}^{N} h(x_s) a(x_s, y_s) f(y_s)| \le (d|A|)^l D_N^*,$$

where d is the mean value of the nonzero elements per row, l is the length of the Markov chain, D_N^* is the star discrepancy of the sequence used, and $\|A\| < 1$.

Let us remind that usually the average number d of nonzero entries per row is much smaller than the size of the matrix n, $d \lessdot n$. Thus the order of the above estimation is the order of D_N^* which is $O((log^l N)N^{-1})$.

Convergence and Complexity

In the Monte Carlo methods there are two kind of errors that controll the convergence: systematic, which comes from the method, and stochastic, which comes from the approximation of the mean value with an averaged sum. The complexity of a Monte Carlo method is a product of the expected value of the length of the corresponding walk (Markov chain), and a number of walks (chains).

For computing an element of the inverse matrix $A^{-1} = C = \{c_{rr'}\}$ the computational complexity is lN, where l is the length of the performed walks (Markov chain) which for MCM is $l = E[l_s]$, and for QMCM l the dimension of the quasirandom sequence; l depends on the spectrum of the matrix A. Let us note that first few steps of a random (quasirandom) walk tend to improve results greatly, whereas many additional steps would be necessary to refine the result to sufficient accuracy. We suggest to use these methods with relatively small l for a quick rough estimation.

The convergence for MCM and QMCM in this case is $O\left(\frac{\|A\|^l \|r^{(0)}\|}{1 - \|A\|} + \sigma N^{-1/2}\right)$ and $O\left(\frac{\|A\|^l \|r^{(0)}\|}{1 - \|A\|} + (log^l N)N^{-1}\right)$ correspondingly.

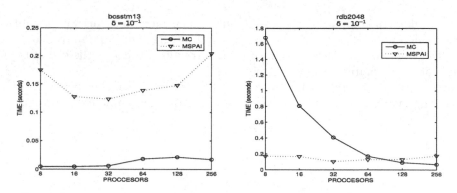

Fig. 1. Scalability properties of hybrid Monte Carlo/BiCGSTAB compared to MSPAI/BiCGSTAB for bcsstm13 and rdb2048.

4 Experiments

The aim in the present study is to compare the performance of the Monte Carlo and quasi-Monte Carlo SPAI preconditioners for computing a rough matrix inverse. To check the possible advantages of each of the approaches we tested the quasi-Monte Carlo approach using scrambled Sobol [4] and modified Halton sequences [3].

As an input for computing the preconditioners, two matrices from the University of Florida Sparse Matrix Collection [19] were used. The matrices we used are the symmetric *Si5H12* of size 19896×19896 and *Si10H16* of size 17077×17077 from the PARSEC matrix group with 875 923 and 735 598 nonzeros entries respectively and are sparse, indefinite with multiple and clustered eigenvalues. Experiments were also run with real, non-symmetric and no positive definite matrices (rdb2048) as well as real, symmetric, positive semidefinite (bcsstm13) matrices from the matrix market (see Fig. 1).

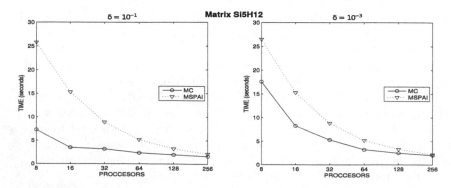

Fig. 2. Scalability properties of hybrid Monte Carlo/BiCGSTAB compared to MSPAI/BiCGSTAB for Si5H12.

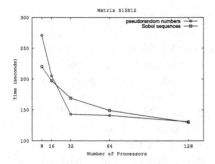

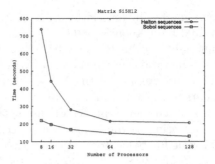

Fig. 3. Scalability properties for the test matrix Si5H12 using pseudorandom and Sobol quasirandom sequences.

Fig. 4. Scalability properties for the test matrix Si5H12 using quasirandom sequences.

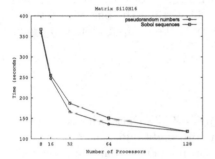

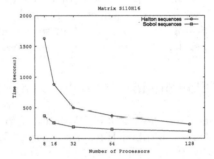

Fig. 5. Scalability properties for the test matrix Si10H16 using pseudorandom and Sobol quasirandom sequences.

Fig. 6. Scalability properties for the test matrix Si10H16 using quasirandom sequences.

The numerical experiments have been executed on the MareNostrum III supercomputer at the Barcelona Supercomputing Center (BSC). It currently consists of 3056 compute nodes that are each equipped with 2 Intel Xeon 8-core processors, 64 GB RAM and are connected via an InfiniBand FDR-10 communication network. The experiments have been run multiple times to account for possible external influences on the results. The computation times for both the preconditioner calculated by MSPAI, as well as our Monte Carlo based results, have been noted. While conducting the experiments, the parameters for probable errors were configured to produce preconditioners with similar properties and therefore producing residuals within similar ranges when used as preconditioners for BiCGSTAB and GMRES. The experiments of hybrid parallel Monte Carlo preconditioner with BiCGSTAB using pseudorandom sequences are depicted on Figs. 1 and 2. The results show clearly the efficiency of the approach.

The numerical experiments have been also executed on the HPC Cluster at IICT-BAS. The technical parameters of this supercomputing facility are the following: HP Cluster Platform Express 7000 enclosures with 36 blades BL 280c with dual Intel Xeon X5560 @ 2.8 Ghz (total 576 cores), 24 GB RAM per blade, 8 controlling nodes HP DL 380 G6 with dual Intel X5560 @ 2.8 Ghz, 32 GB RAM. The total storage capacity available in the three disk systems is 144 TB.

The performance of the Monte Carlo and quasi Monte Carlo SPAI preconditioners with Halton and Sobol sequences is shown in the series of four figures. Maximum number of processors used for the tests are 128. Figures 3 and 5 show that increasing of the number of processors leads to decrease in the computational time for both stochastic and hybrid algorithms. In the case of the bigger matrix (*Si5H12*) Monte Carlo with pseudorandom numbers performs better than the quasi-Monte Carlo with scrambled Sobol achieving less than 150 seconds for more than 32 processors. The quasi-Monte Carlo with scrambled Sobol outperforms the QMC with the modified Halton sequence (Figs. 4 and 6). But there is very interesting behaviour of Halton QMC: fast decrease of computing time with the increased number of CPUs used. With respect to the matrix size the QMC with Halton sequences is much more sensitive than the one with scrambled Sobol showing more than seven times drop in the computational runtime in dependence of the number of processors (Fig. 6).

5   Conclusions and Future Work

Numerical experiments of Monte Carlo and quasi-Monte Carlo with scrambled Sobol and modified Halton sequences for SPAI preconditioning have been performed. It is evident that the Sobol QMC performs better than the modified Halton QMC, but still worse than the standard MC preconditioner. Tests with other scrambled sequences have to be performed.

Additional tests on other supercomputing facilities could contribute significantly in evaluating the preconditioning with stochastic and hybrid approaches.

Acknowledgment. The research work reported in the paper is partly supported by the Bulgarian NSF grant Grant DFNI-I02/8, and second author would like to thank CONACYT-Mexico for supporting potsdoctoral position in BSC.

References

1. Alexandrov, V.N., Karaivanova, A.: Parallel monte carlo algorithms for sparse SLAE using MPI. In: Margalef, T., Dongarra, J., Luque, E. (eds.) PVM/MPI 1999. LNCS, vol. 1697, pp. 283–290. Springer, Heidelberg (1999)
2. Allèon, G., Benzi, M., Giraud, L.: Sparse approximate inverse preconditioning for dense linear systems arising in computational electromagnetics. Numer. Algorithm. **16**(1), 1–15 (1997)
3. Atanassov, E.I., Durchova, M.K.: Generating and testing the modified halton sequences. In: Dimov, I., Lirkov, I., Margenov, S., Zlatev, Z. (eds.) NMA 2002. LNCS, vol. 2542, pp. 91–98. Springer, Heidelberg (2003)

4. Atanassov, E., Karaivanova, A., Ivanovska, S.: Tuning the generation of sobol sequence with Owen scrambling. In: Lirkov, I., Margenov, S., Waśniewski, J. (eds.) LSSC 2009. LNCS, vol. 5910, pp. 459–466. Springer, Heidelberg (2010)
5. Benzi, M., Meyer, C., Tuma, M.: A sparse approximate inverse preconditioner for the conjugate gradient method. SIAM J. Sci. Comput. **5**, 1135–1149 (1996)
6. Branford, S.: The parallel hybrid Monte Carlo algoritm. Master's thesis, Schools of Systems Engineering, The Univerity of Reading (2003)
7. Branford, S.: Hybrid Monte Carlo methods for linear algebra problems. Ph.D. thesis, School of Systems Engineering, The University of Reading, April 2009
8. Branford, S., Weihrauch, C., Alexandrov, V.N.: A sparse parallel hybrid Monte Carlo algorithm for matrix computations. In: Sunderam, V.S., van Albada, G.D., Sloot, P.M.A., Dongarra, J. (eds.) ICCS 2005. LNCS, vol. 3516, pp. 743–751. Springer, Heidelberg (2005)
9. Caflisch, R.: Monte Carlo and quasi-Monte Carlo methods. Acta Numerica **7**, 1–49 (1998)
10. Carpentieri, B., Duff, I., Giraud, L.: Some sparse pattern selection strategies for robust Frobenius norm minimization preconditioners in electromagnetism. Numer. Linear Algebra Appl. **7**, 667–685 (2000)
11. Carpentieri, B., Giraud, L., et al.: Experiments with sparse preconditioning of dense problems from electromagnetic applications. Technical report, CERFACS, Toulouse, France (2000)
12. Fathi, B., Liu, B., Alexandrov, V.N.: Mixed Monte Carlo parallel algorithms for matrix computation. In: Sloot, P.M.A., Tan, C.J.K., Dongarra, J., Hoekstra, A.G. (eds.) ICCS-ComputSci 2002, Part II. LNCS, vol. 2330, pp. 609–618. Springer, Heidelberg (2002)
13. Grote, M., Hagemann, M.: Spai: sparse approximate inversepreconditioner. Spaidoc.pdf paper in the SPAI 3:1 (2006)
14. Huckle, T.: Factorized sparse approximate inverses for preconditioning. J. Supercomput. **25**(2), 109–117 (2003)
15. Huckle, T., Kallischko, A., Roy, A., Sedlacek, M., Weinzierl, T.: An efficient parallel implementation of the MSPAI preconditioner. Parallel Comput. **36**(56), 273–284 (2010). Parallel Matrix Algorithms and Applications
16. Karaivanova, A.: Quasi-Monte Carlo methods for some linear algebra problems. Convergence and complexity. Serdica J. Comput. **4**, 58–72 (2010). ISSN: 1312–6555
17. Niederreiter, H.: Random Number Generation and Quasi-Monte Carlo Methods. SIAM, Philadelphia (1992)
18. Strassburg, J., Alexandrov, V.: Enhancing Monte Carlo preconditioning methods for matrix computations. Procedia Comput. Sci. **29**, 1580–1589 (2014)
19. http://www.cise.ufl.edu/research/sparse/matrices/

Energy Performance Evaluation of Quasi-Monte Carlo Algorithms on Hybrid HPC

E. Atanassov, T. Gurov$^{(\boxtimes)}$, and A. Karaivanova

IICT-BAS, Acad. G. Bonchev Street, Bl. 25A,1113 Sofia, Bulgaria
gurov@bas.bg

Abstract. The increasing demands of scientific applications and the increasing capacity of modern computing systems lead to the need of evaluating energy consumption and, consequently, to the development of energy efficient algorithms. In this paper we study the energy performance of a class of quasi-Monte Carlo algorithms on hybrid HPC systems. These algorithms are applied to solve quantum kinetic integral equations using Sobol and Halton sequences. The energy performance results are compared on a CPU-based computer platform and computer platforms with accelerators like GPU cards and Intel Xeon Phi coprocessors with respect to several metrics. Directions for future work are also given.

Keywords: Quasi-Monte Carlo algorithms · Hybrid HPC systems · Energy efficiency

1 Introduction

The latest developments in the domain of HPC have lead to the deployment of complex extreme-scale systems, based on diverse computing devices (CPU, GPU, accelerators) and posed the question of scalability in the light not only of parallel efficiency, but also in terms of energy consumption. Development of energy-efficient algorithms is becoming more and more important with the growing size of the applied problems that need solution using the power of modern computer systems [1,2]. The FLOPS/WATT (F/W) metric was introduced and successfully used as the de facto standard in measuring the energy efficiency of a computing system (see [13]). Considering the significant probability of an error during a run, some authors [3,4] proposed the *time_to_solution* metric for estimating the algorithm performance. In [6] a new metric has been introduced in order to account also for the initial investment. In this paper we consider all of these metrics in order to analyze the performance of a class of quasi-Monte Carlo (QMC) algorithms with an awareness for their energy use.

Let us remind that a generic approach to improving the convergence of Monte Carlo (MC) methods involved using highly uniform random numbers (called quasirandom sequences - QRNs) in place of the usual pseudo-random numbers. The methods based on QRNs, called QMC methods, are very popular not only

© Springer International Publishing Switzerland 2015
I. Lirkov et al. (Eds.): LSSC 2015, LNCS 9374, pp. 172–181, 2015.
DOI: 10.1007/978-3-319-26520-9_18

for solving multidimensional integrals but also for various Markov chain based applications.

The paper is organized as follows: the next section presents the metric and related work, Sect. 3 describes the studied algorithms, Sect. 4 presents numerical tests and our findings. At the end we present our conclusion.

2 Motivation

Our work has been motivated by the use-cases that we observed during the establishment of a regional high-performance computing (HPC) infrastructure for South-Eastern Europe, taking into account the specific requirements that arise due to the economical and social conditions in the region.

We concentrate to the study energy efficiency of QMC algorithms for integral equations on extreme-scale parallel computing systems. The extreme-scale parallel computing systems are HPC systems with low latency interconnection, equipped with GPGPU devices (manufactured by companies like NVIDIA) and/or co-processors (e.g. those using Intel MIC technology) to speed up the calculations that make use of thousands of processor cores in high-density deployment. Here we point out two important features:

(i) the extensive use of computational accelerators like GPU-computing cards;
(ii) the rapid evolution of hardware in HPC clusters, which leads to a frequent necessity to upgrade in order to meet the challenges of contemporary research.

These points motivate the inclusion of a substantial factor to account for the purchasing price of the equipment, so that the individual optimization efforts at the level of algorithms should lead to a global optimum in the sense of computational results achieved for a given yearly budget. Our experience shows that although the hardware can be operational for a longer period, there is an *efficient lifetime* for a cluster that lasts between 3 to 4 years. On the other hand the x86-based HPC clusters can be upgraded in a more gradual way, presenting the possibility to use savings from energy costs for hardware upgrading. We propose to enhance the formula in [3] in the following way:

$$F(T) \times (E + nCT), \tag{1}$$

where C is the price of one core-hour, excluding energy, n is the number of cores used by the algorithm, T is the time to solution and E is the cost of energy consumed. The price C should be based mainly on the purchasing price of the equipment, divided by the total number of cores and number of hours in the efficient lifetime. Based on the substantial improvements in the computational power of accelerators over time we can postulate the efficient lifetime to be equal to 4 years, because experience shows that after 4 years the same computational results can be achieved by several times less expensive equipment that uses much less energy. We point out that cloud providers offer access to their equipment based on a single price-per-core number. However, a national computational infrastructure provider has more flexibility and can stimulate the development of algorithms that minimize the above function instead. Our formula is not any harder to compute because the purchasing price is readily available.

3   Energy Efficiency Study for a Class of Quasi-Monte Carlo Algorithms

As a case study we consider QMC algorithms for solving Wigner equation for the nanometer and femtosecond transport regime. We use the formulation of the Wigner equation in an inhomogeneous case (in quantum wire - more realistic case) where the electron evolution depends on the energy and space coordinates [10]. Particularly we consider a quantum wire, where the carriers are confined in the plane normal to the wire by infinite potentials. The initial condition is assumed both in energy and space coordinates.

The numerical results that we present for estimating the energy efficiency metrics, are for the inhomogeneous case with applied electric field [10]. We recall the integral form of the quantum-kinetic equation, [11]:

$$f_w(z, k_z, t) = f_w(z - \frac{\hbar k_z}{m}t + \frac{\hbar F}{2m}t^2, k_z, 0) + \int_0^t dt'' \int_{t''}^t dt' \int d\mathbf{q}'_\perp \int dk'_z \times \quad (2)$$

$$\left[S(k'_z, k_z, t', t'', \mathbf{q}'_\perp) f_w \left(z - \frac{\hbar k_z}{m}(t - t'') + \frac{\hbar F}{2m}(t^2 - t''^2) + \frac{\hbar q'_z}{2m}(t' - t''), k'_z, t'' \right) \right.$$

$$\left. - S(k_z, k'_z, t', t'' \mathbf{q}'_\perp) f_w \left(z - \frac{\hbar k_z}{m}(t - t'') + \frac{\hbar F}{2m}(t^2 - t''^2) - \frac{\hbar q'_z}{2m}(t' - t''), k_z, t'' \right) \right]$$

Here, $f_w(z, k_z, t)$ is the Wigner function described in the $2D$ phase space of the carrier wave vector k_z and the position z, and t is the evolution time. The kernel $S(k'_z, k_z, t', t'', \mathbf{q}'_\perp)$ of the integral equation in (2) are well described in [10,11].

In the inhomogeneous case the wave vector (and respectively the energy) and the density distributions are given by the integrals

$$f(k_z, t) = \int \frac{dz}{2\pi} f_w(z, k_z, t); \qquad n(z, t) = \int \frac{dk_z}{2\pi} f_w(z, k_z, t). \quad (3)$$

In this work we investigate algorithms that estimate these quantities, as well as the Wigner function (2), by using a QMC approach [7], where one point of a low-discrepancy sequence is used to sample one numerical trajectory. The computations were performed with both Sobol and modified Halton sequences. The implementations of these sequences are given in [8,9,12].

Parallel Implementation

The QMC algorithms are perceived as computationally intensive, but naturally parallel. Different strategies, based on either static or dynamic load-balancing are possible. The so-called "master-slave" model is usable for dynamic load-balancing, while static load-balancing is sufficient in the case when the variations in the computational load can be controlled. For example, by blocking together a constant amount of samples one achieves decreased variation in the computational load by a factor of square-root of the number of samples in the block. The blocking also decreases the required communication. The partial results are collected and used to assemble an accumulated result with smaller variance. To achieve maximum code re-use with QMC algorithms we follow the same parallelisation approach. Our parallel implementation uses MPI for the CPU-based and Xeon Phi-based computations as well as CUDA for the GPU-based parallelisation. Since in the QMC computations one can not afford to lose computations we use the blocking approach in the generation of the sequences, combined with static load balancing.

Implementation using Hyper-Threading, and GPUs and Xeon Phi

For our test cluster we compared the CPU performance of the parallel code with and without hyper-threading (HT). Although for some codes using hyper-threading does not improve the overall speed of calculations, because the floating point units of the processor are shared between the threads, in our experience substantial gains may be achieved without any additional coding effort by simply using logical instead of physical cores when sizing the launch of the MPI job. For this particular application we observed about 30 % improvement when HT is turned on, which should be considered a good result and also shows that our overall code is reasonably efficient.

For the GPGPU-based implementation we used CUDA for parallel computations. Parallel processing is based upon splitting the computations between grid of threads. We use thread size of 256, which is optimal taking into account the relatively large number of registers. Generators for the scrambled Sobol sequence and modified Halton sequence have been developed and tested in our previous works [8,9].

For the Xeon Phi-based implementation we again used generators that we have developed before. In the case of Xeon Phi the issue of hyper-threading is more interesting than in the case of CPUs. Without using hyper-threading one can not achieve the best possible performance of the Xeon Phi cards, due to some architectural peculiarities. That is why we performed tests with different number of threads. While the number of physical cores of our Xeon Phi cards is 60, we tested with 60, 120, 180 and 240 threads in order to see when the maximum performance will be achieved and also in order to analyse the results from the point of view of energy efficiency.

4   Numerical Results

The numerical results were obtained on the heterogeneous HPC system at the Institute of Information and Communication Technologies, Bulgarian Academy of Sciences [5].

The HPC system combines 3 different computing platforms: (i) HP Cluster Platform Express 7000 enclosures with 36 blades BL 280c (Total 576 CPU cores), 24 GB RAM per blade; 8 controlling nodes HP DL 380 G6 with dual Intel

X5560 @ 2.8 GHz, 32 GB RAM (total 128 CPU cores); (Total CPU peak performance $3.2 Tflops$); (ii) HP ProLiant SL390s G7 4U servers with 16 NVIDIA Tesla M2090 graphic cards (total 8192 GPU cores with 10.64 Tflops in double precision); (iii) HP SL270s Gen8 4U server with 8 Intel Xeon Phi 5110P Coprocessors (total 480 cores, 1920 threads, with 8.088 Tflops of double-precision peak performance) (Table 1).

These 3 computing platforms are connected with 2 InfiniBand Switches and use 3 storage file systems with a total of 132 TB storage;

Taking into account the purchasing price of our equipment, we obtained the following: the cost of a CPU-core is 1.24 euro cents per hour; the cost of 1 GPU card NVIDIA M2090 is 11 euro cents per hour; the cost of 1 Intel Xeon Phi Coprocessor 5110P card is 7 euro cents per hour. The price of energy is assumed to be 8 euro cents per $1KWh$. The energy consumption of n (CPU nodes/Xeon Phi coprocessor) or GPU devices is denoted by W_0 (without jobs execution) while W_n denotes the energy consumption with running jobs. The difference $\Delta W_n = W_n - W_0$ is attributed to the computational workload being run. The function $F(T)$ that penalizes the algorithms is chosen in the following way: $F(T) = \exp(a|T/T_0 - 1|)$. Thus the full metric (1) minimizes the number of blades, GPU cards or Xeon Phi co-processors necessary for the completion of the task (running job) for a fixed time - T_0, leaving the energy cost close to the minimum. In our test $T_0 = 600s$ and $a = 1$. The constant a is chosen so that the results are more easily comparable for all cases. The number of realization of the quasirandom Sobol/Halton sequences is $N = 10$ million. In each table the optimal setup in terms of the metrics is underlined. Looking at the results we see that there is a difference between the Sobol and Halton sequences. For the CPU-based computations we observe that by turning on the hyper-threading in the case of Sobol' sequence one can achieve the same metrics with just 2 blades

Table 1. Test results for the QMC algorithm with Sobol sequence (upper part of the table) and Halton sequence (down part of the table) using CPU devices.

Blades/ Cores/HT	CPU time (s)	ΔW_n	Equipment cost: nCT	Energy cost: E	Total cost: E+nCT	Full metric $F(T) \times (E + nCT)$
1/8/2	1555.49	414	8.572	14.310	22.882	112.48
2/8/2	780.93	543	8.608	9.423	13.031	<u>24.38</u>
4/8/2	388.36	1081	8.561	9.329	17.890	25.46
8/8/2	195.33	1987	8.612	8.625	17.237	33.84
16/8/2	100.56	3689	8.867	8.244	17.111	39.33
1/8/2	1587.63	609	8.750	21.486	30.236	156.84
2/8/2	803.14	712	8.852	12.707	21.559	30.25
<u>4/8/2</u>	403.22	1130	8.889	10.125	19.014	<u>26.39</u>
8/8/2	222.39	1767	9.800	8.728	18.528	34.77
16/8/2	121.20	3624	10.687	9.761	20.448	45.42

Table 2. Test results for the QMC algorithm with Sobol sequence (upper part of the table) and Halton sequence (down part of the table) using CPU devices.

Blades/ Cores	CPU time (s)	ΔW_n	Equipment cost: nCT	Energy cost: E	Total cost: E+nCT	Full metric $F(T) \times (E + nCT)$
1/8	2401.52	272	6.62	14.52	21.14	415.48
2/8	1206.56	415	6.65	11.13	17.78	48.86
4/8	605.12	861	6.67	11.58	18.25	18.41
8/8	307.55	1624	6.78	11.10	17.88	29.11
16/8	160.12	3152	7.06	11.22	18.28	38.05
1/8	1587.63	372	6.89	20.66	27.55	652.73
2/8	803.14	435	6.92	12.13	19.05	56.74
4/8	403.22	951	6.96	13.35	20.31	21.41
8/8	222.39	1622	7.84	12.83	20.67	31.05
16/8	121.20	3073	7.93	12.28	20.21	40.71

Table 3. Test results for the QMC algorithm with Sobol sequence(upper part of the table) and Halton sequence (down part of the table) using GPU devices.

NVIDIA Tesla M2090	GPGPU time (s)	ΔW_n	Equipment cost: nCT	Energy cost: E	Total cost: E+nCT	Full metric $F(T) \times (E + nCT)$
1	893.24	156	2.73	3.10	5.83	9.50
2	456.32	316	2.79	3.20	5.99	7.61
4	246.15	638	3.01	3.49	6.50	11.72
8	155.88	1295	3.81	4.49	8.30	17.40
16	103.97	2419	5.08	5.59	10.67	24.39
1	803.85	158	2.46	2.82	5.28	7.42
2	411.01	307	2.51	2.80	5.31	7.29
4	228.86	614	2.80	3.12	5.92	10.99
8	142.07	1094	3.47	3.45	6.92	14.87
16	98.24	2494	4.80	5.44	10.24	23.65

instead of 4. For the Halton sequence in both cases 4 blades are to be used. The improvement from using hyper-threading is larger in the case of Sobol sequence and in general the Sobol sequence outperforms (Table 2).

In the case of GPGPU computations the optimal number of GPGPU devices used is found to be 2 and the Halton sequence outperforms slightly. In the case of Xeon Phi-based computations the best number of devices seems to be 4 and hyper-threading should be used with 120 threads (two times the number of physical cores). The Sobol's sequence outperforms slightly and the use of 180 threads (three times the number of physical cores) yields similar results

Table 4. Test results with Sobol sequence using Intel Xeon Phi coprocessors.

Threads	Xeon Phi 5110P	CPU time (s)	ΔW_n	Equipment cost: nCT	Energy cost: E	Total cost: E+nCT	Full metric $F(T) \times (E + nCT)$
60	1	2488.86	40	4.84	2.21	7.05	164.21
	2	1252.44	88	4.87	2.45	7.32	21.72
	4	658.17	196	5.12	2.87	7.99	8.80
	8	348.12	208	5.42	1.61	7.03	10.68
120	1	1703.87	56	3.31	2.12	5.43	34.18
	2	855.62	120	3.33	2.28	5.61	8.59
	4	462.58	196	3.60	2.01	5.61	7.05
	8	290.92	372	4.53	2.40	6.93	11.60
180	1	1547.83	68	3.01	2.34	5.35	25.97
	2	778.73	144	3.03	2.49	5.52	7.44
	4	460.33	300	3.58	3.07	6.65	8.39
	8	317.92	544	4.95	3.84	8.79	14.07
240	1	1579.59	72	3.07	2.53	5.60	28.66
	2	848.10	164	3.30	3.09	6.39	9.66
	4	501.91	328	3.90	3.66	7.56	8.90
	8	360.89	596	5.61	4.78	10.39	15.48

Table 5. Test results with Halton sequence using Intel Xeon Phi coprocessors.

Threads	Xeon Phi 5110P	CPU time (s)	ΔW_n	Equipment cost: nCT	Energy cost: E	Total cost: E+nCT	Full metric $F(T) \times (E + nCT)$
60	1	2591.70	48	5.04	2.76	7.80	215.64
	2	1325.97	96	5.16	2.83	7.99	26.79
	4	665.62	185	5.18	2.74	7.92	8.82
	8	345.60	376	5.38	2.89	8.26	12.62
120	1	1840.95	60	3.58	2.45	6.03	47.70
	2	928.91	120	3.61	2.48	6.09	10.54
	4	476.91	268	3.71	2.84	6.55	8.04
	8	270.69	508	4.21	3.06	7.27	12.59
180	1	1705.13	68	3.32	2.58	5.90	37.16
	2	911.06	144	3.54	2.92	6.46	10.85
	4	486.71	300	3.79	3.24	7.03	8.49
	8	276.25	556	4.30	3.41	7.71	13.22
240	1	1998.41	76	3.89	3.38	7.27	74.67
	2	1069.23	156	4.16	3.71	7.87	17.18
	4	532.00	316	4.14	3.74	7.88	8.81
	8	314.80	576	4.90	4.03	8.93	14.36

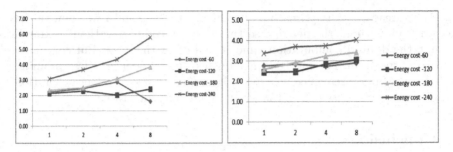

Fig. 1. Comparing the energy costs using Sobol (left picture) and Halton (right picture) sequences in the QMC algorithm on the Xeon Phi coprocessor platform.

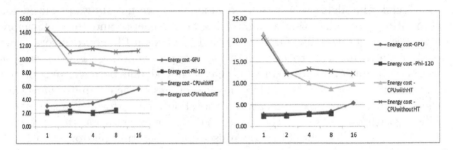

Fig. 2. Comparing the energy costs using Sobol (left picture) and Halton (right picture) sequences in the QMC algorithm on different computing platforms.

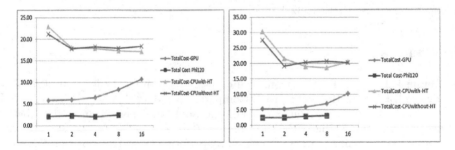

Fig. 3. Comparing the total costs using Sobol (left picture) and Halton (right picture) sequences in the QMC algorithm on different computing platforms.

(Table 3). In all cases of accelerator-based computations (using GPGPU and Xeon Phi) we notice that the use of all devices is not optimal. In part this is due to the relatively high initial startup overhead for the CUDA and MPI implementations. This strengthens the case for Grid computing on these devices, in the sense that it looks beneficial to have a distribution of the computational devices among more than one computational job for maximum efficiency. Figure 1 shows a comparison between energy costs for the Sobol and Halton sequences, where one can evaluate the optimal number of threads to be used. It seems that 120 threads is the optimal number here. In Figs. 2 and 3 we show only the cost function without the penalty, so that one can compare without taking into account the desired computation time. In all cases the Xeon Phi coprocessors seem to outperform. For the Halton sequence this difference is less prominent, while for the Sobol' sequence it is much larger. Of course, if newer generation of graphics cards were used the situation may be different (Table 4).

5   Conclusion

The results of our study show the need of new metrics in order to demonstrate the advantages of different HPC platforms (with GPU cards, Xeon Phi coprocessors, etc.) Although the algorithms under consideration are not fully optimized for the computing platforms, they demonstrate better energy and total efficiency when we use accelerators like GPU cards and Intel Xeon Phi coprocessors. New computing platforms lead to new challenges in programming. That is why it is necessary to optimize codes (in some cases they should be rewritten) of the existing algorithms in order to exploit the advantages of the computer hardware with accelerators. More precise measurements should be performed in the future and the algorithms describing the Sobol and Halton generators, that produce quasirandom sequences, should be optimised for computing platforms with co-processors (Table 5).

Acknowledgments. This work was supported by the National Science Fund of Bulgaria under Grant DFNI-I02/8.

References

1. Demmel, J., et al.: Perfect strong scaling using no additional energy. In: Proceedings of IEEE 27th IPDPS13, IEEE Computer Society (2013)
2. Meswani, M., et al.: Modeling and predicting application performance on hardware accelerators. Int. J. High Perform. Comput. (2012)
3. Bekas, C., Curioni, A.: A new energy aware performance metric. Comput. Sci. Res. Dev. **25**, 187–195 (2010). doi:10.1007/s00450-010-0119-z. Springer
4. Bekas, C., Curioni, A., Fedulova, I.: Low cost high performance uncertainty quantification. In: Workshop on HPC finance, SC 2009, Portland, OR, USA (2009)
5. Atanassov, E., et al.: Tuning for scalability on hybrid HPC cluster. In: Slavova, A. (ed.) Mathematics in Industry, pp. 64–77. Cambridge Scholar Publishing, Newcastle upon Tyne (2014)
6. Atanassov, E., et al.: Energy aware performance study for a class of computationally intensive MC algorithms. J. Comp. Math. Appl. (2015, accepted). Elsevier
7. Atanassov, E., et al.: Ultra-fast semiconductor carrier transport simulation on the grid. Sci. Int. J. Par. Dist. Comp. **11**(2), 137–147 (2010). SCPE
8. Atanassov, E., Karaivanova, A., Ivanovska, S.: Tuning the generation of sobol sequence with owen scrambling. In: Lirkov, I., Margenov, S., Waśniewski, J. (eds.) LSSC 2009. LNCS, vol. 5910, pp. 459–466. Springer, Heidelberg (2010)

9. Atanassov, E.I., Durchova, M.K.: Generating and testing the modified Halton sequences. In: Dimov, I., Lirkov, I., Margenov, S., Zlatev, Z. (eds.) NMA 2002. LNCS, vol. 2542, pp. 91–98. Springer, Heidelberg (2003)
10. Nedjalkov, M., Gurov, T.V., Kosina, H., Vasileska, D., Palankovski, V.: Femtosecond evolution of spatially inhomogeneous carrier excitations part I: kinetic approach. In: Lirkov, I., Margenov, S., Waśniewski, J. (eds.) LSSC 2005. LNCS, vol. 3743, pp. 149–156. Springer, Heidelberg (2006)
11. Gurov, T.V., Atanassov, E.I., Dimov, I.T., Palankovski, V.: Femtosecond evolution of spatially inhomogeneous carrier excitations part II: stochastic approach and grid implementation. In: Lirkov, I., Margenov, S., Waśniewski, J. (eds.) LSSC 2005. LNCS, vol. 3743, pp. 157–163. Springer, Heidelberg (2006)
12. Sobol, I., Asotsky, D., Kreinin, A., Kucherenko, S.: Construction and comparison of high-dimensional Sobol generators. Wilmott J. 2011(56), 64–79 (2011)
13. EEHPCWG_PowerMeasurementMethodology.pdf (2015). http://www.green500.org

Towards RBF Interpolation on Heterogeneous HPC Systems

Gundolf Haase[1]([⊠]), Dirk Martin[2,3], and Günter Offner[3]

[1] Institute for Mathematics and Scientific Computing,
University of Graz, Graz, Austria
gundolf.haase@uni-graz.at
http://www.uni-graz.at/~ghaase
[2] VRVis Research Center, Vienna, Austria
[3] AVL List GmbH, Graz, Austria

Abstract. We present a general approach for the parallelization of the interpolation with radial basis functions (RBF) on distributed memory systems, which might use various shared memory hardware as accelerator for the local subtasks involved. The calculation of an interpolant in general requires a global dense system to be solved. Iterative methods need appropriate preconditioning to achieve reasonable iteration counts. For the shared memory approach we use a special Krylov subspace method, namely the FGP algorithm. Addressing the distributed task we start with a simple block-Jacobi iteration with each block solved in parallel. Adding a coarse representation leads to a two-level block-Jacobi iteration with much better iteration counts and a wider applicability.

1 Introduction

The numerical treatment of many simulation problems in science and industry has to handle changing computational domains originating from the given PDE (system) or from design variables in optimization and optimal control problems. A re-meshing in case of a direct problem, i.e. the PDE, is possible but rather inefficient because all mesh-dependent data have to be reallocated and recomputed. In the context of an optimization problem the re-meshing would destroy the continuous differentiability of the objective functional and therefore we are forced to apply a mesh deformation instead.

We use interpolation with radial basis functions (RBF interpolation) for mesh deformation as proposed in [4]. In this paper we focus on the parallelization of RBF interpolation with its application for mesh deformation in view. Within this context parallelization covers shared memory and distributed memory parallel computing.

Calculating an RBF interpolant requires the solution of a dense system of linear equations. There have been several achievements to overcome the ill-conditioning of the linear system [2,3,16]. Nevertheless a direct solution of the system is inhibited, if the problem size exceeds certain limits, thus iterative methods have to be used. Due to the ill-conditioning of the linear system, some

© Springer International Publishing Switzerland 2015
I. Lirkov et al. (Eds.): LSSC 2015, LNCS 9374, pp. 182–190, 2015.
DOI: 10.1007/978-3-319-26520-9_19

preconditioning has to be applied. One way is to use domain decomposition methods [2,13,19]. We employ a Krylov-subspace method that uses approximate Lagrange functions as preconditioner, namely the Faul-Goodsell-Powell (FGP) algorithm [9,10] for a shared memory solution. Our reasonable approach to a distributed memory solution is applying well known domain decomposition methods.

Due to our application of deforming given computational meshes our data distribution is predetermined. In particular we work on distributed finite volume discretizations with one cell-layer overlap.

The remaining paper is organized as follows. Section 2 gives a short introduction to RBF interpolation. We feature the FGP algorithm and our approach to an efficient shared memory implementation in Sect. 3. Section 4 covers the distributed memory approach. We present some numerical results in Sect. 5 and annotate some conclusions in Sect. 6.

2   Interpolation with Radial Basis Functions

This introduction to RBF interpolation closely follows [6]. Further analysis and treatment of basis functions with compact support can be found in [18].

Given is a set of points $\mathcal{X} = \{x_i\}_{i=1}^{N}$ in a domain $\Omega \subseteq \mathbb{R}^d$. A set of associated real function values $f_i = f(x_i)$ is assigned to these points. The function f is usually unknown, but its existence is postulated for the reasonableness of the interpolation task. Sought is an approximating function $s : \Omega \rightarrow \mathbb{R}$ by interpolation. Restricted on the set $\mathcal{X}$ we request the interpolation condition

$$s|_{\mathcal{X}} = f|_{\mathcal{X}}. \tag{1}$$

In the context of RBF interpolation we seek for an interpolant of the form

$$s(x) = \sum_{i=1}^{N} \lambda_i \phi(\|x - x_i\|) + p(x), \quad \lambda_i \in \mathbb{R}, p \subset \mathbb{P}^M. \tag{2}$$

The polynomial term p is required for the existence and uniqueness of a solution. The required degree M for the existence of a solution depends in the choice of the basis function ϕ, see [6]. We choose $\phi = \sqrt{r^2 + c^2}, c \in \mathbb{R}$ and $M = 0$ for our numerical examples in Sect. 5.

If the basis function requires a polynomial term ($M \geq 0$) then the given set of points $\mathcal{X}$ has to be unisolvent w.r.t. polynomials of degree M, i.e., $p|_{\mathcal{X}} = 0 \Rightarrow p \equiv 0$ for polynomials $p \in \mathbb{P}^M$.

Requiring the interpolation condition (1) in all given points and demanding a side condition on the coefficients of the polynomial term leads to a system of linear equations for the determination of the coefficients λ and π:

$$\sum_{i=1}^{N} \lambda_i \phi(\|x_i - x_k\|) + \sum_{j=1}^{M} \pi_j p_j(x_k) = f(x_k), \qquad 1 \le k \le N,$$

$$\sum_{i=1}^{N} \lambda_i p_l(x_i) = 0, \qquad 1 \le l \le M, \qquad (3)$$

or, in short notation

$$\begin{pmatrix} \Phi & \Pi \\ \Pi^\top & 0 \end{pmatrix} \begin{pmatrix} \boldsymbol{\lambda} \\ \boldsymbol{\pi} \end{pmatrix} = \begin{pmatrix} \boldsymbol{f} \\ \boldsymbol{0} \end{pmatrix}. \qquad (4)$$

3 Accelerated Computation of Interpolants

Analysis of the properties of the linear equations system (4) yields that a direct solution is inhibited, if the number of interpolation points exceeds certain limits. Therefore, we employ an iterative method [9,10] to solve (4).

The FGP algorithm is a Krylov-subspace method. The implemented version of the algorithm can be applied to interpolations with basis functions where constant polynomial terms are sufficient ($M = 0$).

Let X denote the functional space spanned by functions of the form (2). The basis for the algorithm is the semi-inner product

$$\langle s, t \rangle_\phi = -\boldsymbol{\lambda}^\top \Phi \boldsymbol{\mu} \qquad \text{for } s, t \in X$$

with $s(x) = \sum_{i=1}^{N} \lambda_i \phi(\|x - x_i\|) + \alpha$ for $\lambda_i, \alpha \in \mathbb{R}$ and $t(x) = \sum_{i=1}^{N} \mu_i \phi(\|x - x_i\|) + \beta$ for $\mu_i, \beta \in \mathbb{R}$, induced by the radial basis function ϕ. Φ denotes the associated kernel to ϕ. Other basis functions may require a different sign at the definition of the semi-inner product. The semi-inner product $\langle \cdot, \cdot \rangle_\phi$ induces a semi-norm

$$|s|_\phi = \langle s, s \rangle^{1/2} = \left(-\boldsymbol{\lambda}^\top \Phi \boldsymbol{\lambda} \right)^{1/2}.$$

The FGP algorithm uses a linear operator $A : X \to X$. A is chosen such, that for all iterations k the resulting searching direction lies in the subspace of X that is spanned by the functions $A^l s^\star$, $l = 1, \ldots, k$, where $s^\star$ denotes the sought interpolant. In addition to this A performs some preconditioning. A is an approximation of the optimal preconditioning operator A^{opt}. The operator A^{opt} can be derived from the interpolation tasks

$$\hat{u}_j(x) = \sum_{i=1}^{N} \zeta_{j,i} \phi(x - x_i) + \beta, \quad \text{for } x \in \Omega, j = 1, \ldots, N,$$

due to the Lagrange conditions $\hat{u}_j(x_i) = \delta_{ij}$, $i, j = 1, \ldots, N$, where δ_{ij} denotes the Kronecker-delta.

For the construction of A the functions $\hat{u}_j$, $j = 1, \ldots, N$ are determined by solving the interpolation tasks on subsets of $\mathcal{X}$, that contain not more than q

interpolation centers. Generally the relation $q \ll N$ holds. These subsets are called $\mathcal{L}$-sets.

The main computational costs of an implementation lie in the construction of the $\mathcal{L}$-sets and the calculation of matrix-vector products with dense matrices. The authors in [11] show an efficient implementation including a modified setup stage and an approximated matrix-vector product by using a fast multipole method. We use an octree as spatial hierarchy to efficiently construct the $\mathcal{L}$-sets during the setup phase. We also use the same octree structure to employ a multipole-method [1]. The far field series expansion for the multiquadric basis function is described in [7].

Our approach to a shared memory parallelization of the matrix-vector product on CPUs using the OpenMP API [15] applies task based parallelization on octree box level.

We use a slightly different parallelization scheme for acceleration of the matrix-vector product on graphics processors using the CUDA programming model [8]. For the direct calculation of the matrix-vector multiplication the boundary-condition nodes (bc nodes) are organized in blocks. A block of bc nodes correspond to a thread-block on the GPU. Within each block the result for each bc node is calculated by a single thread. Since the hierarchical structure that is used for the multipole approximation is an octree, the far field series expansion of each box in the 'evaluation region' (see [1]) of an octree box has to be evaluated for each node within this box. The thread-blocks on the GPU correspond to the octree boxes. Each thread on the GPU evaluates the polynomial for one node within the box. All threads read the (common) series coefficients to shared memory. If an octree box holds more nodes than the number of threads that are started within a block the box is virtually split in boxes that do not hold more nodes than threads-per-block each. After the completion of our work we learned about the earlier results in [12].

4   Parallel Computation of Interpolants

The referred domain decomposition methods for the computation of RBF interpolants [2,13,19] can be seen as preconditioning methods according to the classification in [17]. Thereby the solution of a large system over the domain Ω is subdivided into P smaller problems over the subdomains $\Omega_s, s = 1, \ldots, P$. The solutions of the smaller problems are used to construct a preconditioner for the solution of the large system.

Our general approach is to use the subdivision predetermined by the application and to apply the methods from Sect. 3 to solve the regarding subproblems. Our given subdivision of the domain Ω consists of overlapping subdomains Ω_s, i.e., $\Omega = \bigcup_{s=1}^{P} \Omega_s$ with $\Omega_i \cap \Omega_j \neq \emptyset$, $i \neq j$. Further let $\tilde{\Omega}_s$ denote the appropriate non-overlapping subdomains such that $\Omega = \bigcup_{s=1}^{P} \tilde{\Omega}_s$ holds with $\Omega_i \cap \Omega_j = \emptyset$, $i \neq j$.

Let R_s denote the restriction matrix projecting a vector $\boldsymbol{x}$ from domain Ω onto a vector $\boldsymbol{x}_s = R_s \boldsymbol{x}$ on subdomain Ω_s. Similarly $\tilde{R}_s$ denotes the restriction matrix which restricts a vector $\boldsymbol{x}$ from domain Ω onto a vector $\boldsymbol{x}_s$ on subdomain $\tilde{\Omega}_s$. For the non-overlapping subdivision a vector over the domain Ω can be composed by applying the transposed mapping operations $\tilde{R}^\top$ on the local vectors $\boldsymbol{x}_s$

$$\boldsymbol{x} = \sum_{s=1}^{P} \tilde{R}^\top \boldsymbol{x}_s.$$

4.1 Block-Jacobi Iteration

The authors in [19] show the utilization of a Restricted Additive Schwarz method (RASM) as preconditioner for the calculation of an RBF interpolant. The proposed method is applicable for positive (negative) definite basis functions, thus no polynomial term is required in (2).

The n-th iteration of the RASM to calculate the solution of (4) is

$$\boldsymbol{\lambda}^{(n+1)} = \boldsymbol{\lambda}^{(n)} + \sum_{i=1}^{P} \tilde{R}_i^\top \Phi_i^{-1} R_i \left(\boldsymbol{f} - \Phi\boldsymbol{\lambda}^{(n)} \right), \tag{5}$$

where Φ_i denotes the system matrix of the subproblem restricted to the domain Ω_i. This can be extended to conditionally positive (negative) basis functions. Let Π_i denote the respective matrix blocks of the restricted system.

Algorithm 1. Block-Jacobi iteration

$\boldsymbol{r}_i^{(n)} \longleftarrow R_i \left(\boldsymbol{f} - \Phi\boldsymbol{\lambda}^{(n)} + \Pi\boldsymbol{\pi}^{(n)} \right)$

for *all domains s* **do**

$\quad$ Solve $\begin{pmatrix} \Phi_s & \Pi_s \\ \Pi_s^\top & 0 \end{pmatrix} \begin{pmatrix} \hat{\boldsymbol{\lambda}}_s \\ \boldsymbol{\pi}_s \end{pmatrix} = \begin{pmatrix} \boldsymbol{r}_s \\ \boldsymbol{0} \end{pmatrix}$

$\hat{\boldsymbol{\lambda}}^{(n)} \longleftarrow \sum_{s=1}^{P} \tilde{R}^\top \hat{\boldsymbol{\lambda}}_s$

Correct $\hat{\boldsymbol{\lambda}}^{(n)}$ such that it fulfills (3) $\longrightarrow \boldsymbol{\lambda}^{(n+1)}$

Compute $\boldsymbol{\pi}^{(n+1)}$

Depending on the size of the subproblem we either use a direct method or employ the FGP algorithm to find a solution for the local subproblems. The authors in [13] show that it is sufficient to calculate an approximative solution of the local subproblem requiring an error relative to the current residual in each iteration.

4.2 Two-Level Block-Jacobi Iteration

The authors in [2] sketched a simplified two-level domain decomposition method for RBF interpolation fitting. The two-level algorithm represented on page 180 requires a coarse grid representation $\mathcal{Y}$ consisting of interpolation centers from each subdomain $\tilde{\Omega}_s$. Let $R_{\mathcal{Y}}$ denote the restriction matrix that restricts a global

Algorithm 2. Two level block-Jacobi iteration

$$r_i^{(n)} \longleftarrow R_i \left(f - \Phi\lambda^{(n)} + \Pi\pi^{(n)} \right)$$

for *all domains s* **do**

$$\text{Solve} \begin{pmatrix} \Phi_s & \Pi_s \\ \Pi_s^\top & 0 \end{pmatrix} \begin{pmatrix} \hat{\lambda}_s \\ \pi_s \end{pmatrix} = \begin{pmatrix} r_s \\ 0 \end{pmatrix}$$

$$\hat{\lambda}^{(n)} \longleftarrow \sum_{s=1}^{P} \tilde{R}^\top \hat{\lambda}_s$$

Correct $\hat{\lambda}^{(n)}$ such that it fulfills (3) $\longrightarrow$ $\check{\lambda}^{(n)}$

Evaluate the residual $\rho_y^{(n)} \longleftarrow R_y \left(f - \Phi\check{\lambda}^{(n)} \right)$.

$$\text{Solve} \begin{pmatrix} \Phi_y & \Pi_y \\ \Pi_y^\top & 0 \end{pmatrix} \begin{pmatrix} \lambda_y \\ \pi_y \end{pmatrix} = \begin{pmatrix} \rho_y \\ 0 \end{pmatrix}.$$

$$\lambda^{(n+1)} \longleftarrow \check{\lambda}^{(n)} + R_y^\top \lambda_y,$$

$$\pi^{(n+1)} \longleftarrow \pi_y$$

vector to the coarse grid interpolation centers and indices y denote restricted vectors and matrices accordingly. We again use either a direct method or the FGP algorithm to solve the subproblems.

5   Numerical Examples

In this section we present timing results for performance tests of the method described in Sect. 3 and iteration counts for the parallel computation of interpolants described in Sect. 4.

The test system is equipped with two Intel Xeon E5-2450 processors with 8 cores each. The OpenMP parallelized C++ code uses 16 threads. The used GPGPU accelerator is an Nvidia Tesla K20Xm. Table 1 lists timings for the brute force computation and an approximation using a multipole method of a single matrix-vector product.

Table 1. Timings for a single matrix-vector product in seconds.

# bc nodes	Brute force		Multipole method			
	CPU [s]	GPU [s]	CPU setup	CPU eval	GPU setup	GPU eval
602	1.09E-02	5.61E-04	2.35E-02	2.67E-02	1.15E-02	2.97E-03
1562	9.01E-03	1.34E-03	2.03E-02	5.19E-02	1.31E-02	3.12E-03
3542	1.06E-02	2.42E-03	2.49E-02	1.43E-01	1.98E-02	3.40E-03
9902	7.03E-02	6.32E-03	2.30E-02	3.48E-01	3.42E-02	4.49E-03
39802	5.59E-01	7.40E-02	8.00E-02	5.54E-01	1.65E-01	2.39E-02
89702	2.84E+00	3.91E-01	1.71E-01	2.15E+00	3.44E-01	4.77E-02

Table 2. Iteration count for the parallel test case.

N	Block-Jacobi			Two-level block-Jacobi							
				$|\mathcal{Y}| = 200$				$|\mathcal{Y}| \approx N/8$			
	# of domains			# of domains				# of domains			
N	2	4	8	2	4	8	16	2	4	8	16
416	12	23	39	3	4	4	5	5	6	8	9
1832	36	178	–	5	6	7	8	5	6	7	8
7472	–	–	–	9	11	13	13	5	6	7	7
30080	–	–	–	30	29	40	39	5	5	6	6

The GPU version is faster by one order of magnitude at least for both methods and the multipole method becomes superior with increasing problem size. Observing the multipole evaluation on the GPU for problem sizes below 10.000 nodes reveals the kernel invocation overhead. In contrast to the CPU implementation the octree boxes in the multipole method have to padded with additional zeros in order to achieve coalesce memory access on the GPU. The fill rate of these octree boxes changes with the tree depth and influences directly the run time such that the observed speed-up varies. This can be seen in Table 1 for the multipole evaluation wherein the GPU timings follow closely the expected run time $\mathcal{O}(N \log N)$ but the CPU performs even better, i.e., on 39802 nodes. The double precision peak performance and memory bandwidth of the Tesla K20Xm is 5 times better than those of the two Xeon processors. This factor is observed for the brute force approach while the multipole evaluation on GPU is considerably better vectorized than on CPU. A detailed discussion of the results can be found in [14].

Table 2 compares the iteration counts for the block-Jacobi and the two-level block-Jacobi preconditioning described in Sect. 4. We imposed test function F3 as in [5] as boundary conditions on N nodes distributed over a sphere. We test the effect of a fixed-size coarse grid ($|\mathcal{Y}| = 200$) versus an adjusted coarse grid size ($|\mathcal{Y}| \approx N/8$), arising as the natural choice for a prospective multi-level preconditioning based on an octree hierarchy. The iteration count indicates clearly that the two-level method is superior even though the parallelization of the coarse grid is still a future work. We expect that the coarse grid parallelization combined with a further recursive coarsening will result in a fastest preconditioner.

6   Conclusion

We showed that an implementation of the FGP algorithm can be adapted to exploit the massive parallelism of GPGPU accelerator cards. The FGP algorithm evinces to be applicable as solution method for the arising sub-problems when the original problem is distributed. Extending the resulting block-Jacobi preconditioning with an additional coarse block results in a two-grid method with constant iteration count. The ideal coarsening factor $N/8$ indicates the reasonableness of a further multilevel preconditioning for the RBF-interpolation based on the (already existing) octree hierarchy in $\mathbb{R}^3$.

This first application with the preconditioner in a simple iteration will be extended to a Krylov subspace method as outer iteration in future. Therein our preconditioner has to be adapted in order to fulfill the requirements of the respective method.

Further research is to utilize a wider set of programming standards, such as OpenACC and OpenMP 4.0, for many-core programming, which hold out the prospect of an incomplex transition to different accelerator hardware (e.g. Intel©Xeon Phi^{TM}).

References

1. Beatson, R.K., Greengard, L.: A short course on fast multipole methods. In: Ainsworth, M., Levesley, J., Light, W., Marletta, M. (eds.) Wavelets, Multilevel Methods and Elliptic PDEs, pp. 1–37. Oxford University Press, Oxford (1997)
2. Beatson, R.K., Light, W., Billings, S.: Fast solution of the radial basis function interpolation equations: domain decomposition methods SIAM. J. Sci. Comput. **22**(5), 1717–1740 (2001)
3. Beatson, R., Levesley, J., Mouat, C.: Better bases for radial basis function interpolation problems. Comput. Appl. Math. **236**, 434–446 (2011)
4. de Boer, A., van der Schoot, M.S., Bijl, H.: Mesh deformation based on radial basis function interpolation. Comput. Struct. **85**(11–14), 784–795 (2007)
5. Bozzini, M.T., Rossini, M.F.: Multivariate approximation and interpolation with applications. In: Testing Methods for 3D Scattered Data Interpolation (Almunecar, 2001), pp. 111–135. Acad. Cienc. Exact.Fs.Qum. Nat., Zaragoza (2002)
6. Buhmann, M.: Radial Basis Functions: Theory and Implementations, Cambridge Monographs on Applied and Computational Mathematics. Cambridge University Press, New York (2003)
7. Cherrie, J.B., Beatson, R.K., Newsam, G.N.: Fast evaluation of radial basis functions: methods for generalized multiquadrics in Rn. SIAM J. Sci. Comput. **23**(5), 1549–1571 (2001)
8. NVIDIA Corporation.: CUDA programming guide 6.5 (2014). http://docs.nvidia.com/cuda/cuda-c-programming-guide/index.html
9. Faul, A.C., Powell, M.J.D.: Krylov Subspace Methods for Radial Basis Function Interpolation. University of Cambridge, DAMP Cambridge (1999)
10. Faul, A.C., Goodsell, G., Powell, M.J.D.: A Krylov subspace algorithm for multiquadric interpolation in many dimensions. IMA J. Numer. Anal. **25**(1), 1–24 (2005)
11. Gumerov, N., Duraiswami, R.: Fast radial basis function interpolation via preconditioned krylov iteration. SIAM J. Sci. Comput. **29**(5), 1876–1899 (2007)
12. Gumerov, N., Duraiswami, R.: Fast multipole methods on graphics processors. J. Comput. Phys. **227**, 8290–8313 (2008)
13. Ling, L., Kansa, E.J.: Preconditioning for radial basis functions with domain decomposition methods. Math. Comput. Model. **40**(13), 1413–1427 (2004)
14. Martin, D., Haase, G.: Interpolation with radial basis functions on GPGPUs using CUDA. Technical Report SFB-Report 2014-04, SFB MOBIS, University of Graz (2014)

15. OpenMP Architecture Review Board.: OpenMP Application Program Interface, Version 3.1 (2011). http://www.openmp.org/mp-documents/OpenMP3.1.pdf
16. Powell, M.J.D.: Some algorithms for thin plate spline interpolation to functions of two variables. Adv. Comput. Math. **4**, 303–319 (1993)
17. Smith, B.F., Bjørstad, P.E., Gropp, W.D.: Domain decomposition: Parallel Multilevel Methods for Elliptic Partial Differential Equations. Cambridge University Press, New York (1996)
18. Wendland, H.: Scatterred Data Approximation: Cambridge Monographs on Applied and Computational Mathematics. Cambridge University Press, New York (2010)
19. Yokota, R., Barba, L.A., Knepley, M.G.: PetRBF A parallel O(N) algorithm for radial basis function interpolation with Gaussians. Comput. Method. Appl. Mech. Eng. **199**(25–28), 1793–1804 (2010)

On the Relation Between Matrices and the Greatest Common Divisor of Polynomials

Nikolai L. Manev[1,2](✉)

[1] University of Structural Engineering & Architecture
"Lyuben Karavelov", Sofia, Bulgaria
[2] Institute of Mathematics and Informatics, BAS, Sofia, Bulgaria
nlmanev@math.bas.bg

Abstract. Following the Barnett's approach to gcd(a(x),b(x)) based on the use of companion matrix we develop an extended algorithm that gives effectively $d(x)$, $u(x)$, $v(x)$, $a_1(x)$ and $b_1(x)$, where $a_1(x) = a(x)/d(x), b_1(x) = b(x)/d(x)$ and $d(x) = u(x)a(x) + v(x)b(x)$. The algorithm is suitable for parallel realization on GPU, FPGA, and smart cards.

Keywords: Greatest common divisor of polynomials · Reduced row echelon form · Gauss elimination

1 Introduction

Let $\mathbb{F}$ be a field or a factorial ring (an integral domain, where each nonzero element admits a unique decomposition into a product of irreducible elements). Any two polynomials $a(x), b(x) \in \mathbb{F}[x]$ have a greatest common divisor (**gcd**): $d(x) = (a(x), b(x))$, that is unique up to an invertible element of $\mathbb{F}$. In the case when $\mathbb{F}$ is a field $d(x)$ is chosen to be a monic polynomial. For $\mathbb{F}$ field and for some rings, like $\mathbb{Z}$ (the ideal $(d) = (a) + (b)$), the following **relation of Bézout** holds:

$$u(x)a(x) + v(x)b(x) = d(x), \qquad (1.1)$$

where $u(x), v(x)$, are uniquely determined if $\deg u(x) < \deg b_1(x)$, $\deg v(x) < \deg a_1(x)$, $a_1(x) = a(x)/d(x)$, and $b_1(x) = b(x)/d(x)$.

Greatest common divisor and Bézout's relation plays an important role in many areas of mathematics as differential equations, linear multivariable control systems, solving algebraic equations (e.g. simple and multiple roots can be separated by $\mathbf{gcd}(f(x), f'(x))$ and $f_1(x)$), etc. For coding theory and cryptography (e.g. Euclid algorithm for decoding cyclic codes and finding the minimal linear generators of a sequence) polynomials $u(x)$ and $v(x)$ are sometimes more interesting.

The classical approach to finding $d(x)$, $u(x)$ and $v(x)$ is based on euclidean polynomial division or on pseudo-division in the case of $\mathbb{F}$ integral domain.

This work was partially supported by the National Science Fund of Bulgaria under Grant DFNI-I02/8.

© Springer International Publishing Switzerland 2015
I. Lirkov et al. (Eds.): LSSC 2015, LNCS 9374, pp. 191–199, 2015.
DOI: 10.1007/978-3-319-26520-9_20

Much more effectively $d(x)$, $u(x)$ and $v(x)$ can be determined by transforming (with elementary row operations) a suitable matrix to its reduced echelon form. The relation between **gcd** and matrices dates back to Sylvester (1840), but the Gaussian elimination approach is manly due to Stephen Barnett ([1,2]). Of course many mathematicians have worked in this direction (MacDuffee [6], Laidacker [3], Gonzales-Vega [4,5], etc.), but the results of Barnett and the others have been mainly motivated by problems in linear control theory and other areas, where the Bézout's relation is not much interesting. However, more attention is payed to the polynomials $u(x)$ and $v(x)$ by some topics in algebra, coding theory and cryptography.

The next section contains the necessary definitions and results. In Sect. 3 we describe the algorithm, which is proved in the forth section. In the last section we discuss realizations and further works.

2   Preliminaries

Let us first agree on notations. Matrices and vectors are denoted by bold capital and small letters, respectively. $\mathbf{I}_n$ is the $n \times n$ identity (unit) matrix having 1's along its principle diagonal and zeros everywhere else. $\mathbf{F}_n$ denotes the matrix obtained from $\mathbf{I}_n$ by flipping, i.e., having 1's along its secondary diagonal. Polynomials are represented by vectors of fixed length (e.g., n) whose rightmost element is the constant term and the leftmost nonzero entry is the leading coefficient of the polynomial.

Definition 1. *Let* $a(x) = x^n + a_1 x^{n-1} + \cdots + a_{n-1} x + a_n \in \mathbb{F}[x]$ *be a polynomial of the variable* x *with coefficients in* $\mathbb{F}$. *The following* $n \times n$ *matrix:*

$$
\mathbf{C}_a = \begin{pmatrix} -a_1 & -a_2 & \ldots & -a_{n-1} & -a_n \\ 1 & 0 & \ldots & 0 & 0 \\ \vdots & \vdots & \vdots & \vdots & \vdots \\ 0 & 0 & \ldots & 0 & 0 \\ 0 & 0 & \ldots & 1 & 0 \end{pmatrix}
$$

is called **companion matrix** *associated with* $a(x)$.

The companion matrix is the matrix of cyclic shifting in the factor ring $\mathbb{F}[x]/a(x)$, i.e., the matrix corresponding to the map

$$
\chi : p(x) \longrightarrow x\, p(x) \quad (\mathrm{mod}\ a(x)),
$$

in the basis $\{x^{n-1}, x^{n-2}, \ldots, 1\}$ ($\chi(p) = \mathbf{p}\mathbf{C}_a$, where $\mathbf{p}$ is the vector representing $p(x)$).

The rank of $\mathbf{C}_a$ is n when $a_n \neq 0$ and it is straightforward to check that its characteristic polynomial is

$$
f_{C_a}(x) = \det(\mathbf{C}_a - x\mathbf{E}) = (-1)^n a(x).
$$

Let $b(x) = b_0 x^m + b_1 x^{m-1} + \cdots + b_{m-1} x + b_m \in \mathbb{F}[x]$. Recall that the value of $b(x)$ at $\mathbf{A}$ is the matrix

$$b(\mathbf{A}) = b_0 \mathbf{A}^m + b_1 \mathbf{A}^{m-1} + \cdots + b_{m-1} \mathbf{A} + b_m \mathbf{E}$$

and according to the Cayley-Hamilton theorem $a(\mathbf{C}_a) = \mathbf{O}$.

In the case when $\mathbb{F}$ is not a field and $a_0 \neq 1$ the **generalized companion matrix** (where 1's are replaced by a_0) is used instead (see [4]). In this paper we restrict ourselves only to the case when $\mathbb{F}$ is a field.

Let

$$\mathbf{b} = (\underbrace{0 \ldots 0}_{n-m-1}\, b_0\, b_1\, \ldots\, b_m)$$

be the vector corresponding to the polynomial $b(x)$. The multiplication of $\mathbf{b}$ by $\mathbf{C}_a$ is equivalent to cyclic shift of $\mathbf{b}$ to the left, i.e.,

$$\mathbf{b}^{(i)} = \mathbf{b}\mathbf{C}_a^i = (\,\underbrace{0 \ldots 0}_{n-m-1-i}\, b_0\, b_1\, \ldots\, b_m\, \underbrace{0 \ldots 0}_{i}), \qquad i = 1, 2, \ldots, n - m - 1$$

Let us denote

$$\mathbf{w}^{(0)} = \mathbf{w} = \mathbf{b}^{(n-m-1)} = (b_0\, b_1\, \ldots\, b_m\, \underbrace{0 \ldots 0}_{n-m-1}),$$

$$\mathbf{w}^{(k)} = (w_1^{(k)}, w_2^{(k)}, \ldots, w_n^{(k)}) \stackrel{def}{=} \mathbf{w}\mathbf{C}_a^k = \mathbf{w}^{(k-1)}\mathbf{C}_a, \qquad k = 1, \ldots, m.$$

Proposition 1 (Barnett [2]). If $b(x) = b_0 x^m + b_1 x^{m-1} + \cdots + b_{m-1} x + b_m \in \mathbb{F}[x]$, $b_0 \neq 0$, then

$$b(\mathbf{C}_a) = \begin{pmatrix} \mathbf{w}^{(m)} \\ \mathbf{w}^{(m-1)} \\ \vdots \\ \mathbf{w}^{(1)} \\ \mathbf{w} \\ \vdots \\ \mathbf{b}^{(1)} \\ \mathbf{b} \end{pmatrix} = \begin{pmatrix} \mathbf{w}\mathbf{C}_a^m \\ \mathbf{w}\mathbf{C}_a^{m-1} \\ \vdots \\ \mathbf{w}\mathbf{C}_a \\ b_0\ b_1\ \ldots\quad b_m \qquad 0 \ldots 0 \\ 0\ b_0\ \ldots \qquad \ldots \qquad b_m \ldots 0 \\ \vdots \quad \vdots \\ 0\ \ 0\ \ldots \qquad b_0 \qquad \ldots\ b_m \end{pmatrix}. \tag{2.1}$$

and $\operatorname{rank}(b(\mathbf{C}_a)) \geq n - m$.

The last $n - m$ rows $\mathbf{b}, \mathbf{b}^{(1)}, \ldots, \mathbf{b}^{(n-m-1)} = \mathbf{w}$ of $b(\mathbf{C}_a)$ are cyclic shift of $\mathbf{b}$. Any of the first m rows can be obtained by n multiplications and n additions. Hence the calculation of $b(\mathbf{C}_a)$ requires $2mn$ operations.

Theorem 1 (Barnett [1]). *Let* $a(x) \in \mathbb{F}[x]$ *have degree* n *and* $\mathbf{C}_a$ *be its companion matrix. The degree of the greatest common divisor* $d(x) = (a(x), b(x))$ *of* $a(x)$ *with an arbitrary polynomial* $b(x) \in \mathbb{F}[x]$ *is*

$$\deg d(x) = n - \operatorname{rank}(b(\mathbf{C}_a)).$$

The coefficients of $d(x)$ *appear as a row after suitable Gauss eliminations on* $b(\mathbf{C}_a)$.

3   Description of the Algorithm

INPUT DATA: $a(x) = x^n + a_1 x^{n-1} + \cdots + a_{n-1} x + a_n \in \mathbb{F}[x]$, and $b(x) = b_0 x^m + b_1 x^{m-1} + \cdots + b_m \in \mathbb{F}[x]$, where $b_0 \neq 0$, $m \leqq n - 1$.
OUTPUT DATA: $\gcd(a(x), b(x)) = d(x) = x^k + d_1 x^{k-1} + \cdots + d_k$,
$u(x)$, $v(x)$, $a_1(x)$, $b_1(x)$.

Step 1. Compute the $n \times n$ matrix

$$
\mathbf{D} = \begin{pmatrix}
b_0 \cdots & & \cdot & & \cdot & \cdots & 0 \\
\vdots & \ddots & \vdots & \vdots & \vdots & \vdots & \vdots \\
0 & \cdots & b_0 & b_1 & b_2 & \cdots & b_m \\
& & & \mathbf{w}^{(1)} & & & \\
& & & \mathbf{w}^{(2)} & & & \\
& & & \vdots & & & \\
& & & \mathbf{w}^{(m)} & & &
\end{pmatrix}.
$$

The matrix $\mathbf{D}$ differs from $b(\mathbf{C}_a)$ only in order of rows. As we mentioned in the previous section only nm multiplications and nm additions are necessary for computing $\mathbf{D}$.

Step 2. Construct the $m \times m$ matrix

$$
\mathbf{W} = -\begin{pmatrix}
0 & 0 & 0 & \cdots & 0 & b_0 \\
0 & 0 & 0 & \cdots & b_0 & w_1^{(1)} \\
\vdots & \vdots & \vdots & \vdots & \vdots & \vdots \\
0 & b_0 & w_1^{(1)} & \cdots & w_1^{(m-3)} & w_1^{(m-2)} \\
b_0 & w_1^{(1)} & w_1^{(2)} & \cdots & w_1^{(m-2)} & w_1^{(m-1)}
\end{pmatrix},
$$

where $w_1^{(k)}$ is the first coordinate of $\mathbf{w}^{(k)} = (w_1^{(k)}, w_2^{(k)}, \ldots, w_n^{(k)})$.
No operations, only shifting is required in this step. Indeed $\mathbf{W}$ can be simultaneously constructed during the process of computing $\mathbf{D}$ in Step 1.

Step 3. Construct the $n \times (2n + m)$ matrix

$$
\mathbf{R} = \left(\mathbf{D} \,\middle|\, \begin{matrix} \mathbf{O}_{(n-m)\times m} \\ \mathbf{W} \end{matrix} \,\middle|\, \begin{matrix} \mathbf{O}_{(n-m)\times m} & \mathbf{E}_{n-m} \\ \mathbf{F}_m & \mathbf{O}_{m\times(n-m)} \end{matrix} \right) \tag{3.1}
$$

Step 4. Carry out elementary row operations on $\mathbf{R}$ in order to transform $\mathbf{D}$ into a row echelon form (trapezium shape), but without interchanging the rows.

Step 5. The last row having nonzero entries in the first n columns has the form

$$
(\underbrace{0 \ldots 0}_{n-k-1} 1\, d_1\, \ldots\, d_k | \mathbf{u} | \mathbf{v}),
$$

where $\mathbf{u}$ and $\mathbf{v}$ represent $u(x)$ and $v(x)$, respectively.

The next row has the form

$$(\underbrace{0\ldots0}_{n}\mid -\mathbf{b_1}\mid \mathbf{a_1})$$

The vector $\mathbf{b_1}$ represents $b_1(x) = b(x)/d(x)$, and $\mathbf{a_1}$ gives the coefficients of $a_1(x) = a(x)/d(x)$.

Example 1. Let $a(x) = x^5+3x^4+3x^3+x^2--4x-4$ and $b(x) = 2x^3+x^2+3x-2$. Their greatest common divisor is $d(x) = x^2 + x + 2$. The companion matrix of $a(x)$ is

$$\mathbf{C}_a = \begin{pmatrix} -3 & -3 & -1 & 4 & 4 \\ 1 & 0 & 0 & 0 & 0 \\ 0 & 1 & 0 & 0 & 0 \\ 0 & 0 & 1 & 0 & 0 \\ 0 & 0 & 0 & 1 & 0 \end{pmatrix}, \quad \Longrightarrow \quad b(\mathbf{C}_a) = \begin{pmatrix} -25 & -23 & -24 & 28 & 48 \\ 12 & 11 & 13 & -12 & -20 \\ -5 & -3 & -4 & 8 & 8 \\ 2 & 1 & 3 & -2 & 0 \\ 0 & 2 & 1 & 3 & -2 \end{pmatrix}.$$

Therefore

$$\mathbf{D} = \begin{pmatrix} 2 & 1 & 3 & -2 & 0 \\ 0 & 2 & 1 & 3 & -2 \\ -5 & -3 & -4 & 8 & 8 \\ 12 & 11 & 13 & -12 & -20 \\ -25 & -23 & -24 & 28 & 48 \end{pmatrix} \qquad \mathbf{W} = -\begin{pmatrix} 0 & 0 & 2 \\ 0 & 2 & -5 \\ 2 & -5 & 12 \end{pmatrix}.$$

In Step 3 we construct the matrix

$$\mathbf{R} = \begin{pmatrix} 2 & 1 & 3 & -2 & 0 & 0 & 0 & 0 & 0\,0\,0\,1\,0 \\ 0 & 2 & 1 & 3 & -2 & 0 & 0 & 0 & 0\,0\,0\,0\,1 \\ -5 & -3 & -4 & 8 & 8 & 0 & 0 & -2 & 0\,0\,1\,0\,0 \\ 12 & 11 & 13 & -12 & -20 & 0 & -2 & 5 & 0\,1\,0\,0\,0 \\ -25 & -23 & -24 & 28 & 48 & -2 & 5 & -12 & 1\,0\,0\,0\,0 \end{pmatrix}.$$

In Step 4 we carry out elementary row operations on $\mathbf{R}$ in order to transform $\mathbf{D}$ into echelon (trapezium shape) form. (To perform the operations in $\mathbb{Z}$ we first multiply the third and fifth rows then third, forth and fifth rows by 2.)

$$\mathbf{R} \sim \left(\begin{array}{rrrrr|rrr|rrr|rr}
2 & 1 & 3 & -2 & 0 & 0 & 0 & 0 & 0 & 0 & 0 & 1 & 0 \\
0 & 2 & 1 & 3 & -2 & 0 & 0 & 0 & 0 & 0 & 0 & 0 & 1 \\
0 & -1 & 7 & 6 & 16 & 0 & 0 & -4 & 0 & 0 & 2 & 5 & 0 \\
0 & 5 & -5 & 0 & -20 & 0 & -2 & 5 & 0 & 1 & 0 & -6 & 0 \\
0 & -21 & 27 & 6 & 96 & -4 & 10 & -24 & 2 & 0 & 0 & 25 & 0
\end{array}\right)$$

$$\sim \left(\begin{array}{rrrrr|rrr|rrr|rr}
2 & 1 & 3 & -2 & 0 & 0 & 0 & 0 & 0 & 0 & 0 & 1 & 0 \\
0 & 2 & 1 & 3 & -2 & 0 & 0 & 0 & 0 & 0 & 0 & 0 & 1 \\
0 & 0 & 15 & 15 & 30 & 0 & 0 & -8 & 0 & 0 & 4 & 10 & 1 \\
0 & 0 & -15 & -15 & -30 & 0 & -4 & 10 & 0 & 2 & 0 & -12 & -5 \\
0 & 0 & 75 & 75 & 150 & -8 & 20 & -48 & 4 & 0 & 0 & 50 & 21
\end{array}\right)$$

$$\sim \left(\begin{array}{rrrrr|rrr|rrrrr}
2 & 1 & 3 & -2 & 0 & 0 & 0 & 0 & 0 & 0 & 0 & 1 & 0 \\
0 & 2 & 1 & 3 & -2 & 0 & 0 & 0 & 0 & 0 & 0 & 0 & 1 \\
0 & 0 & 15 & 15 & 30 & 0 & 0 & -8 & 0 & 0 & 4 & 10 & 1 \\
0 & 0 & 0 & 0 & 0 & 0 & -4 & 2 & 0 & 2 & 4 & -2 & -4 \\
0 & 0 & 0 & 0 & 0 & -8 & 20 & -8 & 4 & 0 & -20 & 0 & 16
\end{array}\right)$$

$$\sim \left(\begin{array}{rrrrr|rrr|rrrrr}
2 & 1 & 3 & -2 & 0 & 0 & 0 & 0 & 0 & 0 & 0 & 1 & 0 \\
0 & 2 & 1 & 3 & -2 & 0 & 0 & 0 & 0 & 0 & 0 & 0 & 1 \\
0 & 0 & 1 & 1 & 2 & 0 & 0 & -8/15 & 0 & 0 & 4/15 & 10/15 & 1/15 \\
0 & 0 & 0 & 0 & 0 & 0 & -2 & 1 & 0 & 1 & 2 & -1 & -2 \\
0 & 0 & 0 & 0 & 0 & -2 & 5 & -2 & 1 & 0 & -5 & 0 & 4
\end{array}\right)$$

The third row shows that

$$d(x) = x^2 + x + 2, \qquad u(x) = -\frac{8}{15}, \qquad v(x) = \frac{1}{15}(4x^2 + 10x + 1).$$

The next (forth) row gives quotients

$$-b_1(x) = -2x + 1, \qquad a_1(x) = x^3 + 2x^2 - x - 2.$$

4   Proof of Correctness of the Algorithm

Lemma 1. *Any linear combination of the rows of $b(\mathbf{C}_a)$ represents a polynomial of the form*

$$f(x)b(x) - g(x)a(x), \tag{4.1}$$

where $\deg f(x) \leq n - 1, \quad \deg g(x) \leq m - 1.$

Proof. Let $w^{(k)}(x)$ be the polynomial corresponding to $\mathbf{w}^{(k)}$. Having in mind the form of $\mathbf{C}_a$ it is easy to check that

$$\mathbf{w}^{(1)}(x) = \mathbf{w}(x) - b_0 a(x) = x^{n-m}b(x) - b_0 a(x),$$
$$\mathbf{w}^{(2)}(x) = x\mathbf{w}^{(1)} - w_1^{(1)}a(x) = x^{n-m+1}b(x) - (b_0 x + w_1^{(1)})a(x),$$
$$\vdots \quad \vdots \quad \vdots \tag{4.2}$$
$$\mathbf{w}^{(m)}(x) = x\mathbf{w}^{(m-1)} - w_1^{(m-1)}a(x) =$$
$$= x^{n-1}b(x) - (b_0 x^{m-1} + w_1^{(1)}x^{m-2} + \cdots + w_1^{(m-1)})a(x).$$

Since the last rows of $\mathbf{C}_a$ represent the polynomials $x^{n-m-1}b(x), \ldots, xb(x), b(x)$, then any linear combination of the rows has the form (4.1). The multipliers at $a(x)$ have degree at most $m-1$, and ones at $b(x)$ have at most $n-1$. Therefore, $\deg f(x) \leq n-1, \quad \deg g(x) \leq m-1$. $\qquad\qquad\qquad\square$

Theorem 2. *Let* $\gcd(a(x), b(x)) = d(x) = x^k + d_1 x^{k-1} + \cdots + d_k$ *and* $u(x)$, $v(x)$, $a_1(x)$, *and* $b_1(x)$ *are defined by (1.1). Then after carrying out the described in the algorithm row operations the* $(n-k)$*th row of* $\mathbf{R}$ *(the last row having nonzero entries in* $\mathbf{D}$*) has the form* $(\mathbf{d} \mid \mathbf{u} \mid \mathbf{v})$,. *The* $(n-k+1)$*th row is* $(\mathbf{o} \mid -\mathbf{b_1} \mid \mathbf{a_1})$.

Proof. Replacing x by $\mathbf{C}_a$ in (1.1) we get

$$v(\mathbf{C}_u)b(\mathbf{C}_u) - d(\mathbf{C}_u).$$

According to Theorem 1 $\operatorname{rank}(b(\mathbf{C}_a)) = \operatorname{rank}(d(\mathbf{C}_a)) = n - k, \quad (v(\mathbf{C}_a)$ is nonsingular). Since the last row of $v(\mathbf{C}_a)$ is $\mathbf{v} = (0, \ldots, 0, v_0, \ldots, v_{n-k-1})$, and that of $d(\mathbf{C}_a)$ is $\mathbf{d} = (0, \ldots, 0, d_0, \ldots, d_k)$, then $\mathbf{v}$ (as a column) is a solution of

$$[b(\mathbf{C}_a)]^\tau \begin{pmatrix} x_1 \\ x_2 \\ \vdots \\ x_n \end{pmatrix} = \mathbf{d}^\tau. \qquad (4.3)$$

The last $n-k$ rows of $b(\mathbf{C}_a)$, i.e., the last $n-k$ columns of $[b(\mathbf{C}_a)]^\tau$, are linear independent. Hence $x_1, \ldots, x_k$ are free unknowns (parameters) and $x_{k+1}, \ldots, x_n$ are determined by them. Setting $x_1 = \cdots = x_k = 0$ we obtain a particular solution, which is obviously $\mathbf{v}$ (the first k entries of $\mathbf{v}$ are zeros) The general solution has the form

$$\mathbf{v} + \mathbf{z},$$

where $\mathbf{z}$ is a solution of the homogeneous system corresponding to (4.3).

The described in Step 4 row operations carrying out up to down are equivalent to such operations on $b(\mathbf{C}_a)$, but down to up and in order to make first k rows only of zeros. Thus, the first m rows of the transformed $b(\mathbf{C}_a)$ are the flipped last m rows of the submatrix $\mathbf{D}$ at the end of Step 4.

Let $\mathbf{P}$ be the triangle matrix (with zeros under the main diagonal), which corresponds to the row operations carried out in Step 4, that is, $\mathbf{P}b(\mathbf{C}_a)$ is with all zeros first k rows and the $(k+1)$-th row is $\mathbf{d}$. Matrix $\mathbf{P}$ has zeros under the main diagonal since the operations are done down to up. If $\mathbf{p}_i$ is the i-th row of $\mathbf{P}$, then

$$\mathbf{p}_{k+1}b(\mathbf{C}_a) = \mathbf{d}.$$

Therefore $\mathbf{p}_{k+1}$ is a solution of (4.3) and its first k entries are zeros. Thus

$$\mathbf{p}_{k+1} = \mathbf{v}.$$

Now let us look at the positions corresponding to matrix $\mathbf{W}$, i.e., from $(n+1)$st to $(n+m)$th column. Lemma 1 and its proof give that the vector $\mathbf{d}$ has also to represent the polynomial $v(x)b(x) - g(x)a(x)$ and the vector in the $(n+1)$st to $(n+m)$th positions has to be $-g(x)$. Hence

$$-g(x) = u(x).$$

The k-th row $\mathbf{p}_k$ satisfies $\mathbf{p}_k b(\mathbf{C}_a) = \mathbf{o}$, thus, it is a solution of the homogeneous system. On the other hand $a_1(\mathbf{C}_a)b(\mathbf{C}_a) = a(\mathbf{C}_a)b_1(\mathbf{C}_a) = \mathbf{O}$ shows that the vector $\mathbf{a}_1$ corresponding to $a_1(x)$ is a solution of the same homogeneous system. But both $\mathbf{p}_k$ and $\mathbf{a}_1$ have zeros in the first $k-1$ positions (i.e., $x_1 = \cdots = x_{k-1} = 0$,). Hence they are proportional and if they are normalized

$$\mathbf{p}_k = \mathbf{a}_1.$$

Similarly, according to Lemma 1 $a_1(x)b(x) - g(x)a(x) = 0$. Taking into account the degree of $g(x)$, we can conclude that the vector in the $(n+1)$st to $(n+m)$th positions has to be $-\mathbf{b}_1$.

Remark. If we do not reduce the matrix $\mathbf{R}$ in prescribed way (e.g., interchanging rows) we will obtained instead $a_1(x)$ a polynomial congruent with it modulo $b(x)$ (it will be of a higher degree). The same is true about $b_1(x)$.   □

5   Conclusion

In terms of O-notation there is no essential difference in the number of the required operations in $\mathbb{F}$, but the described algorithm is much faster in practice.

Also, it is naturally parallelizable. This is important especially in the case when coefficients of polynomials belong to a large finite field (which we are interested in), $\mathbb{Z}$, or special types of fields or factorial rings.

We realized the algorithm on GPUs. To each thread it is assigned a column of the matrix and it carries out all operations with elements in this column.

An important application of the algorithm is its modification to an algorithm for decoding cyclic codes.

The described algorithm can be generalized for the case when F is a factorial ring and for finding **gcd** of several polynomials, but such generalizations are out of the focus of this paper.

References

1. Barnett, S.: Greatest common divisor of two polynomials. Linear Algebra Appl. **3**, 7–9 (1970)
2. Bamett, S.: Matrices: Methods and Applications, Oxford Applied Mathematics and Computing Science Series. Clarendon Press, Oxford (1990)
3. Laidacker, M.: Another theorem relating Sylvester's matrix and the greatest common divisor. Math. Mag. **42**(3), 126–128 (1969)
4. Gonzalez-Vega, L.: An elementary proof of Barnett's theorem about the greatest common divisor of several univariate polynomials. Linear Algebra Appl. **247**, 185–202 (1996)

5. Diaz-Toca, G.M., Gonzalez-Vega, L.: Computing greatest common divisors and squarefree decompositions through matrix methods: the parametric and approximate cases. Linear Algebra Appl. **412**(2–3), 222–246 (2006)
6. MacDuffee, C.: Some applications of matrices in the theory of equations. Amer. Math. Mon. **57**, 154–161 (1950)

Applications of Metaheuristics to Large-Scale Problems

Distributed Evolutionary Computing Migration Strategy by Incident Node Participation

Todor Balabanov[✉], Iliyan Zankinski, and Maria Barova

Institute of Information and Communication Technologies,
Bulgarian Academy of Sciences, Acad. G. Bonchev Str, Block 2, 1113 Sofia, Bulgaria
todorb@iinf.bas.bg
http://www.iict.bas.bg/

Abstract. Distributed evolutionary algorithms are implemented on heterogeneous computing nodes. In a distributed environment it is usual these nodes to differ in operating system and hardware. Such an environment has major problems related to network latency. Some of the evolutionary optimization algorithms are very suitable for distributed computing implementation, because of their high level of parallel scalability. In most cases only fitness function calculation is distributed synchronously (or asynchronously). In this case the population is presented only in the master node. In the next case each separate node has part of the distributed population (it is called island model). The last common model is based on shared memory and each computing node has access to the whole population (it is called fine-grained model). All other models are some hybridization. In the island model there is a common parameter related to migration strategy. The most often used node topology is the ring topology. On a regular basis each node sends its best individual to the next node in the ring. In this paper, a hybrid model, based on incident node participation in star topology, is proposed.

Keywords: Distributed computing · Evolutionary algorithms · Migration strategies

1 Introduction

Evolutionary Algorithms (EAs) are efficient search methods based on principles of natural selection and recombination. They have been applied successfully to find acceptable solutions to problems in business, engineering, and science [1,15]. By using EAs good solutions can be found in reasonable amount of time. When the problem is relatively small, a single computing node can be used. When the problems are bigger or harder, the time to find adequate solutions increases. With the increasing popularity of parallel computations it becomes a promising choice for EAs speed-up. Some distributed evolutionary algorithms (DEAs) are using a single population, while others split the population into several relatively independent subpopulations. A strict classification can be found into the literature [2–4]. In the case of distributed computing, migration strategy is one of the most important aspects of DEAs implementation.

© Springer International Publishing Switzerland 2015
I. Lirkov et al. (Eds.): LSSC 2015, LNCS 9374, pp. 203–209, 2015.
DOI: 10.1007/978-3-319-26520-9_21

In the cases when EA's population is distributed among many different computing nodes, each node has a fraction of the population called subpopulation. In such a case, recombination and fitness function evaluation is done locally. For better optimization convergence, the computing nodes are exchanging individuals. This exchange is called migration. The process of migration is controlled by several parameters and it forms a strategy for information interchange in DEAs. The first parameter is related to traveler selection and corresponds to the strategy of choosing which individual of the subpopulation to be sent. In most of the cases the best individual is selected. The second important parameter is the migration frequency and it determines how often the selected individuals will travel across subpopulations. This parameter is problem specific and usually it is adjusted experimentally. The third and maybe the most important parameter is related to the migration destination. Migration destination means how nodes are organized, as topology, and how each individual will travel in this topology. One option for this parameter is individual broadcasting when each node receives a copy of others nodes selected travelers. Another option is a ring topology where each traveler migrates one neighbor ahead. Grid topology is also possible [5,16], where each node has four neighbors. Many other topologies are also possible as hierarchical, 3D based, hybrid and etc.

In this paper distribution strategy is proposed, based on incident node participation (volunteers are expected to join the project and contribute computing power). Computing nodes are organized as star topology and island model [6,17] for DEAs is applied.

The rest of this paper is organized as follows: Sect. 2 presents the model proposed. Section 3 describes distribution parameters in details. Section 4 is devoted to some experiments and results. Final Sect. 5 concludes and present some ideas for further research.

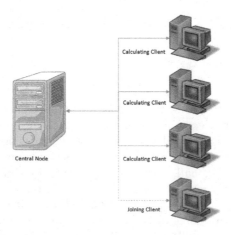

Fig. 1. Incident node participation

2   The Model Proposed

The model proposed in this paper is provoked by the implementation used in VitoshaTrade project [7]. The project is an open source volunteer distributed computing system, which uses EAs for artificial neural networks (ANNs) training. ANNs in VitoshaTrade are used for financial time series forecasting. The goal of optimization is ANN's weights adjustment in order for better forecasting to be achieved. Training of ANN with EAs can be very computationally intensive, because (depending of the ANN topology) it can has more than 400 weights. Because of the time consuming nature of EA based ANN training, the computing nodes in the project are relatively autonomous. The distributed system is organized as star topology with lightweight central node (server) and heavy-loaded remote computing nodes (clients). As DEA implementation, island model is used. There is a global EA population, located in the central node, and many local EA populations distributed on the remote computing nodes. Each remote computing node can join the system and leave the system at any moment of time, asynchronously (Fig. 1). For each joining client the central node sends a subset of the global population. After that there is no more communication between the central node and the remote node. The remote node evolves the local EA population as sequential EA.

The communication between the remote node and the central node is initiated again only if there is a better solution, found on the remote node, which should be reported to the central node. By such organization of the calculation process, each remote node may work for hours, weeks or months before having a need to report something to the central node. In practice, failure of the central node will not affect the performance of the remote nodes. Even in the central node failure situation, the remote nodes will report their results when the central node is available again.

The distribution of the individuals is done only during the remote node joining process. Local best found solutions are reported to the central node and eventually they will migrate to the next remote node that joins. EA based training of ANNs is so slow that such distribution strategy is very efficient. Also the training set in financial time series forecasting is constantly changing, because new data are constantly coming out. In this way the objective function, which should be optimized (ANN's total error), is constantly changed. That is why continuous ANN training should be applied.

3   Distribution Parameters

Distribution is the operator responsible for the exchange of individuals between the nodes in DEA. In relation to distribution, a group of parameters are relevant (such as distribution gap, distribution rate, selection/replacement, topology and heterogeneity). In the case of Distribution Strategy by Incident Node Participation (DSINP) some of these parameters can be applied, but others do not have reasonable meaning. In this section distribution parameters will be discussed according to their applicability to the proposed model.

3.1 Distribution Gap

In the model proposed it is very difficult to define such a parameter as distribution gap. There are two popular approaches described in the literature. The first one is distribution gap on a regular basis (given number of steps) and the second one is probabilistic, on each generation with probability Pm [8–10]. The distribution gap is closer to the probabilistic model. As it was described in the previous sections, distribution is done only once when the computing node joins to the system. In this aspect, distribution gap is probabilistic with exponential distribution. In the model proposed there is no no-effect problem or super-individual problem, because the distribution is relatively rare.

3.2 Distribution Rate

The distribution rate is a problem specific parameter. It determines how many individuals will travel across the local populations. Usually it is expressed as percentage of the population or as absolute value. There are different suggestions how this parameter to be estimated, but the general approach is experimentally [6,11,12]. In the model proposed, this parameter is percentage (a fraction) of the global population (population kept on the central node). As absolute values it is equal to the size of the local population (the size of the population on the remote node).

3.3 Selection and Replacement

There are two general ways to select migrants. The first way is to select the best individuals, the second way is to select random individuals. Of course many other selection operators can be applied similar to those used in genetic algorithms (GAs) [13,18]. In the model proposed, selection is done on the central node. In VitoshaTrade project random selection is used. The number of selected migrants is the same as the size of the remote subpopulation. Practically there is no replacement procedure, because distribution is done only once from the central node to the remote node. In the opposite direction (remote to central), a lot of reported individuals can be sent, but they will participate in other remote subpopulations, according to node joining process.

3.4 Topology

DEAs are divided in two common models (stepping-stone and island). This division depends on whether individuals can freely migrate to any local population or if they are restricted to migrating to geographically nearby islands. Many works exists trying to decide the best topology for a DEA and in most of the cases the most preferred topologies are ring and hyper-cube [2–4,12]. In most of the cases there are problems (parallelization and scalability) with the fully-connected and centralized topologies due to the tight connectivity. In the model proposed, the best suited topology is a star topology with centralized node and relatively independent remote computing nodes. EA based ANN training is a slow process and the remote computing nodes can spend relatively long time without need to communicate with the central node. For this reason the central node is not s risky part of the distributed system. Even more, losing the connection with the central node does not affect local optimization process. Failure of the central node will affect only new nodes trying to join the system. Even with these disadvantages, the model proposed has very high degree of scalability, because the lightweight central node. In the VitoshaTrade project, the central node is presented as PHP based web server.

3.5 Heterogeneity

The model proposed is perfectly suited for heterogeneity. On each remote computing node, a different optimization algorithm can be applied. In this way, a much better balance can be achieved between exploration and exploitation (a well-known trade-off decision in EAs). This parameter is problem dependant. In the case of ANN's weights distribution, it can be implemented very efficiently in such distributed system, because there are very effective gradient based training algorithms.

4 Experiments and Results

DSINP was compared with Ring Topology for the problem of Artificial Neural Network (ANN) training, based on Differential Evolution (DE) weights optimization. All experiments were done in local computer network. As central node,

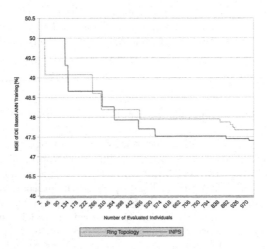

Fig. 2. Artificial Neural Network trained with Differential Evolution for Time Series Forecasting. X - number of evaluated individuals. Y - MSE of ANN. Dotted line is for the ring topology. Solid line is for the Incident Node Participation Strategy (INPS).

Linux (Ubuntu Server 12.04.5 LTS) desktop machine with a 2.33 GHz Intel Core 2 Duo CPU and a 1.95 GB RAM. On the central XAMPP package were used for Apache web server and MySQL database. As computing nodes, four identical Windows 7 Home Premium 64 bits based laptops were used (Acer, CPU Intel U4100 1.30GHz, 4GB RAM).

Because ring topology is pretty different organization, compared with incident node participation strategy, as progress of the computation done (X - axis), number of evaluated individuals was selected (Fig. 2). Respectively, on the Y - axis, ANN Mean Square Error (MSE) is measured. As it can be seen in Fig. 2, DSINP outperforms Ring Topology migration in some stages of the optimization process (around 134 and 486 on X-axis).

5   Conclusions

The model presented suggests efficient distribution strategy in the field of distributed evolutionary algorithms. Current implementation is concentrated on EA based ANN training for financial forecasting. The approach is innovative because the training itself is continuous with constantly changing objective function. In such a distributed computing system it is expected for remote computing nodes to often join and disjoin. Because of all facts presented, incident node participation is the natural way for individuals to migrate.

As further work, it will be interesting hybrid algorithms to be tested. For example, DE based ANN training in combination with Particle Swarm Optimization, Ant Colony Optimization, Simulated Annealing and etc. Also, as described in [14], PicoBlaze FPGAs can be researched for remote computations.

Acknowledgements. This work was supported by the Bulgarian National Scientific Fund under the grants Efficient Parallel Algorithms for Large Scale Computational Problems and InterCriteria Analysis A New Approach to Decision Making.

References

1. Goldberg, D.E.: Genetic and evolutionary algorithms come of age. Commun. ACM **37**(3), 113–119 (1994)
2. Adamidis, P.: Review of parallel genetic algorithms bibliography. Technical Report version 1, Aristotle University of Thessaloniki, Thessaloniki, Greece (1994)
3. Gordon, V.S., Whitley, D.: Serial and parallel genetic algorithms as function optimizers. In: Forrest S. (ed.) Proceedings of the Fifth International Conference on Genetic Algorithms, pp. 177–183. Morgan Kaufmann, San Mateo (1993)
4. Lin, S.-C., Punch, W., Goodman, E.: Coarse-grain parallel genetic algorithms - categorization and new approach. In: Sixth IEEE Symposium on Parallel and Distributed Processing, IEEE Computer Society Press, Los Alamitos, CA (1994)
5. Spiessens, P., Manderick, B.: A massively parallel genetic algorithm. In: Belew, R.K., Booker, L.B. (eds.) Proceedings of the 4th International Conference on Genetic Algorithms, pp. 279–286. Morgan Kaufmann, San Francisco (1991)

6. Tanese, R.: Distributed genetic algorithms. In: Schaffer, J.D. (ed.) Proceedings of the 3rd International Conference on Genetic Algorithms, pp. 434–439. Morgan Kaufmann, San Francisco (1989)

7. Balabanov, T.: VitoshaTrade open source project, (2014). http://code.google.com/p/vitoshatrade/

8. Gorges-Schleuter, M.: ASPARAGOS An asynchronous parallel genetic optimisation strategy. In: Schaffer, J.D. (ed.) Proceedings of the 3rd ICGA, pp. 422–427. Morgan Kaufmann, San Francisco (1989)

9. Munetomo, M., Takai, Y., Sato, Y.: An efficient migration scheme for subpopulation-based asynchronously parallel GAs. Technical Report HIER-IS-9301, Hokkaido University (1993)

10. Voigt, H.M., Santibanez-Koref, I., Born, J.: Hierarchically structured distributed genetic algorithms. In: Manner, R., Manderick, B. (eds.) Proceedings of the International Conference Parallel Problem Solving from Nature, vol. 2, pp. 155–164. North-Holland, Amsterdam (1992)

11. Belding, T.C.: The distributed genetic algorithm revisited. In: Eshelman, L.J. (ed.) Proceedings of the 6th International Conference on GAs, pp. 122–129. Morgan Kaufmann, San Francisco (1995)

12. Mejia-Olvera, M., Cantu-Paz, E.: DGENESIS-software for the execution of distributed genetic algorithms. In: Proceedings of the XX Conferencia Latinoamericana de Informatica, pp. 935–946. Monterrey, Mexico (1994)

13. Baker, J.E.: Reducing bias and inefficiency in the selection algorithm. In: Grefenstette, J.J. (ed.) Proceedings of the Second International Conference on Genetic Algorithms, pp. 14–21. Lawrence Erlbaum Associates Publishers, Hillsdale (1987)

14. Ivanov, V.: An approach for a PicoBlaze system generation. In: Proceedings of Distributed Computer and Communication Networks, pp. 233–241. Moscow (2013)

15. Pappa, G., Ochoa, G., Hyde, M., Freitas, A., Woodward, J., Swan, J.: Contrasting meta learning and hyper-heuristic research: the role of evolutionary algorithms. Genet. Program. Evolvable Mach. 15(1), 3–35 (2014)

16. Krüger, F., Wagner, D., Collet, P.: Massively parallel generational GA on GPGPU applied to power load profiles determination. In: Legrand, P., Corsini, M.-M., Hao, J.-K., Monmarché, N., Lutton, E., Schoenauer, M. (eds.) EA 2013. LNCS, vol. 8752, pp. 227–239. Springer, Heidelberg (2014)

17. Uchida, T., Matsuzawa, T., Inoguchi, Y.: The influence of elitism strategy on migration intervals of a distributed genetic algorithm. In: Proceedings in Adaptation, Learning and Optimization, vol. 2, pp. 363–374 (2015)

18. Lim, T.Y.: Structured population genetic algorithms: a literature survey. Artif. Intell. Rev. 41(3), 385–399 (2014)

Slot Machine RTP Optimization and Symbols Wins Equalization with Discrete Differential Evolution

Todor Balabanov$^{(\boxtimes)}$, Iliyan Zankinski, and Bozhidar Shumanov

Institute of Information and Communication Technologies,
Bulgarian Academy of Sciences, acad. G. Bonchev Str, Block 2, 1113 Sofia, Bulgaria
todorb@iinf.bas.bg
http://www.iict.bas.bg/

Abstract. It is possible to solve slot machine RTP optimization problem by using evolutionary algorithms. In practice this optimization is done by hand adjustment of the symbols placed on the game reels. By arranging symbols positions, it is possible to achieve optimal return to player percentage (RTP). Equalization of the prizes distribution, generated by different win combinations, can be optimized also. In this paper a DE based RTP optimization and prizes equalization is proposed. DE is used in its discrete variation, because the problem of optimal symbols distribution on the reels is in the discrete domain. DE is selected as an alternative to genetic algorithms (GA) because of its faster convergence. The convergence is a key factor in such optimizations, because each fitness value is calculated based on intensive Monte-Carlo simulations. The scope of this paper is focused on the symbols distribution placed on the machine reels in such a way that two common goals to be satisfied - desired RTP and keeping relatively equal levels of the prizes (prizes expressed as amount of money won from combinations with each particular symbol), with relatively good symbol diversity on the reels.

Keywords: Slot machine · Gambling · Discrete Differential Evolution · Return to player · Optimization

1 Introduction

Slot machines are electronic gambling devices, which are popular all over the world. The most popular slot machines consists of five reels. The reels start spinning when the button is pushed. Nowadays slot machines are computerized with PRNG embedded in them. In 1984 Inge Telnaes received a patent for a device titled, "Electronic Gaming Device Utilizing a Random Number Generator for Selecting the Reel Stop Positions" (US Patent 4448419) [1]. In the beginning slot machines have been mechanical. They have had a lever on the side of the machine (because of this lever, machines were known as one armed bandits), which have been used for reels spinning activation. The machine pays

© Springer International Publishing Switzerland 2015
I. Lirkov et al. (Eds.): LSSC 2015, LNCS 9374, pp. 210–217, 2015.
DOI: 10.1007/978-3-319-26520-9_22

off according to symbol patterns, visible on the screen, when the reels stop. Slot machines are the most popular gambling method in casinos and constitute about 70 percent of the average US casino income [2].

A gambler playing a slot machine has credit inserted - cash, by printed ticket or loaded by the attendant. The machine is activated by means of a lever (or a button), or by pressing a touchscreen. The objective of the game is to win money from the machine, which usually involves matching symbols on reels (mechanical or virtual) that spin and stop to reveal one or several symbols. Most games have a variety of winning combinations of symbols. If a player matches a combination according to the given patterns, the slot machine rewards the player.

Each machine has a table that lists the number of credits the player will receive if the symbols listed on the pay table line up on the pay line of the machine. Some symbols are wild and can represent many (or all) of the other symbols to complete a winning line [3]. Symbols are statistically distributed on the reels. Some symbols show up more often than others. Some symbols pay more than others, according to the pay table. Slot machines are usually adjusted to pay out as winnings 75 to 98 percent of the money that is wagered by players. It is known as theoretical payout percentage or RTP (return to player). The minimum RTP varies among jurisdictions and it is subject of law regulations.

The motivation for this research is our previous work related to GA based optimization of slot machine RTP. In this work RTP is optimized again, but prizes equalization is also included, as s second criteria. As third, less important criteria, symbols diversity on the reels is also controlled. The source code, used for this research, is available as open-source project in Github global repository [4]. The rest of this paper is organized as follows: Sect. 2 presents the model proposed. Section 3 is devoted to some experiments and results. Final Sect. 4 concludes and present some ideas for further research.

2 The Model Proposed

The model proposed is based on Discrete Differential Evolution (DDE). DDE is applied over RTP, prizes equalization and symbol diversity, as multi-criteria optimization.

2.1 RTP Optimization

Modern slot machines are computerized. They have virtual reels with symbols distributed on them. Stops of the virtual reels are selected by PRNGs. RTP of the game is directly dependent from symbols distribution on the reels. Usually symbols (and their positions) are selected manually by the mathematicians. In practice, slot machine reels are discrete distribution of symbols. Such distribution can be achieved by discrete optimization, according given constraints like desired RTP, prizes equalization and symbols diversity. From these three criteria, symbol diversity is easily calculated without simulation, but RTP and prizes equalization need Monte-Carlo simulations in order to participate in DDE cost function.

Multi-criteria cost function is converted in single criteria by linear transformation and coefficients given for each of the criteria. Coefficients are selected by the decision maker, according his/her personal preferences. In this research 1 were selected for symbols diversity, 100 for RTP and 10 for prizes equalization. These numbers are selected in correlation with the importance of each criteria. Symbol diversity is a parameter related to how many symbols of the same kind are next to each other in a single reel. Symbol diversity is easily achievable with simple swaps of symbols which are in inappropriate order. Slot machine reels are represented as individuals in the DDE optimization. Symbol diversity can be controlled even before the Monte-Carlo simulation to be executed. The cost function is preferred for this criteria in order to make solution space exploration easier.

2.2   Discrete Differential Evolution

DE is one of the stochastic optimization algorithms. It is used for the following search problem: Minimize an objective function which is a mapping from a parameter vector x in n dimensional real values space into to one dimensional real values space. DE has self-organization for mutation, crossover and selection, but strategy parameters are selected empirically [5]. DE was proposed by Storn and Price [6].

DE shares similarities with traditional evolutionary algorithms. However it does not use binary encoding as a simple genetic algorithm [7] and it does not use a probability density function to self-adapt its parameters as an Evolution Strategy [8]. DE differs in its mutation. Mutation is performed based on the distribution of the solutions in the population. In this way, search directions and possible stepsizes depend on the location of the individuals selected to calculate the mutation values. The most popular model is called DE/rand/1/bin, where DE means Differential Evolution, rand indicates that individuals selected to compute the mutation values are chosen at random, 1 is the number of pairs of solutions chosen and finally bin means that a binomial recombination is used [9].

Similar to other EAs, DE can not deal with constrained optimization. In this research there is a strict constraint which states that slot machine reels should be valid. Slot machine reels validity depends on the game rules. For example, the reels should consists only of valid symbols. This constraint is manually guaranteed after each new individual reproduction. Each invalid symbols is replaced by a randomly selected valid symbol. This correction can be accepted as addition to the mutation operation.

This work uses DDE for the optimization goals. The original DE is modified in such way that weighted difference vector is calculated with discrete values. Instead of regular difference, normalized discrete difference vector is used. This difference vectors consists of three common values (minus one, zero and plus one), as it can been seen at [4].

2.3   Implementation

DDE individuals are presented as 2D arrays of symbols (reels). All symbols are presented as integer numbers. Each DDE individual is a point into solution space (discrete finite space). To be valid each reel should have only integer numbers from the listed (Table 1). In this research symbols used are from 3 to 12. Symbols from 0 to 2 are usually reserved for special symbols called wilds. Symbols between 14 and 16 are usually reserved for special symbols called scatters. The currently presented model of a slot machine does not have wild or scatter symbols.

Table 1. Slot machine pay table. Each column represents the winning of one particular symbol (10 possible symbols in this game). Each row shows the winning of the symbols when the combination is of 3, 4 or 5 symbols.

	SYM03	SYM04	SYM05	SYM06	SYM07	SYM08	SYM09	SYM10	SYM11	SYM12
3 of	250	100	75	50	25	15	9	6	3	1
4 of	500	250	100	75	50	25	15	10	7	3
5 of	750	500	250	150	100	75	50	30	20	10

Free spins were not presented in the model, but a simple bonus game was added. The bonus game imitates a bingo game. There is a bonus prize for bingo line and another bonus prize for bingo.

Initialization of the population is done by manually constructed reels. Some of the individuals are shuffled inside for initial population diversity. The size of the population is a subject of experimental estimation and may vary from several individuals to hundreds or thousands of individuals.

The first step of DDE optimization is selecting a target vector, base vector and two other vectors to be used for weighted difference vector. All of the four vectors are selected randomly (slightly different than original DE algorithm). As a second step, a discrete difference vector is calculated. The only valid values for the discrete difference vector are minus one, zero and plus one. In the third step, a mutation is done. The mutation sums of the difference vector with the base vector. Invalid symbol numbers (in this case 2 or 13) are randomly replaced with valid ones (from 3 to 12). The fourth step is crossover between the base vector and mutated vector. For this operation binomial crossover is used. The final fifth step is related to fitness value calculation and decision which vector to be kept for the next generation (the target vector or the newly recombined one).

This is a multi-criteria problem. By giving weights of the three criteria they are used as linear equation and the problem is converted to a single-criteria problem. The decision maker is responsible for the coefficients selection for each criteria. In this research 1 is used for symbol diversity, 100 for target RTP and 10 for prizes equalization. Monte-Carlo simulation is used for the target RTP and prizes equalization estimation. Symbols diversity is calculated directly from the reels. For better accuracy in Monte-Carlo simulations, 10 times by 1000000 separate slot game runs are executed.

The maximum number of recombinations is used as an optimization termination criteria. Manual observation/termination of the process is also possible. The final solution, found by DDE, is an integer matrix. This matrix is directly applicable as slot machine reels strips. For example, if there is a slot game with 5 reels (visible on the screen as 5 columns and 3 rows), and each reel consists of 63 symbols, the final DDE solution would be an integer matrix of 5×63 values (refer to [4] for more details).

3   Experiments and Results

All experiments have been done on an open-source slot machine simulator [4] (5×3 screen) with a particular pay table (Table 1) and nine winning lines (Table 2).

All winning combinations are paid from left to right. The lowest winning is 1 - for a combination of 3 symbols SYM12 (Table 1). The highest winning is 750 - for a combination of 5 symbols SYM03 (Table 1). There are 10 symbols, which form winning patterns on the screen.

Table 2. Slot machine winning lines. There are nine possible lines. On each of these lines (from left to right) win patterns can appear. Win patterns are formed by 3, 4 or 5 symbols.

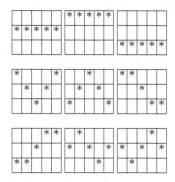

All experiments are done with natural elitism rule embedded in DE, population size of 17, maximum number of recombinations 100, and one million separate simulation game runs, accomplished in ten separate sessions for the fitness value estimation. The DDE strategy is DE/rand/1/bin. Initial solutions use handpicked reels.

Two target RTP were selected for the experiments. Fist set of experiments was done for target RTP of 90, which is the lower legal value for the Bulgarian gambling market. Three independent runs of the algorithm were done and it is visible that fast convergence was achieved in the beginning (4–7 generations). Between 40–80 generations finer convergence was obtained (Fig. 1). In the second

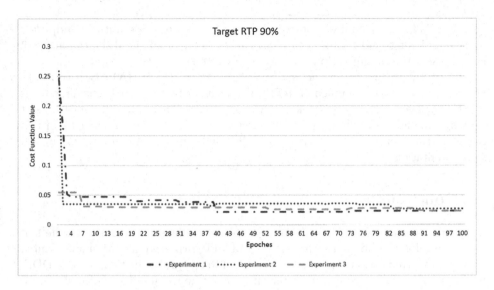

Fig. 1. DDE convergence for target RTP of 90 percent. On the axes: x - number of generations, y - cost function value.

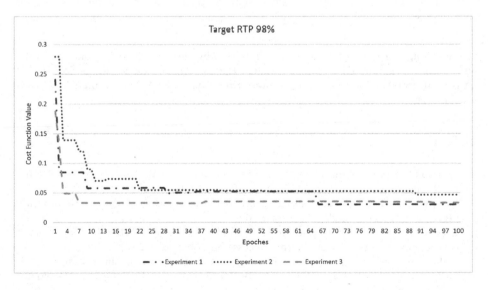

Fig. 2. DDE convergence for target RTP of 98 percent. On the axes: x - number of generations, y - cost function value.

set of experiments, fast convergence can be seen between 4–25 generations and finer convergence between 64–91 generations (Fig. 2). It is visible that DDE convergence is faster in first set of experiments (Fig. 1), because initial handpicked reels are closer to RTP of 90 than 98 (Fig. 2). Because of the discrete nature of the process, the optimization is done in separate stairs like steps.

The results were compared with results presented in NMA14 Borovets [11], which is related to slot machine RTP optimization by GA. DDE convergence is a little bit faster than convergence achieved by GA. There are differences in slot machine models, but the general optimization idea is similar. In the case of this research the cost function is more complex than the cost function used in GA implementation.

4   Conclusions

Experiments show that using DDE may be very efficient and improve the slot games development by better adjustment of RTP, prizes equalization and symbol diversity. Optimization convergence is related to the probabilistic nature of DDE. Even thought, DDE converge faster than other heuristics (like GA), slot RTP estimation is time consuming and slows down the optimization process. Because of the limited scope of this research, multi-objective approach was not tested.

As further research, it could be interesting for DDE to be implemented as distributed computing algorithm. Such distributed implementation is efficiently applicable for the class of evolutionary algorithms between which is DDE. Also, as described in [10], PicoBlaze FPGAs can be researched for faster Mote-Carlo simulations.

Acknowledgements. This work was supported by the Bulgarian National Scientific Fund under the grants DFNI 02/20 Efficient Parallel Algorithms for Large Scale Computational Problems and DFNI 02/5 InterCriteria Analysis A New Approach to Decision Making.

References

1. Inge, S.: Electronic gaming device utilizing a random number generator for selecting the reel stop positions, US 4448419 A (1984). Published on 15 May 1984
2. Cooper, M.: How slot machines give gamblers the business, The Atlantic Monthly Group (2005). Accessed on 21 April 2008
3. Casino Observer: How to play slots. CasinoObserver.com (2013). http://casinoob server.com/how-to-play-slots.htm, Accessed on 06 March 2013
4. Balabanov, T.: Bingo slot discrete differential evolution optimization of RTP and prizes distribution, (2015). http://github.com/TodorBalabanov/BingoSlot DifferentialEvolutionOptimization/
5. Storn, R.: Differential evolution - a simple and efficient heuristic strategy for global optimization over continuous spaces, J. Global Optim., **11**, 341–359. Dordrecht (1997)

6. Price, K.: An introduction to differential evolution. In: Corne, D., Dorigo, M., Glover, F. (eds.) New Ideas in Optimization, pp. 79–108. Mc Graw-Hill, UK (1999)
7. Goldberg, D.: Genetic algorithms in search, Optimization and Machine Learning, Reading. Addison-Wesly Publishing Co., Massachusetts (1989)
8. Schwefel, H.-P. (ed.): Evolution and Optimization Seeking. John Wiley and Sons, New York (1995)
9. Mezura-Montes, E., Velazquez-Reyes, J., Coello, C.: Modified differential evolution for constrained optimization. In: Proceedings of IEEE Congress on Evolutionary Computation, pp. 25–32. Vancouver (2006)
10. Ivanov, V.: On the approach for automatic generation of small embedded PicoBlaze system. In: Proceedings of the 13th International Conference on ACSD, pp. 257–260. Barcelona (2013)
11. Balabanov, T., Zankinski, I., Shumanov, B.: Slot machines RTP optimization with genetic algorithms. In: Dimov, I., Fidanova, S., Lirkov, I. (eds.) NMA 2014. LNCS, vol. 8962, pp. 55–61. Springer, Heidelberg (2015)

Application of Ants Ideas on Image Edge Detection

Stefka Fidanova$^{(\boxtimes)}$ and Zlatolilya Ilcheva

Institute of Information and Communication Technologies,
Bulgarian Academy of Sciences,
Acad. G. Bonchev Str. Bl25A, 1113 Sofia, Bulgaria
stefka@parallel.bas.bg, zlat@isdip.bas.bg

Abstract. The aim of the image edge detection is to find the points, in a digital image, at which the brightness level changes sharply. Normally they are curved lines called edges. Edge detection is a fundamental tool in image processing, machine vision and computer vision, particularly in the areas of feature detection and feature extraction. Edge detection may lead to finding the boundaries of objects. It is one of the fundamental steps in image analysis. Edge detection is a hard computational problem. In this paper we apply a multiagent system. The idea comes from ant colony optimization. We use the swarm intelligence of the ants to search the image edges.

1 Introduction

Edge detection plays an important role in image processing. It is a main stage in image segmentation, pattern recognition and scene analysis [6,7,11]. Edge detection refers to the process of identifying and locating sharp discontinuities in an image. The discontinuities are abrupt changes in pixel intensity which characterize boundaries of objects and textures in a scene. Classical methods of edge detection involve convolving the image with an operator, a 2-D filter, which is constructed to be sensitive to large gradients in the image while at the same time returning values of zero in uniform regions. There is an extremely large number of edge detection operators available, each designed to be sensitive to certain types of edges. The variables involved in the selection of an edge detection operator include: Edge orientation:(i) The geometry of the operator determines a characteristic direction in which it is most sensitive to edges. Operators can be optimized to look for horizontal, vertical, or diagonal edges; (ii) Noise environment: Edge detection is difficult in noisy images, since both the noise and the edges contain high-frequency content. Attempts to reduce the noise result in blurred and distorted edges. Operators used on noisy images are typically larger in scope, so they can average enough data to discount localized noisy pixels. This results in a less accurate localization of the detected edges; (iii) Edge structure: Not all edges involve a step change in intensity. Effects such as refraction or poor focus can result in objects with boundaries defined by a gradual change in intensity. The edge detection operator needs to be chosen to be responsive

© Springer International Publishing Switzerland 2015
I. Lirkov et al. (Eds.): LSSC 2015, LNCS 9374, pp. 218–225, 2015.
DOI: 10.1007/978-3-319-26520-9_23

to such a gradual change in those cases. There are many ways to perform edge detection. However, the majority of different approaches may be grouped into two categories [6,7,9,11]: Gradient based operators (First order edge detection): The gradient approach detects edges by looking for the maximum or minimum in the first derivative of the image. Laplacian based operators (Second-order edge detection): The Laplacian approach searches for zero crossings in the second derivative of the image to find edges. (i) Gradient based edge operators are based on the use of a first order derivative. Roberts, Prewitts and Sobel are classified as typical operators in [6,9]. The Roberts cross operator consists of a pair of 2×2 convolution kernels. One kernel is the other rotated by 90 deg. The Prewitts and Sobel operators consist of a pair of 3×3 convolution kernels as one kernel is the other rotated by 90 deg. The two kernels of the three operators are used for detecting vertical and horizontal edges. These classical operators are easy to operate but highly sensitive to noise. (ii) Laplacian based operators are based on the second order derivative, in particular, on the Laplacian. These operators mark a pixel as an edge at a position where the second derivative of the image function becomes zero [6,9]. The Laplacian of Gaussian (LoG) or Marr-Hildreth edge detector uses both Gaussian and Laplacian operator so that the Gaussian operator reduces the noise and the Laplacian operator detects the sharp edges. The Marr-Hildreth operator, however, suffers from two main limitations. It generates responses that do not correspond to edges, so-called false edges, and its localization error may be rather serious at curved edges. In the field of edge detection, the edge detector of Canny is of particular importance [3]. The method was developed by John F. Canny in 1986. It is an optimal edge detection technique as it provides good detection, clear response and good localization. The process of the Canny edge detection algorithm can be broken down into 5 different steps: 1. Apply Gaussian filter to smooth the image in order to remove the noise; 2. Find the intensity gradients of the image; 3. Apply non-maximum suppression as an edge thinning technique; 4. Apply double threshold to determine potential edges; 5. Track edges by hysteresis: Finalize the detection of edges by suppressing all other edges that are weak and not connected to strong edges. Some disadvantages of the edge detection algorithm of Canny and some means to overcome them are presented in [13].

In our paper we apply multiagent system for image edge detection. The idea comes from Ant Colony Optimization (ACO) methods. ACO approach is applied to solve hard combinatorial optimization problems, which need huge amount of computational resources. We use the swarm intelligence of the ant to look for brightness change and to detect images. We simulate ant behavior, in a similar way as in optimization applications.

The paper is organized as follows. In Sect. 2 we describe the traditional ACO. In Sect. 3 we propose our ACO-based image edge detection algorithm. Experimental results are shown in Sect. 4. In Sect. 5 we draw some conclusions and directions for future work.

2   Ant Colony Optimization Algorithm

The idea for ant algorithm comes from the real ant behavior. They put on the ground chemical substance called pheromone, which help them to return to their nest when they look for a food. The ants smell the pheromone and follow the path with a stronger pheromone concentration. Thus they find shorter path between the nest and the source of the food. The ACO algorithm uses a colony of artificial ants that behave as cooperating agents. With the help of the pheromone they try to construct better solutions and to find the optimal ones. The problem is represented by a graph and the solution is represented by a path in the graph or by tree in the graph. For the successes of the algorithm, it is very important how the graph will be constructed. Ants start from random nodes of the graph and construct feasible solutions. When all ants construct their solution the pheromone values are updated. Ants compute a set of feasible moves and select the best one, according to the transition probability rule. The transition probability p_{ij}, to chose the node j when the current node is i, is based on the heuristic information η_{ij} and on the pheromone level τ_{ij} of the move, where $i, j = 1, \ldots, n$. α and β show the importance of the pheromone and the heuristic information respectively.

$$p_{ij} = \frac{\tau_{ij}^{\alpha} \, \eta_{ij}^{\beta}}{\sum\limits_{k \in \{allowed\}} \tau_{ik}^{\alpha} \, \eta_{ik}^{\beta}} \tag{1}$$

The heuristic information is problem dependent. It is appropriate combination of problem parameter and is very important for ants management. The ant selects the move with highest probability. The initial pheromone is set to a small positive value τ_0 and then ants update this value after completing the construction stage [2,4,5]. The search stops when $p_{ij} = 0$ for all values of i and j.

The pheromone trail update rule is given by:

$$\tau_{ij} \leftarrow \rho\tau_{ij} + \Delta\tau_{ij}, \tag{2}$$

where $\Delta\tau_{ij}$ is a new added pheromone and it depends of the quality of achieved solution.

The pheromone is decreased with a parameter $\rho \in [0; 1]$. This parameter models evaporation in the nature and decreases the influence of old information in the search process. After that, we add the new pheromone, which is proportional to the quality of the solution (value of the fitness function). There are several variants of ACO algorithm. The main difference is the pheromone updating.

3   Image Edge Detection Algorithm Inspired by ACO

The problem of Image edges detection is a problem to find pixels of the image which correspond to edges. The two dimensional image is represented as $N \times M$

matrix. The element (i, j) of the matrix corresponds to the pixel (i, j), and its value is the pixel intensity. The graph of the problem is as follows. The nodes of the graph corresponds to the pixels. The arcs of the graph connect adjacent nodes. Every internal node has 8 adjacent nodes: up and down nodes, left and right nodes and the four diagonal nodes. If the node (i, j) is internal node, the set of its adjacent nodes is $\{(i - 1, j), (i - 1, j - 1), (i, j - 1), (i + 1, j - 1), (i + 1, j), (i + 1, j + 1), (i, j + 1), (i - 1, j + 1)\}$.

The ants start to create the solution starting from random node. When the ACO algorithm solve optimization problem, every ant create its own solution and at the end of the iteration we compare them and choose the best one for the current best solution. On the next iteration the ants try to find better solutions. When we apply ant idea on image edge detection, the ants try to construct part of the image edges and merging the edges detected by individual ant we construct the image edges. In traditional ACO algorithm for optimization problems, ants construct new solutions taking in to account the regions with good solutions according their experience, marked by the pheromone. In our application in every iteration the ants include new edges in current solution, thus they continue to rebuild the image edges.

In our algorithm an ant start, to look for edges, from random internal node (i, j), if the image matrix is $N \times M$, $0 < i < N - 1$ and $0 < j < M - 1$. The ant move to one of the adjacent nodes according transition probability rule:

$$p_{i,j} = \tau_{ij} * \eta_{ij} \tag{3}$$

where $\eta_{ij} = \frac{V_{ij}}{Vmax}$, $Vmax$ is the maximal value of V_{ij} for $0 < i < N - 1$ and $0 < j < M - 1$.

$$V_{ij} = |I_{i-1,j-1} - I_{i+1,j+1}| + |I_{i-1,j} - I_{i+1,j}| + \\ |I_{i-1,j+1} - I_{i+1,j-1}| + |I_{i,j-1}, I_{i,j+1}| \tag{4}$$

An ant stops to add new pixels in the solution when $p_{ij} < \epsilon_1$, where ϵ_1 is a parameter. When all ants finish to create their edge we delete false edges, new edges consisting only one node. At the end of every iteration, the pheromone of the visited pixels is updated according the rule:

$$\tau_{ij} = \rho * \tau_{ij} + (1 - \rho) * (\tau_{init} + \epsilon) \tag{5}$$

where τ_{init} is the initial pheromone value and ϵ is a small number close to 0.

At the next iterations ants again start from random nodes and add new edges on current solution. The algorithm continue till no new edges are detected. The end condition is:

$$|T(k - 1) - T(k)| < \epsilon \tag{6}$$

where $T(k)$ is the sum of the pheromone of all pixels. If the pheromone of the image is not changed, it means that new edge is not detected.

Fig. 1. Lena

4   Experimental Results

In this section we show some experimental results. The experiments are on well know image called Lena and which is used in image processing as a good example, Fig. 1. The image has a size of 256×256 pixels. By the parameter ϵ_1 we control how detailed to be the edges. We can need different level of details for different use of edge detection. On Fig. 2 the ϵ_1 parameter is equal to 0.05 and to 0.25. We run the algorithm on computer with Pentium processor 2.8 GHz. We find the reported image edges for 4 seconds. When the value of parameter ϵ_1 is small the image edges are much more detailed than, when the value of the ϵ_1 is higher. We observe that our algorithm can find the image edges with big details.

The parameters of our algorithm are:

- $\tau_{init} = 0.5$ - initial pheromone
- $\rho = 0.5$ - evaporation parameter
- $A = 50$ - number of ants
- $\epsilon = 0.00001$
- $n = 1000$ - number of iterations

The number of iterations is set to be 1000, but because the end condition, the algorithm performs up to 300 iterations.

We compare our algorithm on other well known test images too as FRUIT-Copy, Fruits, Mandrill and Peppers. We achieved very detailed image edges without beforehand smoothing the image.

Let us compare the image edges achieved by our algorithm with this achieved by ant algorithms, proposed by other authors. Our algorithm finds much more detailed edges comparing with ant based algorithms proposed in [1,10,12]. The image edges detected by the ant algorithm proposed in [8] are detailed as the image edges detected by our algorithm, but their algorithm detects some false edges too (Figs. 3, 4, 5 and 6). Other algorithms for images edge detection are

Fig. 2. Detected edges when $\epsilon_1=0.05$ and $\epsilon_1=0.25$

Fig. 3. FRUIT-Copy

Fig. 4. Fruits

Fig. 5. Mandrill

Fig. 6. Peppers

wavelet based algorithms [14]. They achieves similar or worst edges detection, but the algorithm complexity is higher. These algorithms need smoothing of the original image to find good result. The complexity of the wavelet based algorithms is $O(L \times M \times N)$, where M and N is the image size and L is the length of the used filters. In ACO algorithm we fixed the number of ants to be 50 and every one of the ants starts from random pixel and adds new pixels, in the edge set. Thus our algorithm do not need to perambulate all pixels. The complexity of the proposed algorithm is $A \times n \times L_1$, where A is the number of used ants, n is the number of iterations and L_1 is the number of added pixels by one ant. Thus the algorithm complexity is $O(N + M)$.

5   Conclusion

In this paper an ACO based algorithm for image edge detection is presented. By parameters we can control the how detailed to be the edges. The algorithm deletes the false edges. Experimental results show the feasibilities of achieved results. Comparison with ACO algorithms proposed by other authors show that our algorithm find more detailed image edges and less false edges. We can conclude that achieved results are very encouraging. Our algorithm is fast and achieves good solutions.

Acknowledgments. This work was supported by the Bulgarian National Scientific Fund under the grants DFNI 02/20 "Efficient Parallel Algorithms for Large Scale Computational Problems" and DFNI 02/5 "InterCriteria Analysis. A New Approach to Decision Making" and by EC grant AcomIn.

References

1. Baterina, A.V., Oppus, C.: Image edge detection using ant colony optimization. WSEAS Trans. Sig. Process. **6**(8), 58–67 (2010)
2. Bonabeau, E., Dorigo, M., Theraulaz, G.: Swarm Intelligence: From Natural to Artificial Systems. Oxford University Press, New York (1999)

3. Canny, J.F.: A computational approach to edge detection. IEEE Trans. Pattern Anal. Mach. Intell. (PAMI) **8**(6), 679–697 (1986)
4. Dorigo, M., Stutzle, T.: Ant Colony Optimization. MIT Press, Cambridge (2004)
5. Fidanova, S., Atanasov, K.: Generalized net model for the process of hybrid ant colony optimization. C. R. l'Academie. Bulgare Sci. **62**(3), 315–322 (2009)
6. Gonzalez, R.C., Woods, R.E.: Digital Image Processing. Prentice-Hall Inc., Upper Saddle River (2002)
7. Jain, A.K.: Fundamentals of Digital Image Processing. Prentice-Hall Inc, Upper Saddle River (1989)
8. Jevtic, A., Li, B.: Ant algorithm for adaptive edge detection. In: Abrao, T. (ed.) Search Algorithms for Engineering Optimization, Chapter 2, INTECH publisher (2013)
9. Mlsna, P.A., Rodriguez, J.J.: Gradient and laplacian-type edge detection. In: Bovik, A. (ed.) Handbook of Image and Video Processing, pp. 415–431. Academic Press, San Diego (2000)
10. Nezamabadi-pour, H., Saryazdi, S., Rashedi, E.: Edge detection using ant algorithm. Soft Comput. **10**(7), 623–628 (2006)
11. Pratt, W.K.: Digital Image Processing, 2nd edn. Wiley, New York (1991)
12. Tian, J., Yu, W., Xie, S.: An ant colony optimization algorithm for image edge detection. In: IEEE Congress on Evolutionary Computation, pp. 751–756. Hong Kong (2008)
13. Zhou, P., Ye, W.Q., Wang, Q.: An improved canny algorithm for edge detection. J. Comput. Inf. Syst. **7**(5), 1516–1523 (2011)
14. Zhang, Z., Ma, S., Liu, H., Gong, Y.: An edge detection approach based on directional wavelet transform. J. Comput. Math. Appl. **57**(8), 1265–1271 (2009)

ACD with ESN for Tuning of MEMS Kalman Filter

Petia Koprinkova-Hristova$^{(\boxtimes)}$ and Kiril Alexiev

Institute of Information and Communication Technologies,
Bulgarian Academy of Sciences,
Acad. G. Bonchev Str. Bl.25A, 1113 Sofia, Bulgaria
{pkoprinkova,alexiev}@bas.bg

Abstract. In the present work we designed a neuro-fuzzy approach for on-line optimal tuning of a Kalman filter of a gyroscope within a Micro ElectroMechanical Sensor (MEMS) device. It consists of Adaptive Critic Design (ACD) scheme in which the controller (a Fuzzy Rule Base (FRB) designed to adapt the measurement noise covariance matrix of a Kalman filter) is tuned using only information about the direction to which the estimation error changes (increase or decrease). A novel fast training dynamic neural network structure - Echo state network (ESN) - was used in the role of the critic element. Application to data collected from real MEMS demonstrated the ability of the proposed approach to tune Kalman filter and improve the quality of its estimates in changing working conditions of the MEMS in real time.

Keywords: Adaptive Critic Design · Echo state network · Fuzzy rule base · Kalman filter · Micro electromechanical sensors

1 Introduction

The Kalman filter is a commonly used tool for state estimation of micro electromechanical sensors (MEMS) [15]. The main problem arising in practical applications is due to incomplete a priori information about the covariance matrices of measurement noise and of the estimated process. Nowadays there are developed different intelligent approaches for their estimation. On one hand the fuzzy logic is a powerful tool for description of accumulated knowledge how to tune Kalman filter [5,12,18,19,23,24]. However experts information is too subjective. Another adaptive approach exploits neural networks ability to be trained without exact knowledge about the process model [1,6,13,14,20]. The disadvantage of this technique is the need of big amount of experimental data to tune neural network model as well as the need to re-train it in case of changes in working conditions. Fortunately the application of adaptive algorithms of neural networks for fine tuning of linguistically defined fuzzy rule bases offers a powerful tool that overcomes disadvantages of both approaches [22].

The adaptation of Kalman filter covariance matrix of measurement noise needs to account for MEMS working conditions in real time having minimum

© Springer International Publishing Switzerland 2015
I. Lirkov et al. (Eds.): LSSC 2015, LNCS 9374, pp. 226–233, 2015.
DOI: 10.1007/978-3-319-26520-9_24

available information. Hence reinforcement learning (RL) algorithms are the most proper tool adopted from neural networks [21]. In present study we proposed to tune defined by experts FRB [15] for on-line adaptation of covariance matrix using RL approach called Adaptive Critic Design [21]. In order to apply ACD we need information about quality of estimations as reinforcement signal. In [8–11] GPS signal was used for this aim. Since our device does not contain such sensor we'll use another kind of additional information coming from a camera and assessment of its intentional movement.

The paper is organized as follows: next section describes briefly ACD algorithm, FRB and Kalman filter design; section three describes our MEMS device and experimental set-up; section four presents results and discussion followed by concluding remarks.

2   Methods and Algorithms

2.1   Adaptive Critic Design

The main scheme of RL approach called ACD is shown on Fig. 1. It solves in forward manner the following task: for the given discrete dynamical system with state $x(k)$ find a time profile of control actions $\Delta\sigma(0)$ $\Delta\sigma(1)$ ... $\Delta\sigma(N-1)$ that maximizes (minimizes) given utility function $U(k)$ through all time steps $k = 1 \div N$.

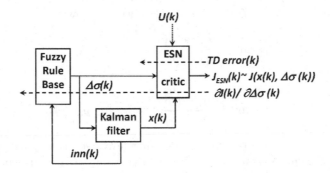

Fig. 1. Adaptive Critic Design scheme.

In our case utility function depends on the difference between predictions of Kalman filter (that is system state $x(k)$) and reference signal $x_{ref}(k)$ coming from additional sensor (here we suppose that this could be movement assessment from the camera) as follows:

$$U(k) = \begin{cases} -0.1 \ if & x(k) - x_{ref}(k) \leq -0.0001 \\ 0 \ \ if & -0.0001 < x(k) - x_{ref}(k) < 0.0001 \\ 0.1 \ if & x(k) - x_{ref}(k) \geq 0.0001 \end{cases} \tag{1}$$

The RL algorithms and their gradient version ACD were created to cope with the "curse of dimensionality" of above task. It consists of training of a model (usually neural network) called "adaptive critic" or briefly critic that approximates Bellman's equation [4]. Training of the critic is done by minimization of the "temporal difference error" (TD error) [25]. Having adequately trained critic allows solving of the above optimization task in forward manner in real time by tuning of the "controller" (in our case it is the FRB that is used to tune Kalman filter) using a gradient algorithm. Here as critic element we used a fast trainable recurrent neural network called Echo state network (ESN). More details about its structure, TD error training algorithm and gradient training of the controller can be found in [16,17].

2.2   Fuzzy Rule Base

Our FRB is adopted from the idea proposed in [18]. It consists of the following three fuzzy rules:

$$IF\ inn(k)\ is\ ZERRO_{inn}\qquad THEN\ \Delta\sigma(k)\ is\ ZERRO_{\Delta\sigma}$$
$$IF\ inn(k)\ is\ POSITIVE_{inn}\quad THEN\ \Delta\sigma(k)\ is\ NEGATIVE_{\Delta\sigma}$$
$$IF\ inn(k)\ is\ NEGATIVE_{inn}\ THEN\ \Delta\sigma(k)\ is\ POSITIVE_{\Delta\sigma}$$

Here ZERO, NEGATIVE and POSITIVE denote the corresponding fuzzy values of the two linguistic variables ($\Delta\sigma(k)$ and $inn(k)$) that are fuzzy numbers defined with triangular membership functions with increasing and decreasing parts (denoted by indexes I and D respectively) as follows:

$$\mu_I(var) = \frac{var - p_1}{p_2 - p_1}, \ var \in [p_1, p_2] \tag{2}$$

$$\mu_D(var) = \frac{p_3 - var}{p_3 - p_2}, \ var \in [p_2, p_3] \tag{3}$$

Here p_1, p_2 and p_3 are parameters defining universe of discourse of the corresponding fuzzy number and var denotes crisp value of the corresponding linguistic variable. Variable inn is the innovation calculated by Kalman filter and $\Delta\sigma(k)$ is the "prescribed" by FRB change of the standard deviation of the measurement noise.

Calculation of crisp output from our FRB is done using minimum as implication function and mean of maxima method for defuzzyfication [17].

Tuning of FRB parameters (these include all p_1, p_2 and p_3 parameters of all fuzzy values) is done by backpropagation of utility that is gradient algorithm as follows:

$$p^{new} = p^{old} - \eta \frac{\partial^* J}{\partial p} \tag{4}$$

Here $0 < \eta < 1$ is learning rate and $\frac{\partial^* J}{\partial p}$ is ordered derivative of the predicted by the critic discounted sum of future utilities with respect to the corresponding vector of parameters:

$$p = \left[p_1^{ZERO_{inn}}, p_2^{ZERO_{inn}}, p_3^{ZERO_{inn}}, \ldots, p_1^{ZERO_{\Delta\sigma}}, p_2^{ZERO_{\Delta\sigma}}, p_3^{ZERO_{\Delta\sigma}} \right]$$

More details about gradient algorithm and calculation of derivatives of J with respect to FRB parameters can be found in [17].

2.3 Kalman Filter

The Kalman filter is a linear minimum variance unbiased estimator [2,3,7], used here to filter gyroscope measurement noise. The state equation in its discrete realization is:

$$x(k+1) = F x(k) + w(k) \tag{5}$$

where k is discrete time; w approximates non-modeled dynamics and parameters uncertainties; the state vector $x = [\omega \; \dot{\omega} \; \ddot{\omega}]$ consists of predicted rotation rate ω and its first and second derivative; F is transition matrix as follows:

$$F = \begin{pmatrix} 1 & \Delta t & \Delta t^2/2 \\ 0 & 1 & \Delta t \\ 0 & 0 & 1 \end{pmatrix} \tag{6}$$

Here Δt is constant sampling interval. The process noise can be approximated or intuitively estimated. In our case a proper choice is:

$$Q = E\left[w(k) w(k)^T \right] = \begin{pmatrix} \Delta t^5/20 & \Delta t^4/8 & \Delta t^3/6 \\ \Delta t^4/8 & \Delta t^3/6 & \Delta t^2/2 \\ \Delta t^3/6 & \Delta t^2/2 & \Delta t \end{pmatrix} \sigma_w^2 \tag{7}$$

Here σ_w is process noise standard deviation. The measurement equation is:

$$y(k+1) = H x(k) + v(k) \tag{8}$$

where $y = [\omega_{measured} \; 0 \; 0]^T$ and $\omega_{measured}$ is received raw data from the gyroscope; the observation matrix H is:

$$H = \begin{pmatrix} 1 & 0 & 0 \\ 0 & 0 & 0 \\ 0 & 0 & 0 \end{pmatrix} \tag{9}$$

and v is additive measurement noise estimated from stand-by mode of the system or taken from catalog sensor data as follows:

$$R = E\left[v(k) v(k)^T \right] = \begin{pmatrix} 1 & 0 & 0 \\ 0 & 10 & 0 \\ 0 & 0 & 100 \end{pmatrix} \sigma_v^2 \tag{10}$$

where σ_v is the standard deviation of the measurement noise.

The recursive steps of the Kalman filter are as follows:

Prediction step: calculates predicted state and state prediction covariance based on their estimates from previous update step:

$$\hat{x}(k+1|k) = F\hat{x}(k|k)$$
$$\hat{P}(k+1|k) = F\hat{P}(k|k)F^T + Q \tag{11}$$

Update step (posteriori step): after receiving of new measurement from the sensor at this step the predicted state and covariance matrix are corrected in accordance with the received new information as follows:

$$S = H\hat{P}(k+1|k)H^T + R$$
$$\hat{y}(k+1|k) = H\hat{x}(k+1|k)$$
$$K = \hat{P}(k+1|k)H^T S^{-1}$$
$$inn(k+1) = y(k+1) - \hat{y}(k+1|k) \tag{12}$$
$$\hat{x}(k+1|k+1) = \hat{x}(k+1|k) + Kinn(k+1)$$
$$\hat{P}(k+1|k+1) = (I - KH)\hat{P}(k+1|k)$$

3   Experimental Set-Up

The gyroscope is an inertial device, for measuring rotation rate. In our experiment it is a part of MEMS that has one gyroscope as well as one accelerometer at each axis in 3-dimensional space. The MEMS is incorporated into a mobile phone. The integral of rotation rate gives the change of orientation of the system.

The raw data were collected during rotation of the mobile device around one of its axes (in our case it was y axis). Since our MEMS has different time stamp for gyroscopes and accelerometers, we preprocessed and approximated raw data so as to have them with constant time step Δt at the same discrete moments in time k.

Initial measurement noise standard deviation σ_v of our gyro sensor was calculated based on the first 70 measured data. Then we apply the ACD scheme from Fig. 1 to adjust σ_v as follows:

$$\sigma_v{}^{new} = \sigma_v{}^{old} - \beta\Delta\sigma(k) \tag{13}$$

Since at this stage we do not have reference signal from the camera, we replaced it with measurement data signal.

4   Results and Discussion

In present study we tested our algorithm to tune described above Kalman filter for one of the three gyroscope sensors (placed on y axis) of our MEMS.

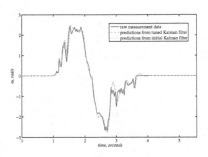 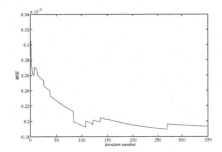

Fig. 2. Predictions of Kalman filter before and after tuning in comparison with raw data (left) and MSE trough training iterations (right).

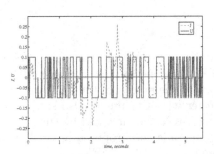 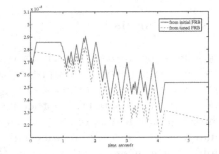

Fig. 3. Predicted by critic utility function (left) and tuned by FRB measurement noise standard deviation (right).

Figure 2(left) represents predictions of Kalman filter before and after tuning in comparison with reference signal. Figure 2(right) presents changes of the mean square error (MSE) during training iterations that is calculated as follows:

$$MSE = \frac{1}{N} \sum_{k=1}^{N} (x(k) - x_{ref}(k))^2 \tag{14}$$

We observed that after optimization of FRB the predictions of Kalman filter become closer to the reference signal that is proved by the decrease of the MSE.

Figure 3(left) represents predictions J of the ESN critic in comparison with the utility function U. It is observed that the critic copes well with the task to predict on time increase or decrease of reinforcement signal. Figure 3(right) represents initial prediction of measurement noise standard deviation σ_v by FRB and assessed by trained FRB measurement noise standard deviation σ_v.

Figure 4 represents initial (solid lines) and tuned (dash lines) membership functions of input (inn) and output $(\Delta\sigma)$ linguistic variables of the FRB.

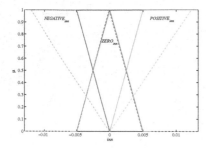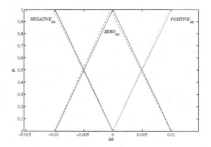

Fig. 4. Initial (solid line) and tuned (dash line) membership functions of input (left) and output (right) linguistic variable of the FRB.

5   Conclusions

Combination between fuzzy logic for description of expert information and reinforcement learning for fine adjusting of FRB demonstrated the power of the proposed algorithm for optimal tuning of covariance matrix of the measurement noise of the Kalman filter. Another parameter whose tuning is also subjective is the covariance matrix of the estimated process. Hence it could be tuned in a similar way. Our further investigations will upgrade the proposed algorithm to simultaneous tuning of both matrices R and Q.

Acknowledgment. The research work reported in the paper is partly supported by the project AComIn, grant 316087, funded by the FP7 Capacity Programme (Research Potential of Convergence Regions).

References

1. Amir, P.: Enhanced SLAM for a mobile robot using unscented Kalman filter and radial basis function neural network. Res. J. Recent Sci. **2**(2), 69–75 (2013)
2. Bar-Shalom, Y., Li, X.R.: Estimation and Tracking: Principles, Techniques and Software. Artech House, Boston (1993)
3. Bar-Shalom, Y., Li, X.R., Kirubarajan, T.: Estimation with Applications to Traching and Navigation: Algorithms and Software for Information Extraction. Wiley, Chichester (2001)
4. Bellman, R.E.: Dynamic Programming. Princeton University Press, Princeton (1957)
5. Choomuang, R., Afzulpurkar, N.: Hybrid Kalman filter/fuzzy logic based position control of autonomous mobile robot. Int. J. Adv. Robot. Syst. **2**(3), 197–208 (2005)
6. Filho, E.A.M., Neto, A.R., Kuga, H.K.: A low cost INS/GPS navigation system integrate with a multilayer feed forward neural network. J. Aeros. Eng. Sci. Appl. **II**(2), 26–36 (2010)
7. Gelb, A. (ed.): Applied Optimal Estimation. The MIT Press, Cambridge (2001)
8. Goodall, C.L.: Improving usability of low-cost INS/GPS navigation systems using intelligent techniques. Ph.D. thesis, University of Calgary (2009)

9. Goodall, C., El-Sheimy, N., Syed, Z.: On-line tuning of an extended Kalman filter for INS/GPS navigation applications. In: ION GNSS 21st International Technical Meeting of the Satellite Division, Savannah, GA, pp. 38–47, 16–19 September 2008
10. Goodall, C., Niu, X., El-Sheimy, N.: Intelligent tuning of a Kalman filter for INS/GPS navigation applications. In: ION GNSS 20th International Technical Meeting of the Satellite Division, Fort Worth, TX, pp. 2121–2128, 25–28 September 2007
11. Guo, H.: Neural network aided Kalman filtering for integrated GPS/INS navigation system. TELKOMNIKA **11**(3), 1221–1226 (2013)
12. Hasan, A.M., Samsudin, K., Ramli, A.R.: GPS/INS integration based on dynamic ANFIS network. Int. J. Control Autom. **5**(3), 1–21 (2012)
13. Jwo, D.-J., Huang, H.-C.: Neural network aided adaptive extended Kalman filtering approach for DGPS positioning. J. Navig. **57**, 449–463 (2004)
14. Kihas, D., Djurović, Ž.M., Kovačević, B.D.: Adaptive filtering based on recurrent neural networks. J. Autom. Control Univ. Belgrade **13**(1), 13–24 (2003)
15. Kramer, J.A.: Accurate localization given uncertain sensors. Thesis Submitted in Partial Fulfillment of the Requirements for the Degree of Master of Science in Computer Science, University of South Florida (2010)
16. Koprinkova-Hristova, P., Oubbati, M., Palm, G.: Heuristic dynamic programming using echo state network as online trainable adaptive critic. Int. J. Adapt. Control Sig. Process. **27**(10), 902–914 (2013)
17. Koprinkova-Hristova, P.: Backpropagation through time training of a neuro-fuzzy controller. Int. J. Neural Syst. **20**(5), 421–428 (2010)
18. Loebis, D., Sutton, R., Chudley, J., Naeem, W.: Adaptive tuning of a Kalman filter via fuzzy logic for an intelligent AUV navigation system. Control Eng. Pract. **12**, 1531–1539 (2004)
19. Matia, F., Jimenez, A., Al-Hadithi, B.M., Rodriguez-Losada, D., Galan, R.: The fuzzy Kalman filter: state estimation using possibilistic techniques. Fuzzy Sets Syst. **157**, 2145–2170 (2006)
20. Petrović, E., Ćojbašić, Ž., Ristić-Durrant, D., Nikolić, V., Ćirić, I., Matić, S.: Kalman filter and NARX neural network for robot vision based human tracking. Facta Univ. Ser. Autom. Control Robot. **12**(1), 43–51 (2013)
21. Prokhorov, D.V.: Adaptive critic designs and their applications. Ph.D. dissertation. Department of Electrical Engineering, Texas Technical University (1997)
22. Si, J., Wang, Y.-T.: On-line learning control by association and reinforcement. IEEE Trans. Neural Netw. **12**(2), 264–276 (2001)
23. Subramanian, V., Burks, T.F., Dixon, W.E.: Sensor fusion using fuzzy logic enhanced Kalman filter for autonomous vehicle guidance in citrus groves. Trans. ASABE **52**(5), 1411–1422 (2009)
24. Wang, L., Wang, F.: Intelligent calibration method of low cost MEMS inertial measurement unit for an FPGA-based navigation system. Int. J. Intell. Eng. Syst. **4**(2), 32–41 (2011)
25. Sutton, R.S.: Learning to predict by methods of temporal differences. Mach. Learn. **3**, 9–44 (1988)

Optimal Discretization Orders for Distance Geometry: A Theoretical Standpoint

Antonio Mucherino[✉]

IRISA, University of Rennes 1, Rennes, France
antonio.mucherino@irisa.fr

Abstract. Distance geometry consists in embedding a simple weighted undirected graph $G = (V, E, d)$ in a K-dimensional space so that all distances d_{uv}, which are the weights on the edges of G, are satisfied by the positions assigned to its vertices. The search domain of this problem is generally continuous, but it can be discretized under certain assumptions, that are strongly related to the order given to the vertices of G. This paper formalizes the concept of optimal partial discretization order, and adapts a previously proposed algorithm with the aim of finding discretization orders that are also able to optimize a given set of objectives. The objectives are conceived for improving the structure of the discrete search domain, for its exploration to become more efficient.

1 Introduction

Let $G = (V, E, d)$ be a simple weighted undirected graph. The Distance Geometry Problem (DGP) [10,16] asks whether an embedding $\sigma : V \longrightarrow \mathbb{R}^K$ of G exists in an Euclidean space having dimension $K > 0$ so that all distances d_{uv}, for each $(u, v) \in E$, are satisfied. The DGP is NP-hard [17], and its search space is, in general, continuous. However, under particular assumptions, it can be discretized, and reduced to a tree [11]. The *discretization* of the DGP does not reduce the problem complexity (which is still NP-hard [7]), but it allows for developing combinatorial algorithms performing a search on a tree [9]. The discretization makes it possible to work with a discrete and finite search space, which is continuous (and hence infinite) otherwise.

This paper focuses on the problem of finding a suitable order for the vertices of G that allows for the discretization. The discretization assumptions, in fact, strongly depend on the vertex order associated to the vertices of G (see Sect. 2). *Discretization orders* can be either total or partial. In chronological order, discretization orders were initially handcrafted for a particular class of DGP instances related to protein conformations [3,8]. These handcrafted orders are total orders. Subsequently, new methods for the identification of discretization orders were proposed [5,6,12]: these methods are able to automatically detect total and partial orders for any DGP, and they were used for constructing orders satisfying different properties for the protein backbones. More recently, in [13], discretization orders were represented as sequences of overlapping cliques of G.

© Springer International Publishing Switzerland 2015
I. Lirkov et al. (Eds.): LSSC 2015, LNCS 9374, pp. 234–242, 2015.
DOI: 10.1007/978-3-319-26520-9_25

In comparison with these previous works, the present paper attempts for the first time to develop a theory for optimal discretization orders, which can be at the basis of novel efficient methods. Depending on the desired additional properties for the searched discretization orders, the ordering problem can actually be either NP-hard or solvable in polynomial time [2]. The use of heuristics might therefore be necessary for finding orders satisfying particular properties. In this work, the focus will be on orders that can be identified in polynomial time.

A discretization order is said to be *optimal* when it optimizes a given set of objectives, which are supposed to be conceived for improving the structure of the search tree, with the aim of making its exploration more efficient. These objectives define multi-level optimization problems, whose solutions are subsets of optimal vertices. An algorithm, that was previously presented in [6,12], is adapted in this work for dealing with partial orders and with several objectives, which imply the definition of several (small) multi-level optimization problems.

This paper is organized as follows. Partial discretization orders are briefly discussed in Sect. 2, with some extended definitions that were previously given for total orders. Section 3 formally introduces optimal partial discretization orders, where a given set of objectives needs to be optimized, and proposes an algorithm for the construction of optimal orders. Section 4 concludes the paper.

2   Partial Discretization Orders

Let $G = (V, E, d)$ be a simple weighted undirected graph representing an instance of the DGP. Let $d : (u, v) \in E \longrightarrow [\underline{d}_{uv}, \bar{d}_{uv}] \subset \mathbb{R}_+$ be the function that associates a weight to each edge $(u, v) \in E$. In general, each weight is a real-valued interval providing the lower and the upper bound on the known distances. When this interval is degenerate, i.e. $d_{uv} = \underline{d}_{uv} = \bar{d}_{uv}$, then the distance d_{uv} is said to be "exact". Let E' be the subset of E related to exact distances. Let S be the set of all subsets $s \subseteq V$.

Definition 2.1. *An ordered partition of V is a function $r : \mathbb{N} \longrightarrow S$ with length $|r| \in \mathbb{N}$ (for which $r_i = \emptyset$ for all $i > |r|$) such that, for each $v \in V$, there exist a non-empty subset $s \in S$ containing v and an index $i \in \mathbb{N}$ such that $r_i = s$.*

Since an ordered partition naturally induces a partial order, the function r will be referred to as a *partial order* on V in the rest of the text. Moreover, the function r also allows for the definition of partitions with repetitions, i.e. of orders where the same vertex can appear more than once. In fact, in Definition 2.1, there is no hypothesis on the intersection of pairs of subsets r_i (which should always be empty to guarantee the absence of repetitions). Repetitions can be necessary for the optimization of some objectives (see for example [13]). However, the possibility to allow repetitions in the orders implies the development of a more complex theory, that cannot be included in this paper for lack of space. It is here supposed therefore that repetitions are not allowed in the partial orders r considered in this work.

If no repetitions are allowed, the function r satisfies the typical properties of partial orders, i.e. the reflexivity, the asymmetry and the transitivity. The indices $i \in \mathbb{N}$ are the *ranks* of the partial order r. An order r is not total in general because the same rank can be associated to more than one vertex (belonging to the same subset r_i such that $|r_i| > 1$). The order becomes total if all subsets r_i have cardinality $|r_i| = 1$.

Let r be a suitable partial order for the vertices of the graph G. For every $v \in r_i$, the two following sets are of intest:

$$\Lambda_\alpha(r_i, v) = \{(u, v) \in E \mid \exists j < i : u \in r_j\},$$
$$\Lambda_\beta(r_i, v) = \{(v, u) \in E \mid \exists j \geq i : u \in r_j\}.$$

All edges in $\Lambda_\alpha(r_i, v)$ are between the vertex v and other vertices that appear earlier in the partial order (lower rank), and therefore suitable positions for them can be supposed to be known when the computation of positions for v is attempted [9]. For this reason, the edges in $\Lambda_\alpha(r_i, v)$ are the *reference distances* for $v \in r_i$. The vertices $u \in V$ such that $(u, v) \in \Lambda_\alpha(r_i, v)$ are named *reference vertices*. Inversely, the edges in $\Lambda_\beta(r_i, v)$ represent distances that serve as a reference for vertices having either the same or a greater rank. The cardinalities of the two sets Λ_α and Λ_β allow to define two important counters related to the partial orders:

$$\alpha(r_i) = \min_{v \in r_i} |\Lambda_\alpha(r_i, v)|, \qquad \beta(r_i) = \max_{v \in r_i} |\Lambda_\beta(r_i, v)|.$$

The choice of considering the minimal cardinality of $\Lambda_\alpha(r_i, v)$ and the maximal cardinality of $\Lambda_\beta(r_i, v)$ will be evident in the following. In order to distinguish between exact distances and nondegenerate distances (related to intervals), the following two counters are also introduced:

$$\alpha_{ex}(r_i) = \min_{v \in r_i} |\Lambda_\alpha(r_i, v) \cap E'|, \qquad \beta_{ex}(r_i) = \max_{v \in r_i} |\Lambda_\beta(r_i, v) \cap E'|.$$

Definition 2.2. *A partial discretization order in dimension K is a partial order* $r : \mathbb{N} \longrightarrow S$ *such that:*

(a) *$G[r_1, \ldots, r_K] \equiv (C, E_C)$ is a clique with $|C| = K$ and $E_C \subset E'$;*
(b) *$\forall i \in \{K+1, \ldots, |r|\}$, $\alpha(r_i) \geq K$ and $\alpha_{ex}(r_i) \geq K - 1$.*

where $G[\cdot]$ is the subgraph induced by a subset of vertices.

The assumptions in Definition 2.2 make it possible to discretize the search domain of a given DGP represented by the graph G [4]. In fact, assumption (a) allows to fix the positions of the first K vertices in the order r, avoiding this way to consider DGP congruent solutions that can be obtained by rotations and translations. For simplicity, it is supposed that reference vertices for a vertex $v \in r_i$ never belong to r_i (even if the counter β counts this kind of distances). This implies the necessity to have a total ordering on the first K vertices, i.e. on the ones that form the clique C. However, it is worth remarking that the internal order for this clique is not relevant, because every total order for the vertices of the clique can be chosen.

Assumption (b) ensures that, for every vertex v that do not belong to the initial clique C, at least K reference vertices exist for v, and that only one of them is related to a distance represented by an interval. Under these assumptions, the search space is reduced to a tree, which can be explored by employing branch-and-prune algorithms [8,9,15]. Once the first K vertices have been positioned by exploiting assumptions (a), the main idea is to explore (and generate) the search tree recursively. New candidate positions for the vertices are computed by using the distances that are ensured by assumptions (b). Additional distances that might be available can be then considered for verifying the feasibility of computed candidate positions. The identification of infeasible positions allows for pruning parts of the search tree.

3   Finding Optimal Discretization Orders

Let $G = (V, E, d)$ be a simple weighted undirected graph representing an instance of the DGP. The simple result presented in [12] for total discretization orders can be extended to partial orders r.

Proposition 3.1. *Necessary condition for G to admit a partial discretization order in dimension K is that, for every suitable order r on V,*

$$\forall i \in \{1, 2, \ldots, |r|\}, \quad \alpha(r_i) + \beta(r_i) \geq K.$$

Proof. Suppose that there exists a rank $i \in \mathbb{N}$, for a certain partial order r, for which $\alpha(r_i) + \beta(r_i) < K$. By definition, there exists $\hat{v} \in r_i$ such that the cardinality of $|\Lambda_\alpha(r_i, \hat{v})|$ is minimal and equal to $\alpha(r_i)$. Since $\beta(r_i)$ is instead a maximal cardinality, $|\Lambda_\beta(r_i, \hat{v})|$ is at most equal to $\beta(r_i)$. Therefore, for the vertex $\hat{v}$, $|\Lambda_\alpha(r_i, \hat{v})| + |\Lambda_\beta(r_i, \hat{v})| < K$, which implies the absence of a sufficient number of edges for this vertex for constructing a discretization order (i.e. no r_i containing $\hat{v}$ can satisfy the two discretization assumptions). □

Notice that a similar necessary condition depends on the counters $\alpha_{ex}(r_i)$ and $\beta_{ex}(r_i)$: for every $i \in \{1, 2, \ldots, |r|\}$, it is necessary that $\alpha_{ex}(r_i) + \beta_{ex}(r_i) \geq K - 1$ in order to discretize.

The algorithm proposed in [6] for dealing with total discretization orders can be extended to partial orders. Algorithm 1 gives a sketch of this algorithm in the case no objectives are to be optimized. The algorithm starts with selecting an initial clique (see assumption (a) in Definition 2.2), and to assign the first K ranks to its vertices. All other sets r_i are generated by removing, from the set of not yet employed vertices (recall that repetitions are not allowed), all the ones that do not satisfy the discretization assumption (b).

For every possible initial clique, Algorithm 1 constructs partial discretization orders, when they exist. For every constructed partial order, a set of different total orders can be defined. Moreover, other partial orders compatible with the

---

Algorithm 1. An algorithm for finding partial discretization orders

---

1: **Searching partial discretization orders**   *in: G out: r*
2:   // *initial clique*
3:   **choose** a K-clique (C, E_C) in V with $C = \{u_1, u_2, \ldots, u_K\}$ and $E_C \subset E'$
4:   **set** $r_1 = u_1$, $r_2 = u_2$, $\ldots$, $r_K = u_K$
5:   **set** $A = C$, $i = K + 1$
6:   // *constructing the rest of the order*
7:   **while** $(A \neq V)$ **do**
8:       $r_i = V \setminus A$
9:       **while** $(\alpha(r_i) < K$ and $r_i \neq \emptyset)$ **do**
10:           $r_i = r_i \setminus \{u\}$, for all $u = \arg\min_{v \in r_i} |\Lambda_\alpha(r_i, v)|$
11:       **end while**
12:       **while** $(\alpha_{ex}(r_i) < K - 1$ and $r_i \neq \emptyset)$ **do**
13:           $r_i = r_i \setminus \{u\}$, for all $u = \arg\min_{v \in r_i} |\Lambda_\alpha(r_i, v) \cap E'|$
14:       **end while**
15:       **if** $(r_i = \emptyset)$ **then**
16:           **break:** no possible orders; **choose** another initial clique
17:       **else**
18:           **let** $A = A \cup r_i$, $i = i + 1$
19:       **end if**
20:   **end while**

---

initial orders found by Algorithm 1 can also be generated. It is important to remark that Algorithm 1 places a given vertex $\hat{v}$ in the subset r_i because it cannot be included in any other r_j with $j < i$, because assumption (b) would otherwise not be satisfied. However, suitable discretization orders may place this vertex $\hat{v}$ in subsets r_j with $j > i$, as far as no other vertex with rank between i and j strictly needs $\hat{v}$ as a reference.

In this work, the aim is not only to find discretization orders, but rather to identify orders that are also able to optimize a given set of objectives. Such objectives are supposed to be conceived for having an impact on the structure of the search tree obtained with the discretization. The main idea is to generate search trees that can be explored in a more efficient way. The interest is in selecting optimal partial orders from the initial orders obtained by Algorithm 1.

The considered objectives are functions $f_\ell : \mathbb{N} \longrightarrow \mathbb{R}$ that associate ranks of a given order r (representing a subset r_i of vertices in the partial order) to a real number. The subscript $\ell \in \mathbb{N}$ is the label associated to every objective, which also gives the priority order for the objective (lower-numbered labels correspond to the objectives that are to be optimized first). The objectives are to be conceived in a way that non-optimal vertices can be unequivocally identified. It is supposed, without losing generality, that all objectives are to be maximized, as an objective f_ℓ can be minimized by maximizing $-f_\ell$.

Definition 3.2. *Given a set of $M > 0$ objectives f_ℓ, with priority levels $\ell \in \{0, 1, \ldots, M-1\}$, an optimal partial discretization order is a partial discretization order where every r_i, with $i > K$, is solution of the multi-level optimization*

Algorithm 2. Code to be added to Algorithm 1 for performing the optimization of the f_ℓ's

---

for (each objective f_ℓ, with $\ell = 0, 1, \ldots, M - 1$) **do**
  $r_i = \{v \in r_i : f_\ell \text{ is optimized}\}$
end for

---

problem

$$\max f_{M-1}(x_{M-1})$$
$$\text{s.t. } x_{M-1} = \arg\max f_{M-2}(x_{M-2})$$
$$\text{s.t} \ldots \tag{1}$$
$$\text{s.t. } x_1 = \arg\max f_0(x_0),$$

where x_0 is the initial set of vertices such that $\alpha(x_0) \geq K$ and $\alpha_{ex}(x_0) \geq K - 1$ (all vertices in x_0 admit rank i).

Multi-level optimization is a class of difficult optimization problems (refer for instance to robust optimization, a survey can be found in [1], while [14] is an example of an application). However, the multi-level optimization problems considered in this context have relatively small search domains (their maximal cardinality is $|V|$), which are discrete and finite. These problems can be therefore solved by a simple exhaustive search.

Suppose that Algorithm 1 is able to find a partial order for a given graph G. Let x_M be the solution to the multi-level problem (1) for a certain r_i obtained by Algorithm 1 (hence r_i originally contains vertices that satisfy assumption **(b)**). If $x_M \neq r_i$, then not all vertices in r_i *optimize* the objectives f_ℓ. In this case, the vertices in $r_i \setminus x_M$ need to be moved to subsequent subsets r_j with $j > i$ (assumption **(b)** would not be satisfied if they were moved to lower-rank subsets). Next step is therefore to include all these "rejected" vertices in r_{i+1}, on which another multi-level problem (1) can be defined and solved. It is evident therefore how the optimization of the objectives can imply an increase on the total number of ranks in the partial orders. The proof of Theorem 3.3 is based on this idea.

The optimization of the objectives can be performed during the execution of Algorithm 1. To this purpose, the code in Algorithm 2 needs to be included between line 17 and 18 of Algorithm 1. The resulting algorithm will be referred to as Algorithms 1+2. In Algorithms 1+2, every set r_i, obtained by removing all vertices v that do not satisfy assumption **(b)**, is progressively filtered by applying the optimization of the objectives, in their priority levels. Notice that the set r_i cannot become empty during this optimization process.

The idea of optimizing the set of objectives during the search of discretization orders was firstly proposed in [5]. The intuition to employ a greedy algorithm comes instead from [6] (the inclusion of Algorithm 2 in Algorithm 1 makes in fact the algorithm a greedy one). Algorithms 1+2 has polynomial complexity and a quadratic worst-case complexity, achieved when the constructed order is actually total. No objectives f_ℓ having as definition domain the sets r_i can increase the complexity of the ordering problem, as far as they satisfy the hypotheses above. For example, the consecutivity assumption (refer to [13]), which makes the ordering problem NP-hard, can be seen as the combination of two objectives, where one of the two imposes the order to be total. When employing this latter objective, however, the non-optimal vertices in every r_i cannot be unequivocally selected, which goes against the above hypotheses.

As already remarked, Algorithm 1 is able to construct partial discretization orders, when they exist. Algorithm 1+2 does not consider anymore all possible orders, but only the ones for which the objectives are optimized. The following theorem proves that, if an order exists where the objectives are not optimized, then an optimal order also exists. This theorem extends a previous result initially presented in [6].

Theorem 3.3. *Let $G = (V, E, d)$ be a simple weighted undirected graph representing an instance of the DGP. When they exist, Algorithms 1+2 is able to construct partial discretization orders for G, which are optimal.*

Proof. By contradiction, consider that a partial discretization order exists for G but this algorithm is not able to identify it. In this hypothesis, there must exist a sequence of non-empty subsets r_i that covers V, which can be found by applying the original Algorithm 1. Therefore, the only reason why the algorithm cannot find the existing order is related to the optimization of the objectives f_ℓ for every r_i with $i > K$.

By hypothesis, non-optimal vertices can be unequivocally identified in the sets r_i: let $\hat{v}$ be a vertex belonging to the set r_i with the smallest rank i for which at least one objective f_ℓ is not optimized. Set $r_i = r_i \setminus \{\hat{v}\}$ (notice that r_i cannot become empty, as all objectives would be optimized in a set containing only one vertex). Let $k > i$ be the smallest rank for which r_k contains a vertex which strictly needs $\hat{v}$ as a reference (in other words, assumption **(b)** would not be satisfied if $\hat{v}$ were included in sets having a rank greater than k). If such a k does not exist, then $k = \infty$. Include then $\hat{v}$ in the set r_j, with $i < j \leq \min(k, |r|)$, such that $\hat{v} \in x_M$ when the corresponding multi-level problem is solved with $x_0 = r_j \cup \{\hat{v}\}$. If no such an r_j exists (all these multi-level problems "reject" the vertex $\hat{v}$), then add a new rank to the order: let $r_{h+1} = r_h$, for all $h \geq k$, and set $r_k = \{\hat{v}\}$.

This procedure can be repeated until there are no longer vertices $\hat{v}$ for which some of the objectives are not optimized. This procedure is able therefore to construct an optimal partial discretization order from a non-optimal one, and it is based on the idea of filtering subsets r_i by optimizing the set of given objectives (see Alg 2). Thus, the obtained order can be found by Algorithms 1+2: contradiction. □

Several objectives can be defined for improving the structure of DGP search trees that are obtained with the discretization. Two objectives that were already considered in previous publications simply correspond to two of the counters that were introduced in Sect. 2. The maximization of the counter α_{ex}, for every r_i that is not part of the initial clique, allows to anticipate the use of exact distances in the search tree, with the aim of reducing its width [5]. The maximization of the counter α, instead, anticipates all kinds of distances (either exact or represented by an interval), which can be useful for pruning the search tree at upper layers [5,6].

The structure of the search tree can be optimized by considering other additional criteria. For example, graph edges (known distances) that "cross" several ranks in the partial order imply the late detection of infeasible candidate vertex positions. It is desirable therefore to minimize the rank difference in edges representing reference distances. Moreover, a *light* consecutivity assumption, which does not impose the order to be total (see above), can be considered for maximizing the number of cliques consisting of reference vertices. The definition of new objectives for these above mentioned criteria, as well as for other novel ones, will be subject of future research.

4   Conclusions

Discretization orders for graphs representing instances of the DGP allow to reduce the search space of the DGP to a discrete set having the structure of a tree. This paper formalizes the concept of optimal partial discretization orders, which are orders that do not only allow for the discretization, but also to optimize a given set of objectives. An algorithm is proposed for constructing optimal partial discretization orders.

Future works will mainly follow these directions: *(i)* the extension of the presented theory to partial orders admitting vertex repetitions; *(ii)* the conception of new objectives, tailored to certain classes of the DGP (e.g. for molecular conformation determination [7,8]); *(iii)* the verification of the "best" orders for the objectives' priority levels, their impact on the search tree, and their relative compatibility; *(iv)* the integration of this methodology in a general branch-and-prune framework for the DGP.

Acknowledgments. I am thankful to Douglas S. Gonçalves and Leo Liberti for the fruitful discussions.

References

1. Bertsimas, D., Brown, D.B., Caramanis, C.: Theory and applications of robust optimization. SIAM Rev. **53**(3), 464–501 (2011)
2. Cassioli, A., Günlük, O., Lavor, C., Liberti, L.: Discretization vertex orders in distance geometry. Discrete Appl. Math. **197**, 27–41 (2015). doi:10.1016/j.dam.2014.08.035
3. Costa, V., Mucherino, A., Lavor, C., Cassioli, A., Carvalho, L.M., Maculan, N.: Discretization orders for protein side chains. J. Global Optim. **60**(2), 333–349 (2014)
4. Gonçalves, D.S., Mucherino, A.: Discretization orders and efficient computation of cartesian coordinates for distance geometry. Optim. Lett. **8**(7), 2111–2125 (2014)
5. Cabalar, P.: Answer set; programming? In: Balduccini, M., Son, T.C. (eds.) Logic Programming, Knowledge Representation, and Nonmonotonic Reasoning. LNCS, vol. 6565, pp. 334–343. Springer, Heidelberg (2011)

6. Lavor, C., Lee, J., Lee-St.John, A., Liberti, L., Mucherino, A., Sviridenko, M.: Discretization orders for distance geometry problems. Optim. Lett. **6**(4), 783–796 (2012)
7. Lavor, C., Liberti, L., Maculan, N., Mucherino, A.: The discretizable molecular distance geometry problem. Comput. Optim. Appl. **52**, 115–146 (2012)
8. Lavor, C., Liberti, L., Mucherino, A.: The interval branch-and-prune algorithm for the discretizable molecular distance geometry problem with inexact distances. J. Global Optim. **56**(3), 855–871 (2013)
9. Liberti, L., Lavor, C., Maculan, N.: A branch-and-prune algorithm for the molecular distance geometry problem. Int. Trans. Oper. Res. **15**, 1–17 (2008)
10. Liberti, L., Lavor, C., Maculan, N., Mucherino, A.: Euclidean distance geometry and applications. SIAM Rev. **56**(1), 3–69 (2014)
11. Liberti, L., Lavor, C., Mucherino, A., Maculan, N.: Molecular distance geometry methods: from continuous to discrete. Int. Trans. Oper. Res. **18**(1), 33–51 (2011)
12. Mucherino, A.: On the identification of discretization orders for distance geometry with intervals. In: Nielsen, F., Barbaresco, F. (eds.) GSI 2013. LNCS, vol. 8085, pp. 231–238. Springer, Heidelberg (2013)
13. Mucherino, A.: A pseudo de Bruijn graph representation for discretization orders for distance geometry. In: Ortuño, F., Rojas, I. (eds.) IWBBIO 2015, Part I. LNCS, vol. 9043, pp. 514–523. Springer, Heidelberg (2015)
14. Mucherino, A., Fuchs, M., Vasseur, X., Gratton, S.: Variable neighborhood search for robust optimization and applications to aerodynamics. In: Lirkov, I., Margenov, S., Waśniewski, J. (eds.) LSSC 2011. LNCS, vol. 7116, pp. 230–237. Springer, Heidelberg (2012)
15. Mucherino, A., Lavor, C., Liberti, L.: The discretizable distance geometry problem. Optim. Lett. **6**(8), 1671–1686 (2012)
16. Mucherino, A., Lavor, C., Liberti, L., Maculan, N. (eds.): Distance Geometry: Theory, Methods and Applications. Springer, New York (2013)
17. Saxe, J.: Embeddability of weighted graphs in k-space is strongly NP-hard. In: Proceedings of 17th Allerton Conference in Communications, Control and Computing, pp. 480–489 (1979)

Sensitivity Analysis of Checkpointing Strategies for Multimemetic Algorithms on Unstable Complex Networks

Rafael Nogueras and Carlos Cotta$^{(\boxtimes)}$

Department Lenguajes Y Ciencias de la Computación,
Universidad de Málaga, ETSI Informática,
Campus de Teatinos, 29071 Málaga, Spain
ccottap@lcc.uma.es

Abstract. The use of volatile decentralized computational platforms such as, e.g., peer-to-peer networks, is becoming an increasingly popular option to gain access to vast computing resources. Making an effective use of these resources requires algorithms adapted to such a changing environment, being resilient to resource volatility. We consider the use of a variant of evolutionary algorithms endowed with a classical fault-tolerance technique, namely the creation of checkpoints in a safe external storage. We analyze the sensitivity of this approach on different kind of networks (scale-free and small-world) and under different volatility scenarios. We observe that while this strategy is robust under low volatility conditions, in cases of severe volatility performance degrades sharply unless a high checkpoint frequency is used. This suggest that other fault-tolerance strategies are required in these situations.

1 Introduction

Distributed computing platforms have been used for running population-based metaheuristics for decades now. This is a direct consequence of the flexibility and adaptability of these techniques whose functioning is intrinsically parallel. Hence they can be naturally deployed on networked computers, cf. [1]. Numerous research works have focused on different design aspects of these techniques and how they affect performance in distributed environments – see, e.g., [2,6,25]. Exploiting efficiently distributed computing resources has become one of the signature weapons of these techniques and is a major factor for boosting their performance. In this sense, it is worth noting how technological advances are reshaping both the underlying computational substrate and the very needs to be addressed in computational terms. Regarding the latter, the problems and their data are becoming increasingly larger and complex [22]. The term *Big Data* [28] is nowadays a hot buzzword used to denote such large collections of data, very much requiring vast computational power in order to harnessed.

While traditional supercomputing techniques (namely, dedicated systems hosting a large array of processors and colossal memory banks) are certainly

© Springer International Publishing Switzerland 2015
I. Lirkov et al. (Eds.): LSSC 2015, LNCS 9374, pp. 243–250, 2015.
DOI: 10.1007/978-3-319-26520-9_26

one of the lines of attack to Big Data problems, the preponderance of computing resources permanently connected to the Internet has led to the emergence of other computational environments such as peer-to-peer (P2P) networks [14] and volunteer computing networks [23]. These are bound to play a key role in this kind of endeavors since they allow the orchestration of enormous decentralized collections of computational nodes. This comes at a cost though: these computational resources are unstable (they are typically contributed by volunteers during their idle time) and this must be taken into account when deploying applications on this kind of environments. Focusing specifically on applications (population-based metaheuristics in our case) running natively on these environment (i.e., being aware of its dynamicity and dealing with it directly), they can either use some fault-management policy for corrective purposes [12] or can self-adapt their behavior/parameterization to cope with it. We aim our attention at the former approach in this work. More precisely we analyze the performance of strategies based on creating restoration checkpoints [16]. This is done within the context of multimemetic algorithms [11], namely memetic algorithms which self-adapt the local search procedure, cf. [21]. These are described next.

2   Fault-Tolerant Model in an Island-Based Multimemetic Algorithm

As stated before we consider the use of multimemetic algorithms (MMAs) on an unstable computational scenario. Our MMA is organized as an island-based algorithm [24, 29], that is, it has a population distributed over a collection of n islands. Each of these islands comprises a panmictic (i.e., unstructured) subpopulation and runs a basic steady-state MMA procedure. This procedure follows the standard pattern of memetic algorithms, namely, selection, recombination, mutation and local search [15] but has a distinctive feature: local search is not done using a predefined strategy but using search patterns (*memes*) embedded in each individual and evolving alongside the latter (note the connection with the concept of memetic computing [20]). Inspired by the model by Smith [26, 27], these memes are expressed as variable-length pattern-based rewriting rules $A \rightarrow B$ (i.e., find A in the genome and change it into B; both $A, B \in \Sigma \cup \{\#\}$ where Σ is the alphabet used for encoding solutions and $\#$ is a wildcard). They evolve via mutation and are transferred from parent to offspring via local selection. We refer to [18] for further details.

The islands are distributed over a network of nodes and perform migration asynchronously (randomly picking an individual from an island and transferring it to another one, where it replaces the worst individual [17]). Two factors define this computational scenario: the interconnection topology and the dynamic model of the network. Regarding the topology, we consider to possibilities:

- Scale-free networks (SF): these are characterized by the existence of a power-law distribution in node degrees, and are often observed in many natural processes. We use the Barabási-Albert (BA) model [3] to generate this kind of

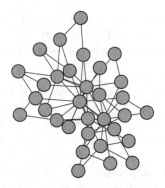
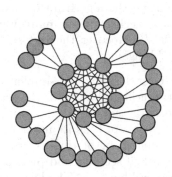

Fig. 1. Example networks for $n = 32$. The left figure is a SF network ($m = 2$) and the right one is a SW network ($M = 61$).

SF networks. This model uses preferential attachment [4] to grow a network by adding a new node at a time. A parameter m determines the number of links each new node gets.

- Small-world networks (SW): these are characterized by very small average distances between nodes (often $O(\log n)$ where n is the number of nodes). We use a variant of the Barmpoutis-Murray (BM) model [5] to create ultra-SW networks. This model takes as a parameter the total number of nodes n and the total number of links M and uses a backtracking procedure to successively build the largest clique that leaves enough links available to connect the rest of the network. In our variant, each of these cliques are then connected using random vertices in the first clique created so as to make the resulting network more resilient.

Figure 1 shows an example of both kinds of network with the same number of nodes and links.

As to the dynamics of the network, it is characterized by the availability patterns of computing nodes. We use the model in [16]: all n nodes are initially available and then their permanence in the system follows a Weibull distribution. This distribution is characterized by a shape parameter η that determines whether failure probability increases with time ($\eta > 1$), decreases with time ($\eta < 1$) or is time-independent ($\eta = 1$), and a scale parameter β determining the mean lifetime for a given shape. Each node has an independent dynamics and will contribute to the so-called *churn* phenomenon, namely the collective effect on the network of computing nodes independently entering and leaving it over time. Churn can have different effects on a distributed population-based metaheuristic, the most obvious being that the current incumbent solution can be lost [9]. Needless to say, the progress of the search will be also affected by the disappearance of whole subpopulations. To tackle this in the context of corrective fault-management policies, we consider two strategies [16]:

- rand: when a node becomes available again, it is initialized from scratch much like in the initialization of the algorithm.

– checkpoint: the algorithm uses some external safe storage in order to create restoration checkpoints, namely periodical backups of the populate state that are used to recover the last state of the population when a node becomes available again.

It is clear that rand is a simpler strategy that has also the potential advantage of reintroducing diversity in the search process. On the other hand, checkpoint has the advantage of not wasting the previous progress of the search, being more amenable to keep its momentum. The negative side of this latter strategy is the requirement of this external safe storage and the associated overhead (particularly if security and privacy concerns are important [13]) introduced by the periodical backups. The latter effect can be somehow ameliorated by tuning the period λ (measured in number of iterations) between checkpoints. The effect of this parameter is studied next.

3   Experiments

We have done experiments using a distributed MMA with $n = 32$ islands. Each island has a population size of $\mu = 16$ individuals. Meme lengths evolve within $l_{min} = 3$ and $l_{max} = 9$, mutating their length with probability $p_r = 1/9$ following [18]. We use crossover probability $p_X = 1.0$ (one-point crossover), mutation probability $p_M = 1/\ell$ (bit-flip mutation), where ℓ is the genotype length, and migration probability $p_{mig} = 1/80$. In order to generate the network topology we use $m = 2$ in the BA model of SF networks, and the corresponding value of $M = nm - m(m+1)/2$ in the BM model of SW networks so that the number of links is the same in both cases. As to node dynamics, we use the shape parameter $\eta = 1.5$ (and hence the probability of failure increases with time), and scale parameters $\beta = -1/\log p(k)$ for $p(k) = 1 - (kn)^{-1}$, $k \in \{1, 2, 5, 10, 20\}$. By doing this, the mean availability stint per node is about $90\% \cdot kn$ iterations. We therefore obtain scenarios ranging from rather low ($k = 20$) churn up to extremely high ($k = 1$) churn. To analyze the sensitivity of the checkpoint strategy we consider values $\lambda \in \{\mu, 10\mu, 100\mu\}$ where μ is the island population size. For comparison purposes we also consider in the experimentation the use of the rand strategy. We have considered four test functions, namely Deb's trap (TRAP) function [7], Watson et al.'s Hierarchical-if-and-only-if (HIFF) and Hierarchical-Exclusive-OR (HXOR) functions [30] and Goldberg et al.'s Massively Multimodal Deceptive Problem (MMDP) [8]. We perform 25 simulations running for a total number of 50 000 evaluations for each value of λ, churn scenario, problem and network topology.

Figure 2 shows the results obtained in terms of deviation with respect to the optimal solution (averaged for the four problems) as a function of the churn rate, separately for each λ value and for each network topology. First of all, it is clear that performance degrades for increasing churn rate. This fact notwithstanding, we can observe that variants using checkpoint reactivation perform notably better than random reactivation. This confirms previous research on this kind of strategies [16] and validates its usefulness on different network topologies.

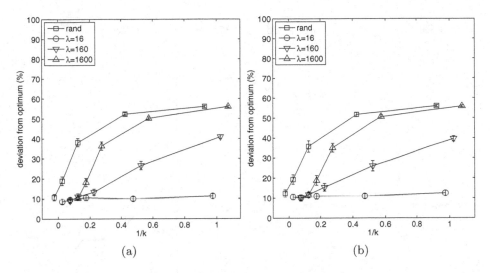

Fig. 2. Deviation from the optimal solution as a function of the churn rate for (a) SF and (b) SW.

Note however that there is a marked performance degradation when the checkpoint frequency is increased. This degradation is shown to be statistically significant according to Quade test (p-value ≈ 0) both globally and when SF and SW are separately analyzed. Subsequently we used Holm test to do a post-hoc analysis. The use of checkpoint with parameter $\lambda = \mu$ is shown to be statistically superior to the remaining techniques at $\alpha = 0.05$ level – see Table 1. This result suggests that less expensive strategies (in terms of requiring less frequent state snapshots) are not capable of dealing with churn (this result also holds if a separate analysis is conducted for SF and SW topologies). A more clear depiction of the behavior of the MMAs is provided by Fig. 3 for low ($k = 20$), high ($k = 5$) and extremely high ($k = 1$) churn. Note that in the most stable scenario the algorithm performs robustly regardless of the frequency of the snapshots (although as seen in Fig. 3f there is a noticeable difference in genetic diversity when the period λ is large). However, as churn increases the difference in fitness turns out to be remarkably higher in favor of $\lambda = \mu$. As seen in Figs. 3d and f, the MMA has convergence problems in these scenarios when λ is high. The less frequent snapshots cannot keep the momentum of the search in such unstable environments.

Table 1. Results of holm test ($\alpha = 0.05$) using $\lambda = 16$ as control parameter.

i	Strategy	z-statistic	p-value	α/i
1	$\lambda = 160$	2.598e+00	4.687e–03	5.000e–02
2	$\lambda = 1600$	7.015e+00	1.151e–12	2.500e–02
3	Rand	8.747e+00	1.097e–18	1.667e–02

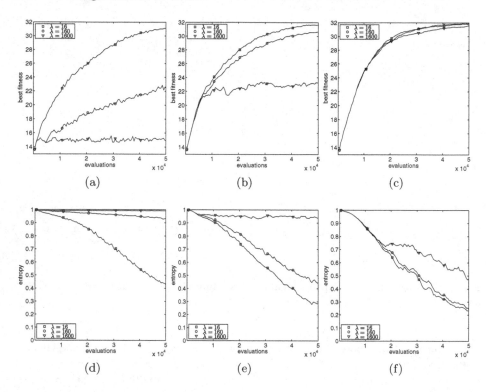

Fig. 3. Best fitness (top) and entropy (bottom) for TRAP with SF topology. From left to right: $k = 1$, $k = 5$ and $k = 20$.

4   Conclusion

Any algorithm directly deployed on an unstable computational environment must be resilient to the volatility of its substrate. Metaheuristics are no exception and, while they are intrinsically resilient to some extent [10], they must be augmented with adequate policies in order to cope with the loss of information associated to computing nodes that become inactive. A classical fault-tolerance technique for this purpose is the creation of periodical backups of the state of these nodes in order to recover from failures. We have performed a sensitivity analysis of this strategy in the context of island-based multimemetic algorithms. It turns out that this approach can be affordable in scenarios with low churn rates. In such a situation, checkpoints need not to be frequent for the algorithm to perform adequately. However, scenarios featuring large churn rates require much more frequent backups in order to cope with node volatility. The additional overhead of such backups together with the need for having access to persistent external storage makes this approach less appealing in such situations, suggesting other approaches –autonomous, self-adaptive and purely local– can be more appropriate. Work is already in progress in this direction [19].

Acknowledgements. This work is partially supported by the MINECO project EphemeCH (TIN2014-56494-C4-1-P), by the Junta de Andalucía project DNEMESIS (P10-TIC-6083) and by the Universidad de Málaga, Campus de Excelencia Internacional Andalucía Tech.

References

1. Alba, E.: Parallel Metaheuristics: A New Class of Algorithms. Wiley-Interscience, New York (2005)
2. Alba, E., Troya, J.M.: Influence of the migration policy in parallel distributed GAs with structured and panmictic populations. Appl. Intell. **12**(3), 163–181 (2000)
3. Albert, R., Barabási, A.L.: Statistical mechanics of complex networks. Rev. Mod. Phys. **74**(1), 47–97 (2002)
4. Barabási, A.L., Albert, R.: Emergence of scaling in random networks. Science **286**(5439), 509–512 (1999)
5. Barmpoutis, D., Murray, R.M.: Networks with the smallest average distance and the largest average clustering. arXiv 1007.4031 [q-bio] (2010)
6. Cantu-Paz, E.: Efficient and Accurate Parallel Genetic Algorithms. Kluwer Academic Publishers, Norwell (2000)
7. Deb, K., Goldberg, D.E.: Analyzing deception in trap functions. In: Whitley, L.D. (ed.) Second Workshop on Foundations of Genetic Algorithms, pp. 93–108. Morgan Kaufmann, Vail (1993)
8. Goldberg, D.E., Deb, K., Horn, J.: Massive multimodality, deception, and genetic algorithms. In: Parallel Problem Solving from Nature - PPSN II, pp. 37–48. Elsevier, Brussels (1992)
9. Hidalgo, J.I., Lanchares, J., Fernández de Vega, F., Lombraña, D.: Is the Island model fault tolerant? In: Proceedings of the 9th Annual Conference Companion on Genetic and Evolutionary Computation, GECCO 2007, pp. 2737–2744. ACM, New York (2007)
10. Laredo, J.J.L., Bouvry, P., González, D.L., de Vega, F.F., Arenas, M.G., Merelo, J.J., Fernandes, C.M.: Designing robust volunteer-based evolutionary algorithms. Genet. Program Evolvable Mach. **15**(3), 221–244 (2014)
11. Krasnogor, N., Blackburne, B.P., Burke, E.K., Hirst, J.D.: Multimeme algorithms for protein structure prediction. In: Guervós, J.J.M., Adamidis, P.A., Beyer, H.-G., Fernández-Villacañas, J.-L., Schwefel, H.-P. (eds.) PPSN 2002. LNCS, vol. 2439, pp. 769–778. Springer, Heidelberg (2002)
12. Lombraña González, D., Jiménez Laredo, J.L., Fernández de Vega, F., Merelo Guervós, J.J.: Characterizing fault-tolerance of genetic algorithms in desktop grid systems. In: Cowling, P., Merz, P. (eds.) EvoCOP 2010. LNCS, vol. 6022, pp. 131–142. Springer, Heidelberg (2010)
13. Mihaljević, M.J., Imai, H.: Security issues of cloud computing and an encryption approach. In: Despotović-Zrakić, M., Milutinović, V., Belić, A. (eds.) Handbook of Research on High Performance and Cloud Computing in Scientific Research and Education, pp. 388–408. IGI Global, Hershey (2014)
14. Milojičić, D.S., Kalogeraki, V., Lukose, R., Nagaraja, K., Pruyne, J., Richard, B., Rollins, S., Xu, Z.: Peer-to-peer computing. Technical report, HPL-2002-57, Hewlett-Packard Labs (2002)
15. Neri, F., Cotta, C., Moscato, P.: Handbook of Memetic Algorithms, Studies in Computational Intelligence, vol. 379. Springer, Heidelberg (2012)

16. Nogueras, R., Cotta, C.: Studying fault-tolerance in island-based evolutionary and multimemetic algorithms. J. Grid Comput. (2015). doi:10.1007/s10723-014-9315-6
17. Nogueras, R., Cotta, C.: An analysis of migration strategies in Island-based multimemetic algorithms. In: Bartz-Beielstein, T., Branke, J., Filipič, B., Smith, J. (eds.) PPSN 2014. LNCS, vol. 8672, pp. 731–740. Springer, Heidelberg (2014)
18. Nogueras, R., Cotta, C.: On meme self-adaptation in spatially-structured multimemetic algorithms. In: Dimov, I., Fidanova, S., Lirkov, I. (eds.) NMA 2014. LNCS, vol. 8962, pp. 70–77. Springer, Heidelberg (2015)
19. Nogueras, R., Cotta, C.: Studying self-balancing strategies in island-based multimemetic algorithms. J. Comput. Appl. Math. **293**, 180–191 (2016). doi:10.1016/j.cam.2015.03.047
20. Ong, Y.S., Lim, M.H., Chen, X.: Memetic computation-past, present and future. IEEE Comput. Intell. Mag. **5**(2), 24–31 (2010)
21. Ong, Y.S., Lim, M.H., Zhu, N., Wong, K.W.: Classification of adaptive memetic algorithms: a comparative study. IEEE Trans. Syst. Man Cybern. Part B: Cybern. **36**(1), 141–152 (2006)
22. Reichhardt, T.: It's sink or swim as a tidal wave of data approaches. Nature **399**(6736), 517–520 (1999)
23. Sarmenta, L.F.: Bayanihan: web-based volunteer computing using java. In: Masunaga, Y., Katayama, T., Tsukamoto, M. (eds.) Worldwide Computing and Its Applications - WWCA 1998. Lecture Notes in Computer Science, vol. 1368, pp. 444–461. Springer, Heidelberg (1998)
24. Schaefer, R., Byrski, A., Smołka, M.: The Island model as a Markov dynamic system. Int. J. Appl. Math. Comput. Sci. **22**(4), 971–984 (2012)
25. Skolicki, Z., Jong, K.D.: The influence of migration sizes and intervals on Island models. In: Genetic and Evolutionary Computation Conference 2005, pp. 1295–1302. ACM, New York (2005)
26. Smith, J.E.: Self-adaptation in evolutionary algorithms for combinatorial optimisation. In: Cotta, C., Sevaux, M., Sörensen, K. (eds.) Adaptive and Multilevel Metaheuristics, Studies in Computational Intelligence, vol. 136, pp. 31–57. Springer, Heidelberg (2008)
27. Smith, J.: Self-adaptative and coevolving memetic algorithms. In: Neri, F. (ed.) Handbook of Memetic Algorithms. SCI, vol. 379, pp. 199–220. Springer, Heidelberg (2011)
28. Snijders, C., Matzat, U., Reips, U.D.: 'Big Data': big gaps of knowledge in the field of internet. Int. J. Internet Sci. **7**, 1–5 (2012)
29. Tanese, R.: Distributed genetic algorithms. In: 3rd International Conference on Genetic Algorithms, pp. 434–439. Morgan Kaufmann Publishers Inc., San Francisco (1989)
30. Watson, R.A., Hornby, G.S., Pollack, J.B.: Modeling building-block interdependency. In: Eiben, A.E., Bäck, T., Schoenauer, M., Schwefel, H.-P. (eds.) PPSN 1998. LNCS, vol. 1498, pp. 97–106. Springer, Heidelberg (1998)

Free Search in Multidimensional Space III

Kalin Penev[(✉)]

School of Media, Art and Technology, Southampton Solent University,
East Park Terrace, Southampton SO14 0YN, UK
Kalin.Penev@solent.ac.uk

Abstract. Various scientific and technological fields, such as design, engineering, physics, chemistry, economics, business, and finance often face multidimensional optimisation problems. Although substantial research efforts have been directed in this area, key questions are still waiting for answers, such as: What limits computer aided design systems on optimisation tasks with high variables number? How to improve capabilities of modern search methods applied to multidimensional problems? What are software and hardware constraints? Approaching multidimensional optimisation problems raises in addition new research questions, which cannot be seen or identified on low dimensional tasks, such as: What time is required to resolve multidimensional task with acceptable level of precision? How dimensionality reflects on the search space complexity? How to establish search process orientation, within multidimensional space? How task specific landscapes embarrass orientation? This article presents an investigation on 300 dimensional heterogeneous real-value numerical tests. The study aims to evaluate relation between tasks' dimensions' number and required for achieving acceptable solution with non-zero probability number of objective function evaluations. Experimental results are presented, analysed and compared to other publications.

Keywords: Free search · 300 dimensional optimisation

1 Introduction

Various scientific and technological fields, such as design, engineering, physics, chemistry, economics, business, and finance often face multidimensional optimisation problems [2]. Multidimensional optimisation problems, however, require sufficient computational resources. In the same time natural life suggests that capability to cope by finite and limited resources with infinite and unlimited environment and problems can be considered as an advanced characteristic of living systems. This article attempts to explore model of similar behaviour. It presents an investigation on 300 dimensional scalable heterogeneous real-value numerical tests optimisation. Due to a specific performance identified in earlier publications [7], optimisation method explored in this study is Free Search (FS) [6] only.

© Springer International Publishing Switzerland 2015
I. Lirkov et al. (Eds.): LSSC 2015, LNCS 9374, pp. 251–257, 2015.
DOI: 10.1007/978-3-319-26520-9_27

The aim of this investigation is to find answers of the questions: How to improve capabilities of modern search methods applied to multidimensional problems? What are software and hardware constraints? The study aims also to investigate specific for multidimensional optimisation research questions such as: What time is required to resolve multidimensional task with acceptable level of precision? How dimensionality reflects on the search space complexity? How to establish search process orientation, within multidimensional space? How task specific landscapes embarrass orientation?

For this purpose five scalable numerical tests are used - Ackley [1], Griewank [4], Michalewicz [5], Rosenbrock [9] and Step [3] test functions.

2   Test Problems

Criteria for tests selection are: - must be scalable for multidimensional format; - must be with heterogeneous landscape. Chosen numerical test are scalable and form different search spaces. All tests are transformed for maximisation.

2.1   Ackley Test

This test [1], know from the literature is widely used for search methods evaluation.

$$f(x) = a \exp\left[-b\left(\frac{1}{n}\sum_{i=1}^{n} x_i^2\right)\right]^{1/2} + \exp\left(\frac{1}{n}\sum_{i=1}^{n}\cos(cx_i)\right) - a - \exp(1) \quad (1)$$

where $a = 20$, $b = 0.2$, $c = 2\pi$. The maximum is $f_{max} = 0$, for $x_i = 0$, $i = 1, \ldots, n$. The search space borders are defined by $x_i \in (-32, 32)$.

2.2   Griewank Test

The test [4], is given by the following analytical expression:

$$f(x) = -\left(1 + \frac{1}{4000}\sum_{i=1}^{n} x_i^2 - \prod_{i=1}^{n}\cos\left(\frac{x_i}{\sqrt{i}}\right)\right) \quad (2)$$

where $x_i \in [-600.0, 600.0]$. The maximum is $f_{max} = 0$, for $x_i = 0$, $i = 1, \ldots, n$.

2.3   Michalewicz Test

Michalewicz test function is described [5] as global optimisation problem. Optimal value is dependent on dimensions number.

$$f(x) = \sum_{i=1}^{n}\sin(x_i)\left(\sin\left(\frac{ix_i^2}{\pi}\right)\right)^{2m} \quad (3)$$

where search space is defined as $x_i \in [0, \pi]$, $i = 1, ..., n$. For 300 dimensions maximum is unknown.

2.4   Rosenbrock Test

This function landscape is smooth flat hill with one optimal solution [9]. The test function is:

$$f(x) = -\sum_{i=1}^{n-1} \left[100(x_{i+1} - x_i^2)^2 + (x_i - 1)^2\right] \tag{4}$$

where $x_i \in [-2, 2]$, $i = 1, \ldots, n$. It has one maximum $f_{max} = 0$, for $x_i = 1$, $i = 1, \ldots, n$.

2.5   Step Test

Step test [3] introduces plateaus to the topology. The search process cannot rely on local correlation. Maximal are all locations, which belong to the plateau $x_i \in [2.0, 2.5)$. The maximum is dependent on dimensions number. The test function is:

$$f(x_i) = \sum_{i=1}^{n} \lfloor x_i \rfloor \tag{5}$$

where $x_i \in [-2.5, 2.5]$. For $n = 300$ maximum is $f_{max} = 600$, for $x_i \in [2.0, 2.5)$, $i = 1, \ldots, n$.

3   Optimization Method

Due to the abilities to produce acceptable results within feasible period of time identified in earlier publications [7,8], optimisation method selected for this study is Free Search (FS) [6] only.

3.1   Free Search

Free Search is adaptive heuristic method [6–8] for real coded optimisation. This section refines the description of some of its essential properties, published earlier. Optimisation process is organised in individual explorations within individuals' neighbour space [6]. In the beginning algorithm has no knowledge about the search space. First exploration is initial trial, which generates knowledge stored in a form of qualitative indicators related with evaluated locations. These indicators facilitate further explorations. Individuals develop sense to the indicators' quality. This sense is an original peculiarity of Free Search, which has no analogue in other methods. Individuals use their sense to locations quality for orientation within the search space.

Although individuals' sensibilities are highly uncertain a review of idealised general states of distribution such as uniform, enhanced and reduced sensibility related with locations' qualities can clarify the search process self-regulation. On initial stage locations quality and sensibility are uniformly distributed among low, medium and high levels. Individuals with low level of sensibility can select for start position any marked location. The individuals with high sensibility can select for start position marked locations with high quality and will ignore locations with low quality.

When marked locations quality highly differs and stochastically generated sensibility produces accidentally high values only, then the individuals will search around the area of the highest quality solutions. Such situations appear naturally. In this manner process converges to high quality locations. External addition of a constant or a variable to the sensibility could lead to an enforced enhancement of the sensibility. In this case all the individuals with enhanced sensibility will select and can differentiate more precisely locations with high quality and will ignore these with low quality. This could accelerate convergence to areas with high quality locations.

Other situation which naturally appears is when marked locations qualities are very similar and randomly generated sensibility is low. In this case individuals can select low quality marked locations with high probability, which indirectly will decrease the probability for selection of high quality marked locations. Subtraction of a constant or a variable from individuals' sensibility could make an enforced sensibility reduction. Individuals with reduced sensibility can select to explore around locations marked with low quality. As far as locations quality is independent on their position within the search space, similar quality locations could be remotely distributed. This facilitates divergence across the entire search space. Sensibility varies across all the individuals and during the optimisation process.

One of the objectives of this study is to evaluate how this manner of orientation performs for multidimensional space. For presented experiments Free Search operates with 10 individuals and explorations are 5 steps, for all experiments. The sense is random in the highest 10 % of the sensibility, and the neighbouring space varies from 0.5 to 1.5 with step 0.1 [6].

4   Experimental Methodology

Experimental Methodology aims to identify method's performance and level of precision for 300 dimensional tests limited to 3.10^8 objective functions evaluations. Each test function is evaluated in one series of 320 experiments, with start from random initial locations different for each experiment. Start locations are defined as:

$$x_{i0} = X_{min} + random_i(X_{max} - X_{min}) \qquad (6)$$

where X_{max} and X_{min} are search space borders and $random_i(X_{max} - X_{min})$ generates random value between X_{max} and X_{min}, $i = 1, \ldots, n$. All variables are 300 dimensional vectors.

Rosenbrock test only is evaluated additionally one more series of 320 experiments limited to 3.10^9 objective functions evaluations.

5   Experimental Results

Achieved results are analysed for maximal and mean values, standard deviation and number of results with precision 0.01 from the maximal value.

On Tables 1, 2, and 3 FE denotes function evaluations number. Time periods in Table 3 are measured on processor Intel i7 3960x at 4.6 GHz and memory G. Skill Trident X at 1866 MHz, motherboard ASUS Rampage VI and solid state disk - SanDisk Extreme SSD SATA III.

6   Discussion

Analysis of experimental results suggests that Ackley, Michalewicz and Step tests can be resolved with 100 % probability with precision 0.001 for 3.10^8 function evaluations. Griewank test can be resolved with 82.81 % probability with precision 0.001 for 3.10^8 function evaluations.

Rosenbrock test cannot be resolved with acceptable level of precision for 3.10^8 function evaluations. Rosenbrock test is evaluated additional for 3.10^9 function evaluations. Rosenbrock test can be resolved with 76.56 % probability with precision 0.001 for 3.10^9 function evaluations, however the period of search is longer.

Comparison of the periods of search for these 300 dimensional tests and 200 dimensional tests publishes earlier [8] suggest that time increases higher than

Table 1. Experimental results from 320 experiments

Test	FE	Maximal results	Mean results	Standard deviation
Ackley	3.10^8	-0.000329198000	-0.000688728	0.000216832
Griewank	3.10^8	-0.000000215366	-0.004886839	0.008172175
Michalewicz	3.10^8	299.603000000000	299.595365600	0.003252990
Rosenbrock	3.10^8	-0.001858470000	-112.786252900	72.692261520
Rosenbrock	3.10^9	-0.000030781900	-0.090739686	1.472556809
Step	3.10^8	600	600	0

Table 2. Number and percentage of the results with precision above 0.01

Test	FE	Successful results	Successful results %
Ackley $\geq$ -0.00	3.10^8	320	100.00 %
Griewank $\geq$ -0.00	3.10^8	265	82.81 %
Michalewicz $\geq$ 299.59	3.10^8	320	100.00 %
Rosenbrock $\geq$ -0.00	3.10^8	4	0.39 %
Rosenbrock $\geq$ -0.00	3.10^9	245	76.56 %
Step $=$ 600	3.10^8	320	100.00 %

Table 3. Periods of time for 3.10^8 objective functions evaluations

Test	FE	Time
Ackley	3.10^8	31 min
Griewank	3.10^8	48 min
Michalewicz	3.10^8	185 min
Rosenbrock	3.10^8	15 min
Rosenbrock	3.10^9	148 min
Step	3.10^8	12 min

linearly and hardware and software speed appears as potential constraints. To improve capabilities of modern search methods time consuming events should be identified and optimised. For further investigation on high dimensional problems hardware speed should be improved. Regarding the time required to resolve multidimensional task with acceptable level of precision, presented on Table 2 results suggest that on used hardware configuration selected 300 dimensional tests could be resolved with high probability for the range of 0.5 to 3.5 hours. For more general conclusion additional experiments with 300 dimensional tests should be done.

Comparison on 100 [7], 200 [8] and 300 dimensional tests performance indicates that:

(1) Complexity of task specific landscapes varies among the tests and for same dimensionality different number of functions evaluations could guarantee successful results. This is illustrated with Tables 1, 2, and 3 with Rosenbrock test function.
(2) Test complexity increases nonlinearly to test dimensionality and higher number of functions evaluations are required to reach the same level of precision and standard deviation.

According to results published earlier on Michalewicz test for 100 dimensions for 10^8 objective function evaluations Free Search reaches 0.00048003 standard deviation [7], for 200 dimensions for 2.10^8 objective function evaluations Free Search reaches 0.001784807 standard deviation [8]. In this investigation for 300 dimensions 3.10^8 objective function evaluations Free Search reaches 0.00325299 standard deviation (Table 4). The results suggests that although the number of objective function evaluations is proportional to the number of dimensions, achieved standard deviation tends to decrease. For higher precision additional objective function evaluations are required.

In summary presented results suggest that search process orientation based on heuristic trial and error could cope with multidimensional space. For more general conclusion additional investigation should be done.

Table 4. Performance on Michalewicz test for 100, 200 and 300 dimensions

Michalewicz test	Function evaluations	Maximal	Mean	Deviation
100 dimensions	100 000 000	99.6191	99.618175	0.000480030
200 dimensions	200 000 000	199.612	199.608409	0.001784807
300 dimensions	300 000 000	299.603	299.595365	0.003252990

7   Conclusion

This article presents experimental evaluation of Free Search on 300 dimensional tests. Identified are specific issues related with multidimensional optimisation. Experimental results are also summarized and analysed. Further investigation could focus on evaluation and measure of the time and computational resources sufficient for completion of other multidimensional tasks using parallel processing systems or parallel implementation of the method, which uses several processor cores in parallel or apply accelerated processing based on Graphics Processing Unit (GPU). Algorithms analysis and improvement could be also subject of future research.

Acknowledgements. I would like to thank to my students Adel Al Hamadan, Asim Al Nashwan, Dimitrios Kalfas, Georgius Haritonidis, and Michael Borg for the design, implementation and overclocking of desktop PC used for completion of the experiments presented in this article.

References

1. Ackley, D.H.: A Connectionist Machine for Genetic Hillclimbing. Kluwer, Boston (1987)
2. Censor, Y.: Optimisation Methods, Encyclopedia of Computer Science. Nature Publishing Group, London (2000). pp. 1339–1341
3. De Jung, K.A.: An analysis of the behaviour of a class of genetic adaptive systems. Ph.D thesis, University of Michigan, USA (1975)
4. Griewank, A.O.: Generalized decent for global optimization. J. Optim. Theor. Appl. **34**, 11–31 (1981)
5. Michalewicz, Z.: Genetic Algorithms + Data Structures = Evolution Programs. Springer, Heidelberg (1992)
6. Penev, K.: Free search of real value or how to make computers think. St. Qu, UK (2008). ISBN 978-0-9558948-0-0
7. Penev, K.: Free search – comparative analysis 100. Int. J. Metaheuristics **3**(2), 118–132 (2014)
8. Penev, K.: Free search in multidimensional space II. In: Dimov, I., Fidanova, S., Lirkov, I. (eds.) NMA 2014. LNCS, vol. 8962, pp. 103–111. Springer, Heidelberg (2015)
9. Rosenbrock, H.H.: An automate method for finding the greatest or least value of a function. Comput. J **3**, 175–184 (1960)

Speeding up Parallel Combinatorial Optimization Algorithms with Las Vegas Method

Bogdan Zavalnij[✉]

Institute of Mathematics and Informatics, University of Pecs, Pecs, Hungary
bogdan@ttk.pte.hu

Abstract. In this paper we introduce a new method for speeding up parallel run times of discrete optimization problems which can be used for different problems. We propose that the variant of the Monte Carlo method, the Las Vegas method can be used for overcoming some special barriers that can occur in the course of dividing such problems. Especially the problem of maximum clique and k-clique is examined, and the new algorithm with the relevant measurements is presented.

Keywords: Las Vegas method · Parallel algorithms · Maximum clique

1 Introduction

There arise two major problems when one tries to create a parallel algorithm for a combinatorial optimization problem. First, the search space for the divided different subproblems differ in several magnitudes. This makes the scheduling of the subproblems hard, and the resulting algorithm often inefficient. Second, the problem of the equality of the sum of divided subproblems and the original problem. Because of very specific heuristics used in the algorithm, solving the subproblems sometimes tend to do much more, and sometimes much less work than the original problem. The special effect of super linear speedup that can be observed with some back tracking algorithms also related to this problem.

We would like to deal with this second problem, and propose a parallel randomized method, that *tends* to reformulate the original problem into a simpler one. The nice property of this method is, that it will result a reorganization that could have been done without the randomization, if we would have been given the information that comes out of the randomization. That means, that the proposed method is robust and clear cut. Another nice result of this method is, that it tends to produce more even distribution of the subproblems, thus also addressing the first problem mentioned above.

The problem we concentrate on is the maximum clique problem, although the concept described in this paper applies to other problems in the field of discrete optimization as well. The maximum clique problem can be formulated in the following way. Given a finite simple graph $G = (V, E)$, where V represents

© Springer International Publishing Switzerland 2015
I. Lirkov et al. (Eds.): LSSC 2015, LNCS 9374, pp. 258–266, 2015.
DOI: 10.1007/978-3-319-26520-9_28

the nodes and E represents the edges. We call Δ a clique of G if the set of vertices of Δ is subset of V; Δ is an induced subgraph of G; and Δ is an all connected graph, that is all its nodes connected to all the other nodes. We call Δ a maximum clique if no other clique of G is bigger than that. The size of Δ is called the clique number of the graph, and denoted by ω. The maximum clique problem is a discrete optimization problem to find a maximum clique and determine its size, and it is a well known NP-hard problem. A variation of this problem is the k-clique problem, which is a problem in the NP-complete class. This problem states the question that if given a graph G, and a positive integer k, is there a clique of size k in the graph. To answer the question either we must present a k-clique of the graph or prove that there is none in the graph.

While the maximum clique and the k-clique problems are related and similar, there are important differences. One difference arises from the NP-completeness of the k-clique problem. It is easy to show a 'Yes' answer if we can provide the clique, but it is hard to state 'No'. On the other hand the actual running times of the maximum clique problem can vary greatly, depending on how fast we can find an actual maximum clique, which will produce better bounds later [2]. While these problems are easy to understand and to present well, it is hard to find a good example for showing actual results of a new algorithm. Because of these problems we performed my tests for k-clique problem, where k was such that it was larger by one as the size of a maximum clique. This question is the hard part of the k-clique problem, and it eliminates the chance factor of finding the maximum clique fast or slowly as in the maximum clique problem.

2 Background

Parallel Implementations of the Maximum Clique Problem. In the literature of discrete optimization there are several papers describing the possibility of parallelization of such problems. In the field of maximum clique search [3] and [2] among others made interesting contributions. It is important to note that, alas, most works examine parallelization on small number of processors. We would especially point out the latter work of McCreesh and Prosser. Their contribution is important because they clearly pointed out the effect of possibility of super linear speedup for maximum clique search. This effect takes place because there is a larger possibility for one thread to find an actual maximum clique, and this helps the other threads, so they can use a better bound. This paper describes the implementation of the k-clique problem, so since we do make the search for k one too large, this effect is not considered my work.

Las Vegas Algorithms. The Las Vegas algorithms, first described by Babai [1], is a variation of the Monte Carlo randomized algorithm. Formally, we call an algorithm a Las Vegas algorithm if for a given problem instance the algorithm terminates returning a solution, and this solution is guaranteed to be a correct solution; and for any given problem instance, the run-time of the algorithm applied to this problem is a random variable.

The variance in the running time of a Las Vegas algorithm led Truchet, Richoux and Codognet to implement an interesting way of parallelization of the algorithms for some NP-complete discrete optimization problems [9]. The authors note that the algorithm implementation for those problems heavily depends on the "starting point" of the algorithm, as it starts from a random incorrect solution and constantly changes it to find a real solution. Depending on the incorrect starting solution the convergence of the algorithm may be very fast or slow. The idea behind the Las Vegas parallel algorithm was to start several instances of the sequential algorithm from different starting points and let them run independently. The first instance that finds the solution shuts down all the other instances and the parallel algorithm terminates. As the running time of the different instances vary, some will terminate faster, thus ending the procedure in shorter time. The article describes the connection of the variance of the running times and the possible speedup when using k instances and found that for some problems a linear speedup could be achieved.

Effect of Subproblem Sequence. We should point out one more effect of discrete optimization problems. The usual Branch-and-Bound method is sensitive to the sequence of subproblems in a branch. This was shown for SAT problems [8], and could be shown for clique search problems as well. This effect will be used by my algorithm, as it eventually finds a better sequence for solution, and thus reduces the search space.

3   Parallel Las Vegas Algorithms

We choose, as an example algorithm, a parallel algorithm for k-clique problem proposed by Szabo [6]. The basic step of this algorithm is the removing of an edge from the graph. Given an edge v_i, v_j. If one can prove that this edge is not part of any k-clique, then this edge can be freely deleted from the graph without altering the answer to the k-clique question. The proof takes the subgraph of G spanned by the nodes $N(v_i) \cap N(v_j)$, and examines whether there is a $(k - 2)$-clique in it. ($N(v)$ denotes the neighboring nodes of node v.) If the answer is 'Yes', then we found a 'Yes' answer for the original question. If the answer is 'No', then we can delete this edge, as it cannot be an edge of a k-clique.

The algorithm uses the concept of disturbing edges. Given some set of edges, if we can delete them, then the original question can be answered by 'No', and this answer is trivial. The actual algorithm enumerates disturbing edges by a quasi coloring, and by deleting these edges we get a graph which can be colored by $k - 1$ colors, thus there can be no k-clique in it. Actually this method of reducing the graph by taking the neighbors of *two* nodes is closely related to the two level branching detailed in [2].

The subproblems, which can be denoted by an edge, are independent, but overlapping. In order to eliminate the overlapping parts let us consider a fixed sequence of the disturbing edges. Should we solve the problems in a sequential manner, then after solving the first one we can delete this edge from all the other problems to be solved. And so on for all the edges, one by one. The resulting problems are free of overlapping. Because these problems can still be solved independently, we can run them parallelly, deleting these edges *a priory* even before solving the specific problem.

We propose two other methods. First, which we call the Las Vegas edge deletion, starts the original overlapping problems without any edge deletion parallelly. The problems will be of different complexity, thus some of them will run fast, and others slowly. If one problem (a fast one) finishes, then the edge denoted by this problem shall be deleted from all the other problems *in the course they are being solved*. (Obviously this can be done only if the sequential program can allow this.) This method will do some surplus calculation because of overlapping search spaces, but can profit by reordering the sequence of edge deletion. As we already pointed out previously the sequence in which the problems are solved can affect the size of the search space. The structure of the program:

Master	Slave
Get report / get request / asked for deleted edges	Report / request for new task (edge)
IF found: exit	Construct the task from edge
IF not found: delete the edge	Solve the task
IF asked: give deleted edges	(repeatedly ask for deleted edges)
IF requested: give new task (edge)	

Second version, which we call the Las Vegas edge deletion with restart, do the same as the previous one, but uses an other technique, as well. If no work left to be given out (the number of subproblems fall below the number of processors), then we give out an edge, which is already given out to another process, and so two threads do the same calculation. It may seem redundant, but the restarting process can start from scratch, and with some already deleted edges, can produce better preconditioning of the subproblem. This method is well known in SAT solving community, but my proposal is somehow different, as given spare processors we run the original, half solved subproblem, along with the newly started. On the other hand, if the already long-running solver is near to the finish, we do not need to throw it out.

As it will be seen from the evaluation, both methods are really strong. Also, there is an interesting effect of reproducibility. While different runs will delete edges in different order, one can save the actual order of a given run. It is clear, that given this special order of edges we can run the original *a priory* edge deletion algorithm with the same, or even better speed.

Morphology of the Proposed Algorithm. The method described in this paper can be used for parallelization of different combinatorial optimization problems such as SAT or set covering. In order to utilize the algorithm it is important, that the original problem can be divided into several independent (and actually overlapping) subproblems. Also, the result of a subproblem should be usable to help other subproblems – it will reduce the search tree by the overlapping part.

The usage can be done in three different ways. First, the "help" can be given *a priory* in a chosen sequence. In fact, this is the usual method for parallelizing discrete optimization problems. Second, start the independent subproblems, and if one finished give this information to all the other threads. This is my method of Las Vegas edge deletion. Third, if some threads have no work to do, then restart a subproblem, but with the information gained from the previous results. The third algorithm uses this technique together with the edge deletion.

It is clear, that this method can be used, for example, in parallel implementation of a SAT problem.

4   Tests

We programmed the described algorithms in c++ and MPI, and performed several tests on different graph sets. Table 1 below summarize the running times for sequential and different parallel algorithms for $k = \omega + 1$. The columns labeled as follows. "N" denotes the size of the graph; "%" denotes the edge density; "ω" denotes the size of the maximum clique. The "parts" indicates the number of disturbing edges, thus the number of different subproblems that will be started. The rest of the columns denote the number of processors we used, where "1" means the sequential run time. The "seq" denotes the original algorithm where the edges deleted in the given sequential order, *a priori*. The "lv" denotes the Las Vegas method where the edge deletion is performed after a subproblem was solved. The "rest" denotes that, apart from the Las Vegas edge deletion, the problems also restarted when free processors were available, so the same problem run on several processors. The running times are in seconds, and for really big figures we used "k" for denoting thousand. All tests, including the sequential runs, were performed on the same supercomputer.

The first set consists of random graphs, the second set is the DIMACS set of graph problems [4], the third set is graphs of hard problems of monotonic matrices [7] and of deletion code [5]. The data presented here is partial, as we left only those instances, where the run-times are big enough to be of any interest.

The time limit for sequential and 16 processor runs was 12 hours, while for 64 and 512 processor runs 72 hours. The symbol "*" denotes run times exceeding the time limit, and "-" means that we have not run the test.

5   Evaluation

The test runs lead to several conclusions. First, it is clear, that the original idea of disturbing edges by Sandor Szabo enables quite good parallel speedups even for large number of cores. Second, for some (harder) problems there is a limit for the original algorithm. For other problems, such as random graphs, the speedup

Table 1. Test runs

	N	%	ω	parts	1	16 seq	16 lv	64 seq	64 lv	64 rest	512 seq	512 lv	512 rest
rand 200p9	200	90	40	152	480	27	35	22	30	20	22	30	32
rand 300p8	300	80	29	540	754	41	45	13	16	17	12	19	19
rand 300p9	300	90	47	341	*	*	*	*	*	*	*	16k	14k
rand 500p6	500	60	17	2478	48	9	9	2	2	2	0	0	0
rand 500p7	500	70	22	2231	3069	167	179	40	45	46	9	15	15
rand 500p8	500	80	32	1664	*	*	*	15k	17k	17k	5703	3543	3377
rand 800p5	800	50	14	7296	71	26	26	6	7	7	1	1	1
rand 800p6	800	60	19	6345	2371	158	166	38	41	41	5	6	6
rand 800p7	800	70	25	5587	*	*	*	5999	6571	6578	830	968	967
rand 900p5	900	50	15	8729	132	40	41	10	10	10	3	2	2
rand 900p6	900	60	19	8215	6493	398	423	96	103	104	13	14	15
rand 1000p5	1000	50	15	10955	282	65	66	16	16	16	2	2	2
rand 1000p6	1000	60	20	9823	14k	798	847	192	206	207	25	30	29
brock800_3	800	65	25	4888	6383	357	373	86	91	91	12	14	15
brock800_4	800	65	26	4592	4899	280	291	68	71	71	9	11	11
latin_sq_10	900	76	90	380	4053	91	121	40	61	59	40	55	54
keller5	776	75	27	420	4071	369	382	118	198	166	115	114	133
sanr200_0.9	200	90	42	128	297	15	25	14	33	33	14	29	28
sanr400_0.7	400	70	21	1408	351	22	23	5	6	6	1	2	2
p_hat700-2	700	50	45	826	740	36	59	27	48	44	27	82	84
p_hat300-3	300	74	36	297	198	9	15	7	14	14	7	15	15
p_hat500-3	500	75	50	657	*	*	*	9645	6020	4960	9653	4328	2253
monoton-8	512	82	23	590	2123	367	266	367	165	145	367	154	114
monoton-9	729	84	28	932	*	-	-	137k	39k	23k	137k	27k	6377
deletion-9	512	93	52	375	*	-	-	-	-	-	*	*	67k

Table 2. Test runs for `monoton-9`

	N	%	ω	parts	1	64 seq	64 lv	64 rest	512 seq	512 lv	512 rest
monoton-9	729	84	28	932	~1150k	137k	39k	23k	137k	27k	6377
hours:					~320h	38h	11h	6h	38h	8h	2h
speed-up:					1x	8x	30x	50x	8x	43x	180x
average run:									1232	1717	909
minimum run:									1	11	3
maximum run:									137k	27k	5k

is nearly linear. Actually, in my opinion, this indicates not as much the goodness of the algorithm, as it is rather shows the problem of testing with random graphs.

The Las Vegas methods are also performing well, and while being a bit behind for smaller and easier problems, they make a big difference for bigger and harder problems. Actually the lower performance for the easy problems are less interesting: one would not use a supercomputer for those problems! The difference between the simpler edge deletion Las Vegas algorithm and the restarting algorithm is similar. For very small problems the second can be a little slower – it takes time to shut down the threads that are not needed. For most problems they run for the same time, and for really hard problems the restarting version makes one more huge leap in performance.

We would point out the problems of `monoton-9` and `deletion-9` These problems are extremely hard, and only few achieved solving them with the aid of reducing the problem by finding symmetries, and in the second case, with the aid of semi definite programming. My method is using none of these.

Details of One Problem. Let us take a close look at one special problem, the `monoton-9`. Table 2 presents the `monoton-9`, problem alone, and we indicated also the running times in hours, the speedup and the average run time for the subproblems as well. The sequential run time is calculated by summing up the run times of the 512-seq subproblems. This is obviously not the same, as we would run the sequential program, but it is as if the *a priory* edge deleted problems would be run in sequential order. This figure, in my opinion, should be close enough to the real one.

Evaluating the results, one can clearly see, that the *a priory* edge deletion method is dominated by one subproblem, and that is why it cannot speed up using more processors. The edge deletion Las Vegas method is much better, but it also has some limits of speedup. The restarting Las Vegas method on the other hand scales very well. A scale up is usually considered good if by doubling the cores the running time reduced by a factor of 1.5. In case of this hard combinatorial optimization problem we could achieve a better scale up using 512 processors. This indicates that the method possibly can be scaled up for several thousands processors.

Also, the average run times of the subproblems are quite interesting. The edge deletion Las Vegas method has a bigger average run time than the *a priory* edge deletion – and one would expect this, as this method is certainly making more calculations in order to eliminate the dominating subproblem. More interesting is that the average time of the restarting algorithm is less. That indicates, that this method even possibly can achieve super linear speedups in some lucky problems one day. Obviously then the whole question of super linearity should be examined and rethinked, because it depends on which sequential algorithm we compare it to. There certainly will be a faster sequential algorithm, but we cannot find it without using the Las Vegas randomization method.

Finally, we would present the graphs of actual running times of the subproblems. In the first graph on Fig. 1 we reordered the problems by the magnitude of

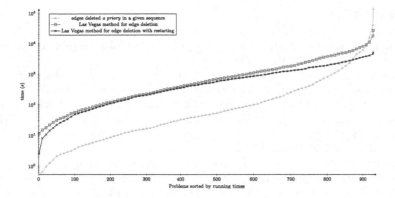

Fig. 1. The sorted running times of the `monoton-9` subproblems.

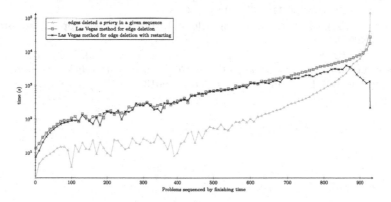

Fig. 2. The time sequence of running times of the `monoton-9` subproblems.

the times. Be aware, that the time scale is logarithmic, so the actual differences are of several magnitudes! In this graph one can see, that all three algorithms are dominated by the longest subproblem, although the restarting Las Vegas can smooth out this problem the best, reducing the variance of running times by more than 2 magnitudes.

The second graph on Fig. 2 is the running times sorted by finishing times. To the left the time passes while we run the parallel algorithm, and the finished subproblem times denoted on the x axis. The graph is smoothed, to have less 'noise' of big variance of running times. It could clearly be seen, that the restarting Las Vegas algorithm helps at the very end by reducing the dominating problems exactly where it needs it the most.

Acknowledgments. We would like to thank the HPC Europe grant for the fruitful visit to Helsinki, to the Finish Computer Science Center which hosts the supercomputer Sisu on which the computations were performed. (https://research.csc.fi/sisu-supercomputer)

References

1. Babai, L.: Monte-Carlo algorithms in graph isomorphism testing. Université de Montréal, D.M.S. No.79-10 (1979)
2. McCreesh, C., Prosser, P.: The shape of the search tree for the maximum clique problem, and the implications for parallel branch and bound. ACM Trans. Parallel Comput. **2**(1), 1–27 (2015). http://dx.doi.org/10.1145/2742359, Article No.: 8
3. Depolli, M., Konc, J., Rozman, K., Trobec, R., Janežič, D.: Exact parallel maximum clique algorithm for general and protein graphs. J. Chem. Inf. Model. **53**(9), 2217–2228 (2013). doi:10.1021/ci4002525
4. DIMACS (2014). ftp://dimacs.rutgers.edu/pub/challenge/graph/
5. Sloan, N.: (2015) http://neilsloane.com/doc/graphs.html
6. Szabo, S.: Parallel algorithms for finding cliques in a graph. J. Phys. Conf. Ser. **268**, 012030 (2011)
7. Szabo, S.: Monotonic matrices and clique search in graphs. Annales Univ. Sci. Budapest., Sect. Comp. **41**, 307–322 (2013)
8. Ouyang, M.: How good are branching rules in DPLL? Discrete Appl. Math. **89**(1–3), 281–286 (1998)
9. Truchet, C., Richoux, F., Codognet, P.: Prediction of parallel speed-ups for Las Vegas algorithms. In: 42nd International Conference on Parallel Processing, ICPP 2013, pp. 160–169, Lyon, France, 1–4 October 2013. IEEE Computer Society (2013)

Computational Microelectronics — From Monte Carlo to Deterministic Approaches

Optimization of the Deterministic Solution of the Discrete Wigner Equation

Johann Cervenka$^{(\boxtimes)}$, Paul Ellinghaus, Mihail Nedjalkov, and Erasmus Langer

Institute for Microelectronics, TU Wien, Vienna, Austria
{cervenka,ellinghaus,nedjalkov,langer}@iue.tuwien.ac.at

Abstract. The development of novel nanoelectronic devices requires methods capable to simulate quantum-mechanical effects in the carrier transport processes. We present a deterministic method based on an integral formulation of the Wigner equation, which considers the evolution of an initial condition as the superposition of the propagation of particular fundamental contributions.

Major considerations are necessary, to overcome the memory and time demands typical for any quantum transport method. An advantage of our method is that it is perfectly suited for parallelization due to the independence of each fundamental contribution. Furthermore, a dramatic speed-up of the simulations has been achieved due to a preconditioning of the resulting equation system.

To evaluate this deterministic approach, the simulation of a Resonant Tunneling Diode, will be shown.

1 Introduction

To describe the carrier transport processes in novel nanoelectronic devices the effects of quantum mechanics have to be considered. The Wigner formulation of quantum mechanics challenges deterministic methods due to difficulties in the discretization of the diffusion term in the differential equation. Even high-order schemes show very different output characteristics because of rapid variations of the Wigner function in the phase-space [1]. However, the high precision of this methods makes them a desirable approach in cases where physical quantities vary over many orders of magnitude. To overcome these problems, an adaptive momentum discretization scheme has been proposed [2]. Alternatively, the developed approach, shown here, uses an integral formulation of the Wigner equation so that the differentiation can be avoided.

We consider the evolution of an initial condition described by a phase-space superposition of particular fundamental solutions. To calculate the distribution at desired time-steps, the Wigner equation has to be solved for each such solution and all "fundamental evolutions" have to be summated.

Unfortunately, the usual approach to solve at sequential time-steps is not practical due to the huge memory consumption: during the time evolution the complete history of all fundamental solutions in phase-space has to be stored in

© Springer International Publishing Switzerland 2015
I. Lirkov et al. (Eds.): LSSC 2015, LNCS 9374, pp. 269–276, 2015.
DOI: 10.1007/978-3-319-26520-9_29

parallel. To overcome this drawback, the calculation order is modified in such a way that for each solution its specific time evolution is calculated separately.

As the particular calculations are independent from each other, this method is well suited for parallelization using MPI and OpenMP.

2   The Deterministic Approach

The Wigner equation [3,4]

$$\frac{\partial f(x,k,t)}{\partial t} - \frac{\hbar k}{m^*} \frac{\partial f(x,k,t)}{\partial x} = \sum_m V_w(x, k - k') \, f(x, k', t), \tag{1}$$

describes the evolution of the function $f(x,k,t)$ under the action of the Wigner potential $V_w(x, \Delta k)$ which is obtained as a Wigner-Weyl-transform of the electrostatic potential [5].

Our approach uses the integral formulation of the Wigner equation. The integral form of (1) is obtained [6,7] by considering the characteristics of the Liouville operator on the left-hand-side of the equation, which are the Newtonian trajectories $x(\cdot, t)$ initialized with x', k', t' [8]:

$$x(x', k', t', t) = x' + \frac{\hbar k'}{m^*}(t - t'). \tag{2}$$

A weak formulation of the numerical task is used

$$f_\Theta(\tau) = \int_0^\tau dt \int dx \sum_m f_i(x, k) \, e^{-\int_0^t \gamma(x_i(y))dy} \, g_\Theta(x_i(t), k, t), \tag{3}$$

which calculates the mean value f_Θ – the integral of the solution inside a particular domain with indicator Θ. τ is the evaluation time, f_i the initial condition, $x_i(t)$ is the trajectory, initialized by $(x, k, 0)$, and g_Θ is the forward solution of the adjoint integral equation:

$$g_\Theta(x', k', t') = \Theta(x', k') \, \delta(t', \tau) +$$

$$+ \int_{t'}^\tau dt \sum_m e^{-\int_{t'}^t \gamma(x(y))dy} \, \Gamma(x(t), k, k') \, g_\Theta(x(t), k, t). \tag{4}$$

Within (4) $\gamma(x) = \sum_k V_w^+(x, k)$,

$$\Gamma(x, k, k') = V_w^+(x, k - k') + V_w^+(x, k' - k) + \gamma(x) \, \delta(k, k'),$$

and $x(t)$ initialized by (x, k, t). The time integration in Eq. (3) can be carried out, delivering the new equation system

$$f_\Theta(\tau) = \int dx \sum_m f_i(x, k) \, p_\Theta(x, k, 0), \tag{5}$$

$$p_\Theta(x', k', t') = e^{-\int_{t'}^{\tau} \gamma(x'(y))\mathrm{d}y}\, \Theta(x'(\tau), k')+$$

$$+ \int_{t'}^{\tau} \mathrm{d}t \sum_m e^{-\int_{t'}^{t} \gamma(x'(y))\mathrm{d}y}\, \Gamma(x'(t), k, k')\, p_\Theta(x'(t), k, t) \tag{6}$$

without time dependency in (5). The trajectories $x'(y)$ are initialized by (x', k', t').

3   Discretization

The numerical procedure is developed by first discretizing the variables of the equation by:

$$x = n\Delta x,\ n \in [0, N];\quad k = m\Delta k,\ m \in [-M/2, M/2];\quad t = l\Delta t,\ l \in [0, L].$$

In the same way, the considered domains are discretized and correspond to a point in phase-space (u, v). Also the trajectories $x'(t)$ are replaced by a discrete version, depicted by $N'(l)$, delivering the new equation system:

$$f_{u,v}(l_\tau) = \sum_n \sum_m f_i(n, m)\, q_{u,v,l_\tau}(n, m, 0), \tag{7}$$

$$q_{u,v,l_\tau}(n', m', l') = e^{-\sum_{j=l'}^{l_\tau} \gamma(N'(j))\Delta t\, \omega_j}\, \delta(N'(\tau), u)\, \delta(m, v)+$$

$$+ \sum_{l=l'}^{l_\tau} \Delta t\, \omega_l \sum_m e^{-\sum_{j=l'}^{l} \gamma(N'(j))\Delta t\, \omega_j}\, \Gamma(N'(l), m, m')\, q_{u,v,l_\tau}(N'(l), m, l) \tag{8}$$

with the discrete trajectory $N'(j)$ initialized by (n', m', l').

The obtained discrete equation system brings several challenges in its implementation, which will be solved in the following.

Re-insertion of Old Values. At each time step l_τ only the new values $q(n', m', 0)$ have to be calculated. The values for different l' can be reused from the previous calculations. In this case, the main computation time shifts from solving the equation system to assembling the equation system. However, by elimination of the time integration it is also possible to reinsert the already calculated values as initial values for the new calculations. In this case the equation system (8) changes to:

$$q_{u,v,l_\tau}(n', m', 0) = e^{-\sum_{j=l'}^{l_\tau} \gamma(x'(j))\Delta t\, \omega_j}\, q_{u,v,l_\tau}(N'(T), m', T)+$$

$$+ \sum_{l=0}^{T} \Delta t \sum_m e^{-\sum_{j=l'}^{l} \gamma(N'(j))\,\Delta t\, \omega_j}\, \Gamma(N'(l), m, m')\, q_{u,v,l_\tau}(N'(l), m, l)\, \omega_l, \tag{9}$$

where $q_{u,v,l_\tau}(N'(T), m', T)$ is the solution T time-steps ago.

Time Integration. The time integration from (6) has to be approximated by numerical integration. This may be done by several methods, which is depicted by the different weights ω_j and ω_l in equation (9). A detailed examination of the summation shows that the first term with $l = l'$ contributes to the unknowns $q(n', m', 0)$ – the system matrix of the equation system. Left-handed or right-handed approximations of the integration lead to a big under- or overestimation of the results. The iteration may result in an unstable system behavior; at least a trapezoidal approximation of the integration has to be performed.

Interpolation. The proper discretization of the trajectory presents a big challenge. As the space coordinate can only take discrete values, the trajectory may be discretized by

$$N'(n', m', l', j) = n' + int\left[\frac{\hbar m \Delta k \Delta t}{m^* \Delta x}(j - l')\right]. \tag{10}$$

For common device dimensions, the contribution in the $int[\ldots]$ expression stays nearly constant for a long number of time-steps. Especially for low m the shifting value is always 0, which results in a non-moving distribution. This aspect can be accounted for by manipulation of the integer contribution depending on the accumulated error of the trajectory discretization.

Another issue can be identified by examining the first part in Eq. (9)

$$q_{u,v,l_\tau}(n', m', 0) = \ldots\ q_{u,v,l_\tau}(N'(T), m', T) + \ldots, \tag{11}$$

which shows a difficulty in the discretization of the initial condition. Even with the proposed manipulation of the calculation of N', the discrete values of N' stay constant for a wide range of l_τ, which results in a stepwise moving wavefront. As a consequence this stepwise movement may cause increasing amplifying oscillations in the solution, especially near corners in the potential distribution.

A correction is introduced by interpolation of the initial conditions between the left-side and right-side integer values N'_{left} and N'_{right}:

$$N'_{left} \leq N'_{left} + \Delta N' = N'(n', m', 0, T) \leq N'_{left} + 1 = N'_{right} \tag{12}$$

and insertion into (9)

$$q_{u,v,l_\tau}(n', m', 0) = \ldots\ \left[q_{u,v,l_\tau}(N'_{left})(1 - \Delta N') + q_{u,v,l_\tau}(N'_{right})\Delta N'\right] + \ldots. \tag{13}$$

4   Parallelization Issues

A direct implementation of the algorithm is to assemble and solve the equation system (9) of rank $N \cdot M$ (the number of points in phase-space) and then to back-insert the solution into (7).

This requires

– the assembly of (9) with effort $\mathcal{O}(M \cdot T \cdot T)$,

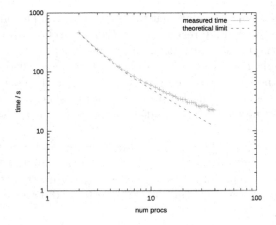

Fig. 1. Used time for simulation runs in dependency of the number of parallel processes. The values are compared to the theoretical limit without an overhead $t_{used} \sim 1/n$.

- solving a system with rank $N \cdot M$,
- back-insertion in (7) with effort $\mathcal{O}(N \cdot M)$,
- the storage of all $q_{u,v,l_\tau}(n', m', l')$, and
- the temporary storage for the equation system (9).

This equation system has to be computed for all time steps for each particular indicator, leading to solve $N \cdot M \cdot L$ times Eq. (9).

Concerning memory and computation time demands this offers special possibilities for parallelization purposes.

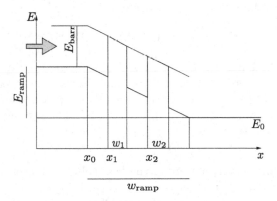

Fig. 2. The considered RTD device is specified as follows: $x_1 = 55\,\text{nm}$, $x_2 = 65\,\text{nm}$, drift region from x=40 nm to 70 nm. The ramp height varies between 0 and 0.2 eV.

If the solution has no feedback to the Wigner potential, the different equation systems are independent from each other and, therefore, they are very well suited for parallelization. The different tasks for (u, v) can be split over the computation nodes. Only the final results $f(u, v, l_\tau)$ have to be transmitted. An MPI parallelization without communication can be used.

Also on a single node it seems feasible, not to parallelize the solver, but to share the common resources to split the different tasks (u, v) in parallel by OpenMP on the nodes, which also does not require synchronization. Even the system matrix is common for all equation systems and may also be assembled in parallel on the nodes.

In Fig. 1 the relation between execution times and number of parallel processes is shown. They are compared to the theoretical limit. The differences to this value at higher number of processes arises due to a nearly constant overhead of calculating the Wigner potential, which is performed on each machine.

5   Preconditioning and Inversion of the System Matrix

Analyzing the computational costs of the method, it can be seen that the order is higher than $(N \cdot M)^2$. Looking at the system matrix of the equation system, which has to be solved,

$$\mathbf{A} \cdot \mathbf{q} = \mathbf{b}, \tag{14}$$

the matrix $\mathbf{A}$ can be expressed as:

$$A(n', m', n', m) = \begin{cases} 1 - \Delta t\, \gamma(n')\, \omega_0, & m' = m \\ \Delta t\, \Gamma(n', m, m')\, \omega_0, & m' \neq m. \end{cases} \tag{15}$$

The matrix is sparse and, for sufficiently small Δt, of good dominance and the solutions may be calculated iteratively. A main speedup is achieved by Jacobi Over-relaxation Methods like

$$\mathbf{q}_{i+1} = (\mathbf{I} - \epsilon \mathbf{D}^{-1}\mathbf{A})\mathbf{q}_i + \epsilon \mathbf{D}^{-1}\mathbf{b} \tag{16}$$

with $\mathbf{D}$ the diagonal part of the matrix $\mathbf{A}$ and ϵ the over-relaxation factor. For the resulting equation systems this method shows better performance than common gradient based techniques. The simulation procedure implies solving $N \cdot M \cdot T$ times an equation system of rank $N \cdot M$. The solving method (16) is used for each calculation

$$\mathbf{A} \cdot \mathbf{q}_i = \mathbf{e}_i, \tag{17}$$

with $\mathbf{e}_i$ the i-th unit vector, obtaining the final solution vector

$$\mathbf{q} = \sum_i \mathbf{q}_i\, b_i = \mathbf{Q}^T \mathbf{b}. \tag{18}$$

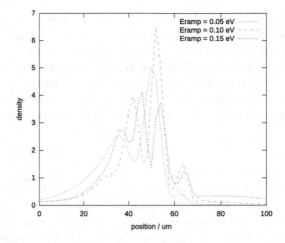

Fig. 3. The wave package evolving after 300fs for different bias voltages.

6   Application to a Resonant Tunneling Diode

As an application we consider a resonant tunneling diode (RTD) [2] schematically shown in Fig. 2. The device consists of two 3 nm wide 0.1eV high potential barriers as described in the figure caption. Depending on the bias, the transmission of electrons through the barriers is influenced [9]. The transmitted part of the packet has a maximum if its mean energy coincides with the resonant energy of the structure.

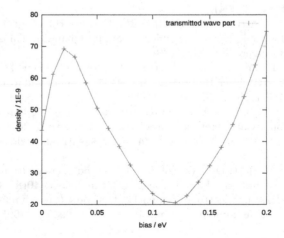

Fig. 4. Wave package passing through a double barrier. The transmitted portion as dependent on the bias potential demonstrates the typical for RTDs dipping region.

The initial wave-packet is accelerated by the voltage drop [10]. The dependency of the density distribution on the applied bias is shown in Fig. 3. To calculate the amount of passed signal, the distribution in the right device area with 200 nm length is integrated. The portion of the transmitted part as depending on the bias is shown in Fig. 4. The typical region for RTD devices can be observed. This gives rise to negative differential resistance which can be utilized as negative feedback in transistor circuits, like Terahertz oscillators.

7   Conclusion

In this paper the technique for a deterministic Wigner solver in its integral formulation has been shown. A modified simulation approach was discussed regarding scalability and the performance due to optimization was investigated. The method is capable to correctly simulate physical effects of typically quantum devices.

References

1. Kim, K.Y., Lee, B.: On the high order numerical calculation schemes for the wigner transport equation. Solid-State Electron. **43**, 2243–2245 (1999)
2. Dorda, A., Schürrer, F.: A WENO-solver combined with adaptive momentum discretization for the wigner transport equation and its application to resonant tunneling diodes. J. Comput. Phys. **284**, 95–116 (2015)
3. Griffiths, D.: Introduction to Quantum Mechanics. Pearson Prentice Hall, Upper Saddle River (2005)
4. Kosik, R.: Numerical challenges on the road to nanoTCAD. Ph.D. thesis, Institut für Mikroelektronik (2004)
5. Nedjalkov, M., Querlioz, D., Dollfus, P., Kosina, H.: Wigner function approach. Nano-electronic Devices; Semiclassical and Quantum Transport Modeling, pp. 289–358. Springer, New York (2011)
6. Nedjalkov, M., Kosina, H., Selberherr, S., Ringhofer, C., Ferry, D.K.: Unified particle approach to Wigner-Boltzmann transport in small semiconductor devices. Phys. Rev. B **70**, 115319 (2004)
7. Sellier, J.M.D., Nedjalkov, M., Dimov, I., Selberherr, S.: A Benchmark study of the Wigner Monte Carlo method. Monte Carlo Meth. Appl. **20**(1), 43–51 (2014)
8. Dimov, I.T.: Monte Carlo Methods for Applied Scientists. World Scientific, Singapore (2008)
9. Sudiarta, I.W., Geldart, D.J.W.: Solving the Schrödinger equation using the finite difference time domain method. J. Phys. A: Math. Theor. **40**(8), 1885 (2007)
10. Fu, Y., Willander, M.: Electron wave-packet transport through nanoscale semiconductor device in time domain. J. Appl. Phys. **97**(9), 094311 (2005)

The Influence of Electrostatic Lenses
on Wave Packet Dynamics

Paul Ellinghaus[(⊠)], Mihail Nedjalkov, and Siegfried Selberherr

Institute for Microelectronics, TU Wien, Vienna, Austria
{ellinghaus,nedjalkov,selberherr}@iue.tuwien.ac.at

Abstract. The control of coherent electrons is becoming relevant in emerging devices as (semi-)ballistic transport is observed within nanometer semiconductor structures at room temperature. The evolution of a wave packet – representing an electron in a semiconductor – can be manipulated using specially shaped potential profiles with convex or concave features, similar to refractive lenses used in optics. Such electrostatic lenses offer the possibility, for instance, to concentrate a single wave packet which has been invoked by a laser pulse, or split it up into several wave packets. Moreover, the shape of the potential profile can be dynamically changed by an externally applied potential, depending on the desired behaviour. The evolution of a wave packet under the influence of a two-dimensional potential – the electrostatic lens – is investigated by computing the physical densities using the Wigner function. The latter is obtained by using the signed-particle Wigner Monte Carlo method.

1 Introduction

Analogies often serve as a source of inspiration to advance research in science and technology. An example is the electrostatic lens, inspired by concepts from geometrical optics, which can be used to steer and control coherent electrons. The term *electrostatic lens* refers to a specially shaped potential with convex/concave features, as found in optical lenses, used to steer electron waves. The concept was first demonstrated experimentally in 1990 in [1,2], in low-temperature, high-mobility semiconductors, which ensured that the coherent electrons had a sufficiently long mean free path to conduct experiments with structures made with the lithographic capabilities at the time. The astounding decrease of the feature sizes in semiconductor devices, along with novel materials like graphene, has made (semi-)ballistic electron transport applicable at room temperatures [3]. This has sparked new interest in applying concepts from optics in semiconductors: electrons can be guided in a channel using total internal reflection as in optical fibres [4] or focused towards the centre of nanowires, using electrostatic lenses, to increase their mobility by avoiding rough interfaces [5].

Scanning probe microscopy allows the flow of coherent electrons in semiconductor structures to be measured and visualized with a subnanometer resolution [6,7], however, a concurrent temporal resolution to visualize dynamics on the femtosecond time scale still remains out of reach [8]. Computer simulations

© Springer International Publishing Switzerland 2015
I. Lirkov et al. (Eds.): LSSC 2015, LNCS 9374, pp. 277–284, 2015.
DOI: 10.1007/978-3-319-26520-9_30

can provide insight into the temporal dynamics of wave packets, which capture the physics of single electrons in mesoscopic structures. Here, we apply the two-dimensional (2D) Wigner Monte Carlo method to demonstrate its suitability as a simulation tool to investigate wave packet dynamics in the context of electrostatic lenses (and beyond).

The Wigner formalism [9] has re-emerged in recent times as a convenient formalism to consider quantum mechanical phenomena on the mesoscopic scale, since semi-classical transport models can be augmented to the coherent quantum evolution. Multi-dimensional simulations have been made computationally feasible by the signed-particle Wigner Monte Carlo method, as described in Sect. 2. Section 3 shows examples of electrostatic lenses, and their influence on the behaviour of wave packets.

2   Wigner Monte Carlo Method

The Wigner formalism expresses quantum mechanics, normally formulated with the help of wave functions and operators, in terms of functions and variables defined in the phase-space. This reformulation in the phase-space facilitates the reuse of many classical concepts and notions.

The Wigner transform of the density matrix operator yields the Wigner function, $f_w(x, p)$, which is often called a quasi-probability function as it retains certain properties of classical statistics, but the negative values which appear demand a different interpretation than the classical probability [10]. The associated evolution equation for the Wigner function follows from the von Neumann equation for the density matrix, which for the illustrative, one-dimensional case is written as

$$\frac{\partial f_w}{\partial t} + \frac{p}{m^*} \frac{\partial f_w}{\partial x} = \int dp' V_w (x, p - p') f_w (x, p', t). \tag{1}$$

If a finite coherence length is considered, the implications and interpretation of which is discussed in [11,12], the semi-discrete Wigner equation result, the momentum values are quantized by $\Delta k = \frac{\pi}{L}$, and the integral is replaced by a summation. Henceforth, the index q refers to the quantized momentum, i.e. $p = \hbar (q \Delta k)$.

Equation (1) is reformulated as an adjoint integral equation (Fredholm equation of the second kind) and is solved stochastically using the particle-sign method [13]. The latter associates a $+$ or a $-$ sign to each particle, which carries the quantum information of the particle. Furthermore, the term on the right-hand side of (2) gives rise to a particle generation term in the integral equation; the statistics governing the particle generation are given by the Wigner potential (i.e. the kernel of the Fredholm equation), which is defined here as

$$V_w (x, q) \equiv \frac{1}{i \hbar L} \int_{-\frac{L}{2}}^{\frac{L}{2}} ds \, e^{-i 2q \Delta k \cdot s} \{ V (x + s) - V (x - s) \}. \tag{2}$$

A generation event entails the creation of two additional particles with complementary signs and momentum offsets q' and q'', with respect to the momentum q of the generating particle. The two momentum offsets, q' and q'', are determined by sampling the probability distributions $V_w^+(x, q)$ and $V_w^-(x, q)$, dictated by the positive and negative values of the Wigner potential in (2), respectively:

$$V_w^+ (x, q) \equiv \max (0, V_w) ; \qquad (3)$$

$$V_w^- (x, q) \equiv \min (0, V_w) . \qquad (4)$$

The generation events occur at a rate given by

$$\gamma (x) = \sum_q V_w^+ (x, q) , \qquad (5)$$

which typically lies in the order of $10^{15}\,\mathrm{s}^{-1}$ in numerical experiments where potential differences in the order of $100\,\mathrm{meV}$ are encountered. This rapid increase in the number of particles makes the associated numerical burden become computationally debilitating, even for simulation times in the order of femtoseconds.

The notion of particle annihilation is used to counteract the exponential increase in the number of particles, due to particle generation [14]. This concept entails a division of the phase space into many cells – each representing a volume $(\Delta x \Delta k)$ of the phase space – within which particles of opposite sign annihilate each other. This is justified since particles of opposite sign, within the same cell, have the same probabilistic future – their contribution to the calculation of any physical quantity would be equal in magnitude, yet opposite in sign.

3   Electrostatic Lenses

The following experiments consider a minimum-uncertainty wave packet, which captures both the particle- and wave-like physical characteristics of an electron. The associated Wigner function representing this initial condition is given by

$$f_w (\mathbf{x}, \mathbf{q}) = \mathcal{N} e^{-\frac{(x-x_0)^2}{\sigma^2}} e^{-(q\Delta k - k_0)^2 \sigma^2} , \qquad (6)$$

where $\mathbf{x_0}$ and $\mathbf{k_0}$ are two-dimensional vector quantities representing the mean position and the mean wavevector, respectively; σ is the standard spatial deviation and $\mathcal{N}$ represents a normalization constant. The wave packet travels in a two-dimensional plane towards a potential barrier, which forms the electrostatic lens. This setup is representative of a physical system where a 2D electron gas is formed at the interface between two semiconductors, e.g. GaAs/AlGaAs; the potential barrier can be induced by an appropriately shaped gate contact at the surface of the semiconductor (parallel to the interface).

A law of refraction, equivalent to Snell's law in optics, can be derived for electrostatic lenses by considering the principle of energy conservation. A particle with a wavevector $\mathbf{k}$ has a kinetic energy

$$E_k = \frac{\hbar^2 |\mathbf{k}|^2}{2m^*}, \qquad (7)$$

which is reduced as the particle transverses the potential step, while its potential energy increases. The change in kinetic energy is attributed only to a change of the momentum component normal to the interface; the momentum component parallel to the lens interface (potential step) is maintained, therefore

$$|\mathbf{k}|\sin\theta = |\mathbf{k}'|\sin\theta', \tag{8}$$

where θ (θ') is the angle of incidence (transmission) with respect to the normal (of the interface) and $\mathbf{k}'$ is the wavevector of the particle within the lens region. The law of refraction follows:

$$\frac{\sin\theta'}{\sin\theta} = \frac{|\mathbf{k}|}{|\mathbf{k}'|} = \frac{\sqrt{E_k}}{\sqrt{E_k'}}. \tag{9}$$

Therefore, the square root of the kinetic energy of a particle is analogous to the refractive index used in geometrical optics.

The interaction of a wave packet with two different electrostatic lens shapes will be investigated in the following.

3.1   Wave Packet Focusing with a Double-Concave Lens

Optical lenses typically operate in a medium (air) with a lower refractive index, where the familiar double-convex shaped lens is used to focus light. A positive potential step is used here, however, making the refractive index of the electrostatic lens lower than the surrounding regions (due to a decrease in kinetic energy). Therefore, a double-concave shaped profile is needed to form a converging lens for focusing the wave packet. The potential shape used to form the electrostatic lens is shown in Fig. 1 along with the corresponding generation rate. The free evolution of a wave package is compared to the case where it interacts with the proposed lens in Fig. 2. The electrostatic lens has a peak potential of 0.04 eV and the wave packet is initialized with a kinetic energy of

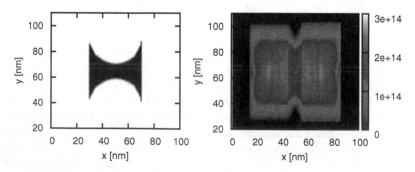

Fig. 1. Two-dimensional potential (represented in black) with a double-concave shape forming a converging electrostatic lens for electrons propagating in the y-direction. The potential value of the lens is constant; it has no three-dimensional features. The corresponding particle generation rate γ is shown on the right.

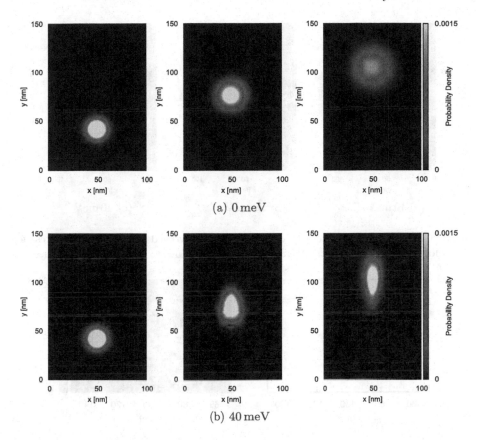

Fig. 2. Wave packet evolving freely (top sequence) and interacting with a double con-cave electrostatic lens (bottom sequence); the time steps (from left to right) correspond to 40 fs, 100 fs and 150 fs.

0.18 eV, moving upwards. The electrostatic lens clearly focuses the wave packet (density) after 150 fs of evolution, compared to the case without a lens. If such a lens is added within a quantum wire (say), the focused wave packet suffers less from the surface roughness at the boundaries when compared to the spread-out wave packet in the case without a lens.

The refractive index of the lens, and thereby its focal length, can be modified by varying the magnitude of the potential with which it is formed. Figure 3 compares the effect of different potential values: It can be clearly seen that the higher potential focuses the wave packet more sharply (at the distance observed at the time instance shown). The applied potential can thereby control, when and at which distance, a wave packet is focused (on a detector, for instance).

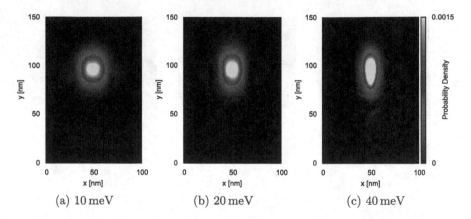

(a) 10 meV   (b) 20 meV   (c) 40 meV

Fig. 3. Comparison of the density of a wave packet evolved for 135 fs in the presence of an electrostatic lens (Fig. 1) at various potential values to show different focussing.

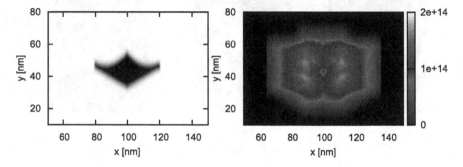

Fig. 4. Two-dimensional potential (left) with rhomboid-like shape and concave-shaped rear edges, forming an electrostatic lens to scatter an electron wave packet in various directions. The potential value of the lens is constant; it has no three-dimensional features. The corresponding particle generation rate γ is shown on the right.

3.2   Wave Packet Splitting with a Rhomboid-Like Potential

An electrostatic lens can also be used to split a wave packet into parts. Figure 4 shows a rhomboid-like potential, along with the corresponding generation rate, which forms a lens to perform such a splitting. Figure 5 illustrates the effect of the lens at different potential values. It should be noted that the electron is not split; it is a single electron in an entangled state. The density peaks indicate regions with a higher probability to find an electron. In Fig. 5a (peak potential 70 meV) the wave packet is almost fully transmitted and split into two parts. The same lens shape, but with a potential of 120 meV, splits the wave packet into four parts (Fig. 5b): The front edges splits off a portion of the wave packet by reflection, while the concave-shaped rear edges focus the transmitted parts again. In the first case, with two peaks (the most-probable components of the state), the y-component of the wavevector remains positive, whereas for the second case,

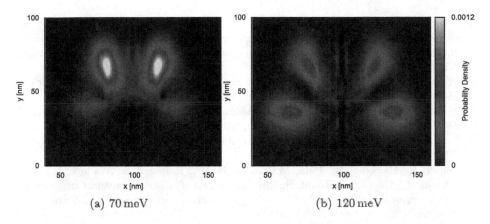

Fig. 5. Wave packet is split either (a) into two or (b) into four parts, after 90 fs evolution, by a rhomboid-like potential profile with concave rear edges with a peak potential of 70 meV and 120 meV, respectively.

at a higher potential, the wavevector of the scattered state also has a negative y-component. This example clearly illustrates how specially shaped potentials can be used to influence the scattering pattern of an electron wave packet. By varying the potential the electron can be guided in a certain direction with a controllable probability. This can be of use in the field of quantum computing to generate a (modifiable) entangled state and direct it to other computing elements.

4   Conclusion

It has been shown that 2D Wigner Monte Carlo simulations, using the signed-particle method, can be applied to investigate the dynamics of wave packets interacting with electrostatic lenses formed in mesoscopic semiconductor structures. The concept of electrostatic lenses enables the control of coherent electrons by focusing or splitting wave packets in a controllable fashion. This ability can be utilized in emerging mesoscopic semiconductor devices, where (semi-)ballistic transport at room temperature becomes feasible.

Acknowledgement. This work was partially supported by the Bulgarian NSF under the grant DFNI 02/20.

References

1. Spector, J., Stormer, H.L., Baldwin, K.W., Pfeiffer, L.N., West, K.W.: Electron focusing in two-dimensional systems by means of an electrostatic lens. Appl. Phys. Lett. **56**(13), 1290–1292 (1990)
2. Sivan, U., Heiblum, M., Umbach, C.P., Shtrikman, H.: Electrostatic electron lens in the ballistic regime. Phys. Rev. B **41**, 7937–7940 (1990)

3. Wang, R., Liu, H., Huang, R., Zhuge, J., Zhang, L., Kim, D.W., Zhang, X., Park, D., Wang, Y.: Experimental investigations on carrier transport in Si nanowire transistors: ballistic efficiency and apparent mobility. IEEE Trans. Electron Devices **55**(11), 2960–2967 (2008)
4. Williams, J.R., Low, T., Lundstrom, M.S., Marcus, C.M.: Gate-controlled guiding of electrons in graphene. Nat. Nanotechnol. **6**(4), 222–225 (2011)
5. Muraguchi, M., Endoh, T.: Size dependence of electrostatic lens effect in vertical MOSFETs. Jpn. J. Appl. Phys. 53(4S), 04EJ09 (2014)
6. LeRoy, B.J.: Imaging coherent electron flow. J. Phys. Condens. Matter **15**(50), R1835 (2003)
7. Sellier, H., Hackens, B., Pala, M.G., Martins, F., Baltazar, S., Wallart, X., Desplanque, L., Bayot, V., Huant, S.: On the imaging of electron transport in semiconductor quantum structures by scanning-gate microscopy: successes and limitations. Semicond. Sci. Technol. **26**(6), 064008 (2011)
8. Loth, S., Burgess, J.A.J., Yan, S.: Scanning probe microscopy: close-up on spin coherence. Nat. Nanotechnol. **9**(8), 574–575 (2014)
9. Wigner, E.: On the quantum correction for thermodynamic equilibrium. Phys. Rev. **40**, 749–759 (1932)
10. Leibfried, D., Pfau, T., Monroe, C.: Shadows and mirrors: reconstructing quantum states of atom motion. Phys. Today **51**(4), 22 28 (1998). Print edition
11. Ellinghaus, P., Nedjalkov, M., Selberherr, S.: Implications of the coherence length on the discrete Wigner potential. In: 2014 International Workshop on Computational Electronics (IWCE), pp. 1–3 (2014)
12. Nedjalkov, M., Vasileska, D.: Semi-discrete 2D Wigner-particle approach. J. Comput. Electron. **7**(3), 222–225 (2008)
13. Nedjalkov, M., Schwaha, P., Selberherr, S., Sellier, J.M., Vasileska, D.: Wigner quasi-particle attributes - an asymptotic perspective. Appl. Phys. Lett. **102**(16), 163113 (2013)
14. Sellier, J.M., Nedjalkov, M., Dimov, I., Selberherr, S.: The role of annihilation in a Wigner Monte Carlo approach. In: Lirkov, I., Margenov, S., Waśniewski, J. (eds.) LSSC 2013. LNCS, vol. 8353, pp. 186–194. Springer, Heidelberg (2014)

Evaluation of Spin Lifetime in Thin-Body FETs: A High Performance Computing Approach

Joydeep Ghosh$^{(\boxtimes)}$, Dmitry Osintsev, Viktor Sverdlov,
Josef Weinbub, and Siegfried Selberherr

Institute for Microelectronics, TU Wien, Vienna, Austria
{ghosh,osintsev,sverdlov,weinbub,selberherr}@iue.tuwien.ac.at

Abstract. Silicon, the prominent material of microelectronics, is perfectly suited for spin-driven applications because of the weak spin-orbit interaction resulting in long spin lifetime. However, additional spin relaxation on rough interfaces and acoustic phonons may strongly decrease the spin lifetime in modern silicon-on-insulator and trigate transistors. Because of the need to perform numerical calculation and appropriate averaging of the strongly scattering momenta depending spin relaxation rates, an evaluation of the spin lifetime in thin silicon films becomes prohibitively computationally expensive. We use a highly parallelized approach to calculate the spin lifetime in silicon films. Our scheme is based on a hybrid parallelization approach, using the message passing interface MPI and OpenMP. The algorithm precalculates wave functions and energies, and temporarily stores the results in a file-based cache to reduce memory consumption. Using the precalculated data for the spin relaxation rate calculations drastically reduces the demand on computational time. We show that our approach offers an excellent parallel speedup, and we demonstrate that the spin lifetime in strained silicon films is enhanced by several orders of magnitude.

1 Introduction

For almost half a century Moore's law has successfully predicted the persistent miniaturization of semiconductor devices, such as the transistor feature size in microprocessors and unit cell in magnetic storage disks and random access memories. However, as devices are scaled down to the nano-scale, fundamental physical limitations will hinder further improvements in device performances in the upcoming years. Employing spin as an additional degree of freedom is promising for boosting the efficiency of future low-power integrated electronic circuits. Silicon is characterized by a weak spin-orbit interaction and long spin lifetime. It is therefore an attractive material for spin-driven applications. A long spin transfer distance of conduction electrons has been shown experimentally [1]. However, a large experimentally observed spin relaxation in electrically-gated silicon films could become an obstacle in realizing spin-driven devices [2]. Henceforth, a deeper understanding of the fundamental spin relaxation mechanism in silicon MOSFETs is required. We consider the surface roughness (SR) and the

© Springer International Publishing Switzerland 2015
I. Lirkov et al. (Eds.): LSSC 2015, LNCS 9374, pp. 285–292, 2015.
DOI: 10.1007/978-3-319-26520-9_31

longitudinal (LA) and transversal (TA) acoustic phonons to cause the prominent spin relaxation mechanisms. In (001) silicon films the conduction electrons are positioned close to the minima of the pair of valleys near the edges of the Brillouin zone along the axis. Each state is described by the subband index, the in-plane wave vector $\mathbf{k}$, and the spin orientation (spin-up and spin-down) on a chosen axis. The subband wave functions and the rates are obtained by the perturbative $\mathbf{k} \cdot \mathbf{p}$ model [3]. The spin relaxation time is calculated from the obtained rates by thermal averaging [4,5]:

$$\frac{1}{\tau} = \frac{\int \frac{1}{\tau(\mathbf{K_1})} \cdot f(E)(1 - f(E))d\mathbf{K_1}}{\int f(E)d\mathbf{K_1}}, \tag{1}$$

$$f(E) = \frac{1}{1 + \exp(\frac{E - E_F}{K_B T})}, \int d\mathbf{K_1} = \int_0^{2\pi} d\phi \cdot \int_0^\infty \frac{|\mathbf{K_1}(\phi, E)|}{|\frac{\partial E(\mathbf{K_1})}{\partial \mathbf{K_1}}|} dE \tag{2}$$

E is the electron energy, $\mathbf{K_1}$ is the in-plane after-scattering wave vector, K_B is the Boltzmann constant, T is the temperature, and E_F is the Fermi level.

The SR-limited spin relaxation rate is expressed as:

$$\frac{1}{\tau_{SR}(\mathbf{K_1})} = \frac{4\pi}{\hbar(2\pi)^2} \sum_{i,j=1,2} \int_0^{2\pi} \pi \triangle^2 L^2 \cdot \frac{1}{\epsilon_{ij}^2(\mathbf{K_2} - \mathbf{K_1})} \cdot \frac{\hbar^4}{4m_l^2} \cdot \frac{|\mathbf{K_2}|}{|\frac{\partial E(\mathbf{K_2})}{\partial \mathbf{K_2}}|} \cdot$$
$$[(\frac{d\psi_{i\mathbf{K_{1\sigma}}}}{dz})^*(\frac{d\psi_{j\mathbf{K_{2-\sigma}}}}{dz})]^2_{z=\pm\frac{t}{2}} \cdot \exp(\frac{-(\mathbf{K_2} - \mathbf{K_1})^2 L^2}{4})d\phi \tag{3}$$

$\mathbf{K_2}$ and $\mathbf{K_1}$ are the in-plane wave vectors before and after scattering, ϕ is the angle between $\mathbf{K_1}$ and $\mathbf{K_2}$, ϵ is the dielectric permittivity, L is the autocorrelation length, $\triangle$ is the mean square value of the SR-fluctuations, $\psi_{i\mathbf{K_{1\sigma}}}$ and $\psi_{j\mathbf{K_{2\sigma}}}$ are the wave functions, $\sigma = \pm 1$ is the spin projection to the [001] axis, and m_l is the longitudinal effective mass. The TA-phonon induced intravalley spin relaxation rate can be written as:

$$\frac{1}{\tau_{TA}(\mathbf{K_1})} = \frac{\pi K_B T}{\hbar \rho \nu_{TA}^2} \sum \int_0^{2\pi} d\phi \cdot \frac{|\mathbf{K_2}|}{|\frac{\partial E(\mathbf{K_2})}{\partial \mathbf{K_2}}|} \left[1 - \frac{|\frac{\partial E(\mathbf{K_2})}{\partial \mathbf{K_2}}|f(E(\mathbf{K_2}))}{|\frac{\partial E(\mathbf{K_1})}{\partial \mathbf{K_1}}|f(E(\mathbf{K_1}))}\right] \cdot$$
$$\int_0^t \int_0^t \exp(-\sqrt{q_x^2 + q_y^2}|z - z'|)[\psi_{\mathbf{K_{2\sigma}}}^\dagger(z)M\psi_{\mathbf{K_{1-\sigma}}}(z)]^*[\psi_{\mathbf{K_{2\sigma}}}^\dagger(z')M\psi_{\mathbf{K_{1-\sigma}}}(z')] \cdot$$
$$[\sqrt{q_x^2 + q_y^2} - \frac{8q_x^2 q_y^2 - (q_x^2 + q_y^2)^2}{q_x^2 + q_y^2}|z - z'|]dzdz' \tag{4}$$

$\rho = 2329 \, \text{kg/m}^3$ is the silicon density, $\nu_{TA} = 5300 \, \text{m/s}$ is , t is the film thickness, $(q_x, q_y) = \mathbf{K_1} - \mathbf{K_2}$, and M written in the two valley plus two spin projection basis is ($D = 14 \, \text{eV}$: shear deformation potential) $\begin{bmatrix} 0 & 0 & \frac{D}{2} & 0 \\ 0 & 0 & 0 & \frac{D}{2} \\ \frac{D}{2} & 0 & 0 & 0 \\ 0 & \frac{D}{2} & 0 & 0 \end{bmatrix}$.

The intravalley spin relaxation rate due to LA-phonons is:

$$\frac{1}{\tau_{LA}(\mathbf{K_1})} = \frac{\pi K_B T}{\hbar \rho \nu_{LA}^2} \sum \int_0^{2\pi} d\phi \cdot \frac{|\mathbf{K_2}|}{\left|\frac{\partial E(\mathbf{K_2})}{\partial \mathbf{K_2}}\right|} \left[1 - \frac{\left|\frac{\partial E(\mathbf{K_2})}{\partial \mathbf{K_2}}\right| f(E(\mathbf{K_2}))}{\left|\frac{\partial E(\mathbf{K_1})}{\partial \mathbf{K_1}}\right| f(E(\mathbf{K_1}))} \right]$$

$$\cdot \int_0^t \int_0^t \exp(-\sqrt{q_x^2 + q_y^2}|z - z'|)[\psi_{\mathbf{K_{2\sigma}}}^\dagger(z) M \psi_{\mathbf{K_{1-\sigma}}}(z)]^* [\psi_{\mathbf{K_{2\sigma}}}^\dagger(z') M \psi_{\mathbf{K_{1-\sigma}}}(z')] \cdot$$

$$\frac{4 q_x^2 q_y^2}{(q_x^2 + q_y^2)^{3/2}} [\sqrt{q_x^2 + q_y^2}|z - z'| + 1] dz dz'$$

$$(5)$$

$\nu_{LA} = 8700\,\mathrm{m/s}$ is the longitudinal phonon velocity.

The intervalley spin relaxation rate due to acoustic phonons contains Elliot and Yafet contributions and is calculated as:

$$\frac{1}{\tau_{LA}(\mathbf{K_1})} = \frac{\pi K_B T}{\hbar \rho \nu_{LA}^2} \sum \int_0^{2\pi} d\phi \cdot \frac{|\mathbf{K_2}|}{\left|\frac{\partial E(\mathbf{K_2})}{\partial \mathbf{K_2}}\right|} \left[1 - \frac{\left|\frac{\partial E(\mathbf{K_2})}{\partial \mathbf{K_2}}\right| f(E(\mathbf{K_2}))}{\left|\frac{\partial E(\mathbf{K_1})}{\partial \mathbf{K_1}}\right| f(E(\mathbf{K_1}))} \right] \quad (6)$$

$$\cdot \int_0^t [\psi_{\mathbf{K_{2\sigma}}}^\dagger(z) M' \psi_{\mathbf{K_{1-\sigma}}}(z)]^* [\psi_{\mathbf{K_{2\sigma}}}^\dagger(z) M' \psi_{\mathbf{K_{1-\sigma}}}(z)] dz$$

M' is
$$\begin{bmatrix} D' & 0 & 0 & D_{SO}(q_y - iq_x) \\ 0 & D' & D_{SO}(q_y - iq_x) & 0 \\ 0 & D_{SO}(-q_y + iq_x) & D' & 0 \\ D_{SO}(q_y + iq_x) & 0 & 0 & D' \end{bmatrix}$$ with

$D' = 12\,\mathrm{eV}$, $D_{SO} = 15\,\mathrm{meV/k_0}$ ($k_0 = 0.15 \frac{2\pi}{a}$, a being silicon lattice constant).

2   Simulations

In order to evaluate the spin relaxation time, a multi-dimensional integral (2) over the energy E and angle ϕ must be computed. The spin relaxation matrix elements are characterized by very narrow and sharp peaks (so-called spin hot spots [6]). To resolve these sharp features, the mesh in the $\mathbf{K}$ space has to be precise. In our application we have determined that the energy step value $\triangle E$ should be upper-bounded by 0.5meV. The lower limit of the integral over E is zero, and we have also identified that it is sufficient to set the corresponding upper limit to be 0.7eV. This particular simulation setup requires almost 1400 points. The lower and the upper limits of the integral over ϕ are 0° and 360° respectively. The step value for ϕ, or $\triangle \phi$, is set to be smaller than 0.5°. Hence, the inner integral over ϕ on before- and after-scattering directions at fixed energies require almost 1000 points each. Thus, the scattering matrix elements and the Jacobians (the derivative of the dispersion energy over the wave vector) must be calculated numerically for almost 1,400,000 times. To compute the matrix elements, the eigenfunction problems for the 4×4 Hamiltonian matrix must be solved for the two wave vectors before and after scattering for a broad range of parameters, which makes the numerical spin relaxation time calculation pro-hibitively expensive: When utilizing a standard adaptive integration technique, we found that a month of calculations on 20 cores, or 15000 core-hours total, was required to evaluate a single data point of τ as a function of stress ε_{xy}. In order to improve the calculation time, we have divided the entire computation into two levels. The first level calculates and archives all static wave functions and energy data to a binary file (a file-based cache technique), and the second level performs the spin lifetime calculations by loading those data in memory. Both of these levels individually perform the calculations in parallel by using the message passing interface MPI and OpenMP.

2.1   Parallelization Algorithm for Spin Relaxation Rate Calculations

In the following our two-level computation algorithm is outlined.
Level 1

- (**1**) Divide the range of angle ϕ into sub-domains for each MPI process.
- (**1.1**) Divide the range of energy E into sub-domains for each OpenMP thread.
- (**1.1.1**) Calculate the derivatives at the interface $(\frac{d\psi}{dz})_{z=\pm\frac{t}{2}}$, and $\frac{|\mathbf{K}|}{|\frac{\partial E(\mathbf{K})}{\partial \mathbf{K}}|}$ in parallel (MPI, OpenMP).
- (**2**) Collect all the cached values at the master MPI process.
- (**3**) Archive the cache to a binary file.

Level 2

- (**4**) Load archived cache by the master MPI process.
- (**5**) Divide the range of ϕ into sub-domains for each MPI process.
- (**5.1**) Divide the range of E into sub-domains for each OpenMP thread.
- (**5.1.1**) Calculate (1) for a given range of values in parallel (MPI, OpenMP).
- (**6**) Collect all calculated relaxation rates into the final relaxation rate.

The performance is measured on the Vienna Scientific Cluster (VSC-2) [7], which consists of 1285 nodes with 2 processors each (AMD Opteron 6132 HE, 2.2 GHz and 8 cores) and 32 GB main memory on each node. We investigate the calculation time, memory consumption, and utilization of computational resources for each level.

In the first level we scrutinize the limitations of our file-based cache calculation technique. We have examined different configurations of MPI and OpenMP with a fixed number of cores 96 (i.e. number of nodes is 6). Using our discretization scheme, we have found that the memory consumption increases by a factor of around three by using a pure MPI approach, compared to a hybrid MPI-OpenMP (i.e. 6 MPI processes, each using 16 threads) configuration. However, even very accurate calculations with a pure MPI approach require less than 10 GB of memory per node, hence memory limitations are not an issue considering a modern supercomputer and this particular simulation setup. In contrast, we find that the total calculation time is reduced by 30 % with a pure MPI configuration, compared to a hybrid MPI-OpenMP scheme. The performance decrease of the hybrid approach is due to data locality issues arising in shared-memory

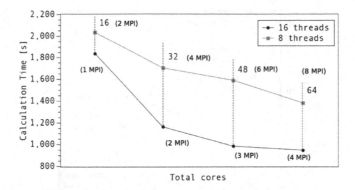

Fig. 1. Dependence of the total calculation time on total cores for a fixed number of threads. 1, 2, 3, and 4 nodes are used respectively. For 16(8)-threaded application, we use one(two) MPI job(s)/node.

techniques. Therefore, we conclude that our cache approach is most efficient for a pure MPI configuration.

In the second level, we compute τ by using the archived data in parallel. For our discretization parameters, the size of the archived cache is up to 3 GB. The even smaller energy steps ($\triangle E = 0.2\,\mathrm{meV}$) improve the computational accuracy but increase the required cache size to more than 7.5 GB. This archived data is to be loaded into the memory while calculating τ by each MPI job, and thus the number of parallel MPI jobs on a single node becomes strictly limited. Theoretically only three processes can work together on a single node on our supercomputer, thus leading to a significant loss of computational resources. Under these conditions, it becomes inevitable to use a hybrid MPI-OpenMP configuration, albeit its execution performance limitations. Figure 1 shows the dependence of the total calculation time on the number of cores and the number of threads. This confirms that increasing the total number of cores at a fixed number of threads decreases the demand on computing time, which is further reduced when the number of threads is increased. This approach was tested with 416 cores and requires only around 40 min for a single relaxation time data point (around 280 core-hours). Hence, we conclude that our suggested two-level computation technique tremendously reduces the overall computational time.

3   Results

First we assume that the spin is injected along the perpendicular OZ-direction, and we investigate the dependence of τ on shear strain $\varepsilon_{\mathrm{xy}}$. It is observed in Fig. 2 that the spin relaxation rate with $\varepsilon_{\mathrm{xy}}$ is dramatically reduced, and the corresponding τ is increased by orders of magnitude. The figure confirms that at higher temperature the phonon scattering rate increases to reduce τ as compared to that at lower temperature, for all values of $\varepsilon_{\mathrm{xy}}$. Figure 2 also describes the inter- and intravalley scattering spin relaxation components, and we notice

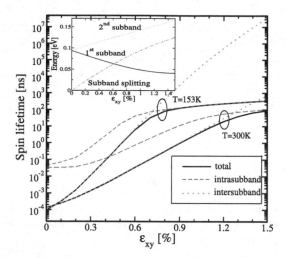

Fig. 2. Total spin relaxation time with components at two distinct temperatures. $t = 1.36\,\mathrm{nm}$, $N_S = 10^{12}\,\mathrm{cm}^{-2}$.

that at low to intermediate strain the major contribution to spin relaxation comes from the intersubband processes, whereas at higher stress the intrasubband component becomes significant. The physical reason for the spin lifetime enhancement by shear strain lies in the ability to completely remove the degeneracy between the two [001] valleys in a confined electron systems by ε_{xy} (inset of Fig. 2). The enhanced valley splitting makes the intervalley spin relaxation much less pronounced which results in a giant spin lifetime enhancement.

Next we study the spin lifetime dependence on the spin injection orientation. The surface roughness scattering matrix elements (M_{SR}), taken to be proportional to the product of the subband wave function derivatives at the interface, are shown in Fig. 3 for several injection orientations, when the additional valley splitting in unstrained films [6] is included. The spin injection orientation is described by the polar angle θ measured from the perpendicular OZ-axis towards the XY-plane. We find that, when θ increases, M_{SR} decreases (Fig. 3 inset). The dependence of M_{SR} on θ can be expressed as:

$$M_{SR}^2(\theta) \propto 1 + \cos^2\theta \cdot \left(\frac{p_x}{p_y}\right)^2 \tag{7}$$

(p_x, p_y) is the in-plane momentum.

Figure 4 describes the variation of τ with ε_{xy}. Contrary to Fig. 2 we find that the increase of τ with ε_{xy} becomes less pronounced, but even in this case τ is boosted by almost two orders of magnitude. In accordance with Fig. 3 we also find in Fig. 4 that the spin relaxation rate (time) decreases (increases), when θ increases, thus τ reaches its minimum (maximum) value for an in-plane spin injection. The inset of Fig. 4 highlights the variation of τ with θ at a fixed stress point ($\varepsilon_{xy} = 0.5\,\%$). We note that the ratio of τ computed for two different

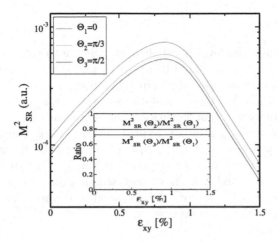

Fig. 3. Variation of intersubband spin relaxation matrix elements with ε_{xy}, $t = 1.36\,\mathrm{nm}$, $k_x = 0.5\,\mathrm{nm}^{-1}$, $k_y = 0.8\,\mathrm{nm}^{-1}$. The inset shows the ratio.

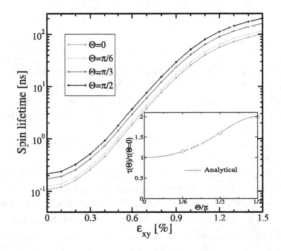

Fig. 4. Variation of τ with ε_{xy} with the spin injection orientation (θ) as parameter. $t = 1.36\,\mathrm{nm}$, $N_S = 10^{12}\,\mathrm{cm}^{-2}$, $T = 300\,\mathrm{K}$. The inset shows the ratio, dots are simulation points.

injection directions at the same stress value does not depend on shear strain. An analytical expression describing the variation of τ with θ for a fixed ε_{xy} can be deduced from (7) by averaging M_{SR}^2 over the in-plane momentum direction, and can be expressed as, $\frac{1}{\tau(\theta)} \propto 1 + \cos^2 \theta$. The simulated results and the analytical expression show a perfect agreement. We point out that a similar dependence of spin lifetime on the injection direction was also reported in bulk silicon [8], indicating that the spin lifetime only depends on the spin injection orientation relative to the valley orientation it is injected to. We conclude that the spin lifetime can be further increased, when spin is injected in-plane.

4   Summary

We have described a two-level parallelization scheme to calculate the spin life-time in silicon. The proposed algorithm precalculates wave functions and energies (first level), and computes the spin relaxation rate by using the precalculated data (second level). These calculations are performed in parallel. We have analyzed the memory and computation time requirements for different parallelization configurations (pure MPI, hybrid MPI-OpenMP), and found that the precalculation step is best performed through a pure MPI scheme, whereas the spin relaxation calculations are efficiently performed by a hybrid approach due to memory demands. Finally, we have depicted that shear strain can boost the spin lifetime by orders of magnitude in thin silicon films. The spin lifetime is further enhanced, once spin is injected in-plane.

Acknowledgements. This work is supported by the European Research Council through the grant #247056 MOSILSPIN. The computational results presented have been partly achieved using the Vienna Scientific Cluster (VSC).

References

1. Huang, B., Monsma, D.J., Appelbaum, I.: Coherent spin transport through a 350 micron thick silicon wafer. Phys. Rev. Lett. **99**, 177209 (2007)
2. Li, J., Appelbaum, I.: Modeling spin transport in electrostatically-gated lateral-channel silicon devices: role of interfacial spin relaxation. Phys. Rev. B **84**, 165318 (2011)
3. Li, P., Dery, H.: Spin-orbit symmetries of conduction electrons in silicon. Phys. Rev. Lett. **107**, 107203 (2011)
4. Fischetti, M.V., Ren, Z., Solomon, P.M., Yang, M., Rim, K.: Six-band $\mathbf{k} \cdot \mathbf{p}$ calculation of the hole mobility in silicon inversion layers: dependence on surface orientation, strain, and silicon thickness. Phys. Rev. Lett. **94**, 1079 (2003)
5. Song, Y., Dery, H.: Analysis of phonon-induced spin relaxation processes in silicon. Phys. Rev. B **86**, 085201 (2012)
6. Osintsev, D., Sverdlov, V., Neophytou, N., Selberherr, S.: Valley splitting and spin lifetime enhancement in strained thin silicon films. In: Proceedings IWCE (2014) ISBN:9781479954346
7. Vienna Scientific Cluster: http://www.vsc.ac.at/systems/vsc-2/
8. Dery, H., Song, Y., Li, P., Zutic, I.: Silicon spin communication. Appl. Phys. Lett. **99**, 082502 (2011)

Free Open Source Mesh Healing for TCAD Device Simulations

Florian Rudolf[1]([⊠]), Josef Weinbub[1], Karl Rupp[1,2], Peter Resutik[1],
Andreas Morhammer[3], and Siegfried Selberherr[1]

[1] Institute for Microelectronics, TU Wien, Vienna, Austria
{rudolf,weinbub,resutik,selberherr}@iue.tuwien.ac.at
[2] Institute for Analysis and Scientific Computing, TU Wien, Vienna, Austria
rupp@iue.tuwien.ac.at
[3] Christian Doppler Laboratory for Reliability Issues in Microelectronics, Institute
for Microelectronics, TU Wien, Vienna, Austria
morhammer@iue.tuwien.ac.at

Abstract. Device geometries in technology computer-aided design
processes are often generated using parametric solid modeling computer-
aided design tools. However, geometries generated with these tools often
lack geometric properties, like being intersection-free, which are required
for volumetric mesh generation as well as discretization methods. Contributing to this problem is the fact, that device geometries often have
multiple regions, used for, e.g., assigning different material parameters.
Therefore, a *healing* process of the geometry is required, which detects
the errors and repairs them. In this paper, we identify errors in multi-
region device geometries created using computer-aided design tools. A
robust algorithm pipeline for *healing* these errors is presented, which has
been implemented in ViennaMesh. This algorithm pipeline is applied on
complex device geometries. We show, that our approach robustly *heals*
device geometries created with computer-aided design tools and is even
able to handle certain modeling inaccuracies.

1 Introduction

Many commercial parametric solid modeling computer-aided design (CAD)
tools, like AutoCAD [1], are available and also various free open source tools, like
FreeCAD [2], are used in many applications. Some technology computer-aided
design (TCAD) simulation suites have modules for CAD processing, but these
modules are usually not as powerful as their standalone counterparts. For example, Synopsys Sentaurus TCAD provides a structure editor for modeling device
geometries [3]. In contrast, many free open source TCAD tools, like DEVSIM [4],
lack the CAD processing module and, therefore, they require a ready-to-simulate
input mesh representing the device geometry.

Regardless of the utilized CAD tools, the task of generating a ready-to-
simulate mesh based on a CAD-based geometry is challenging. The finite difference method is particularly attractive whenever the domain can be represented

© Springer International Publishing Switzerland 2015
I. Lirkov et al. (Eds.): LSSC 2015, LNCS 9374, pp. 293–300, 2015.
DOI: 10.1007/978-3-319-26520-9_32

well with a structured grid, possibly taking additional smooth geometric transformations into account to allow for more complicated domains [5]. In other cases, the finite element method or the finite volume method are popular choices. However, these methods require the mesh to be conforming and valid [6]. Additionally, simulations often require the mesh to be partitioned into several regions, which can, for example, be used to locally assign material properties.

To generate a volumetric mesh which can be used in a simulation, the device geometry created with the CAD tool has to be exported. Usually, a widely supported geometry representation is chosen for this export in order to have a high degree of freedom when selecting the volumetric mesh generation tool. Popular geometry representation formats, like the standard for the exchange of product model data (STEP, ISO 10303) [7], provide a rich feature set, but they are not supported by a variety of popular open source mesh generation tools, like Tetgen [8]. On the other hand, triangular hull geometry representations, like StereoLithography (STL) [9], do not provide a high level of flexibility, but are supported by a large number of mesh generation tools. Triangular hull geometry representations, however, might have topological issues like duplicate elements, or geometrical errors like self-intersections, gaps, or holes. These errors have to be *healed* before a mesh generation can be performed. Additionally, STL and similar geometry representations lack support for multiple mesh regions.

In this work we present an algorithm pipeline which robustly *heals* errors in multi-region triangular hull geometries of complex device structures exported by CAD tools. Section 2 discusses possible errors in triangular hull geometry representations and provides an overview of research in mesh *healing*. The algorithm pipeline for *healing* the triangular hull geometry and generating a mesh is presented in Sect. 3. This algorithm pipeline is implemented in the free open source meshing tool ViennaMesh [10]. The pipeline is applied to example devices created with FreeCAD in Sects. 4 and 5 summarizes the work and gives an outlook for future work.

2   Background and Related Work

When CAD tools export geometries as triangular hulls, a discretization of the geometry has to be performed for the non-planar surfaces. Due to inaccuracies during the modeling process or different discretizations of interfaces, the resulting triangular hull might have topological or geometrical errors. Relevant triangular hull errors are listed below and visualized in Fig. 1 [11,12].

- **Duplicate vertices and elements** are vertices or elements which occur more than once in the mesh.
- **Isolated and dangling elements** are vertices and lines which are not edges or vertices of any hull triangle.
- **Singular edges** occur, when an edge is shared by more than two triangles.
- **Singular vertices** occur, when a vertex is shared by two unrelated sets of triangles.

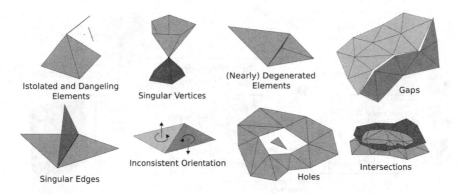

Fig. 1. Visualization of all triangular hull mesh errors except duplicate vertices and elements.

- **Inconsistent orientation** occurs, when two neighboring triangles have different vertex orientations.
- **Nearly degenerated elements** are triangles which are degenerated or nearly degenerated, meaning their surface area is very small compared to their edge lengths.
- **Holes** occur, when a hull is not fully closed.
- **Gaps** occur, when two different hulls are not topologically connected to each other.
- **Intersections** occur, when a triangle intersects another triangle.

Duplicate vertices can be *healed* by merging them and duplicate elements can safely be removed. Isolated and dangling elements can be identified and removed using topological operations. In many file formats, like STL, isolated and dangling vertices and lines cannot occur because vertices and lines are not stored explicitly. Due to the fact that many mesh *healing* algorithms originate in the field of computer graphics, the definition of a valid *healed* mesh is different to the definition of a *healed* mesh for volumetric ready-to-simulate mesh generation. In particular, a singular edge might be valid for multi-region geometries, because it can be an interface edge between two different mesh regions. Similarly, a singular vertex might also be valid, but can lead to numerical issues during the simulation. Therefore, singular vertices must be detected using topological operations and split up into multiple new vertices, one for each triangle set. If the volumetric mesh generation algorithm requires consistent orientations, like most advancing front mesh generation algorithms [13] do, they can be fixed by vertex index swapping of triangles with wrong orientation. Degenerated or nearly degenerated triangles can be fixed by either performing an edge collapse [14], if two vertices are close to each other, or re-meshing the area of the degenerated triangle and its three neighbors.

The other three types of errors, being holes, gaps, and intersections, are much more challenging to *heal*. Many different algorithms have been developed, which address these types of errors [11,12,15]. Several open and closed source mesh

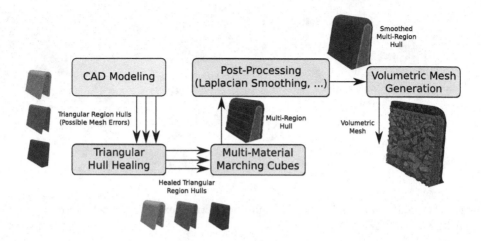

Fig. 2. Each region of a device (represented by one arrow) is exported individually by the CAD tools to a triangular hull and *healed* (cf. Sect. 3.1). A multi-material marching cubes algorithm re-samples these *healed* triangular hulls (cf. Sect. 3.2) and creates a valid multi-region hull. After a post-processing step, a volumetric mesh is generated based on the re-sampled geometry (cf. Sect. 3.3).

healing tools, implementing some of these algorithms, are freely availably [16]. However, most of the algorithms originate in the field of computer graphics and are therefore not able to handle multiple regions properly, which is highly relevant for the field of TCAD.

3   Mesh Healing and Generation Pipeline

In this section, a mesh *healing* and generation pipeline is presented, an overview of which is given in Fig. 2. The pipeline consists of three parts as described in the following subsections.

3.1   CAD Interface and Triangular Hull Healing

After modeling the device in a CAD tool, each region is exported on its own using a triangular hull representation. These triangular region hulls will only be used in the re-sampling step to test if a point is inside the region hull. Therefore, region interfaces do not need to be compatible and each region hull can be treated individually. To ensure stable point inclusion tests, triangular hull errors are *healed* using the mesh *healing* tool Polymender [17].

3.2   Re-Sampling

After *healing* the region hulls, a volumetric re-sampling is performed by creating a regular three-dimensional grid covering the entire device geometry.

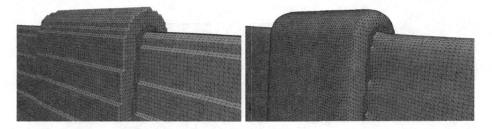

Fig. 3. A re-sampled triangular hull geometry before and after the smoothing process.

For each grid point the corresponding region is determined by using a point-in-hull test on each *healed* region hull. Due to possibly different geometry discretizations at interfaces, regions might intersect or form holes. If a grid point is in more than one mesh region, a user-created priority list resolves the ambiguity. Afterwards, a three-dimensional version of a dilatation and an erosion operation avoids holes in the re-sampled geometry [18]. This regular grid is then used by the multi-material marching cubes algorithm [19] in order to create a triangular hull with multiple regions and valid region interfaces. The entire re-sampling step has been implemented in ViennaMesh.

3.3 Post Processing and Volumetric Mesh Generation

Due to the nature of the marching cubes algorithm, the re-sampled triangular hull has a stair-stepped characteristic. A modified version of the Laplacian smoothing algorithm is applied to mitigate these characteristics [19]. Figure 3 visualizes the re-sampled triangular hull before and after Laplacian smoothing. Like the re-sampling step, the Laplacian smoothing algorithm has also been implemented in ViennaMesh. Depending on the application, chosen grid resolution during the re-sampling step, and required volumetric mesh resolution, a refinement or coarsening algorithm suitable for multi-region triangular hulls is applied on the smoothed mesh. Finally, a mesh generation software, like Tetgen, is used to create a volumetric ready-to-simulate mesh based on the resulting triangular hull.

4 Examples

In this section we apply our algorithm to a bulk silicon trigate transistor [20] and a FlexFET [21], which have been modeled with the free open source CAD tool FreeCAD.

The exported geometry of the bulk silicon trigate transistor has 22 volumetric holes and 24 intersections, visualized in Fig. 4. By applying our algorithm pipeline, we obtain a valid multi-region triangular hull geometry, where all errors of the exported input geometry have been successfully eliminated.

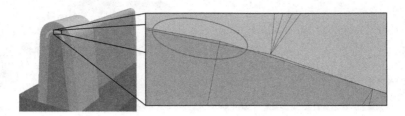

Fig. 4. Hole (marked green) and intersection errors (marked blue) in the bulk silicon trigate transistor due to different discretizations of neighboring regions. The green area on the left indicates the areas, where these errors occur (color figure online).

Fig. 5. The bulk silicon trigate transistor: The original geometry modeled in FreeCAD (left) and a clipped visualization of the generated volumetric mesh (right).

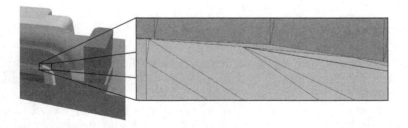

Fig. 6. A hole in the exported FlexFET geometry (visualized in green) which stems from modeling inaccuracies (color figure online).

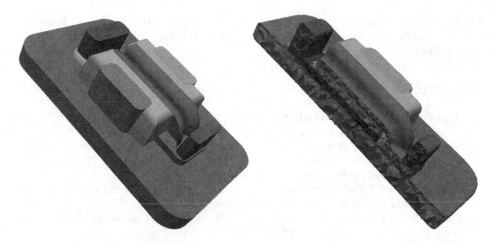

Fig. 7. The FlexFET: The original geometry modeled in FreeCAD (left) and a clipped visualization of the generated volumetric mesh (right).

This *healed* triangular hull geometry is used by ViennaMesh's Tetgen module to create a volumetric ready-to-simulate mesh. The modeled device geometry and the volumetric mesh are visualized in Fig. 5.

The exported FlexFET geometry has a total of 39 errors, being 21 volumetric holes and 18 intersections. In contrast to holes caused by different surface discretizations, the FlexFET geometry has a hole which stems from modeling inaccuracies (cf. Fig. 6). Again, applying our algorithm pipeline successfully *heals* all errors and generates a valid multi-region triangular hull geometry. If, in case of a high re-sampling resolution, holes are not closed, the kernel size of the three-dimensional dilatation and erosion operation during the re-sampling step has to be increased. ViennaMesh's Tetgen module is used to create a volumetric ready-to-simulate mesh based on the *healed* triangular hull. The modeled FlexFET geometry and the volumetric mesh is shown in Fig. 7.

5   Summary and Future Work

We presented an algorithm pipeline for automatic and robust *healing* of CAD geometries for further processing by volumetric mesh generation tools. In contrast to mesh and geometry *healing* algorithms used in the field of computer graphics, our approach supports meshes with multiple regions. We show, that our algorithm pipeline reliably generates volumetric ready-to-simulate meshes based on geometries of complex semiconductor devices modeled in CAD tools.

To further improve the stability for handling big holes which are not closed by dilatation and erosion operations, other filling algorithms, like the flood fill algorithm [18], should be investigated in the future.

Acknowledgements. This work has been supported by the European Research Council (ERC), grant #247056 MOSILSPIN and by the Austrian Science Fund FWF, grant P23598.

References

1. AutoCAD: http://www.autodesk.de/products/autocad/overview/
2. FreeCAD: http://www.freecadweb.org/
3. Synopsys Sentaurus Structure Editor: http://www.synopsys.com/Tools/TCAD/Pages/StructureEditor.aspx
4. DEVSIM: https://github.com/devsim/devsim/
5. Strikwerda, J.C.: Finite difference schemes and partial differential equations, 2nd edn. SIAM, Philadelphia (2004) ISBN: 978-0-89871-567-5
6. Cheng, S.W., Dey, T.K., Shewchuk, J.R.: Delaunay Mesh Generation. CRC Press, Boca Raton (2013) ISBN: 978-1584887300
7. Pratt, M.J.: Introduction to ISO 10303 - the STEP standard for product data exchange. J. Comput. Inf. Sci. Eng. **1**(1), 102–103 (2001). doi:10.1115/1.1354995
8. Si, H.: TetGen a quality tetrahedral mesh generator and three-dimensional delaunay triangulator, Version 1.4, User Manual (2006). http://wias-berlin.de/software/tetgen/files/tetgen-manual.pdf
9. Szilvási-Nagy, M., Mátyási, G.: Analysis of STL files. J. Math. Comput. Model. **38**(7–9), 945–960 (2003). doi:10.1016/S0895-7177(03)90079-3
10. ViennaMesh: http://viennamesh.sourceforge.net/
11. Attene, M., Campen, M., Kobbelt, L.: Polygon mesh repairing: an application perspective. ACM Comput. Surv. **45**(2), 1–33 (2013). doi:10.1145/2431211.2431214
12. Chong, C., Kumar, A.S., Lee, H.: Automatic mesh-healing technique for model repair and finite element model generation. J. Finite Elem. Anal. Des. **43**(15), 1109–1119 (2007). doi:10.1016/j.finel.2007.06.009
13. Frederick, C., Wong, Y., Edge, F.: Two-dimensional automatic mesh generation for structural analysis. Int. J. Numer. Meth. Eng. **2**, 133–144 (1970). doi:10.1002/nme.1620020112
14. Hoppe, H.: Progressive meshes. In: Proceedings of the 23rd Annual Conference on Computer Graphics and Interactive Techniques, pp. 99–108. New York (1996). doi:10.1145/237170.237216
15. Ju, T.: Robust repair of polygonal models. ACM Trans. Graph. **23**(3), 888–895 (2004). doi:10.1145/1015706.1015815
16. Mesh Repairing Software on the Web: http://meshrepair.org/
17. Polymender: http://www1.cse.wustl.edu/taoju/code/polymender.htm
18. Burger, W., Burge, M.J.: Digital Image Processing - An Algorithmic Introduction Using Java. Texts in Computer Science, 1st edn. Springer-Verlag, London (2008)
19. Wu, Z., Sullivan, J.M.: Multiple material marching cubes algorithm. Int. J. Numer. Meth. Eng. **58**(2), 189–207 (2003). doi:10.1002/nme.775
20. Agrawal, N., Kimura, Y., Arghavani, R., Datta, S.: Impact of transistor architecture (bulk planar, trigate on bulk, ultrathin-body planar SOI) and material (silicon or III-V semiconductor) on variation for logic and SRAM applications. IEEE Trans. electron devices **60**(10), 3298–3304 (2013). doi:10.1109/TED.2013.2277872
21. Modzelewski, K., Chintala, R., Moolamalla, H., Parke, S., Hackler, D.: Design of a 32nm independently-double-gated FlexFET SOI transistor. In: Proceedings of the 17th Biennial University/Government/Industry Micro/Nano Symposium, region hulls, a volumetric pp. 64–67 (2008) doi:10.1109/UGIM.2008.24

A Non-Equilibrium Green Functions Study of Energy-Filtering Thermoelectrics Including Scattering

Mischa Thesberg[1]([✉]), Mahdi Pourfath[1], Neophytos Neophytou[2], and Hans Kosina[1]

[1] Institute for Microelectronics, TU Wien, Vienna, Austria
{thesberg,pourfath,kosina}@iue.tuwien.ac.at
[2] Warwick University, Coventry, UK
n.neophytou@warwick.ac.uk

Abstract. Thermoelectric materials can convert waste heat into usable power and thus have great potential as an energy technology. However, the thermoelectric efficiency of a material is quantified by its *figure of merit*, which has historically remained stubbornly low. One possible avenue towards increasing the figure of merit is through the use of low-dimensional nanograined materials. In such a system scattering, tunnelling through barriers and other low-dimensional effects all play a crucial role and thus a quantum mechanical treatment of transport is essential. This work presents a one-dimensional exploration of the physics of this system using the Non-Equilibrium Green's Function (NEGF) numerical method and include carrier scattering from both acoustic and optical phonons. This entirely quantum mechanical treatment of scattering greatly increases the computational burden but provides important insights into the physics of the system. Thus, we explore the relative importance of nanograin size, shape and asymmetry in maximizing thermoelectric efficiency.

1 Introduction

Waste heat is created everywhere; in manufacturing, in the engine of an automobile, in the production of power, in the operation of computer chips, and so forth. A thermoelectric material is able to drive a current when an external temperature gradient is applied. Thus, such materials have great potential in turning our abundant heat losses into energy gains. However, the field of thermoelectrics lies on a sort of precipice; although many theoretical schemes for engineering efficient thermoelectrics exist, commercially available thermoelectrics are still too inefficient for most applications.

The efficiency of a thermoelectric material can be effectively encapsulated within a simple quantity, the figure of merit:

$$ZT = \frac{S^2 G}{\kappa} T$$

© Springer International Publishing Switzerland 2015
I. Lirkov et al. (Eds.): LSSC 2015, LNCS 9374, pp. 301–308, 2015.
DOI: 10.1007/978-3-319-26520-9_33

where T is the temperature, G is the conductance, S is the Seebeck coefficient, to be described later, and κ is the total heat conductivity, having both electron, κ_e, and lattice, κ_L, components. A material with a high figure of merit is a good thermoelectric. The state of the art in terms of commercial thermoelectrics lies at around ~ 1 with research devices lying in the range 1.5–1.8 [1]. However, it's generally accepted that for thermoelectric technology to find application in places beyond niche industries like arctic and space exploration, values of $ZT > 2$ are required [1]. Thus, current methods for increasing the figure of merit are insufficient should thermoelectric technology ever hope to reach its potential.

The figure of merit, ZT, can be decomposed into two crucial aspects, the denominator, which is the thermal conductivity, and the numerator $S^2 G$ which is collectively called the *power factor*. Thus far, the bulk of improvements to ZT have resulted from schemes which minimize the conductivity of phonons, κ_L, without overly harming the electrical conductivity. This approach then seeks to minimize the denominator of the ZT function. It is much rarer to find schemes which seek to maximize the power factor. The reason for this is that the Seebeck coefficient S and the conductance G are highly interdependent and inversely related, and generally when one improves one they tend to erode the other. However, this need not always be the case.

The Seebeck coefficient is defined as $S = -\Delta V / \Delta T$ and thus intuitively encapsulates the ability of a material to separate charge given a certain temperature gradient. Within linearized transport theory it can be written as

$$S = \int \left(-\frac{\partial f_{FD}(E,T)}{\partial E} \right) T(E) \left(\frac{E - \epsilon_F}{k_B T} \right) dE$$

where f_{FD} is the Fermi-Dirac distribution, T is a conductance or transmission function which is also related to the density of states and ϵ_F is the Fermi level. Similarly, the conductance is defined as

$$G = \int \left(-\frac{\partial f_{FD}(E,T)}{\partial E} \right) T(E) dE.$$

Thus, the key difference between the two components of the power factor is only the factor of $(E - \epsilon_F)/k_B T$. This factor means that symmetric transmission above and below the Fermi level acts to cancel each other out and produce a Seebeck coefficient of zero. Thus, to enhance the power factor one seeks to maximize conductance at energies above the Fermi level but minimize conductance at energies below it. In light of this it was generally considered that insulators/semiconductors with asymmetric valence and conduction bands and a Fermi level lying between them were good materials for thermoelectric applications. Similarly, a typical band in three dimensions has a density of state whose shape is approximately $\propto \sqrt{E - E_0}$ where E_0 is the band edge, thus metals whose Fermi level lies deep in a band where the density of states, and thus the transmission, is effectively constant/flat, were deemed to be poor thermoelectrics.

Contrary to this initial assessment, it has been argued [4,5] that a metal can be made a good thermoelectric if a sequence of potential barriers can be added. If the system has a series of potential barriers whose height is near the height of the Fermi level above the conduction band (or below the valence band) then the material will have a high conductivity, as it is a metal, for energies above the Fermi level, and for energies below the Fermi level transport is effectively blocked by the barriers. Further work along this line [7] would argue that scattering with optical and acoustic phonons, which allowed carriers to mix energy and momentum in the regions between barriers, would further heighten this filtering effect.

This system of barrier filtering is not just an abstract contrivance but does indeed occur in real systems. Two prime examples of this are superlattices, where layer-by-layer varying material properties can create a series of transport barriers, and nanograined systems (i.e. polycrystalline systems where the grain size is on the order of nanometers) where grain boundaries can act as potential barriers [6]. However in terms of modelling and theory, such systems offer considerable challenges. As potential barriers and scattering play a large role in the physics, both quantum tunnelling and phonon scattering must be included if any physical model is to be accurate. The standard method for numerically describing quantum transport is the non-equilibrium Green's function method (NEGF). However, the inclusion of electron-phonon scattering within this model results in a substantial computational burden. Thus, the implementation of such a method is necessarily an exercise in high-performance algorithm design.

The issue of ideal barrier shape has been addressed before by authors of this paper, though at the semi-classical level in a system where ionized impurity scattering was the primary source of scattering [6]. An NEGF calculation taking into account scattering to the issue of energy filtering has also been done before [3] with an eye for optimal barrier spacing and height. In this work we consider a fully quantum mechanical study of the effect of barrier shape on the thermoelectric power factor.

The purpose of this work is to use an NEGF algorithm with electron-phonon scattering incorporated to study the effect of barrier height, width and shape on the power factor. A plausible potential barrier shape will be posited and their forms generalized. The ideal barrier shape will be determined as well as the potential loss for non-ideal shapes quantified.

2   Methods

The NEGF method is a robust and accurate method for simulating quantum transport [2]. The numerical crux of the method involves an inversion of a matrix of the form

$$G(E) = [E \pm i\epsilon - H - \Sigma(E, E')]^{-1}$$

where H is the systems Hamiltonian and Σ is a matrix which accounts for both the effect of the contacts and for electron-phonon scattering. The Green's function must be evaluated for every value of the energy considered (typical ~ 1000 values). If scattering is not considered then the Green's functions at each energy are independent. In such a case the computation can be parallelized in energy provided each processing node has sufficient memory to hold both the Hamiltonian (which is sparse) and two versions of the Green's function; the retarded Green's function G^R which holds static information and the greater-than Green's function $G^>$ which holds dynamic information (see [2]).

When scattering is included into the NEGF method then the Σ matrix acquires a dependence on other energies. This is because the process of inelastic scattering causes a carrier and a phonon with an energy E and E' to scatter and end with energies of E'' and E''' where $E + E' = E'' + E'''$. It is important to note that only the energy of the carrier, and not the phonon, is tracked. Thus carriers can scatter between energies and the Green's functions are no longer independent. This means that all Green's functions must then be determined self-consistently. The parallelization of the NEGF method then becomes less straightforward. However, optical phonon energies are often approximated to be constant ($E = \hbar\omega$) and thus the mixing of energy only occurs between energies that are $\pm\hbar\omega$. This allows one to still parallelize in energy space with only minimal communication between nodes.

The simulations performed in this paper were done using the effective mass model with $m_{eff} = m_0$, a lattice constant of 0.5 nm and a channel length of 120 nm. The channel contained six barriers with a 20 nm separation between each. The optical and acoustic phonon coupling strengths are taken to be the same with a value of $1.6 \times 10^{-3} \text{eV}^2$. The Fermi level was chosen to maximize ballistic conductance and was fixed at 0.14eV above the conduction band. These parameters, though plausible for a silicon-like structure, are effectively arbitrary and though quantitative behaviour will be dependent on them it is believed that qualitative insights will be general.

3   Results

The type of barrier explored in this work is an exponentially decaying square barrier. Such a barrier profile may realistically appear in structures where a square barrier is intended but some level of diffusion, possibly due to annealing, has occurred and thus are a good prototypical shape. In a reality a perfectly square barrier cannot be obtained and such tails are thus a more natural shape. The first part of the considered barrier shape is a square barrier of height h_b and width w_b. At the edges of the square portion the potential then decays exponentially according to the function

$$f(x) = \exp\left(-C_b x\right)$$

where C_b is the *curvature* of the exponential decay. A sample set of barriers can be seen in Fig. 1. In addition to the properties of the barrier one must also specify the separation between barriers. A cursory exploration of both barrier separation and barrier height found that the optimal values for an entirely square barrier were determined to be 20 nm and 0.17 eV respectively. The fact that the

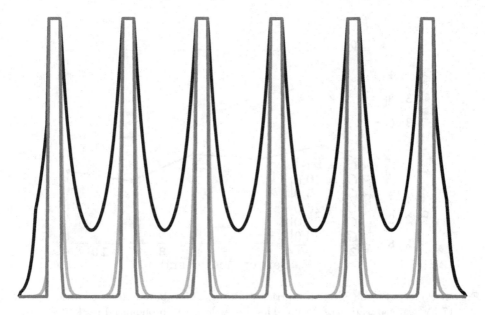

Fig. 1. A schematic diagram of an exponentially decaying square well. The parameter which describes the overall shape is the curvature, C_b. A small C_b corresponds to a very broad, bowl-like, structure (see black line) where a very large C_D corresponds to a square well (see blue line). For reference the curve in red has a curvature of 25.0 (color figure online).

optimal barrier separation corresponds to the specified mean free path suggests the approach taken here is valid.

With the barrier separation and height fixed the power factor dependence on barrier width and curvature were explored. Figure 2 shows the dependence of barrier width on power factor for various curvatures. It is clear that there is a pronounced peak in the power factor at a value of 3 nm. However, the value of the peak is merely an artifact of the chosen system parameters, of greater interest is the loss in power factor associated with deviation from the ideal. For barriers thinner than 3 nm the power factor can be reduced by up to $\sim 31\,\%$. The reason for this is fairly intuitive, as the barriers become thinner the amount of tunnelling through the barriers increases. If the purpose of the barriers is to maximize the power factor by blocking conductance through the barrier than tunnelling erodes the power factor gains resulting from this blockage. It is also interesting to note that barriers thicker than the ideal can cause losses of approximately $\sim 15\,\%$ if one ignores the red and black data (curvatures of 5 and 10), a point to be discussed later. Though this loss is less than the previous case it is still significant. This loss likely results from the poor conductivity of the barrier regions themselves. Since a barrier height of 0.17eV is above the Fermi level of 0.14eV the barrier region is approximately an insulating region. Thus, the thicker the barriers the greater the fraction of the total system volume is comprised of "insulating" material. Thus, conductance drops.

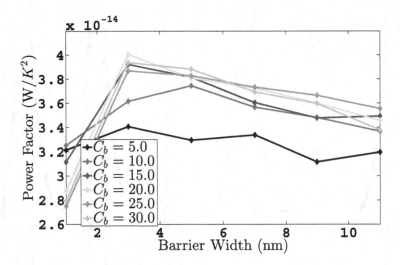

Fig. 2. Power Factors (S^2G) vs. Barrier Width w_b: The above reflects a barrier height of 0.17 eV and separation of 20 nm. There is a clear and pronounced peak at a width of 3 nm with deviations from the ideal causing a loss of $\sim 30\%$ in the power factor. Additionally, for curvatures greater than ~ 15.0 the affect on power factor is minimal.

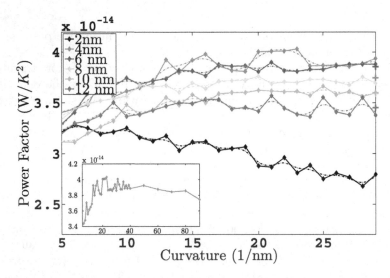

Fig. 3. Power Factor (S^2G) vs. Curvature: The above reflects a barrier height of 0.17 eV and separation of 20 nm. Power factor can be seen to plateau at high curvatures. Lower curvatures have power factors of approximately $\sim 18\%$ lower. The inset shows the power factor of a barrier of ideal width, 3 nm, for curvatures tending towards infinity.

Looking at Fig. 2 it is clear that the low curvature data (the red and black lines) has much worse thermoelectric behaviour than the higher curvature data. A plot of power factor versus curvature can be found in Fig. 3. Looking at this figure it is important to stress that higher values of the curvature reflect increasingly square like wells with an infinite curvature being a square well. Thus, the black line in Fig. 3 represent a barrier of width 1 nm and confirms the earlier statement about tunnelling eroding transport with larger curvatures resulting in thinner barriers. However, for all other widths it is clear that overly curved well shapes have worse power factors than square wells, the difference being $\sim 18\,\%$. At curvatures (C_b) of approximately 15.0 it appears that the power factor effectively saturates. From this it can be concluded that the square barrier is the ideal barrier shape.

4 Conclusions

In this work an NEGF study including electron-phonon scattering of the effect of barrier shape and width on the thermoelectric power factor was explored. It was determined that a square barrier of optimized width is ideal for maximizing energy filtering. Furthermore, it was determined that deviations from ideal width can erode the power factor by $\sim 31\,\%$ and deviations from ideal shape can erode by $\sim 18\,\%$.

Acknowledgements. This work was supported by the Austrian Science Fund (FWF), contract P25368-N30. The computational results presented have been achieved using the Vienna Scientific Cluster (VSC).

References

1. Biswas, K., He, J., Blum, I.D., Seidman, D.N., Dravid, V.P., Kanatzidis, M.G.: High-performance bulk thermoelectrics with all-scale hierarchical architectures. Nature **489**, 414–418 (2012). doi:10.1038/nature11439
2. Datta, S.: Quantum Transport: Atom to Transistor. Cambridge University Press (2005), https://books.google.ca/books?id=Yj50EJoS224C, ISBN: 9780521631457
3. Kim, R., Lundstrom, M.S.: Computational study of energy filtering effects in one-dimensional composite nano-structures. J. Appl. Phys. **111**(2), 024508 (2012). http://scitation.aip.org/content/aip/journal/jap/111/2/10.1063/1.3678001, doi:10.1063/1.3678001
4. Mahan, G.D., Woods, L.M.: Multilayer thermionic refrigeration. Phys. Rev. Lett. **80**, 4016–4019 (1998). http://link.aps.org/doi/10.1103/PhysRevLett.80.4016. doi:10.1103/PhysRevLett.80.4016
5. Moyzhes, B., Nemchinsky, V.: Thermoelectric figure of merit of metal-semiconductor barrier structure based on energy relaxation length. Appl. Phys. Lett. **73**(13), 1895–1897 (1998). http://scitation.aip.org/content/aip/journal/apl/73/13/10.1063/1.122318, doi:10.1063/1.122318

6. Neophytou, N., Kosina, H.: Optimizing thermoelectric power factor by means of a potential barrier. J. Appl. Phys. **114**(4), 044315 (2013). http://scitation.aip.org/content/aip/journal/jap/114/4/10.1063/1.4816792, doi:10.1063/1.4816792

7. Vashaee, D., Shakouri, A.: Improved thermoelectric power factor in metal-based superlattices. Phys. Rev. Lett. **92**, 106103 (2004). http://link.aps.org/doi/10.1103/PhysRevLett.92.106103, doi:10.1103/PhysRevLett.92.106103

Parallelization of the Two-Dimensional Wigner Monte Carlo Method

Josef Weinbub[⊠], Paul Ellinghaus, and Siegfried Selberherr

Institute for Microelectronics, TU Wien, Vienna, Austria
{weinbub,ellinghaus,selberherr}@iue.tuwien.ac.at

Abstract. A parallelization approach for two-dimensional Wigner Monte Carlo quantum simulations using the signed particle method is introduced. The approach is based on a domain decomposition technique, effectually reducing the memory requirements of each parallel computational unit. We depict design and implementation specifics for a message passing interface-based implementation, used in the Wigner Ensemble Monte Carlo simulator, part of the free open source ViennaWD simulation package. Benchmark and simulation results are presented for a time-dependent, two-dimensional problem using five randomly placed point charges. Although additional communication is required, our method offers excellent parallel efficiency for large-scale high-performance computing platforms. Our approach significantly increases the feasibility of computationally highly intricate two-dimensional Wigner Monte Carlo investigations of quantum electron transport in nanostructures.

1 Introduction

The Wigner formalism [11] provides an attractive alternative to the non-equilibrium Green's function formalism, as it provides a reformulation of quantum mechanics - usually defined through operators and wave functions - in the phase space using functions and variables [6]. Thereby, the Wigner formalism provides a more intuitive description which also facilitates the reuse of many classical concepts and notions. Several methods have been applied to solve the Wigner equation of which the stochastic Wigner Monte Carlo method, using the signed-particle technique [4,5], has emerged as probably the most promising approach: it has made multi-dimensional Wigner simulations viable for the first time [8]. An efficient distributed parallel computation approach is the next crucial step to facilitate the use of Wigner simulations to investigate actual devices.

Wigner Monte Carlo simulations have been made computationally feasible by the annihilation step, required to counterbalance the continuous generation of particles [7]. However, the memory demand of the annihilation algorithm itself is proportional to the dimensionality and resolution of the phase space represented in the simulation, which can lead to exorbitant requirements.

All in all, Wigner Monte Carlo quantum simulations suffer not just from compute intensive operations but also from vast memory demands; the latter

© Springer International Publishing Switzerland 2015
I. Lirkov et al. (Eds.): LSSC 2015, LNCS 9374, pp. 309–316, 2015.
DOI: 10.1007/978-3-319-26520-9_34

is much more severe, as more realistic simulations are beyond reach on single workstations with their limited memory.

Conventional parallelization approaches for Monte Carlo methods, using domain replication, are *embarrassingly parallel* [1]. The particle ensemble is split amongst computational units, where each sub-ensemble is treated completely independently. This necessitates domain replication, when working in a distributed-memory context (as is the de facto standard for large-scale parallel computations) to avoid additional communication. Such an approach offers excellent parallel efficiency, however, domain replication is not feasible for the Wigner Monte Carlo method due to the huge memory demands associated with the annihilation algorithm, quickly exceeding the typically available memory on a single computational unit. Further contributing to the challenge of implementing scalable Wigner Monte Carlo simulations is the fact that particle annihilation must be performed in unison across the global simulation domain [2]; a synchronization step between each individual time step is required, impeding parallel efficiency.

We present a message passing interface (MPI)-based domain decomposition approach for two-dimensional problems, which avoids domain replications and thus drastically reduces the memory requirements for each parallel computational unit. This work extends previous investigations of one-dimensional problems [2] to two-dimensional scenarios. Our approach to partition the simulation domain and to accelerate the overall simulation process is discussed for the Wigner Ensemble Monte Carlo simulator, which is part of the free open source ViennaWD simulation package [10]. The parallel efficiency is evaluated based on the execution times of a representative time-dependent, two-dimensional example using randomly placed point charges. With our approach we not only tackle the challenge of reducing simulation times, but much more importantly, we enable to conduct highly memory-intensive Wigner Monte Carlo quantum simulations in the first place.

2   Parallel Algorithm for Two-Dimensional Problems

The parallelization strategy for two-dimensional problems is based on previous investigations regarding one-dimensional problems [2]. The domain decomposition approach entails splitting up the spatial domain amongst processes. Each process represents a subdomain (i.e. a part of the global domain) and only treats particles, which fall within its own subdomain. Thereby, the memory requirements to represent the phase-space, and all other space-dependent quantities, are scaled down with the number of processes (subdomains) used. As the particle ensemble evolves, the particles travel between subdomains. This necessitates an inter-MPI process communication layer, representing spatially neighboring subdomains; a centralized communication where all worker processes transfer data via a single master process is avoided, which would significantly limit parallel scalability.

The applied domain decomposition technique assigns each MPI process to a unique subdomain by splitting the simulation domain uniformly, as illustrated

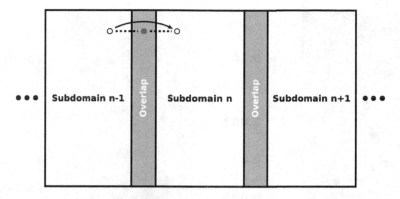

Fig. 1. The simulation domain is splitted uniformly. Each subdomain is assigned to a separate MPI process. If a particle (black circle) enters the overlap area (red), it is transmitted to the neighboring subdomain (Color figure online).

in Fig. 1. A so-called *slab* or one-dimensional decompositioning method is used to partition the simulation domain, meaning that one direction is partitioned, whereas the second direction (in a two-dimensional setting) is kept untouched. Although such a partitioning technique theoretically tends to limit the parallelization efficiency (e.g. the maximum number of utilizable MPI processes is limited to the number of grid elements in the direction of the partitioning, as one MPI process has to be at least responsible for one grid element), the method provides more than enough parallel processing potential for today's relevant problem scenarios (cf. Sect. 3). This is even more so, when the communication is aligned with the partitioning, meaning that the majority of particles primarily propagate in the unpartitioned direction, minimizing the need for communication which in itself further increases parallel scalability.

The subdomains are assigned to MPI processes in a sequential order, inherently providing an MPI/subdomain neighbor-identification mechanism. The primary MPI communication consists of non-blocking direct neighbor communication. After each time step a lightweight message is used to globally trigger an annihilation step within each MPI process. The transfer (communication) of particles between processes only occurs once at the end of each time-step. This necessitates a small overlap between adjacent subdomains, which serves a similar purpose as a *ghost layer* used in conventional domain decomposition techniques [3]. The overlap is used to identify particles traveling towards a neighboring subdomain, which ultimately get transferred to the respective neighbor subdomain at the end of every time step. A larger overlap between subdomains simplifies the transfer of particles between processes (as particles are required to be transferred less often), however, this introduces a larger data redundancy, which negatively affects parallel efficiency. The exact extent of the overlap is a simulation parameter which it should consider the maximum distance a particle can travel within the chosen time-step as well as its direction of travel.

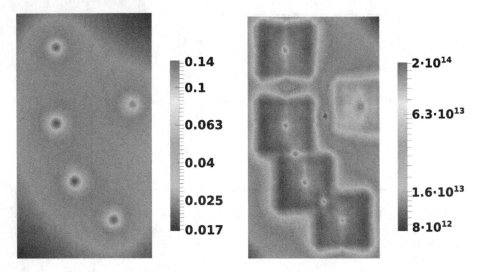

Fig. 2. The potential barrier profile V [eV] (**left**) and the corresponding particle generation rate γ [s^{-1}] (**right**) is shown for the simulation domain (Color figure online).

3   Results

This section investigates a time-dependent, two-dimensional problem (rectangular simulation domain, spatial dimensions are 70 nm × 128 nm) with respect to parallel execution performance. The total number of particles is limited to $32 \cdot 10^7$ particles; the simulation is initialized with $3 \cdot 10^3$ particles. Reflective boundary conditions are used for all boundaries, meaning that no particles leave the simulation domain. The coherence length is 30 nm and the lattice temperature is 300 K. The system is simulated for 200 fs using a 0.5 fs time step.

Five point charges are spread over the simulation domain, each giving rise to particle generation, concentrated within half of the coherence length around it (Fig. 2). The simulation is parallelized via 16, 32, 64, and 128 MPI processes using the VSC-2 supercomputer [9]. One VSC-2 computational node provides 16 cores (two 8-core AMD Opteron Magny Cours 6132HE 2.2 GHz) and 32 GB of system memory; the nodes are connected via an InfiniBand QDR network.

Figure 3 shows the parallel execution performance of our approach which gives an almost perfect, linear parallel scalability. For this setup, a simulation which would otherwise take around 9.3 h (extrapolated, assuming linear scaling relative to 16 MPI processes), takes about 35 min when using 16 MPI processes or around 5 min when using 128 MPI processes. This fact clearly shows the significance of our parallelization approach as it drastically accelerates the simulation process, allowing to considerably increase the pace of research.

In comparison to earlier investigations regarding one-dimensional problems [2], our domain decompositioning approach works even better for two-dimensional

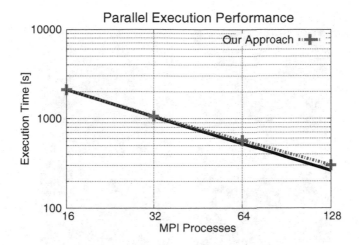

Fig. 3. The execution performance of our two-dimensional parallelization approach (dashed, blue line), shown relative to ideal scalability (black line), offers an almost perfect parallel scaling behavior (Color figure online).

cases and increased particle numbers. The workload per MPI process drastically increases, outweighing any potential communication overhead. In our two-dimensional investigations, we capped the total number of particles at $32 \cdot 10^7$ particles, as compared to $8 \cdot 10^6$, $16 \cdot 10^6$, and $32 \cdot 10^6$ particles in our previous one-dimensional investigations. We, therefore, allow around an order of magnitude more particles to take part in the simulation. An increase in the number of particles is also required in two dimensions to increase the statistical confidence, since the phase space is bigger. We compute the maximum number of particles for each MPI process by dividing the maximum size by the number of MPI processes. Therefore, in the case of using 128 MPI processes each process is responsible for at most $25 \cdot 10^5$ particles, which offers enough workload per process to outweigh the communication overhead required after each time step. Also, the presence of several point charges gives rise to a very high generation rate; the entire simulation domain is rather quickly populated with particles, inherently increasing the load balance over all MPI processes.

Figure 4 depicts the total number of particles (the sum of positively and negatively signed particles) for different time steps as computed via a parallelized simulation. Over time, the entire simulation domain is flooded with particles, however, local maxima occur around the point charges. The maxima take on a rectangular shape (spatial dimensions correspond to the coherence length) due to the taken implementation of the Wigner potential calculation for a given point. While the total number of particles gives an indication of the computational load, the positive and negative particles compensate each other when calculating physically meaningful quantities, like the density shown in Fig. 5.

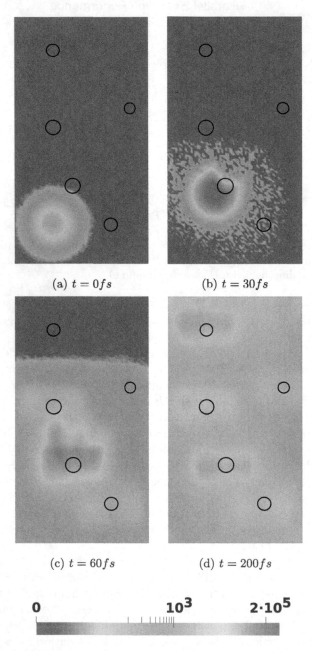

(a) $t = 0 fs$ (b) $t = 30 fs$

(c) $t = 60 fs$ (d) $t = 200 fs$

Fig. 4. Number of particles (positive+negative) for different time steps (a-d). The initial wave package (a) propagates upwards and slightly to the right. Reflecting boundary conditions are used for all four boundaries. Black circles indicate point charges (Color figure online).

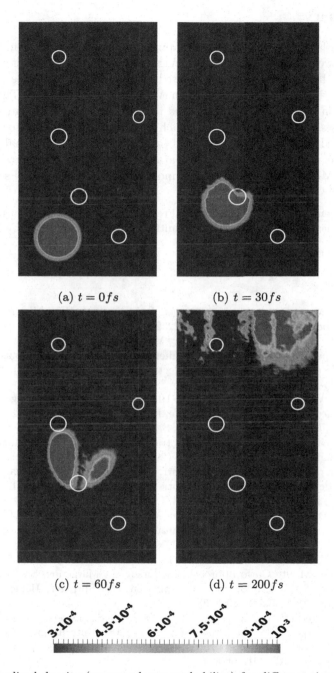

(a) $t = 0fs$

(b) $t = 30fs$

(c) $t = 60fs$

(d) $t = 200fs$

Fig. 5. Normalized density (expressed as a probability) for different time steps (a-d). The initial wave package (a) propagates upwards and slightly to the right. Reflecting boundary conditions are used for all four boundaries. Black circles indicate point charges (Color figure online).

4   Summary

Our approach for parallelizing computationally highly demanding time-dependent, two-dimensional quantum Wigner Monte Carlo simulations has been presented in the context of the MPI-based Wigner Ensemble Monte Carlo simulator, part of the free open source ViennaWD simulation package. The approach uses a domain decomposition technique to distribute the workload among the MPI processes. The conceptual approach for the parallelization technique has been discussed as well as the setup and results of a two-dimensional simulation example. A benchmark depicting the parallel execution performance for 16, 32, 64, and 128 MPI processes shows an almost perfect, linear parallel scalability.

Acknowledgements. The computational results presented have been achieved using the Vienna Scientific Cluster (VSC). The authors thank Mihail Nedjalkov for valuable feedback.

References

1. Dimov, I.: Monte Carlo Methods For Applied Scientists. World Scientific Publishing, Singapore (2008). ISBN 9789810223298
2. Ellinghaus, P., Weinbub, J., Nedjalkov, M., Selberherr, S., Dimov, I.: Distributed-Memory parallelization of the Wigner Monte Carlo method using spatial domain decomposition. J. Comput. Electron. **14**(1), 151–162 (2015). doi:10.1007/s10825-014-0635-3
3. Hager, G., Wellein, G.: Introduction to High Performance Computing for Scientists and Engineers. CRC Press, Boca Raton (2010). ISBN 9781439811924
4. Nedjalkov, M., Kosina, H., Selberherr, S., Ringhofer, C., Ferry, D.K.: Unified particle approach to Wigner-Boltzmann transport in small semiconductor devices. Phys. Rev. B **70**, 115319-1–115319-16 (2004). doi:10.1103/PhysRevB.70.115319
5. Nedjalkov, M., Schwaha, P., Selberherr, S., Sellier, J.M., Vasileska, D.: Wigner quasi-particle attributes - an asymptotic perspective. Appl. Phys. Lett. **102**(16), 163113-1–163113-4 (2013). doi:10.1063/1.4802931
6. Querlioz, D., Dollfus, P.: The Wigner Monte-Carlo Method for Nanoelectronic Devices: Particle Description of Quantum Transport and Decoherence. Wiley, New York (2010). ISBN 9781848211506
7. Sellier, J.M., Nedjalkov, M., Dimov, I., Selberherr, S.: The role of annihilation in a Wigner Monte Carlo approach. In: Lirkov, I., Margenov, S., Waśniewski, J. (eds.) LSSC 2013. LNCS, vol. 8353, pp. 186–193. Springer, Heidelberg (2014). doi:10.1007/978-3-662-43880-0_20
8. Sellier, J., Nedjalkov, M., Dimov, I., Selberherr, S.: Two-dimensional transient Wigner particle model. In: Proceedings of the 18th International Conference on Simulation of Semiconductor Processes and Devices (SISPAD). pp. 404–407 (2013). doi:10.1109/SISPAD.2013.6650660
9. Vienna Scientific Cluster: VSC-2. http://vsc.ac.at/
10. ViennaWD: Wigner Ensemble Monte Carlo Simulator. http://viennawd.sourceforge.net/
11. Wigner, E.: On the quantum correction for thermodynamic equilibrium. Phys. Rev. Lett. **40**, 749–759 (1932). doi:10.1103/PhysRev.40.749

Large-Scale Models: Numerical Methods, Paralel Computations and Applications

A Splitting Numerical Method for Primary and Secondary Pollutant Models

Tatiana Chernogorova[1][(✉)], Ivan Dimov[2], and Lubin Vulkov[3]

[1] Faculty of Mathematics and Informatics, University of Sofia, Sofia, Bulgaria
chernogorova@fmi.uni-sofia.bg
[2] Institute of Information and Communication Technologies,
Bulgarian Academy of Sciences, Sofia, Bulgaria
ivdimov@bas.bg
[3] Faculty of Natural Sciences and Education, University of Rousse, Rousse, Bulgaria
lvalkov@uni-ruse.bg

Abstract. A splitting numerical method is proposed to study the effect of gravitational settling velocity and chemical reaction on pollutants emitted from a point source on the boundary of an urban area. The governing ultra-parabolic equation degenerates on the part of the boundary and we apply a fitted finite volume scheme in order to resolve the degeneration and to preserve the positivity property of the solution (concentration of the pollutant). Computational experiments illustrate the efficiency of our numerical method.

Keywords: Primary and secondary pollutant models · Ultra-parabolic equation · Degeneracy · Finite volume method

1 Introduction

The basic governing equations of primary and secondary pollutants can be written in a general form as [9]

$$\frac{\partial \bar{c}}{\partial \bar{t}} + \bar{u}_z \frac{\partial \bar{c}}{\partial \bar{x}} = \frac{\partial}{\partial \bar{z}}\left(\bar{k}_z \frac{\partial \bar{c}}{\partial \bar{z}}\right) - \bar{w}_s \frac{\partial \bar{c}}{\partial \bar{z}} + \bar{v}_g \bar{k}\bar{c}, \ (\bar{x},\bar{z},\bar{t}) \in (0,\bar{X})\times(0,\bar{H})\times(0,\bar{T}), \ (1)$$

where $\bar{c} = \bar{c}(\bar{x},\bar{z},\bar{t})$ is the ambient mean concentration of pollutant species, $\bar{u}_z$ is the mean wind speed in $\bar{x}$ - direction, $\bar{k}_z$ is the turbulent eddy diffusivity in $\bar{z}$ - direction, $\bar{w}_s$ is the gravitational settling velocity, $\bar{v}_g$ is the mass ratio while $\bar{k}$ is the first order chemical reaction rate coefficient.

Typical meteorological expressions of K-eddy diffusivity are the following:

$$\bar{k}_z(\bar{z}) = \beta\bar{u}\bar{z}e^{-4\bar{z}/\bar{H}} \ \text{ or } \ \bar{k}_z(\bar{z}) = \beta\bar{u}\bar{z}e^{-b\eta}/(a+b\bar{z}), \tag{2}$$

where $\bar{u}$ is the friction velocity and β, a, b, η are meteorological parameters.

The Eq. (1) equipped with one of the diffusion coefficients in (2) is an *ultra-parabolic equation with degeneration*. A typical example of ultra-parabolic equations *without degeneration* is $u_t + xu_y + u_{xx} = f(x,y,t)$, introduced in the work of

© Springer International Publishing Switzerland 2015
I. Lirkov et al. (Eds.): LSSC 2015, LNCS 9374, pp. 319–326, 2015.
DOI: 10.1007/978-3-319-26520-9_35

A. N. Kolmogorov in 1934 for the description of non-isotropic processes. Later equations of such type appeared in mechanics, physics, biology and in other branches of science. One of the first papers on the existence and uniqueness of solution for ultra-parabolic equations is [5]. The well-posedness in special weighted Sobolev spaces of the problem in this paper can be investigated in a similar way as it is done for the parabolic problem in [4]. Finite element and finite difference approximations of ultra-parabolic equations without degeneration are discussed in [1,2,11].

To write the problem under consideration in dimensionless variables, we introduce the new variables $x = \bar{x}/\bar{H}$, $z = \bar{z}/\bar{H}$, $h = \bar{h}/\bar{H}$, $l = \bar{l}/\bar{H}$, $w_s = \bar{w}_s/u_z$, $u = \bar{u}/u_z$, $v_{dp} = \bar{v}_{dp}/u_z$, $t = \bar{t}u_z/H$, $c = \bar{c}u_z/Q$, $k = \bar{k}H/u_z$, $Q_1 = \bar{Q}_1/Q$. Then, in dimensionless variables, the problem under consideration is:

$$\frac{\partial c}{\partial t} + \frac{\partial c}{\partial x} = \frac{\partial}{\partial z}\left(k_z \frac{\partial c}{\partial z}\right) - w_s \frac{\partial c}{\partial z} + k v_g c, \quad x \in (0, X), \ z \in (0, 1), \ t \in (0, T), \quad (3)$$

$$k_z(z) = \beta u z e^{-4z},$$

$$c(x, z, 0) = 0, \quad x \in (0, X), \quad z \in (0, 1), \quad (4)$$

$$c(0, z, t) = Q_1 \delta(z - h), \quad h < 1, \quad z \in (0, 1), \quad t \in (0, T), \quad (5)$$

$$k_z \frac{\partial c}{\partial z}\Big|_{(x,0,t)} = v_{dp} c(x, 0, t) - \mu, \quad \mu = \begin{cases} 1, & 0 < x \leq l, \\ 0, & l < x \leq X, \end{cases} \quad t \in (0, T), \quad (6)$$

$$k_z \frac{\partial c}{\partial z}\Big|_{(x,1,t)} = \mu_1, \quad x \in (0, X), \quad t \in (0, T). \quad (7)$$

Various numerical methods for air-pollution problems described by parabolic equations are developed in [3,4,6–8]. The difficulties that arise at the numerical treatment of the model are: *1-st order degeneracy of the ultra-parabolic equations (1) on the boundary $z = 0$, see (2); Dirac-delta point source; small diffusion or convection domination that cause boundary layers, etc.* In the next section, we describe a hyperbolic-parabolic splitting. In Sect. 3 we construct an exponential fitted finite volume difference scheme for the parabolic subproblems to remove the degeneracy and possible boundary layers. Computational examples are presented in Sect. 4.

2   The Splitting Method

We introduce the non-uniform mesh in time $0 = t_1 < t_2 < \ldots < t_n < t_{n+1} < \ldots < t_{P+1} = T$, $\tau_n = t_{n+1} - t_n$, $\tau = \max_{1 \leq n \leq P} \tau_n$. Starting from the initial condition (4) we solve sequentially [10] on each subinterval $(t_n, t_{n+1}]$, $n = 1, 2, \ldots, P$:

Parabolic Problem: For given $c(x, z, t_n)$ find the solution

$$\psi(x, z, t), \ (x, z, t) \in (0, X) \times (0, 1) \times (t_n, t_{n+1/2}], \quad x - \text{fixed},$$

of the problem

$$\frac{1}{2}\frac{\partial \psi}{\partial t} = \frac{\partial}{\partial z}\left(p(z)z\frac{\partial \psi}{\partial z} + q(z)\psi\right) + r(z)\psi, \tag{8}$$

$$p(z) = \beta u/e^{4z}, \quad q(z) = -w_s, \quad r(z) = v_g k,$$

$$\psi(x,z,0) = c(x,z,0) = 0, \quad \psi(x,z,t_n) = c(x,z,t_n), \quad n = 2,3,\ldots,P, \tag{9}$$

$$k_z\frac{\partial \psi}{\partial z}\bigg|_{(x,0,t)} = v_{dp}\psi(x,0,t) - \mu, \quad \mu = \begin{cases} 1, & 0 < x \leq l, \\ 0, & l < x \leq X, \end{cases} \quad t \in (0,T), \tag{10}$$

$$k_z\frac{\partial \psi}{\partial z}\bigg|_{(x,H,t)} = \mu_1, \quad x \in (0,X), \quad t \in (0,T); \tag{11}$$

Hyperbolic Problem: For given $\psi(x,z,t_{n+1/2})$ find the solution

$$c(x,z,t), \quad (x,z,t) \in (0,X) \times (0,1) \times (t_{n+1/2},t_{n+1}], \quad z - \text{ fixed},$$

of the problem

$$\frac{1}{2}\frac{\partial c}{\partial t} + \frac{\partial c}{\partial x} = 0, \quad (x,z,t) \in (0,X) \times (0,1) \times (t_{n+1/2},t_{n+1}], \tag{12}$$

$$c(x,z,t_{n+1/2}) = \psi(x,z,t_{n+1/2}), \tag{13}$$

$$c(0,z,t) = Q_1\delta(z-h), \quad h < 1, \quad z \in (0,1), \quad t \in (0,T). \tag{14}$$

3 Discretization

3.1 Parabolic Problem

We will implement the S. Wang difference scheme [12]. To write the space discretization, we divide the interval $[0,X]$ into N subintervals $I_{x,i} = [x_i, x_{i+1}]$, $i = 1,2,\ldots,N$ by the nodes $0 = x_1 < x_2 < \ldots < x_N < x_{N+1} = X$. Next we divide the interval by z $[0,1]$ into M subintervals $I_{z,j} = [z_j, z_{j+1}]$, $j = 1,2,\ldots,M$ by the nodes $0 = z_1 < z_2 < \ldots < < z_{D-1} < z_D = h < z_{D+1} < \ldots < z_M < z_{M+1} = 1$. Let $h_{x,i} = x_{i+1} - x_i$ for $i = 1,2,\ldots,N$, $h_x = \max\limits_{1 \leq i \leq N} h_{x,i}$, $h_{z,j} = z_{j+1} - z_j$ for $j = 1,2,\ldots,M$, $h_z = \max\limits_{1 \leq j \leq M} h_{z,j}$. We introduce the secondary mesh $z_{j-1/2} = 0.5(z_{j-1} + z_j)$, $z_{j+1/2} = 0.5(z_j + z_{j+1})$ for each $j = 2,3,\ldots,M$.

A. Internal Grid Nodes. We integrate Eq. (8) on the interval $[z_{j-1/2}, z_{j+1/2}]$, $j = 2,3,\ldots,M$ to get

$$\int\limits_{z_{j-1/2}}^{z_{j+1/2}} \frac{1}{2}\frac{\partial \psi}{\partial t}dz = \int\limits_{z_{j-1/2}}^{z_{j+1/2}} \frac{\partial}{\partial z}\left(p(z)z\frac{\partial \psi}{\partial z} + q(z)\psi\right)dz + \int\limits_{z_{j-1/2}}^{z_{j+1/2}} r(z)\psi dz. \tag{15}$$

For all integrals except for the second one in (15) we use the quadrature formula of the central rectangles:

$$\frac{1}{2}\frac{\partial\psi}{\partial t}\Big|_{(x,z_j,t)} \hbar_{z,j} = \left(p(z)z\frac{\partial\psi}{\partial z} + q(z)\psi\right)\Big|_{(x,z_{j+1/2},t)}$$

$$-\left(p(z)z\frac{\partial\psi}{\partial z} + q(z)\psi\right)\Big|_{(x,z_{j-1/2},t)} + \hbar_{z,j}\, r(z)\psi|_{(x,z_j,t)}, \tag{16}$$

where $\hbar_{z,j} = z_{j+1/2} - z_{j-1/2}$. We rewrite Eq. (16) in the form

$$\frac{1}{2}\frac{\partial\psi_j}{\partial t}\hbar_{z,j} = \rho_{i+1/2} - \rho_{i-1/2} + \hbar_{z,j}r_j\psi_j, \tag{17}$$

where

$$\rho(\psi) = p(z)z\frac{\partial\psi}{\partial z} + q(z)\psi, \quad \frac{\partial\psi_j}{\partial t} = \frac{\partial\psi}{\partial t}\Big|_{(x,z_j,t)}, \quad \psi_j = \psi|_{(x,z_j,t)}, \quad r_j = r|_{(x,z_j,t)}.$$

In order to obtain an approximation of the flux $\rho(\psi)$ at $z_{j+1/2}$, $j = 2,3,\ldots,M$ for fixed x and t, we consider the following auxiliary problem:

$$\left(p_{j+1/2}zv' + q_{j+1/2}v\right)' = 0, \quad z \in I_{z,j}, \tag{18}$$

$$v(z_j) = \psi_j, \qquad v(z_{j+1}) = \psi_{j+1}, \tag{19}$$

where $p_{j+1/2} = p(z_{j+1/2})$, $q_{j+1/2} = q(z_{j+1/2})$.

An integration of (18) leads to the linear ODE of first order

$$p_{j+1/2}zv' + q_{j+1/2}v = M_1 = const.$$

The general solution of this equation is

$$v = M_2 z^{-\alpha_j} + \frac{M_1}{q_{j+1/2}}.$$

Using the boundary conditions (19) we get

$$\rho_{j+1/2} = M_1 = q_{i+1/2}\frac{z_{j+1}^{\alpha_j}\psi_{j+1} - z_j^{\alpha_j}\psi_j}{z_{j+1}^{\alpha_j} - z_j^{\alpha_j}}, \quad j = 2,3,\ldots,M. \tag{20}$$

In analogical way we approximate $\rho_{j-1/2}$ in (17) for $j = 3,4,\ldots,M$:

$$\rho_{j-1/2} = q_{j-1/2}\frac{z_j^{\alpha_{j-1}}\psi_j - z_{j-1}^{\alpha_{j-1}}\psi_{j-1}}{z_j^{\alpha_{j-1}} - z_{j-1}^{\alpha_{j-1}}}. \tag{21}$$

In order to find an approximation of the flow flux at $z_{3/2}$, we consider the problem

$$\left(p_{3/2}zv' + q_{3/2}v\right)' = M_1, \quad z \in I_{z,1}, \quad v(z_1) = \psi_1, \quad v(z_2) = \psi_2.$$

In this case one gets

$$v = \psi_1 + \frac{\psi_2 - \psi_1}{z_2} z,$$

and the approximation for $\rho_{3/2}$ is

$$\rho_{3/2} = \frac{1}{2} \left[(p_{3/2} + q_{3/2}) \psi_2 - (p_{3/2} - q_{3/2}) \psi_1 \right]. \tag{22}$$

B. Boundary Grid Nodes. For *the left vertical boundary* we integrate Eq. (8) over the interval $[z_1, z_{3/2}]$ to get

$$\frac{1}{2} \frac{\partial \psi_1}{\partial t} \frac{h_{z,1}}{2} = \rho_{3/2} - \rho_1 + \frac{h_{z,1}}{2} r_1 \psi_1.$$

For $\rho_{3/2}$ we have the approximation from (22). For ρ_1 we have

$$\rho_1 = p(z_1) z_1 \frac{\partial \psi_1}{\partial z} + q(z_1)\psi_1 = v_{dp}\psi(x, 0, t) - \mu + q(z_1)\psi_1 = \psi_1 (v_{dp} + q_1) - \mu.$$

For *the right vertical boundary* $z = 1$ we integrate Eq. (8) on the interval $[z_{M+1/2}, z_{M+1}]$ to get

$$\frac{1}{2} \frac{\partial \psi_{M+1}}{\partial t} \frac{h_{z,M}}{2} - \rho_{M+1} - \rho_{M+1/2} + \frac{h_{z,M}}{2} r_{M+1}\psi_{M+1}.$$

For $\rho_{M+1/2}$ we have the approximation (20) at $j = M$. For ρ_{M+1} we have

$$\rho_{M+1} = p(z_{M+1}) z_{M+1} \frac{\partial \psi_{M+1}}{\partial z} + q(z_{M+1})\psi_{M+1} = \mu_1 + q_{M+1}\psi_{M+1}.$$

Taking all approximations above, we obtain the semi-discretization of the problem (8)–(11):

$$\frac{1}{2} \frac{\partial \psi_1}{\partial t} \frac{h_{z,1}}{2} = \frac{1}{2} \left[(p_{3/2} + q_{3/2}) \psi_2 - (p_{3/2} - q_{3/2}) \psi_1 \right] - \psi_1 (v_{dp} + q_1) + \mu$$

$$+ \frac{h_{z,1}}{2} r_1 \psi_1;$$

$$\frac{1}{2} \frac{\partial \psi_2}{\partial t} h_{z,2} = q_{5/2} \frac{z_3^{\alpha_2} \psi_3 - z_2^{\alpha_2} \psi_2}{z_3^{\alpha_2} - z_2^{\alpha_2}} - \frac{1}{2} \left[(p_{3/2} + q_{3/2}) \psi_2 - (p_{3/2} - q_{3/2}) \psi_1 \right]$$

$$+ h_{z,2} r_2 \psi_2;$$

$$\frac{1}{2} \frac{\partial \psi_j}{\partial t} h_{z,j} = q_{j+1/2} \frac{z_{j+1}^{\alpha_j} \psi_{j+1} - z_j^{\alpha_j} \psi_j}{z_{j+1}^{\alpha_j} - z_j^{\alpha_j}} - q_{j-1/2} \frac{z_j^{\alpha_{j-1}} \psi_j - z_{j-1}^{\alpha_{j-1}} \psi_{j-1}}{z_j^{\alpha_{j-1}} - z_{j-1}^{\alpha_{j-1}}}$$

$$+ h_{z,j} r_j \psi_j; \ j = 3, 4, \ldots, M;$$

$$\frac{1}{2} \frac{\partial \psi_{M+1}}{\partial t} \frac{h_{z,M}}{2} = \mu_1 + q_{M+1}\psi_{M+1} - q_{M+1/2} \frac{z_{M+1}^{\alpha_M} \psi_{M+1} - z_M^{\alpha_M} \psi_M}{z_{M+1}^{\alpha_M} - z_M^{\alpha_M}}$$

$$+ \frac{h_{z,M}}{2} r_{M+1}\psi_{M+1}.$$

With respect to time we construct an implicit scheme:

$$-\left(\frac{h_{z,1}}{2\tau_k} + \frac{1}{2}\left(p_{3/2} - q_{3/2}\right) + v_{dp} + q_1 - \frac{h_{z,1}}{2}r_1\right)\bar\psi_1 + \left(\frac{1}{2}\left(p_{3/2} + q_{3/2}\right)\right)\bar\psi_2$$

$$= -\frac{h_{z,1}}{2\tau_k}\psi_1 - \mu,$$

$$\frac{1}{2}\left(p_{3/2} - q_{3/2}\right)\bar\psi_1 - \left(\frac{\hbar_{z,2}}{\tau_k} + \frac{q_{5/2}z_2^{\alpha_2}}{z_3^{\alpha_2} - z_2^{\alpha_2}} + \frac{1}{2}\left(p_{3/2} + q_{3/2}\right) - \hbar_{z,2}r_2\right)\bar\psi_2$$

$$+ \frac{q_{5/2}z_3^{\alpha_2}}{z_3^{\alpha_2} - z_2^{\alpha_2}}\bar\psi_3 = -\frac{\hbar_{z,2}}{\tau_k}\psi_2,$$

$$\frac{q_{j-1/2}z_{j-1}^{\alpha_{j-1}}}{z_j^{\alpha_{j-1}} - z_{j-1}^{\alpha_{j-1}}}\bar\psi_{j-1} - \left(\frac{\hbar_{z,j}}{\tau_k} + \frac{q_{j+1/2}z_j^{\alpha_j}}{z_{j+1}^{\alpha_j} - z_j^{\alpha_j}} + \frac{q_{j-1/2}z_j^{\alpha_{j-1}}}{z_j^{\alpha_{j-1}} - z_{j-1}^{\alpha_{j-1}}} - \hbar_{z,j}r_j\right)\bar\psi_j$$

$$+ \frac{q_{j+1/2}z_{j+1}^{\alpha_j}}{z_{j+1}^{\alpha_j} - z_j^{\alpha_j}}\bar\psi_{j+1} = -\frac{\hbar_{z,j}}{\tau_k}\psi_j, \quad j = 3,4,\ldots,M,$$

$$\left(\frac{q_{M+1/2}z_M^{\alpha_M}}{z_{M+1}^{\alpha_M} - z_M^{\alpha_M}}\right)\bar\psi_M - \left(\frac{h_{z,M}}{2\tau_k} - q_{M+1} - \frac{h_{z,M}}{2}r_{M+1} + \frac{q_{M+1/2}z_{M+1}^{\alpha_M}}{z_{M+1}^{\alpha_M} - z_M^{\alpha_M}}\right)\bar\psi_{M+1}$$

$$= -\frac{h_{z,M}}{2\tau_k}\psi_{M+1} - \mu_1,$$

were $\bar\psi$ is the approximate solution on the $n + 1/2$-th time level and ψ is the approximate solution on the n-th time level. The truncation error of the constructed scheme is of order $O(\tau + h_z)$.

3.2 Hyperbolic Problem

We first approximate the δ-function in the boundary condition, namely we replace the δ-function by a δ-like with support 2Δ in the following way:

$$\delta_\Delta = \begin{cases} \frac{\Delta - |z - z^*|}{\Delta^2}, & z \in [z^* - \Delta, z^* + \Delta], \\ 0, & z \notin [z^* - \Delta, z^* + \Delta]. \end{cases}$$

Then we approximate (14) by $c_1 = Q_1\delta_\Delta$. For (12) we construct a backward difference scheme:

$$\frac{\hat c_i - \psi_i}{\tau_n} + \frac{\hat c_i - \hat c_{i-1}}{h_{x,i-1}} = 0, \quad i = 2,3,\ldots,N.$$

This scheme is unconditionally stable and its truncation error is of order $O(\tau + h_x)$. For $\hat c_i$ we get

$$\hat c_i = \frac{h_{x,i-1}\psi_i + \tau_n\hat c_{i-1}}{h_{x,i-1} + \tau_n}, \quad i = 2,3,\ldots,N.$$

4   Numerical Experiments

In order to observe the error behaviour of the difference method we used the analytical solution $c_{an} = (x^2 + z^2)t$ and compared it with the numerical solution when $X = T = 1$, $\beta = 0.4$, $u = 10^{-2}$, $w_s = 8 \cdot 10^{-5}$, $v_g = 1.5$, $k = 10^{-3}$, $\tau = \tau_n = 10^{-4}$. The space meshes are regular and the rate of convergence (RC) is calculated using the double mesh principle. Results are presented in Table 1.

To show the efficiency and usefulness of the discretization method, various test problems with different choices of parameters were solved. For all examples, presented in this paper, we use the following fixed values of the parameters: $\bar{X} = 6000$ m, $\bar{H} = 600$ m, $\bar{T} = 10000$ s, $\bar{u}_z = 5$ m/s, $\bar{Q} = 1$ mg/m^{2}s, $\beta = 0.4$, $\bar{l} = \bar{X}/2$.

As a first example we perform numerical experiments for the following values of the other parameters in the problem under consideration (they are at the lower bound of the interval we investigate): $\bar{u} = 0.05$ m/s, $\bar{w}_s = 4 \cdot 10^{-4}$ m/s, $\bar{k} = 8.3 \cdot 10^{-6}$ 1/s, $\bar{Q}_1 = 0$ mg/m^{2}s, $\bar{v}_{dp} = 5 \cdot 10^{-3}$ m/s, $v_g = 1.5$. The graphics of the numerical solution of the problem under consideration at $\bar{t} = \bar{T}$ is presented in Fig. 1.

For the second example we choose the values of the other parameters at the upper bound of the interval we investigate: $\bar{u} = 0.5$ m/s, $\bar{w}_s = 1 \cdot 10^{-3}$ m/s, $\bar{k} = 2.2 \cdot 10^{-5}$ 1/s, $\bar{Q}_1 = 0$ mg/m^{2}s, $\bar{v}_{dp} = 7 \cdot 10^{-3}$ m/s, $v_g = 4.43$. The graphics of the solution at $\bar{t} = \bar{T}$ is presented in Fig. 2.

For the third example we use the values of the parameters from the second example, but with $\bar{Q}_1 = 0.2$ mg/m^{2}s and $\bar{h}_1 = 60$ m. The graphics of the solution in this case at $\bar{t} = \bar{T}$ is presented in Fig. 3.

From Figs. 1, 2 and 3 one can see the influence of the variable parameters of the problem under consideration on the shape of the solution and its maximal value at $\bar{t} = \bar{T}$.

Table 1. Rate of convergence

N × M	10 × 10	20 × 20	40 × 40	80 × 80	160 × 160	320 × 320
Relative C-norm of the error	2.385 E-2	1.211 E-2	6.058 E-3	2.984 E-3	1.439 E-3	6.715 E-4
RC	-	0.977	0.999	1.021	1.052	1.100

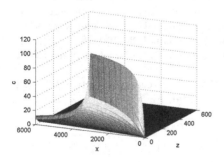
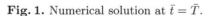

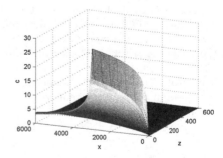

Fig. 1. Numerical solution at $\bar{t} = \bar{T}$.   **Fig. 2.** Numerical solution at $\bar{t} = \bar{T}$.

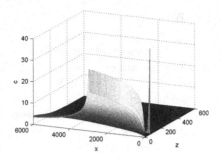

Fig. 3. Numerical solution at $\bar{t} = \bar{T}$.

Acknowledgments. This work was partially supported by the Bulgarian Fund of Sciences under Grants No. FNI I 02/9-2014 and No. FNI I 02/20-2014.

References

1. Akrivis, G., Grouzlix, M., Thomee, V.: Numerical methods for ultraparabolic equations. CALCOLO **31**, 179–190 (1996)
2. Ashyralyev, A., Yilmaz, S.: Modified Crank-Nicholson difference schemes for ultraparabolic equations. Comput. Math. Appl. **64**, 2756–2764 (2012)
3. Chernogorova, T., Vulkov, L.: Finite volume difference scheme for a model of settling particle dispersion from an elevated source in a open-channel flow. Comput. Math. Appl. **67**, 2099–2111 (2014)
4. Chernogorova, T., Vulkov, L.: Fitted finite volume positive difference scheme for a stationary model of air pollution. Numer. Algorithms **70**(1), 171–189 (2015). doi:10.1007/s11075-014-9940-y
5. Gencev, T.: Ultraparabolic equations. Dokl. Akad. Nauk. SSSR **151**, 265–268 (1963)
6. Dang, Q.A., Ehrhardt, M.: Adequate numerical solution of air pollution problems by positive difference schemes on unbounded domains. Math. Comput. Model. **44**, 834–856 (2006)
7. Dimov, I., Farago, I., Havasi, A., Zlatev, Z.: Different splitting techniques with applications to air pollution models. Int. J. Environ. Pollut. **32**(2), 174–199 (2008)
8. Dimov, I., Zlatev, Z.: Computational and Numerical Challenges in Environmental Modeling. Elsevier, Amsterdam (2006)
9. Lakshminarayanachari, K., Suresha, C.M., Prasad, M.S., Pandurangappa, C.: Numerical model of air pollutant emitted from an area source of primary and secondary pollutants with chemical reaction and gravitational settling with point source on the boundary. Int. J. Res. Environ. Sci. Technol. **3**(1), 9–18 (2013)
10. Samarskii, A.A.: The Theory of Difference Schemes. Marcel Dekker, New York (2001)
11. Vabishchevich, P.N.: The numerical simulation of unsteady convective-diffusion processes in a countercurrent. Zh. Vychisl. Mat. Mat. Fiz. **35**(1), 46–52 (1995)
12. Wang, S.: A novel fitted finite volume method for Black-Scholes equation governing option pricing. IMA J. Numer. Anal. **24**, 699–720 (2004)

Snow Cover Assessment with Regional Climate Model - Problems and Results

Hristo Chervenkov$^{(\boxtimes)}$, Todor Todorov, and Kiril Slavov

National Institute of Meteorology and Hydrology, Bulgarian Academy of Sciences,
Tsarigradsko Shose 66, 1784 Sofia, Bulgaria
hristo.tchervenkov@meteo.bg

Abstract. Climate modelling, either global or regional, is usually treated as a typical large-scale scientific computational problem. The regional climate model RegCM, well-known within the meteorological community, is applied in the study to estimate quantitatively the snow water equivalent, which is the most consistent snow cover parameter. Multiple runs for a time window of 14 consecutive winters with different model configurations, in particular with various initial and boundary conditions, have been performed, in an attempt to obtain most adequate representation of the real snow cover. The results are compared with stations' measurements from the network of the National Institute of Meteorology and Hydrology. Generally all runs yield similar results, where the overall (i.e. over the whole time span) biases are acceptable, but, however, with large discrepancies in the day-by-day comparisons, which is typical for climate modelling studies.

Keywords: Regional climate modelling · RegCM4 · Snow water equivalent · Verification

1 Introduction

The snow is a very important component of the climate system which controls surface energy and water balances and is the largest transient feature of the land surface [17]. It has an effect on atmospheric circulation through changes to the surface albedo, thermal conductivity, heat capacity and aerodynamic roughness. The snow properties of surface water storage control the availability of water in many ecosystems and to a sixth of the world's population [3]. Therefore it is vital that snow is properly represented in geophysical models if we want to understand and make predictions of weather, climate, the carbon cycle, flooding and drought.

The various properties characterizing snow are highly variable and so have to be determined as dynamically active components of climate. These include the snow depth (h_s) snow water equivalent (SWE), density, and snow cover area (SCA). The snow water equivalent is a measure of the amount of water contained in snow pack. It can be considered as the depth of water that would

© Springer International Publishing Switzerland 2015
I. Lirkov et al. (Eds.): LSSC 2015, LNCS 9374, pp. 327–334, 2015.
DOI: 10.1007/978-3-319-26520-9_36

theoretically result if the whole snow pack instantaneously melts. SWE is the product of snow depth and snow density. Unfortunately, from the four snow metrics listed above, only extent (i.e., snow cover area (SCA)) is easily monitored using satellites. SCA, however, is only an indirect measure of the world's snow water resources (e.g. [2]). To fully understand global snow water trends, the most fundamental metric to assess is SWE, with h_s a close second. However, on large spatial scales the properties of snow are not easily quantified either from modelling or observations. For example, station based snow measurements often lack spatial representativeness, especially in regions where the topography, vegetation and overlaying atmosphere produce considerable heterogeneity of the snow-pack distribution [12].

Of the two fundamental parameters, depth is quicker and easier to measure than SWE. No detailed estimates of the total number of depth and SWE measurements made worldwide is available, but what is available suggests that considerably more depths are collected than SWE measurements. So, for example, following the directives of the World Meteorological Organization (WMO), h_s is measured in every station of the network of the Bulgarian National Institute of Meteorology and Hydrology every day, at 06 UTC and SWE - usually only five times monthly. Thus, despite the weaknesses of the land surface models, the quantitative assessment of the snow properties by the means of the numerical simulation is reasonable approach for obtaining of spatial and temporal consistent picture of the snow-pack distribution.

The paper aims to present some, preliminary indeed, results from multiple runs with different model configurations, which are performed, in attempt to obtain a most adequate representation of the real snow cover.

2   Concept and Methodology

Regional climate models (RCMs) have been developed and extensively applied in the recent decade for dynamically downscaling coarse resolution information from different sources, such as global circulation models (GCMs) and reanalyses, for different purposes including past climate simulations, as in the presented study and future climate projection. This widely used and productive approach is used here. The main simulation tool is the freely and on-line available from the web-site of the maintaining institute, the International Center of Theoretical Physics in Italy (ICTP, http://gforge.ictp.it/gf/project/regcm/) newest version 4 of the Regional climate model RegCM. RegCM4 is a 3-dimensional, sigma-coordinate, primitive equation RCM with dynamical core based (version 2 and later) on the hydrostatic version of the NCAR-PSU Mesoscale Model 5 (MM5) [8]. The radiative transfer package is taken from the Community Climate Model v. 3 (CCM3) [11]. The large-scale cloud and precipitation computations are performed by Subgrid Explicit Moisture Scheme (SUBEX, [15]) and the land surface physics are according to the Biosphere-Atmosphere Transfer Scheme (BATS, [5]). The adopted convective scheme for the RCM simulations in the present study is the Grell scheme [7] with the Arakawa and Schubert [1] closure assumption.

The model is flexible, portable and easy to use. It can be applied to any region of the World, with grid spacing of up to about 10 km (hydrostatic limit), and for a wide range of studies, from process studies to paleoclimate and future climate simulation. There are a number of previous studies that evaluated the model performance around the world ([6] and references therein).

Main manifestation of the flexibility of the modern RCM, including RegCM4, is the possibility for selection among different initial and boundary conditions datasets (ICBC), parameterization schemes/modules within the model, various constants and closure assumptions, etc., combining them in practically count-less model setups. Obviously the simulation output from such model setups will differ from one another, and, more or less, from the "reality". Investigation of the influence of the changes of some parameter on the output, often considered as "sensitive study" is out of the scope of our work. There is, however, overall agreement that the ICBC plays the most important role for the model per-formance. Although there are numerous tests with different reanalysis data, which are considered as better ICBC compared to those produced by GCMs, there is no single reanalysis data set yielding the best results in every region and/or every season. We have performed simulations with the two most popular and widely used reanalysis datasets - The ERA-Interim of the European Centre for Medium-Range Weather Forecasts (ECMWF) [4] with horizontal resolution $1.5° \times 1.5°$ for RegCM simulations, noted further as EIN15 and the reanalysis 2 of the USA National Centers for Environmental Predictions and the National Center for Atmospheric Research (NCEP/NCAR) [10] with horizontal resolution $2.5° \times 2.5°$, noted further as NNRP2. It is physically reasonable also to expect, that the module, which describes the surface processes and the interactions with the under- and overlaying soil and atmospheric layers, namely the land surface model, plays relevant role especially in the numerical treatment of the snow cover. A major addition to RegCM4 is the option to use the Community Land Model (CLM), version 3.5. Compared to BATS, CLM is a more advanced package (and as a result computationally heavier), which is described in detail in [13,14]. It uses a series of biogeophysically-based parameterizations to describe the land-atmosphere exchanges of energy, momentum, water, and carbon.

3   Performed Computations and Results

The model domain is centered over Bulgaria, includes the whole Balkan penin-sula, as shown in Fig. 1 and consists of 72×77 20 km $\times$ 20 km gridcells. The simulation period is from 1th November till 31th March (hereafter: winter) for 14 consecutive years between 2000 and 2014. Model output is the gridded dis-tribution of the SWE on 6-hourly basis (i.e. at 00, 06, 12 and 18 UTC).

As in many similar studies, part of the difficulty in exploring the dynamic downscaling ability issue is rooted in the lack of validation data for small scale features and reliable measurements. Additionally, due to the shorter (in comparison with higher latitudes) and discontinuous duration of the snow cover in the region, the data series of the station measurements are sparse and also

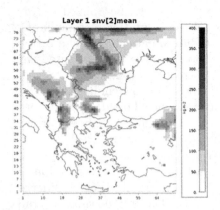

Fig. 1. Average SWE (unit: mm) for February 2012

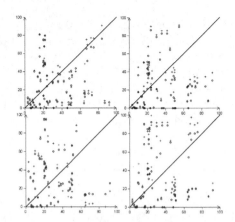

Fig. 2. Scatter plot for Haskovo ($N = 75$, units: mm SWE) for EIN15/BATS (upper left pane), NNRP2/BATS (upper right pane), EIN15/CLM (lower left pane) and NNRP2/CLM (lower right pane). The measured values are plotted along the abscissas. The comparisons measured - nearest gridnode value are marked with crosses and the comparisons measured - interpolated value - with circles.

relatively short. Specific feature of the snow cover is, due to practical absence of horizontal mixture processes, the relatively high (in comparison to the atmospheric lower-level parameters) heterogeneity. Thus, even on small distances, considerable differences in the snow properties can be observed, respectively modelled. In an attempt to compensate this, not only comparison between the point observation and the nearest gridpoint, which is the most commonly used approach, but also comparison between the point observation and the interpolated to this point from the four nearest surrounding gridpoints' values, is performed.

Observational time series with acceptable length for the period and domain under consideration are available only for four Bulgarian stations, but only two of them, Haskovo and Rasgrad, are located over relatively flat terrain. The selected resolution of 20 km, and, more generally, the hydrostatic limit of 8–10 km, do not allows to resolve smaller scale features and thus is not suitable over mountainous orography, where the other stations are deployed. Station Rasgrad is located in the northern, and station Haskovo - in the southern half of the country and they can be roughly considered as representatives of these climatic regions. A traditional way to contrast graphically the measured vs. modelled values is to present them as scatter plot, as shown in Figs. 2 and 3.

The degree of agreement of the time series of the observed (measured) values of SWE O_i and their modelled correspondents M_i, is estimated with the frequently

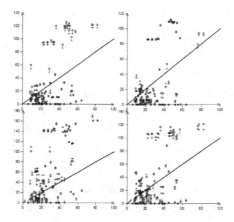

Fig. 3. Same as Fig. 2, but for station Rasgrad ($N = 124$)

used statistical quantities, namely the root mean square error (RMSE), the correlation coefficient (also termed the pearson correlation coefficient, R), the index of agreement (IA) and the mean bias (BIAS). Explicit formulas for the first two will not be given due to their popularity, and the last two are equal accordingly to:

$$IA = 1 - \frac{\sum_{i=1}^{N}(O_i - M_i)^2}{\sum_{i=1}^{N}(|M_i - \overline{O}| + |O_i - \overline{O}|)^2} \tag{1}$$

$$BIAS = \frac{1}{N}\sum_{i=1}^{N}(M_i - O_i) \tag{2}$$

The summation is along the time series length N and the overlines notes time-averaging. The (dimensionless) index of agreement condenses the differences between modelled and observed values for a given time period into one statistical quantity. It provides a measure of the match between the departure of each prediction from the observed mean and the departure of each observation from the observed mean. IA has a theoretical range of 0 to 1, with a value of 1 suggesting "perfect" agreement. The mean bias is simply the average bias between the modelled and observed values.

The result of the statistical comparison is summarized in Table 1.

It is obvious that these observational data are too scarce, respectively the calculated statistics insufficient, to obtain results about each models run performance with satisfying representativeness and confidence. So far, however, some basic facts seem clear.

Generally (and roughly) speaking, no model configuration, among the four, performs significantly better than the others. For both stations, in terms of the BIAS, the performance of the set NNRP2/CLM looks worst. Similar, at the first site, a puzzling fact (due to the above mentioned theoretical superiority of CLM over BATS) is found by Steiner et al. [16] in significantly deeper study of the CLM applicability in the RegCM.

Table 1. Statistical indexes for the comparison for Haskovo (upper half) and Rasgrad (lower half). The first number in each cell is for the comparison measured - nearest gridnode value and the second - for the comparison measured - interpolated value

	RMSE (mm)	R (corr. coeff.)	IA	BIAS (mm)
EIN15/BATS	31.97 31.84	0.17 0.11	0.54 0.51	−6.56 −8.34
NNRP2/BATS	31.89 32.77	0.15 0.09	0.48 0.45	−7.90 −10.26
EIN15/CLM	49.38 41.06	0.17 0.21	0.44 0.50	12.44 10.28
NNRP2/CLM	37.25 35.46	0.22 0.17	0.52 0.51	3.37 −0.65
EIN15/BATS	32.72 34.42	0.58 0.59	0.62 0.60	2.09 2.10
NNRP2/BATS	30.78 30.35	0.52 0.53	0.61 0.62	−1.59 −1.53
EIN15/CLM	50.80 48.77	0.57 0.58	0.48 0.49	23.73 21.45
NNRP2/CLM	31.67 33.51	0.55 0.55	0.61 0.60	3.90 6.13

The interpolation between the gridnodes to the point of the station does not lead to any remarkable convergence of the computed values to the measured ones. The weaknesses in the model physics, rather than the displacement between the gridnode and the station point are the most natural explanation therefore.

As shown in Figs. 2 and 3 the disagreement in the individual pairs is relatively high - up to a factor of 5, whose quantitative expression are the high values of RMSE and low ones of the R and IA. The BIAS-es, however, except the NNRP2/CLM-case, are "acceptably" small. Therefore, it can be concluded that no systematic model under- or overestimation is detected. Keeping in mind also that the BIAS is the difference between the time-averaged modelled and observational values, the model behaviour for the whole period can be estimated as good.

4   Comments and Conclusion

The obvious main reasons for the detected discrepancies between the model results on the one hand and the measurements on the other, are the weaknesses of the parameterization schemes and/or their combination in the selected model setups. Other experiments with different settings have to be performed. Thus, due to the expected minimal influence of the convection during most of the winter season in the domain, the replacement of the standard large-scale (i.e. non-convective) precipitation scheme SUBEX, with the Tompkins's one, can be the subject of further efforts.

As emphasized in many studies (see, for instance, [18]), the "climate downscaling" concept mainly refers to "climate" statistics based on averages (or sums as shown in Fig. 1) of the climate system over periods of a month or more. RCMs can and have been applied for various temporal scales, but as a whole, in contrast to the weather prediction models, they are not intended to fore- or hindcast the state for periods shorter than 1–2 weeks. Thus, strictly speaking, such day-by-day comparisons and any judgments based upon them, are methodologically not correct. Nevertheless, however, such procedure is often treated in similar numerical experiments as a necessary (first) step in verification/model performance evaluation.

Satellite earth snow observation products have the needed spatial and temporal consistency, which allows comparisons with model output over continuous area and time frames. The absence of this consistency of the point measurements is an inherent weakness of every statistical evaluation procedure based upon them and thus utilizing satellite data is a significant step ahead in the quantitative snow cover assessment. Although among the other products SWE has proven more problematic [9], especially for wet snow and during melt, which is a typical case in the region due to its climate, the common treatment of SWE satellite data and model results is already a planned continuation of the presented work.

The model RegCM is constantly developed and, respectively, its simulation capabilities are steadily increasing. The expected in the near future transition to a version with non-hydrostatic dynamical core will improve its downscaling possibilities. Then it will be possible to perform simulations with resolutions below 10 km, which is significant, for instance, for studies as the one presented here, where the representation of the physical processes over the topographic heterogeneous terrain of the Balkan peninsula in proper spatial scale is crucial. This concerns even to a greater extent the snow pack assessment studies, due to the fact that, at given geographical latitude, the mountainous snow covers are, generally speaking, those with the longest duration and thickness, with all hydrological, ecological and socio-economical consequences.

Acknowledgments. Deep gratitude to the organizations and institutes (ICTP, ECMWF, NCEP-NCAR, Unidata, MPI-M and all others), which provides free of charge software and data. Without their innovative data services and tools this study would be not possible.

References

1. Arakawa, A., Schubert, W.H.: Interaction of a cumulus cloud ensemble with the large-scale environment, Part I. J. Atmos. Sci. **31**, 674–701 (1974)
2. Braun, R.D., Walker, A.E., Goodison, B.: Seasonal snow cover monitoring in Canada: an assessment of Canadian contributions for global climate monitoring. In: Proceedings of the 57th Eastern Snow Conference, Syracuse, NY, pp. 1–11 (2000)
3. Clifford, D.: Global estimates of snow water equivalent from passive microwave instruments: history, challenges and future developments. Int. J. Remote Sens. **31**, 3707–3726 (2010)
4. Dee, D.P., Uppala, S.M., Simmons, A.J., Berrisford, P., Poli, P., Kobayashi, S., Andrae, U., Balmaseda, M.A., Balsamo, G., Bauer, P., Bechtold, P., Beljaars, A.C.M., van de Berg, L., Bidlot, J., Bormann, N., Delsol, C., Dragani, R., Fuentes, M., Geer, A.J., Haimberger, L., Healy, S.B., Hersbach, H., Hólm, E.V., Isaksen, L., Kållberg, P., Köhler, M., Matricardi, M., McNally, A.P., Monge-Sanz, B.M., Morcrette, J.-J., Park, B.-K., Peubey, C., de Rosnay, P., Tavolato, C., Thépaut, J.-N., Vitart, F.: The ERA-Interim reanalysis: configuration and performance of the data assimilation system. Q. J. R. Meteorol. soc. **137**, 553–597 (2011)

5. Dickinson, R., Henderson-Sellers, A., Kennedy, P.J.: Biosphere-Atmosphere Transfer Scheme, BATS: version1E as coupled to the NCAR community climate model. NCAR Tech Note NCAR/TN-387+ STR. National Center for Atmospheric Research, Boulder CO (1993)

6. Giorgi, F., Coppola, E., Solomon, F., Mariotti, L., Sylla, M.B., Bi, X., Elguindi, N., Diro, G.T., Nair, V., Giuliani, G., Turuncoglu, U.U., Cozzini, S., Guttler, I., O'Brien, T.A., Tawfic, A.B., Shalaby, A., Zakey, A.S., Steiner, A.L., Stordal, F., Sloan, L.C., Brankovic, C.: RegCM4: model description and preliminary tests over multiple CORDEX domains. Clim. Res. **52**, 7–29 (2012)

7. Grell, G.A.: Prognostic evaluation of assumptions used by cumulus parametrizations. Mon. Weather Rev. **121**, 764–787 (1993)

8. Grell, G.A., Dudhia, J., Stauer, D.R.: A description of the fifth-generation Penn State/NCAR mesoscale model (mm5). Technical report NCAR/TN-398+ STR, National Center for Atmospheric Research, Boulder CO (1994)

9. Hancock, S., Baxter, R., Evans, J., Huntley, B.: Evaluating global snow water equivalent products for testing land surface models. Remote Sens. Environ. **128**, 107–117 (2013)

10. Kanamitsu, M., Ebisuzaki, W., Woollen, J., Yang, S.K., Hnilo, J.J., Fiorino, M., Potter, G.L.: NCEP-DOE AMIP- II reanalysis (R-2). Bull. Am. Meteorol. Soc. **83**, 1631–1643 (2002)

11. Kiehl, J.T., Hack, J.J., Bonan, G.B., Boville, B.A., Briegleb, B.P., Williamson, D., Rasch, P.: Description of the NCAR community climate model (ccm3). Technical report NCAR/TN-420+ STR. National Center for Atmospheric Research, Boulder CO (1996)

12. Liston, G.E.: Representing subgrid snow cover heterogeneities in regional and global models. J. Clim. **17**, 1381–1397 (2004)

13. Oleson, K.W., Dai, Y., Bonan, G., Bosilovich, M., et al.: Technical description of the Community Land Model. National Center for Atmospheric Research Tech Note NCAR/TN-461+ STR, NCAR, Boulder CO (2004)

14. Oleson, K.W., Niu, G.Y., Yang, Z.L., Lawrence, D.M., et al.: Improvements to the Community Land Model and their impact on the hydrologic cycle. J. Geophys. Res. **113**, G01021 (2008). doi:10.1029/2007JD000563

15. Pal, J.S., Small, E.E., Eltahir, E.A.B.: Simulation of regional-scale water and energy budgets: representation of subgrid cloud and precipitation processes within RegCM. J. Geophys. Res. **105**, 29579–29594 (2000)

16. Steiner, A.L., Pal, J., Giorgi, F., Dickinson, R.E., Chameides, W.L.: Coupling of the Common Land Model (CLM0) to a regional climate model (RegCM). Theoret. Appl. Climatol. **82**(3–4), 225–243 (2005)

17. Yang, F., Kumar, A., Wang, W., Juang, H.-M.H., Kanamitsu, M.: Snow-albedo feedback and seasonal climate variability over North America. J. Clim. **14**, 4245–4248 (2001)

18. Xue, Y., Janjic, Z., Dudhia, J., Vasic, R., De Sales, F.: A review on regional dynamical downscaling in intraseasonal to seasonal simulation/prediction and major factors that affect downscaling ability. Atmos. Res. **147–148**(2014), 68–85 (2014)

Input Data Preparation for Fire Behavior Fuel Modeling of Bulgarian Test Cases
(Main Focus on Zlatograd Test Case)

Georgi Dobrinkov[2] and Nina Dobrinkova[1($\boxtimes$)]

[1] Institute of Information and Communication Technologies,
Bulgarian Academy of Sciences, Sofia, Bulgaria
nido@math.bas.bg

[2] Institute of Mathematics and Informatics, Bulgarian Academy of Sciences,
Sofia, Bulgaria
g.dobrinkov@gmail.com

Abstract. Since the year 2000 Bulgaria is facing progressive increase of wildland fire occurrence. That is caused mainly because of human mistakes in having fire camps or agricultural land processing after crop harvesting. At the moment Bulgaria has no working mechanism to spot such fires before they become a threat, however the team from Bulgarian Academy of Sciences is working on fire behaviour modelling issues since 2007 and in this work the first attempts for Bulgarian forestry data classification will be presented according to the existing 53 Fire Behavior Fuel Models (FBFMs), and estimations where custom fuels has to be prepared for better representation of the potential fire spread. Calibrations with FARSITE (Fire Area Simulator) runs have been performed for the area of Zlatograd Forestry Department (Bulgaria) and the results are compared with Harmanli (Bulgaria) WRF-Fire/S-Fire simulations. The differences in the fire behaviour fuel models estimations reflect in the final simulated burned area which is presented in the conclusions. Fire behaviour fuel modeling based on both simulation approaches in Zlatograd and Harmanli areas gives future application of the presented work for Bulgarian test cases.

1 Introduction

FARSITE runs with the standard or custom Fire Behaviour Fuel Models in Bulgaria has never been done until our first attempts for this on the test cases selected randomly for the period 2011–2012 on the territory of Zlatograd Forestry Department. In the previous works on Bulgarian test case nearby the area of Harmanli town has been used CORINE categories [4,5]. The approach in Harmanli test case gives CORINE species division adapted only to the thirteen classes of Anderson [1]. However this approach has some weaknesses, because fuel load parameters description with satellite images only is hard to be well justified. The current work present chronologically how the available forestry data

© Springer International Publishing Switzerland 2015
I. Lirkov et al. (Eds.): LSSC 2015, LNCS 9374, pp. 335–342, 2015.
DOI: 10.1007/978-3-319-26520-9_37

from the Zlatograd Forestry department is prepared for the FARSITE runs from turning the list of biological species into FBFMs according to the FARSITE input instructions. Most of the collected data has been provided as paper maps which processing into digital GIS (Geographic Information Systems) layers had to meet the requirements of FARSITE. The standard FBFMs [1,8] has been taken into consideration with their parameters for fuel load (1-hr, 10-hr, 100-hr, live herbaceous and live woody), compared to the test cases according to the best collected data. This work is giving both Harmanli and Zlatograd FBFMs methodologies and how they have been implemented in the WRF-Fire/S-Fire and FARSITE runs [3,5]. The conclusions give comparison on the achieved results with some plans for future refinements.

2   Summary of Harmanli Test Case Data Preparation

In the Harmanli test case WRF-Fire/S-Fire computer based tool has been used, which is a combination between the mesoscale atmospheric code WRF-ARW [9] with a fire spread module, based on the Rothermel model [6] implemented by the level set method.

The semi-empirical fire propagation model imposes the fire spread rate directly, replaces the leading edge of the combustion wave by instantaneous ignition and replaces the fuel depletion rate by an imposed one. We consider fire burning in the domain $\Omega = \Omega(t)$ in the (x,y) plane, with the boundary $\Gamma = \Gamma(t)$ called the fire line, and with outside normal $\overrightarrow{n} = \overrightarrow{n}(x,y,t)$, $(x,y) \in \Gamma(t)$. The time of ignition $t_i(x,y)$ at a point $(x,y) \in \Omega(t)$ is defined as the time when the point (x,y) is at the fire line, that is, $(x,y) \in \Gamma(t_i(x,y))$. The fire-line propagation model postulates that the fire line evolves with a given spread rate $S = S(x,y,t)$ in the normal direction. The spread rate S is a function of the fuel properties, the wind $\overrightarrow{v}$ and the terrain gradient ∇z in the normal direction $\overrightarrow{n}$ to the fire line, that is, $S = S(\overrightarrow{v} \cdot \overrightarrow{n}, \nabla z \cdot \overrightarrow{n})$, and is given by the modified Rothermel's formula

$$S = \begin{cases} 0, & \text{if } \tilde{S} < 0, \\ S_{max}, & \text{if } \tilde{S} > S_{max}, \\ \tilde{S}, & \text{otherwise,} \end{cases} \tag{1}$$

$$\tilde{S} = min\{B_0, R_0 + \phi_W + \phi_S\},$$

where $S_{\max}$ is the maximum rate of fire spread, R_0 is the spread rate in the absence of wind, $\phi_W = a(\overrightarrow{v} \cdot \overrightarrow{n})^b$ is the wind correction, $\phi_S = d\nabla z \cdot \overrightarrow{n}$ is the terrain correction a, b and d are constants and B_0 is the backing rate that is the minimal fire spread rate even against the wind. A small backing rate of spread must be specified, since fires are known to creep upwind on their upwind edge due to radiation.

In the burning area, the model postulates that the fuel decreases exponentially from the ignition time, that is,

$$F(x,y,t) = \begin{cases} F_0(x,y)e^{-\frac{(t-t_i(x,y))}{W(x,y)}}, & \text{if } (x,y) \in \Omega(t) \\ F_0(x,y), & \text{otherwise} \end{cases} \tag{2}$$

where $F_0(x,y)$ is the initial fuel supply and $W(x,y)$ is the time constant of the fuel. The heat flux from the fire to the atmosphere is determined from the amount of fuel burned by

$$\Phi = -A(x,y)\frac{\partial}{\partial(t)}F(x,y,t) \tag{3}$$

which gives the modifications to the local weather conditions in the area where the fire has occurred. The so called "fire weather" is accounted with this equation. The coefficients B_0, R_0, S_{max}, a, b, d, W, and A, characterize the fuel load. The fire model input data consists of the fuel category array, which is integrated in the WRF input data and can be alternatively set from the namelist for testing.

A simulation with WRF-Fire/S-Fire requires input data from a variety of sources from meteorological initial and boundary conditions to static surface properties. For the meteorological inputs the U.S. National Center for Environmental Protection (NCEP) gives an 1 degree resolution grid covering the entire globe with 6 hour reanalysis cycle. The data is freely available and can be downloaded automatically over HTTP by using a simple script. Creating simulation also requires a number of static data fields describing the surface properties of the area. All such data is available as part of a standard global dataset for WRF. The fields in this dataset are available at various resolutions ranging from about 1 Km to 10 km, which is sufficient for most mesoscale weather modeling purposes. Each field is stored in a unique format consisting of a series of simple binary files described by a text file. A geogrid utility in the WRF preprocessor (WPS) interpolates the data in these files onto the model grid and produces an intermediate NetCDF file used in further preprocessing steps. While the standard geogrid dataset is sufficient for most weather forecasting applications, it lacks two high resolutions fields. These fields are surface topography and fuel information. Both are essential for modeling fire behavior because they directly affect the rate of spread of the fire front inside the model. Topography at a resolution of about 90 m for the area of Harmanli is used from the Shuttle Radar Topography Mission (SRTM) at http://eros.usgs.gov. The data received from the server is a GIS raster format (DTED), which is processed and converted to geogrid binary data format. The final piece of surface data needed for input into geogrid is a categorical field describing the properties of the fuels. In the U.S., this data is readily available from the USGS, however, no such data exists for the Harmanli or any other Bulgarian region. Instead data for this field is used from the Corine Landcover Project (financed by the European Environment Agency and the member states). This project provides landcover data for Bulgaria with 100 m resolution with a 25 ha minimum mapping unit http://www.eea.europa. eu/data-and-maps/data/corine-land-cover-2006-raster.

The downloaded satellite data along with orthophoto data from the geoportal of the Ministry of Regional Development (MRD) of Bulgaria can be used to estimate the fuel types of the domain like conifer or deciduous woods. All rivers, lakes, villages and forest areas can be vectorized using the orthophoto images combined with CORINE2006 into a GIS vector shape file. The vectorized file provides very high accuracy of representation for non burning areas like rivers

Table 1. Fuel categories from satellite imagery and CORINE code (in parentheses).

Category	Description
1	Artificial, non-agricultural vegetated areas (141,142)
2	Sport Complex,Irrigated Cropland and Pasture,Bare Ground Tundra, Arable land (211,212,213), Open spaces with little or no vegetation (331,332,333,334,335)
3	Cemeteries, Dryland Cropland and Pasture, Grassland, Permanent crops(221,222,223), Pastures (231), Heterogeneous agricultural areas(241,242,243,244), Scrub and/or herbaceous vegetation associations (321,322,323,324)
4	Herbaceous Tundra, Parks
5	Wooded Wetland
6	Wooded Tundra, Orchard
7	Mixed Forest
8	Deciduous Needleleaf Forest, Forests (311,312,313)
9-13	N/A
14	Urban fabric (111,112), Industrial, commercial, and transport units (121,122,123,124), Mine, dump and construction sites (131,132,133), Wetlands(411,412,421,422,423), Water bodies (511,512,521,522,523)

and lakes. Table 1 is used for the areas with woods, where a description of the fuel categories for the Harmanli simulation corresponds to the Anderson thirteen classes [1] with additional one class for the non burning areas. This fuel level data combined with the vectorized landcover areas gives us a final shape file with attributes for each polygon fuel level. The resulting input files contain all the standard WRF fields along with several additional variables generated from the high resolution topography and fuel categories. However no fuel load description about 1-hr, 10-hr, 100-hr, live herbaceous and live woody parameters can be estimated with such approach. The problem here is that the satellite images from CORINE project can not provide high resolution with all species information on the land cover. The ortophotos can provide the canopy cover, but not the fuel load description and in these cases the best solution is to find Bulgarian Forestry department plans, which contain information that can be used for the fuel load estimations and refinements in the Table 1 categories. Such approach is described in the Zlatograd test case area.

3   Summary of Zlatograd Test Case Data Preparation

In the Zlatograd test case fifteen wildland fires have been taken into consideration. They occur on the territory of Forestry Department Zlatograd, that includes the municipal areas of Zlatograd, Madan and Nedelino towns. General table with fire occurrence by territory, date and duration is given by Table 2.

The fires that are taken into consideration for calibration are only surface fires. The modelling has been done through the FARSITE (Fire Area Simulator)

Table 2. Fire information provided by the Zlatograd Forestry Department for the period 2011–2012.

Fire No.	Vegetation type	Burned area in decares	Date of occurrence	Hour of start	Hour of end
1	Durmast	3.0	25 March 2012	1330	1530
2	Beechwood	5.0	29 March 2012	1400	1800
3	Scotch pine	1.0	16 June 2012	1500	1700
4	Scotch pine	7.0	6Aug. 2012	1640	1950
5	Scotch pine	5.0	6 Aug. 2012	1710	2130
6	European black pine	4.0	27Aug. 2012	1200	1600
7	Scotch pine	3.0	5 Sept. 2012	1400	2030
8	Scotch pine	6.0	6 Sept. 2012	1400	1930
9	Scotch pine	2.0	6 Oct. 2012	1600	2320
10	Scotch pine	1.0	16 March 2011	1310	1400
11	Scotch pine	1.0	5 April 2011	1715	1900
12	Scotch pine	1.0	10 April 2011	1130	1530
13	Grassland	3.0	30 Aug. 2011	1400	1800
14	Scotch pine	4.0	12 Sept. 2011	1230	1900
15	Scotch pine	1.0	15 Sept. 2011	1600	1830

tool which combines fire behavior models in cases of surface and crown fires, fire acceleration, spotting fires. The base of the surface fire modelling is the Rothermel Rate Of Spread (ROS) equation, which presents in nominator the Heat Source and in denominator the Heat sink for the fire propagation. The equation representation is given by the formula:

$$R = \frac{HeatSource}{HeatSink} = \frac{I_{xig} + \int_{-\infty}^{0} \left(\frac{\partial I_z}{\partial z}\right)_{Z_c} dx}{\rho_{be} Q_{ig}}. \tag{4}$$

where

R - is parameter for fire spread or the so called ROS (rate of spread),

I_{xig} - is the horizontal spread of the heat absorbed by the burning materials evaporating their water content,

ρ_{be} - is the density of the burning materials which are heated until the fire start,

Q_{ig} - is the absorbed energy by the burning materials while they are evaporating their water content,

$\frac{\partial I_z}{\partial z}$ - is the gradient of vertical intensity in the plane, where the energy is released.

Horizontal and vertical coordinates are x and z [6]. All parameters are in English system, where the spread rate is in ch/m, the fuels are in lb/ft^3 and lb.

The first step in preparing data to run spatial fire behaviour analyses was to determine suitable fuel models for fire locations in the Zlatograd test area. This was done by using BehavePlus [2]. BehavePlus is a point fire behaviour prediction system that can be used to analyze fire growth and behaviour for homogeneous vegetation with static weather data. Using a number of standard fuel models developed for the United States [1,8], we evaluated which fuel models were best able to produce estimates of fire behaviour and growth in BehavePlus similar to those observed on each of the fifteen fires. In addition to fuel model, BehavePlus requires inputs for weather, fuel moisture, slope, and duration of the burning period. Weather data was obtained for each fire from TV Met, a private company in Bulgaria, which provided calculated fine dead fuel moisture values [7]. The weather stations in the area of interest are quite sparse and that let to estimations for some of the zones of fires. Estimations on live herbaceous and live woody fuel moisture values were based on the expected phrenological stage for the time of year that the fire occurred. To estimate slope, we used 30 m resolution digital elevation model (DEM) from the National Institute of Geophysics, Geodesy, and Geography in Bulgaria, then subsequently calculated the average slope for each fire using standard geospatial processing in ArcGIS (ESRI 2010). Burn period length for each fire was obtained from the Zlatograd forestry department data. Based on initial BehavePlus results using standard fuel models, custom fuel models were developed for some vegetation types not well represented by the US fuel models. Custom fuel models were developed for native durmast oak and grass as well as one of the Scotch pine sites by modifying fuel loading parameters to better match local vegetation and reflect the lack of woody debris in the understory, as it is collected as firewood by the local population. The custom fuel model developed for grass has a much lower rate of spread and flame length than any of the standard grass fuel models. Following evaluation of fuel models with BehavePlus, we then performed analyses in FARSITE, a spatial fire growth system that integrates fire spread models with a suite of spatial data and tabular weather, wind and fuel moisture data to project fire growth and behavior across a landscape. It was defined test landscapes using a 500 m buffer zone around each of the fifteen Zlatograd fires in order to comprise the extent of the spatial analysis for each individual wildfire. Input for FARSITE consists of spatial topographic, vegetation, and fuels parameters compiled into a multi-layered "landscape file" format. Topographic data required to run FARSITE include elevation, slope, and aspect. Using the aforementioned 30 m DEM, we calculated an aspect layer, and then clipped elevation, aspect, and slope rasters to the extent of the fifteen test landscapes. Required vegetation data include fuel model and canopy cover. Fuel models within the 500 m buffered analysis area for each individual fire were assigned based on the BehavePlus analyses; fuel model assignments were tied to the dominant vegetation for each polygon based on the Zlatograd forestry departments vegetation data. Canopy cover values were visually estimated from orthophoto images and verified with stand data from the Zlatograd forestry department. Additional canopy variables (canopy base height, canopy bulk density, and canopy height) that may be included in the landscape file were omitted, as these variables are most important for calculating crown fire spread or the potential for a surface fire to transition to a crown fire. None of the fifteen fires analyzed experienced crown fire. Tabular weather and wind files for FARSITE were compiled using the weather and wind data from TV Met, Bulgarian meteorological company that included hourly records. Tabular fuel moisture files were created using the fine dead fuel moisture values (the wetness in the grass, srubs and small branches, which are on the ground from previous seasons or have been cut) calculated for the BehavePlus analyses for 1-hr timelag fuels. The 10-hr fuel moisture value was estimated by adding 1 percent to the 1-hr fuel moisture and the 100-hr fuel moisture was generally calculated by adding 3 percents to the 1-hr fuel moisture. The live fuel moisture (the wetness in the grass, srubs and trees, which are alive from the current season) values previously estimated for BehavePlus analyses were used to populate live herbaceous and live woody moisture values. All simulations performed in FARSITE used metric data for inputs and outputs. An adjustment value was not used to alter rate of spread for standard fuel models, rather custom fuel models were created. Crown fire, embers from torching trees, and growth from spot fires were not enabled.

4 Conclusion

With the performed simulations we observed that when it comes to Fire Behaviour Fuel Modelling for Bulgarian vegetations the general assumptions which can be taken are as follows:

1. In cases of Scots pine (Pinus sylvestris) the best proxis from the available models from the thirteen of Anderson [1] and forty of Scott-Burgan [8] are 188 and modified 183 (these classes come from 2005 classification).
2. In cases of Black pine/Acacia (Pinus Nigra/Acacia)the best proxis from the available models from the thirteen of Anderson [1] and forty of Scott-Burgan [8] are 161 and 183 with modification (these classes come from 2005 classification).
3. In cases of Beechwood (Fagus sylvatica) the best proxis from the available models from the thirteen of Anderson [1] and forty of Scott-Burgan [8] are 182/186 in cases of dormant season fire and 161 in growing season fire (these classes come from 2005 classification).
4. In cases of Durmast (Quercus dalechampii) the best proxis from the available models from the thirteen of Anderson [1] and forty of Scott-Burgan [8] are 182/186 in cases of dormant season fire and 161 in growing season fire (these classes come from 2005 classification)
5. In cases of Grasslands the best proxis from the available models from the thirteen of Anderson [1] and forty of Scott-Burgan [8] are 101 in cases of grazed pasture 102 in cases of ungrased pasture and Custom FBFMs with lower ROS and fuel load than the 101 (these classes come from 2005 classification).

In rare cases also the available thirteen models from the classification done by Anderson in 1982 are observed as good match, however the propagation of the simulated runs was faster than expected.

Acknowledgements. This work was supported by the National Science Fund of the Bulgarian Ministry of Education and Science under Grant FNI I02/20.

References

1. Anderson, H.E.: Aids to determining fuel models for estimating fire behavior. USDA Forest Service, Intermountain Forest and Range Experiment Station, Research Report INT-122 (1982). http://www.fs.fed.us/rm/pubsint/intgtr122.html
2. Andrews, P.L.: BehavePlus fire modeling system: past, present, and future. In: Proceedings of Seventh Symposium on Fire and Forest Meteorology, 23–25 October 2007, Bar Harbor, ME. American Meteorological Society, Boston (2007)
3. Dobrinkova, N., Hollingsworth, L., Heinsch, F.A., Dillon, G., Dobrinkov, G.: Bulgarian fuel models developed for implementation in FARSITE simulations for test cases in Zlatograd area. In: Wade. D.D., Fox, R.L. (eds.), Robinson, M.L. (comp.) (E-proceeding) Proceedings of 4th Fire Behavior and Fuels Conference, 18–22 February 2013, Raleigh, NC and 1–4 July 2013, St. Petersburg, Russia, pp. 513–521. International Association of Wildland Fire, Missoula (2014)
4. Dobrinkova, N., Jordanov, G., Mandel, J.: WRF-fire applied in Bulgaria. In: Dimov, I., Dimova, S., Kolkovska, N. (eds.) NMA 2010. LNCS, vol. 6046, pp. 133–140. Springer, Heidelberg (2011). ISSN 0302-9743
5. Jordanov, G., Beezley, J.D., Dobrinkova, N., Kochanski, A.K., Mandel, J., Sousedík, B.: Simulation of the 2009 Harmanli fire (Bulgaria). In: Lirkov, I., Margenov, S., Waśniewski, J. (eds.) LSSC 2011. LNCS, vol. 7116, pp. 291–298. Springer, Heidelberg (2012). ISSN 0302-9743
6. Rothermel, R.C.: A mathematical model for predicting fire spread in wildland fuels. Research Paper INT-115. Forest Service, Intermountain Forest and Range Experiment Station, pp. 1–40. US Department of Agriculture, Ogden, UT (1972)
7. Rothermel, R.C.: How to predict the spread and intensity of forest and range fires. Technical Report, USDA Forest Service, Intermountain Forest and Range Experiment Station General Technical Report INT-GTR-143, Ogden, UT (1983)
8. Scott, J.H., Burgan, R.E.: Standard fire behavior fuel models: a comprehensive set for use with Rothermel's surface fire spread model. Technical report, USDA Forest Service, Rocky Mountain Research Station General Technical Report RMRS-GTR-153. Fort Collins, CO (2005)
9. Skamarock, W.C., Klemp, J.B., Dudhia, J., Gill, D.O., Barker, D.M., Duda, M.G., Huang, X.Y., Wang, W., Powers, J.G.: A description of the Advanced Research, WRF version 3. NCAR Technical Note 475 (2008). http://www2.mmm.ucar.edu/wrf/users/docs/arw_v3.pdf

Supervised 2-Phase Segmentation of Porous Media with Known Porosity

Ivan Georgiev, Stanislav Harizanov$^{(\boxtimes)}$, and Yavor Vutov

Institute of Information and Communication Technologies,
Bulgarian Academy of Science, Sofia, Bulgaria
sharizanov@parallel.bas.bg

Abstract. Porous media segmentation is a nontrivial and often quite inaccurate process, due to the highly irregular structure of the segmentation phases and the huge interaction among them. In this paper we perform a 2-class segmentation of a gray-scale 3D image under the restriction that the number of voxels within the phases are a priori fixed. Two parallel algorithms, based on the graph 2-Laplacian model [1] are proposed, implemented, and numerically tested.

1 Introduction

Porous materials are of current interest within a wide range of applications, where their properties strongly depend on various measurements such as absolute porosity, average pore size, size and shape of individual pores. Therefore, accurate segmentation of a 3D reconstruction image of the corresponding specimen is crucial in practice. Due to the highly irregular structure of the segmentation phases and the presence of noise in the image, such a task is nontrivial and sometimes impossible, unless additional information on the data is provided. In particular, the volume (thus, the cardinality) of the solid phase can be determined from its density and weight.

We consider a 2-phase segmentation that satisfies an equality solid phase volume constraint. Graph 2-Laplacian is used for the mathematical model [1–4]. The derived constraint optimization problem is NP-hard [5]. Hence, we propose two different relaxations of the problem that can be efficiently solved. The paper is organized as follows. In Sect. 2, notation is fixed and the 2-Laplacian model is introduced. The two relaxed modifications of the original optimization problem, together with algorithms for solving them, are described in Sect. 3. In Sect. 4, three numerical examples are considered and the different algorithms are compared. Conclusions are drawn in Sect. 5.

2 Mathematical Formulation of the Problem

Let us first give some preliminary definitions and fix the notation. We consider 3D gray-scale images $\bar{u} : \Omega \rightarrow [0, \nu]$, where Ω is a discrete box domain of dimensions n_1, n_2, and n_3, respectively, while ν is the maximal intensity of

© Springer International Publishing Switzerland 2015
I. Lirkov et al. (Eds.): LSSC 2015, LNCS 9374, pp. 343–351, 2015.
DOI: 10.1007/978-3-319-26520-9_38

the image. For a simpler matrix-vector notation, we assume the image to be column-wise reshaped as a vector $\bar{u} \in [0, \nu]^{\mathbf{n}}$, with $\mathbf{n} = \mathrm{card}(\Omega) = n_1 n_2 n_3$. We keep the same notation $\bar{u}$ for the vectorized image and it will be clear from the context which representation we consider. We denote via $\mathbb{I_n} := \{1, \ldots, \mathbf{n}\}$ the voxel index set. The discrete segment membership vector $v \in \{0, 1\}^{\mathbf{n}}$ is used for image segmentation and for every $i \in \mathbb{I_n}$, it indicates to which class the i-th voxel belongs to ("air" if $v(i) = 0$ or "metal" if $v(i) = 1$). The index set is split into two disjoint subsets $\mathbb{I_n} = L \cup U$ of labeled and unlabeled points, respectively. Without loss of generality (after re-numeration) we consider $L = \{1, \ldots, 2\ell\}$, $U = \{2\ell+1, \ldots, \mathbf{n}\}$, and we split $v = (v_L, v_U)^T$. Furthermore, $L = L_0 \cup L_1$, where $L_0 := \{i \in L | v_L(i) = 0\} = \{1, \ldots, \ell\}$, $L_1 := \{i \in L | v_L(i) = 1\} = \{\ell + 1, \ldots, 2\ell\}$.

The indicator function ι_C of a nonempty set C is given by

$$\iota_C(x) = \begin{cases} 0 & \text{if } x \in C, \\ +\infty & \text{otherwise.} \end{cases}$$

Finally, we denote by e the ones vector $(1, \ldots, 1)^T$ of the appropriate dimension.

2.1   Graph 2-Laplacian Model

Starting with some labeled voxels ($L \neq \emptyset$) we want to segment the unlabeled ones, using their similarities/differences to the former and among themselves. Following [1], we do so via minimizing

$$F(v) := \langle \triangle v, v \rangle = \frac{1}{2} \sum_{i,j=1}^{\mathbf{n}} w_{i,j} \big(v(i) - v(j)\big)^2,$$

with respect to v_U, where $\triangle$ denotes the (graph) 2-Laplacian [6,7]

$$(\triangle v)(i) = \sum_{j=1}^{\mathbf{n}} w_{i,j} \big(v(i) - v(j)\big).$$

The weights are chosen similar to [1]. Let $\mathcal{N}_i^{geo} = \{j : \|j - i\|_1 = 1\}$ be the 1-neighborhood of $i \in U$. Then, for $j \in \mathcal{N}_i^{geo}$ we take $w_{i,j}^{geo} = \frac{1}{6}$. Our feature function f is a weighted average of the intensities of the voxel i and its neighbors

$$f(i) = \frac{1}{12}\Big(6\bar{u}(i) + \sum_{j \in \mathcal{N}_i^{geo}} \bar{u}(j)\Big).$$

We use it as a similarity measure to compute the other two types of weights

$$w_{i,j}^{pho} := \begin{cases} a_i e^{-(f(i)-f(j))^2} & \text{if } j \in \mathcal{N}_i^{pho}, \\ 0 & \text{otherwise,} \end{cases} \qquad w_{i,j}^{lab} := \begin{cases} b_i e^{-(f(i)-f(j))^2} & \text{if } j \in L, \\ 0 & \text{otherwise.} \end{cases}$$

The constants a_i, b_i normalize the weights, so that they sum up to 1 within each group. $\mathcal{N}_i^{pho}$ consists of the 4 voxels j in the $5 \times 5 \times 5$ cube $\|i - j\|_\infty \leq 2$ that minimize $|f(i) - f(j)|$. Finally, $W = \max\left\{\tilde{W}, \tilde{W}^T\right\}$, where

$$W^* = \frac{1}{1 + \nu^{\text{pho}}}W^{geo} + \frac{\nu^{\text{pho}}}{1 + \nu^{\text{pho}}}W^{pho}, \quad \tilde{W} = \max\left\{\frac{\nu^{\text{lab}}}{1 + \nu^{\text{lab}}}W^{lab}, \frac{1}{1 + \nu^{\text{lab}}}W^*\right\}.$$

W is non-negative, symmetric. The parameters $\nu^{\text{pho}}, \nu^{\text{lab}}$ are positive. Let

$$W := \begin{pmatrix} W_{LL} & W_{LU} \\ W_{UL} & W_{UU} \end{pmatrix}; \quad D := \text{diag}(d_i)_{i=1}^{n}, \quad d_i := \sum_{j=1}^{n} w_{i,j}, \ \forall i \in \mathbb{I_n}.$$

Note that $W_{UL} = W_{LU}$ due to symmetry and $F(v) = \frac{1}{2}v^T(D - W)v$. W_{UU} is sparse, and (almost) row-normalized via $d_i \approx 1$, $\forall i \in U$. Since $\nu^{\text{lab}} > 0$, $\sum_{j \in L} w_{i,j} > 0$, $\forall i \in U$, $D_{UU} - W_{UU}$ is strictly diagonally dominant with non-positive non-diagonal entries, thus an M-matrix, and the problem

$$\underset{0 \leq v \leq 1}{\text{argmin}} \, F(v) \quad \text{subject to} \quad v_L(i) = \begin{cases} 0, i \in L_0, \\ 1, i \in L_1, \end{cases} \tag{1}$$

admits a unique solution $\bar{v}$, given by (see [1, Theorem 3.2.] for details)

$$\underbrace{(D_{UU} - W_{UU})}_{Q} \bar{v}_U = \underbrace{W_{UL}v_L}_{q}. \tag{2}$$

Since $Q^{-1}, q \geq \mathbf{0}, \mathbf{0} \leq \bar{v} \leq \mathbf{1}$. To ensure $\bar{v} \in \{0,1\}^{\mathbf{n}}$, hard thresholding with respect to the middle value 0.5 is typically used.

Such segmentation methods work fine for well-separated, smooth phases, but their performance is unclear in the presence of big interaction. In (homogeneous) porous media, the "air" consists of multiple, non-structured, possibly not even connected pores of various size and shape, that "cut" through the material. Combined with the inevitable noise and blur the input image possesses, segmentation (2) is often poor and unreliable. In this paper, we assume that the input is a 3D reconstruction of a given specimen, which volume is a priori known. Thus, the number of the solid phase voxels N is given and can be used as a constraint in the mathematical model. We consider the following problem

$$\underset{v \in \{0,1\}^{\mathbf{n}}}{\text{argmin}} \, F(v) \quad \text{subject to} \quad v_L(i) = \begin{cases} 0, i \in L_0, \\ 1, i \in L_1; \end{cases} \quad \|v\|_0 = N. \tag{3}$$

The ℓ_0 pseudo-norm is non-convex and the problem (3) is NP-Hard [5]. In the binary case, $\|v\|_0 = \|v\|_1 = e^T v$, thus we rewrite the problem accordingly

$$\underset{v \in \{0,1\}^{\mathbf{n}}}{\text{argmin}} \, F(v) \quad \text{subject to} \quad v_L(i) = \begin{cases} 0, i \in L_0, \\ 1, i \in L_1; \end{cases} \quad e^T v = N, \tag{4}$$

and apply convex optimization techniques to this convex reformulation.

3   Convex Optimization Algorithms

We propose 2 different algorithms for dealing with (4) - one direct, and one iterative. Both algorithms does not solve (4) itself, but further modifications of the problem. The algorithms' results are compared on several different inputs.

3.1   Equality Constrained Quadratic Optimization

If we forget about $v \in \{0,1\}^n$, denote by $N_1 := N - \ell$, and use the notation from (2), we derive the equality constrained quadratic optimization problem

$$\operatorname*{argmin}_{v_U} \frac{1}{2} v_U^T Q v_U - q^T v_U \quad \text{subject to} \quad e^T v_U = N_1. \tag{5}$$

Thus, the minimizer $\bar{v}_U$ of (5) is the solution of

$$\begin{pmatrix} Q & e \\ e^T & 0 \end{pmatrix} \begin{pmatrix} \bar{v}_U \\ \lambda \end{pmatrix} = \begin{pmatrix} q \\ N_1 \end{pmatrix}, \tag{6}$$

where λ is Lagrange multiplier. If $s := -e^T Q^{-1} e$ is the Schur complement, then (6) can be rewritten as

$$\begin{pmatrix} Q & \\ e^T & s \end{pmatrix} \begin{pmatrix} I & Q^{-1} e \\ & 1 \end{pmatrix} \begin{pmatrix} \bar{v}_U \\ \lambda \end{pmatrix} = \begin{pmatrix} q \\ N_1 \end{pmatrix} \quad \Rightarrow \quad \begin{vmatrix} \lambda & = (N_1 - e^T Q^{-1} q)/s, \\ \bar{v}_U & = Q^{-1} q + \lambda Q^{-1} e. \end{vmatrix}$$

The matrix Q is sparse and positive definite. We use conjugate gradient (CG) method [8] for solving the linear systems $Qx = q$ and $Qy = e$. For the segment membership vector we take $(v_L, \hat{v}_U)^T$, where the N_1 largest elements of $\bar{v}_U$ are set to 1 (in case of equality, randomness is applied), while the rest are set to 0.

Since $(Qe)(i) = \sum_{j \in L} w_{i,j} \approx \frac{\nu^{\text{lab}}}{1+\nu^{\text{lab}}}$ (the value may slightly differ only for close neighbors $i \in U$ of L), we have that $Q^{-1} e \approx \frac{1+\nu^{\text{lab}}}{\nu^{\text{lab}}} e$. Hence $\bar{v}_U$ is basically a shift of the solution of (1), and the segmentation based on (5) coincides with the N-segmentation of (2).

We also consider a slight generalization of (6)

$$\begin{pmatrix} Q - 2\mu I & e \\ e^T & 0 \end{pmatrix} \begin{pmatrix} \bar{v}_U \\ \lambda \end{pmatrix} = \begin{pmatrix} q \\ N_1 \end{pmatrix}, \tag{7}$$

that corresponds to the penalized version

$$\operatorname*{argmin}_{v_U, \lambda} \frac{1}{2} v_U^T Q v_U - q^T v_U + \lambda(e^T v_U - N_1) - \mu(v_U^T v_U - N_1)$$

of the Lagrange formulation of (5). The penalizer $\mu > 0$ aims at sparsifying the solution $\bar{v}_U$, because $0 \leq v_U \leq 1$, together with $e^T v_U \leq N_1$ and $v_U^T v_U \geq N_1$, guarantee $v \in \{0,1\}^n$, $e^T v = N$. In our experiments, we take $\mu = \nu^{\text{pho}}/(3(1 + \nu^{\text{pho}})) = \frac{1}{3} \min_{i \in U} \sum_{j \in L} w_{i,j}$ to assure that $Q - 2\mu I$ remains an M-matrix and CG solver for (7) still converges (fast).

3.2 Fully Constrained Convex ℓ_2-norm Minimization

Let $\mathbf{n_1 = n} - 2\ell$. As in (5), we start by projecting (4) onto U

$$\underset{v_U \in \{0,1\}^{n_1}}{\mathrm{argmin}} \; \frac{1}{2} v_U^T Q v_U - q^T v_U \quad \text{subject to} \quad e^T v_U = N_1.$$

For segment membership vectors $v_U^2 = v_U$, and the problem is equivalent to

$$\underset{v_U \in \{0,1\}^{n_1}}{\mathrm{argmin}} \; \Big\langle \underbrace{(Q - 2\mathrm{diag}(q))}_{\bar{Q}} v_U, v_U \Big\rangle \quad \text{subject to} \quad e^T v_U = N_1. \tag{8}$$

Here $\mathrm{diag}(q)$ is the diagonal matrix, generated by q. Since

$$q_i = \sum_{j \in L} w_{i,j} u_L(j) = \sum_{j \in L_1} w_{i,j} \implies \bar{Q} = \bar{D}_{UU} - W_{UU}.$$

$$\bar{D}_{UU} = \mathrm{diag}(\bar{d}), \quad \bar{d}(i) = \sum_{j \in L} (-1)^{u_L(j)} w_{i,j} + \sum_{j \in U} w_{i,j}, \; \forall i \in U.$$

We consider the following constrained ℓ_2-norm optimization problem:

$$\underset{v_U \in \{0,1\}^{n_1}}{\mathrm{argmin}} \; \|\bar{Q} v_U\|_2^2 \quad \text{subject to} \quad e^T v_U = N_1. \tag{9}$$

The relation between (8) and (9) is given by the Cauchy-Schwarz inequality.

$$\langle \bar{Q} v_U, v_U \rangle \leq \|\bar{Q} v_U\|_2 \|v_U\|_2 = \sqrt{N_1} \, \|\bar{Q} v_U\|_2.$$

For the equality, we used that if $v_U \in \{0,1\}^{n_1}$, $(v_U - e)^T v_U = 0$. Next, we relax both the constraints in (9) so that the problem becomes convex

$$\underset{0 \leq v_U \leq 1}{\mathrm{argmin}} \, \|\bar{Q} v_U\|_2^2 \quad \text{subject to} \quad e^T v_U \geq N_1. \tag{10}$$

If the halfspace $H := \{x \in \mathbb{R}^{n_1} \mid e^T x \geq N_1\}$ does not contain any zeroes of $\bar{Q}$, the minimizer $\bar{v}_U$ lies on its border and satisfies $e^T \bar{v}_U = N_1$. Similar conclusion cannot be drawn for the box constraint, and once again we take the N_1 largest entries of $\bar{v}_U$ to be "metal", while the rest we set as "air".

Following [10], we rewrite (10) into its equivalent form

$$\underset{v_U \in \mathbb{R}^{n_1}, x \in \mathbb{R}^{3n_1}}{\mathrm{argmin}} \; \big\{ \langle 0, v_U \rangle + \iota_H(x_1) + \|x_2\|_2^2 + \iota_{[0,1]^{n_1}}(x_3) \big\} \; \text{s.t.} \begin{pmatrix} I \\ \bar{Q} \\ I \end{pmatrix} v_U = \begin{pmatrix} x_1 \\ x_2 \\ x_3 \end{pmatrix},$$

and apply the alternating direction methods of multipliers (ADMM).

Algorithm (ADMM): Initialization: $q_{1,2,3}^{(0)} = 0$, $x_{1,3}^{(0)} = e$, $x_2^{(0)} = \bar{Q}e$, $\gamma \in (0,1)$.
For $k = 0, 1, \ldots$ repeat until a stopping criterion is reached

1. $v_U^{(k+1)} = (\bar{Q}^T \bar{Q} + 2I)^{-1} \left(\left(x_1^{(k)} - q_1^{(k)} \right) + \bar{Q}^T \left(x_2^{(k)} - q_2^{(k)} \right) + \left(x_3^{(k)} - q_3^{(k)} \right) \right)$

2. $x_1^{(k+1)} = \begin{cases} q_1^{(k)} + v_U^{(k+1)}, & \left(q_1^{(k)} + v_U^{(k+1)} \right) \in H, \\ q_1^{(k)} + v_U^{(k+1)} + \frac{N_1 - e^T \left(q_1^{(k)} + v_U^{(k+1)} \right)}{n_1^2} e, & \text{otherwise.} \end{cases}$

3. $x_2^{(k+1)} = \gamma \left(q_2^{(k)} + \bar{Q} v_U^{(k+1)} \right)$

4. $x_3^{(k+1)} = \min \left(1, \max \left(0, q_3^{(k)} + v_U^{(k+1)} \right) \right)$

5. $q_i^{(k+1)} = q_i^{(k)} + v_U^{(k+1)} - x_i^{(k+1)}$, $i = 1, 3$, $q_2^{(k+1)} = q_2^{(k)} + \bar{Q} v_U^{(k+1)} - x_2^{(k+1)}$.

Step 1 is solved (implicitly) via parallelized CG, since $\bar{Q}^T \bar{Q} + 2I$ is sparse and positive definite. Steps 3–5 are all componentwise, thus are parallelized, too. The complete splitting of the constraints, due to the introduction of $\langle 0, v_U \rangle$ in the cost function, leads to fast convergence rate of the algorithm.

4   Numerical Examples

In this section, we demonstrate by numerical examples the performance of our algorithms, implemented in C++. The code is parallelized using OpenMP [11]. The assembly of the matrix, matrix-vector, and vector operations is distributed among the available threads. We have tested artificially polluted bone part, based on [9]; a real 3D reconstruction of an Aluminum (AlSi10Mg) metal foam, obtained via industrial CT scan; and a binary image of a sphere inside a unit cube. We take $\ell = 3$, where L_0 (L_1) consists of the indices of the three voxels that minimize (maximize) f. When computing the weights, we use mirror boundary conditions. For the ADMM algorithm, we set $\gamma = 0.3$. In all the examples, the CG method converges fast, so no preconditioning is needed/used. Different segment vectors $v_{1,2} \in \{0,1\}^n$ are compared via both their voxel difference $\|v_1 - v_2\|_1$ and their 2-sided Hausdorf distance, based on the 3D sup-norm $\| \cdot \|_\infty$.

The bone part image has size $64 \times 64 \times 64$. 50604 of its voxels are bone material (porosity 80.7 %). The image was convoluted with a Gaussian kernel

Fig. 1. From left to right: Segmented bone part (binary image), noisy and blurry input image $\bar{u}$, direct N-segmentation, segmentation based on (10).

Fig. 2. 3 slices of the Aluminum foam reconstruction (left) and its segmentation via (10) (right).

($\sigma = 2$). Then, 10 % white (Gaussian) noise was added to derive the input image $\bar{u}$ from Fig. 1. For the weight matrix W we used $\nu^{\text{pho}} = 10$, $\nu^{\text{lab}} = 0.1$.

As shown on Fig. 1 and in Table 1, simply taking the 50604 voxels of $\bar{u}$ highest intensity as the solid phase leads to poor segmentation, both visually and quantitatively. Our algorithms perfectly denoise $\bar{u}$, but are not able to completely overcome its blur. This results in thickening parts of the bone structure at the expense of loosing structure information elsewhere. Unlike the original image, the segmented one is not connected.

The AlSi10Mg foam reconstruction has size $680 \times 680 \times 680$ with sampling distance (voxel size) 0.0272 mm. The specimen has cylindrical shape with diam=14.94 mm and height=16.55 mm (see Fig. 2). Its weight is 2.7070g and $\rho(AlSi10Mg) = 2.6687g/cm^3$. Porosity is computed to be 83.97 %, thus $N = 50405948$. Different segmentations are compared in Table 2. The ADMM algorithm converges fast (less than 70 iterations as shown on Fig. 3) as well as the CG solver within each step (less than 15 iterations per time).

With our last example (Fig. 4) we want to stress that our segmentation algorithms may still produce meaningful results when N differs from the real porosity. The input image is $64 \times 64 \times 64$ big, and the centered sphere takes 28 % of its volume. We execute our algorithms with $N = 131072$ (half the volume). All the outputs are simply connected and visually resemble the original object.

Table 1. Comparison among the original bone, its direct segmentation, and our algorithms. Above the diagonal: voxel difference, below: 2-sided Hausdorff distance.

hausd\\#voxels	Original	Direct	λ-QP (6)	$\mu\lambda$-QP (7)	ℓ_2-CP (10)
Original	*	21 796	16 048	16 156	15 524
Direct	25 547	*	10 476	10 412	11 346
λ-QP (6)	19 417	14 218	*	278	1 486
$\mu\lambda$-QP (7)	19 601	14 182	289	*	1 722
ℓ_2-CP (10)	18 857	15 140	1 509	1 764	*

Table 2. Comparison among different segmentations of the Aluminum metal foam. Above the diagonal: voxel difference, below: 2-sided Hausdorff distance.

hausd\#voxels	Direct	λ-QP (6)	$\mu\lambda$-QP (7)	ℓ_2-CP (10)
Direct	*	2 789 328	2 850 756	2 659 200
λ-QP (6)	2 847 631	*	128 492	378 686
$\mu\lambda$-QP (7)	2 917 305	130 926	*	493 402
ℓ_2-CP (10)	2 702 613	384 633	503 522	*

5   Conclusions

Based on the 2-Lagrangian model in [1], we proposed two different relaxations for 2-phase image segmentation of a porous media of known porosity. The algorithms are implemented in a parallel way, allowing us to work with large 3D images of high resolution. The conducted numerical experiments showed significantly improved results, compared to non-supervised N-constrained segmentation. The outputs, based on (6), (7) and (10) are quite similar and there are no visible differences among them. Assuring connectivity of the solid phase is not achieved at this moment, and remains a task for future work.

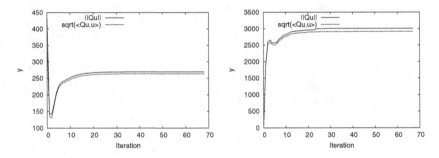

Fig. 3. Relation between (8) and (9): The graphs of $\sqrt{\langle \bar{Q}v_U^{(k)}, v_U^{(k)} \rangle}$ and $\|\bar{Q}v_U^{(k)}\|_2$ from the ADMM algorithm as functions of k. Left: Bone part. Right: Aluminum foam.

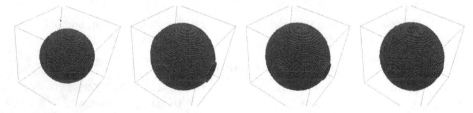

Fig. 4. From left to right: Original, segmented sphere inside a cube (volume $= 28\,\%$). The output of (6), (7) and (10), respectively, for volume $= 50\,\%$ of the cube's.

Acknowledgements. The research is partly supported by the project AComIn "Advanced Computing for Innovation", grant 316087, funded by the FP7 Capacity Program.

References

1. Kang, S.H., Shafei, B., Steidl, G.: Supervised and transductive multi-class segmentation using p-Laplacians and RKHS methods. J. Vis. Commun. Image Represent. **25**(5), 1136–1148 (2014)
2. Shi, J., Szalam, J.: Normalized cuts and image segmentation. IEEE Trans. Pattern Anal. Mach. Intell. **22**(8), 888–905 (2000)
3. von Luxburg, U.: A tutorial on spectral clustering. Stat. Comput. **17**(4), 395–416 (2007)
4. Law, N.Y., Lee, H.K., Ng, M.K., Yip, A.M.: A semisupervised segmentation model for collections of images. IEEE Trans. Image Process. **21**(6), 2955–2968 (2012)
5. Natarajan, B.K.: Sparse approximate solutions to linear systems. SIAM J. Comput. **24**(2), 227–234 (1995)
6. Amghibech, S.: Eigenvalues of the discrete p-Laplacian for graphs. Ars Comb. **67**, 283–302 (2003)
7. Bühler, T., Hein, M.: Spectral clustering based on the graph p-Laplacian. In: Proceedings of the 26th Annual International Conference on Machine Learning, pp. 81–88 (2009)
8. Axelsson, A.: Iterative Solution Methods. Cambridge University Press, Cambridge (1994)
9. Beller, G., Burkhart, M., Felsenberg, D., Gowin, W., Hege, H.-C., Koller, B., Prohaska, S., Saparin, P., Thomsen, J.: Vertebral Body Data Set ESA29-99-L3. http://bone3d.zib.de/data/2005/ESA29-99-L3/
10. Teuber, T., Steidl, G., Chan, R.H.: Minimization and parameter estimation for seminorm regularization models with I-divergence constraints. Inverse Prob. **29**, 1–28 (2013)
11. Dagum, L., Menon, R.: OpenMP: an industry-standard API for shared-memory programming. IEEE Comput. Sci. Eng. **5**(1), 46–55 (1998)

Image Processing Methods in Analysis of Component Composition and Distribution of Dust Emissions for Environmental Quality Management

Andrew Kokoulin[1]([✉]), Irina May[2], and Anastasija Kokoulina[2]

[1] Perm National Research Polytechnic University, 614000 Perm, Russian Federation
a.n.kokoulin@at.pstu.ru
http://pstu.ru
[2] Federal Scientific Center for Medical and Preventive Health Risk
Management Technologies, 614000 Perm, Russian Federation

Abstract. In this article we consider a novel fast approach based on wavelet transform for edge detection and simplified variant of active contour method - active primitives for the image processing in the research on dust emissions from industrial enterprises. The objective of the dust emissions analysis is to determine their component composition and the fine particle size distribution (PM10 and PM2.5). The scanning electronic microscope of high resolution was used to obtain the large-scale images of dust particles. We use the images of particles in different scales as the entire imagery data.

Keywords: Image segmentation · Active contour · Edge detection · Dust emissions · PM10 · PM2.5

1 Introduction

Dust emissions of industrial plants contain a considerable quantity of different chemical components. These include ash, soot, smoke, sulfates, nitrates, oxides of metals and other solid components. The size of dust particles emitted by enterprises are defined by the process environment and the composition of raw materials used. Particle fractions less than 10 microns (PM10) and less than 2.5 microns (PM2.5) can penetrate to the upper and lower respiratory tract. Both short and long-term exposures of fine dust particles on human health leads to respiratory and cardiovascular diseases [1–3].

Our research center uses the following equipment to study the fractional composition of dusts: - laser particle analyzer Microtrac S3500 (covering the particle size range from 20 nm to 2000 microns) to determine the distribution of dust emissions. - scanning electronic microscope of high resolution (the degree of increase - from 5 to 300 000 times) with X-ray fluorescence attachment S3400N HITACHI for microscopy of dust to determine the particle shape and the component composition of dust emission. Results of microscopy are obtained as a series of images of dust specimen with different scales as shown on the Fig. 1.

© Springer International Publishing Switzerland 2015
I. Lirkov et al. (Eds.): LSSC 2015, LNCS 9374, pp. 352–359, 2015.
DOI: 10.1007/978-3-319-26520-9_39

Fig. 1. Tiled microscope images of sample dust probe.

Automation of morphological analysis of dust particles and study of component composition and distribution of dust emissions is the primary task of our laboratory of environmental quality management. This analysis consists of segmentation problem and feature extracting of imagery data. The image segmentation is the basic task defining the quality of whole processing. Original datasets can be large-scaled, so the efficient and fast image processing algorithms required. There is a great amount of papers, methods and software on the image recognition problem and many of them offers good approaches suitable for any particular case. In our research we tried to combine several methods to operate in one toolchain for our subject: processing of dust particles images in multiple scales. We tried to make this toolchain more efficient to operate with large-scale images and made some enhancements resulting in a good computing rate. We consider a fast image processing approach based on wavelet transform for edge detection and novel simplified variant of active contour method (active primitives) for objects enumeration on microscopy image set.

2   Image Analysis: Active Contour Model

Active contour methods have found application in a wide range of problems including visual tracking and image segmentation. The basic idea is to allow a contour (a snake) to evolve so as to minimize a given energy functional in order to produce the desired segmentation. An initial model of active contour was proposed by Kass et al. [7] and named snakes suitable to the appearance of contour evolution. Solving the problem of snakes is to locate the contour C that minimize the total energy term E with the certain set of weights α, β and λ. Two main categories exist for active contours: edge-based and region-based. The complete 'state of art' on active contours is given in [4] with explanation of advantages and disadvantages of all encountered methods.

There are many advantages of region-based approaches when compared to edge-based methods including robustness against initial curve placement and insensitivity to image noise [5]. However, techniques that attempt to model regions using global statistics are usually not ideal for segmenting heterogeneous objects. In cases where the object to be segmented cannot be easily distinguished in terms of global statistics, region-based active contours may lead to erroneous segmentations [6]. Consider the image in Fig. 1. Here, we see a situation when the foreground and background are heterogeneous and share nearly the same statistical model. The construction of this image causes it to be segmented improperly by a standard region-based algorithm, but correctly by an edge-based algorithm.

Theory and practice of active contours. Snake parametric representation [7]: v(s)=(x(s),y(s))

$$E_{snake} = \int_0^1 E_{int}(v(s)) + E_{image}(v(s)) + E_{con}(v(s))ds \qquad (1)$$

E_{int} is internal energy due to bending. Serves to impose piecewise smoothness constraint.

E_{image} are image forces pushing the snake toward image features (edges).

E_{con} are external constraints responsible for putting the snake near the desired local minimum. It may come from higher level interpretation, user interaction, etc.

E_{int} The snake is a controlled continuity spline

$$E_{int} = (\alpha(s)|v_s(s)|^2 + \beta(s)|v_{ss}(s)|^2)/2 \qquad (2)$$

Derivative $v_s(s)$ makes the spline act like a membrane (elasticity, E_{cont}).

Derivative $v_{ss}(s)$ makes it act like a thin-plate (rigidity,E_{bal}). $\alpha(s)$ and $\beta(s)$ controls the relative importance of membrane and thin-plate terms

E_{image} Attracts the snake to features (data term)

$$E_{image} = w_{line}E_{line} + w_{edge}E_{edge} + w_{term}E_{term} \qquad (3)$$

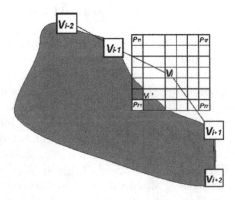

Fig. 2. Shake evolution: estimating V_i.

E_{line}: in practice simplest functional is the image intensity: $E_{line} = I(x, y)$ Depending on the sign of w_{line}, the snake will be attracted to the lightest or darkest nearby contour

Edges attracts the snake to large intensity gradients: we can set $E_{edge} = -|\bigtriangledown I(x, y)|$

E_{term} Attracts the snake toward termination of line segments and corners

Combining E_{edge} and E_{term}, we can create a snake attracted to edges and terminations. The shape of the snake between the edges and lines in the illusion is completely determined by the spline smoothness term (Fig. 2).

In practice we apply the discrete curve approximation and use any of optimization methods to estimate (1) using the equations shown in [8].

Each equation is evaluated on every iteration of contour evolution, leading to long time durations of segmentation process. In the 3rd section we will consider the novel approach of active contour definition and life cycle providing high-speed processing of microscopy image sets of dust probes.

3   The Proposed Segmentation Method: Active Primitives

Our first ideas of active contour method modification were inspired by the article of Yankowitz et al. [9]. This method was proposed in 1988; it provides good results when we need to take into account spatial variations due to uneven background and illumination conditions, i.e. conditions as in our particular case, see Fig. 1. The method uses the gradient map of the image to point at well-defined portions of object boundaries in it. Both the location and gray levels at these boundary points make them a good choice for local threshold. These point values are then interpolated, yielding the threshold surface (see Fig. 3). This idea can be adopted to be used in active contours processing. First of all we can significantly reduce the computations by elimination of several terms in (1). We can use the edge detection results or gradient maps to start the snake from and to stop on. Also we can substitute the contour of varying shape with the series of independently resizing fixed-shaped figures (primitives) with starting points, anchored to edge curves. These shapes can be spirals or circles (bubbles) with growing radius and fixed decay and segments number. Primitive shapes can grow only in one direction according to edges map and the intense map. The initial size is chosen to be two pixels, anchored to the edge line. The benefit is elimination of E_{cont}, E_{bal} and E_{edge}:

- E_{cont} is the contour energy. The minimum value of the $E_{cont}(v_i)$ elements corresponds to the $p_{jk}(v_i)$ closest to a perfect circle, passing through two adjacent vertices v_{i-1} and v_{i+1}. Assuming the elementary spiral (or bubble) has constant shape, the significance of term E_{cont} is obsolete;
- E_{bal} makes the contour grow in defined direction. Elementary shapes grow in one direction, defined by normal to the edge line. So E_{bal} is insignificant;
- E_{edge} is the gradient energy matrix. Gradient map is undoubtedly useful, but we already use it as the starting and termination points for our elementary shapes.

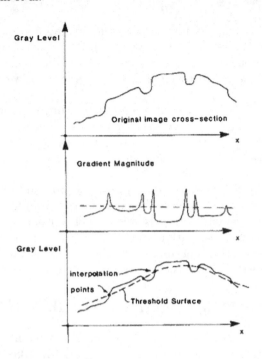

Fig. 3. Description of the process of determining the adaptive threshold surface: (top) cross section in the original image, showing objects on an uneven backgroung, (middle) cross section of the gradient magnitude image, (bottom) peak points in the gradient point at interpolation values, that determine the threshold surface

E_{line} is the only matrix to take into account.

Imagery Data Preprocessing: Contrast Adjustment and Noise Reduction. Noise reduction and smoothing of image set is performed by Gaussian filter. Gaussian filter is a filter which impulse response is a Gaussian function (or an approximation to it). Gaussian filters do not overshoot to a step function input while minimizing the rise and fall time. These methods were chosen and their parameters were adjusted after the series of experiments with the microscopy images for our particular task. Other environments may require their own experimental or theoretical justification.

Edge detection is performed using wavelet transform. If we keep the details of the image obtained with simple Haar transform, remove the coarse-grained low frequency component and perform image reconstruction, we obtain the edges of the objects present in the image. The result is stored on separate image layer. We recommend using sequential transform of original image and transform of diagonally-oriented image (H3 and H4 high-pass blocks on Fig. 4(b)) and their combination since the smooth diagonal edges are hidden otherwise. Also we can use another wavelets such as LeGall 5/3 with fast wavelet transform due to

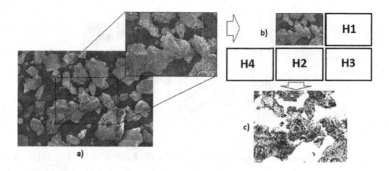

Fig. 4. Wavelet-based edge detection: (a) every image in series is photo of the same surface but with different scale; (b) low-pass and high-pass filtered image; (c) resulting edge map.

elimination of floating-point multiplications as shown by Huing J.K. in A Parallel Algorithm for the Biorthogonal Wavelet Transform Without Multiplication [10].

After the edge detection we can use the dilate/erode combination to produce more continuous edges instead of dotted edge. All the images are ordered by scale. Each image in series is processed starting with topmost and then combined with the lowest (see Fig. 4(a)). Image with a highest resolution is a central tile of lower resolution image, thus we can process them separately and then combine the edge maps. Finally we obtain the edge map of the lowest image. We can expect that at least edges in it's center to be well handled.

Clusterization for pyramidal structure is perfomed in similar manner: from topmost image down to low-quality image, combining the results. The resulting image is used for edges classification and correction. We have to distinguish the background clusters, the particle clusters and intermediate ones (gray shades which belongs to both background or particles depending on unusual texture). All the edges (or parts of the edge lines) have to be classified as internal, thresholding or external, according to intense map: internal, when the particle points are situated on both sides of the edge; external - if background is on both sides; threshold - if background is on one side and particle points on another. If the edge is not continuing we have to find nearest edge according to adjustable parameter of maximum gap. Also we have to use the adjustable parameter minimum particle size to eliminate noise and textures influence.

Active primitives (Spiral contours) combines all the spirals owned by the closed contour. Spiral shapes have different radius depending on particle size, the roundness of contour and texture features. As shown on Fig. 5. the shape grows in normal direction while the area under the spiral consists of points of particle (inside) or intermediate clusters, and stops when it reaches the edge or outside clusters. Finally we get the closed areas covered with spirals. Sometimes afterwards we have to fill holes. If we use an image representation called the integral image,

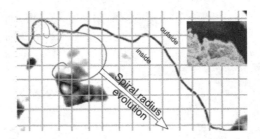

Fig. 5. Spiral evolution: starting from edge points growing until other edge is crossed.

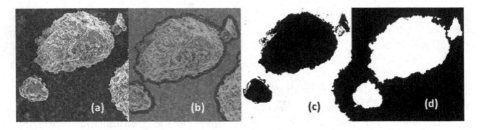

Fig. 6. Segmentation results: Fragment of original image (a); masks obtained by Fiji plugins (b) and (c); mask obtained by active primitives (d).

intense can be evaluated in constant time, and this will give a considerable speed advantage.

The remaining processing: analysis of particle sizes distribution, particles classification is performed using well-known algorithms with the obtained particle contours (masks) (Fig. 6).

4   Results

Test measurements have shown the better rate results compared with same processing using Fiji toolboxes with satisfactory segmentation quality. The image set of 4 grayscale photos 2950 × 1790 was segmented by our active primitives method in 180 s median time, which is at least two times better in comparison with active contours method. The quality of segmentation is the same (each particle was found, 5–10 %% of particles were glued due to overlapping).

5   Conclusion

A fast image processing approach based on wavelet transform for edge detection and novel simplified variant of active contour method for objects enumeration on microscopy image set was considered. The fast segmentation method of active primitives presented in this paper was tested to be successful. The model is not based on single closed active contour (snake) but on sequence of growing spirals or circles - active primitives. This fact eliminates several terms in active contour equations. Future work is to adopt this method to work with very smooth or noisy images and to develop a strict mathematical model of this method.

References

1. Jimoda, L.A.: Effects of particulate matter on human health, the ecosystem, climate and materials: a review. Facta Univ. Ser. Work. Living Environ. Prot. **9**(1), 27–44 (2012)
2. Cormier, S., Lomnicki, S., Backes, W., Dellinger, B.: Origin and health impacts of emissions of toxic by-products and fine particles from combustion and thermal treatment of hazardcomponentous wastes and materials. Environ. Health Perspect. **114**(6), 810–817 (2006)
3. May, I.V., Zagorodnov, S.Y., Maks, A.A.: Fractional and component composition of dust in the working area of machine building enterprise. Occup. Med. Ind. Ecol. **12**, 12–16 (2012)
4. Baswaraj, D., Govardhan, A., Premchand, P.: Active contours and image segmentation: the current state of art. Glob. J. Comput. Sci. Technol. **12**(1), 1–12 (2012). https://globaljournals.org/GJCST_Volume12/1-Active-Contours-and-Image-Segmentation.pdf
5. Chan, T., Vese, L.: Active contours without edges. IEEE Trans. Image Process. **10**, 266–277 (2001)
6. Lankton, S., Tannenbaum, A.: Localizing region-based active contours. IEEE Trans. Image Process. **17**(11), 2029–2039 (2008)
7. Kass, M., Witkin, A., Terzopolous, D.: Snakes: active contour models. Int. J. Comput. Vis. **1**, 321–331 (1988)
8. Petrov, V., Privalov, O.: The modification of active contour algorithm for the interactive segmentation of the raster images of foundry defects. Mod. Prob. Sci. Educ. **6**, 14–19 (2008)
9. Yanowitz, S., Bruckstein, A.: A new method for image segmentation. Comput. Vis. Graph. Image Process. **46**, 82–95 (1989)
10. Hyung, J.K.: A parallel algorithm for the biorthogonal wavelet transform without multiplication. In: Liew, K.-M., Shen, H., See, S., Cai, W., Fan, P., Horiguchi, S. (eds.) PDCAT 2004. LNCS, vol. 3320, pp. 297–300. Springer, Heidelberg (2007)

Fully Implicit Time-Stepping Schemes for a Parabolic-ODE System of European Options with Liquidity Shocks

Miglena N. Koleva[✉] and Lubin G. Vulkov

Faculty of Natural Science and Education, University of Ruse,
8 Studentska Street, 7017 Rousse, Bulgaria
{mkoleva,lvalkov}@uni-ruse.bg

Abstract. We consider the numerical valuation of European options in a market subject to liquidity shocks. Natural boundary conditions are derived on the truncated boundary. We study the fully implicit scheme for this market model, by use of different algorithms, based on the Newton and the Picard iterations at each time step. To validate the efficiency of the time-stepping and the theoretical results, various appropriate numerical experiments are performed.

1 Introduction

We are interested in implementation of implicit finite difference schemes for solving the following system of coupled PDE and ODE which is suggested by M. Ludkowski and Q. Shen [5] in European option pricing with liquidity shocks:

$$
\begin{aligned}
&p_t + \frac{1}{2}\sigma^2 S^2 p_{SS} - \frac{\nu_{01}}{\gamma}\frac{F_1}{F_0}e^{-\gamma(q-p)} + \frac{d_0 + \nu_{01}}{\gamma} - \frac{1}{\gamma}\frac{F_0'}{F_0} = 0, \\
&q_t - \frac{\nu_{10}}{\gamma}\frac{F_0}{F_1}e^{-\gamma(p-q)} + \frac{\nu_{10}}{\gamma} - \frac{1}{\gamma}\frac{F_1'}{F_1} = 0,
\end{aligned}
\tag{1}
$$

with terminal conditions

$$
p(T,S) = q(T,S) = h(S), \quad S > 0.
\tag{2}
$$

Here $p(t,S)$ and $q(t,S)$ are indifference buyer's prices and depend on the current market of the underlying asset, S, and the remaining time t, $0 < t \leq T$, $h(S)$ denotes the terminal payoff of a contingent claim, σ is the volatility of the underlying, ν_{01}, ν_{10} are transition intensities, $d_0 = \mu^2/2\sigma^2$ and μ is a drift rate. The functions $F_0(t), F_1(t)$ are given by

$$
F_0(t) = c_1 e^{\lambda_1 t} + c_2 e^{\lambda_2 t},
$$

$$
F_1(t) = \frac{1}{\nu_{01}}\left(c_1(d_0 + \nu_{01} - \lambda_1)e^{\lambda_1 t} + c_2(d_0 + \nu_{01} - \lambda_2)e^{\lambda_2 t}\right),
$$

where $\quad \lambda_{1,2} = \frac{1}{2}\left(d_0 + \nu_{01} + \nu_{10} \pm \sqrt{(d_0 + \nu_{01} + \nu_{10})^2 - 4d_0\nu_{10}}\right),$

$$
c_1 = \frac{\lambda_2 - d_0}{\lambda_2 - \lambda_1}e^{-\lambda_1 T} \quad \text{and} \quad c_2 = \frac{\lambda_1 - d_0}{\lambda_1 - \lambda_2}e^{-\lambda_2 T}.
$$

© Springer International Publishing Switzerland 2015
I. Lirkov et al. (Eds.): LSSC 2015, LNCS 9374, pp. 360–368, 2015.
DOI: 10.1007/978-3-319-26520-9_40

The presence of liquidity shocks is a source of non-traded risk and makes the market uncomplete. M. Ludkowski and Q. Shen investigated expected utility maximization with exponential utility function $U(x) = e^{-\gamma x}$, where γ is the investor's risk aversion parameter. Standard stochastic control methods and the properties of the exponential utility function imply that the value functions $\widehat{U}^i$, $i = 1, 2$ can be presented by

$$\widehat{U}^1(t, X, S) = e^{-\gamma X} e^{-\gamma p(t, S) + \ln F_0(t)}, \qquad \widehat{U}^2(t, X, S) = e^{-\gamma X} e^{-\gamma q(t, S) + \ln F_1(t)},$$

where X is the wealth process [5].

The well-posedness of the problem (1)–(2) is discussed in [3]. Also, it is proved a minimum principle for the Cauchy problem (1)–(2): *the functions p(t,S) and q(t,S) are non-negative if the payoff h(S) $\geq$ 0*. Numerical solution of the corresponding to (1)–(2) stochastic problem is discussed in [7]. Two implicit-explicit (IMEX) schemes that reproduce the positivity of the differential problem solution are constructed and analyzed in [8]. In the present paper we discuss a fully implicit difference scheme for (1)–(2).

Our article is structured as follows: in Sect. 2, using changes of independent and dependent variables, we obtain a simpler formulation of the main problem (1)–(2). The implicit difference scheme is detailed in Sect. 3. Newton's and Picard's iterative procedure for the solution of the corresponding non-linear system of algebraic equations are described in Sect. 4. To illustrate the applicability of the implicit scheme and to compare the techniques for treating the exponential non-linearities, in Sect. 5 we present results of various experiments. Finally, in the last section some conclusions are formulated.

2 The Auxiliary Differential Problem

By making the substitutions

$$\tau = T - t, \quad u = \gamma p - \ln F_0(t), \quad v = \gamma q - \ln F_1(t), \tag{3}$$

the system (1) becomes

$$\begin{aligned}
L^p(u, v) &\equiv u_\tau - \frac{1}{2}\sigma^2 S^2 u_{SS} + a e^u e^{-v} - b = 0, \\
L^0(u, v) &\equiv v_\tau + c e^v e^{-u} - c = 0,
\end{aligned} \tag{4}$$

where $a = \nu_{01}, b = d_0 + \nu_{01}, c = \nu_{10}$. According to (2) and (3), taking into account that $F_0(T) = F_1(T) = 1$, we set the initial conditions

$$u(0, S) = \gamma h(S) = u^0(S), \quad v(0, S) = \gamma h(S) = v^0(S). \tag{5}$$

Denoting by E the exercise price, in the case of European Call option, the following terminal condition is imposed

$$h(S) = \max(S - E, 0). \tag{6}$$

The functions $u(t, S)$ and $v(t, S)$ can take negative values even the initial data $u^0(S)$, $v^0(S)$ is prescribed to be non-negative. However, it is proved in [3] that, if the terminal conditions $p(T, S)$ and $q(T, S)$ are non-negative, then the prices $p(t, S)$ and $q(t, S)$ remain non-negative for $0 \leq t \leq T$. Now, from the comparison principle for (p, q), see Corollary 2.3 in [3], follows the comparison principle for (u, v):

Proposition 1. Let $(\overline{u}, \overline{v}), (\underline{u}, \underline{v}) \in C([0, T) \times (0, +\infty)) \cap C^{2,1}((0, T) \times (0, +\infty))$ be two pairs of classical solutions of (4)–(5) corresponding to the initial data $h = \overline{h}$ and $h = \underline{h}$ and the following inequalities are fulfilled

$$L^p(\overline{u}, \overline{v}) \geq L^p(\underline{u}, \underline{v}), \quad L^0(\overline{u}, \overline{v}) \geq L^0(\underline{u}, \underline{v}) \quad and \quad \overline{h} \geq \underline{h},$$

then $\overline{u} \geq \underline{u}, \quad \overline{v} \geq \underline{v}$.

Further we develop fully implicit scheme to solve the coupled semi-linear parabolic-ordinary problem (4)–(6).

We obtain the left boundary condition for u, taking $S = 0$ in the first equation in (4). At right boundary, we use the linear (natural) conditions

$$u(\tau, S) = u_1(\tau)S + u_2(\tau), \quad v(\tau, S) = v_1(\tau)S + v_2(\tau), \tag{7}$$

applying the common financial assumption that $p_{SS} \to 0$, $q_{SS} \to 0$ and therefore $u_{SS} \to 0$, $v_{SS} \to 0$ as $S \to \infty$, see [6].

Note that from the payoff (5), (6) we have $u_1(0) = v_1(0) = \gamma$, $u_2(0) = v_2(0) = -\gamma E$ for $S \to \infty$. Next, substituting (7) into (4) and differentiating the resulting equations two times with respect to S and then (after determination u_1 and v_1) solving the ODE system of equations for u_2 and v_2 we get for $S \to \infty$

$$
\begin{aligned}
u(\tau, S) &= \gamma(S - E) + b\tau - \frac{aB}{Q} \ln |e^{Q\tau} - D| - \frac{aA}{Q} \ln |1 - De^{-Q\tau}| + \ln |1 - D|, \\
v(\tau, S) &= \gamma(S - E) + c\tau - \frac{c}{Q}\left(\frac{\ln |Be^{Q\tau} + P|}{B} - \frac{\ln |B + Pe^{-Q\tau}|}{A} + G \ln |B + P|\right),
\end{aligned}
\tag{8}
$$

where $\rho = (c - b)/(2a)$, $M = \rho^2 + c/a$, $A = \sqrt{M} + \rho$, $B = \sqrt{M} - \rho$, $Q = 2a\sqrt{M}$, $D = (1 - B)/(1 + A)$, $P = AD$, $G = (A + B)/(AB)$.

3   The Discrete Problem

In this section we present a second order accurate in space and first order accurate in time fully implicit difference scheme for the problem discussed in Sect. 2. There are many numerical schemes to solve non-linear parabolic equations, however, very few dealt with exponential non-linear term.

The problem (4)–(8) will be solved on truncated, large enough computational interval $[0, L]$, $L > 0$. We define uniform mesh in space and time $\overline{\omega}_{S\tau} = \overline{\omega}_S \times \overline{\omega}_\tau$:

$$
\begin{aligned}
\overline{\omega}_S &= \{S_i = ih, \quad i = 0, \dots, N, \quad h = L/N\}, \\
\overline{\omega}_\tau &= \{\tau^n = n\Delta\tau, \quad n = 0, \dots, N_\tau, \quad \Delta\tau = T/N_\tau\}.
\end{aligned}
$$

The numerical solutions of (4)–(8) at point (τ^n, S_i) are denoted by $U_i^n = U(\tau^n, S_i)$ and $V_i^n = V(\tau^n, S_i)$. For clarity of the exposition we use the notations $\widehat{W}_i := W_i^{n+1}$, $W_i := W_i^n$, $W_{t_i} = (\widehat{W}_i - W_i)/\triangle\tau$, $W_{\overline{S}S,i} = (W_{i+1} - 2W_i + W_{i-1})/h^2$ for the derivative approximations of the mesh function W_i^n [9]. On the discrete domain $\overline{\omega}_{S\tau}$ we approximate the problem by fully implicit finite difference scheme

$$L^p(\widehat{U}_i, \widehat{V}_i) = U_{t_i} - \frac{1}{2}\sigma^2 S_i^2 \widehat{U}_{\overline{S}S,i} + ae^{\widehat{U}_i}e^{-\widehat{V}_i} - b = 0,$$
$$L^0(\widehat{U}_i, \widehat{V}_i) = V_{t_i} + ce^{-\widehat{U}_i}e^{\widehat{V}_i} - c = 0, \quad i = 0, \ldots, N-1, \quad n = 0, \ldots, N_\tau, \quad (9)$$
$$U_i^0 = u^0(S_i), \quad V_i^0 = v^0(S_i), \quad i = 0, \ldots, N,$$
$$\widehat{U}_N = u(\tau^{n+1}, L), \quad \widehat{V}_N = v(\tau^{n+1}, L) \text{ given by } (8), \quad n = 0, \ldots, N_\tau.$$

We have the following convergence result at the pointwise level:

Theorem 1 (Convergence). *Let u, $v \in C((0,L)\times[0,T))\cap C^{4,2}((0,L)\times(0,T))$ are classical solutions of (4)–(6), (8) and U, V are solutions of (9). Then for sufficiently small h and $\triangle\tau$ the following error estimate holds:*

$$\|u - U\|_\infty + \|v - V\|_\infty \leq C(\triangle\tau + h^2),$$

where the constant C doesn't depend on h and $\triangle\tau$.

The following assertion is a discrete analog of Theorem 2.1 in [3]:

Theorem 2 (Comparison principle). *Let the assumptions of Theorem 1 are fulfilled, $(\overline{U}, \overline{V})$, $(\underline{U}, \underline{V})$ are grid functions, defined on $w_{S\tau}$ and*

$$L^p(\overline{U}, \overline{V}) \geq L^p(\underline{U}, \underline{V}), \quad L^0(\overline{U}, \overline{V}) \geq L^0(\underline{U}, \underline{V}),$$
$$\overline{U}_0^0 \geq \underline{U}_0^0, \quad \overline{V}_i^0 \geq \underline{V}_i^0, \quad i = 0, \ldots, N,$$
$$\overline{U}_0^n \geq \underline{U}_0^n, \quad \overline{U}_N^n \geq \underline{U}_N^n, \quad \overline{V}_0^n \geq \underline{V}_0^n, \quad n = 0, \ldots, N_\tau.$$

Then for sufficiently small h and $\triangle\tau$ we have

$$\overline{U}_i^n \geq \underline{U}_i^n, \quad \overline{V}_i^n \geq \underline{V}_i^n, \quad i = 0, \ldots, N, \quad n = 0, \ldots, N_\tau.$$

4   Techniques for Treating the Non-linearities

In order to compute the numerical solution, we discuss different iteration methods for solving the system of non-linear algebraic equations (9).

Suppose that the solutions of (9) after k iterations are denoted by $\widehat{U}_i^{(k)}$ and $\widehat{V}_i^{(k)}$, $i = 0, \ldots, N$. To obtain $\widehat{U}_i^{(k+1)} = \widehat{U}_i^{(k)} + \triangle_{U_i}^{(k+1)}$ and $\widehat{V}_i^{(k+1)} = \widehat{V}_i^{(k)} + \triangle_{V_i}^{(k+1)}$, we apply Newton-like and Picard-like methods, presenting $L^p(\widehat{U}_i, \widehat{V}_i)$ and $L^0(\widehat{U}_i, \widehat{V}_i)$ in different manner.

Classical Newton method (NM):

$$L^e(\widehat{U}_i, \widehat{V}_i) = L^e(\widehat{U}_i^{(k)}, \widehat{V}_i^{(k)}) + \frac{\partial L^e}{\partial U}(\widehat{U}_i^{(k)}, \widehat{V}_i^{(k)})\triangle_{U_i}^{(k+1)} + \frac{\partial L^e}{\partial V}(\widehat{U}_i^{(k)}, \widehat{V}_i^{(k)})\triangle_{V_i}^{(k+1)},$$

where $e = \{p, 0\}$. In this case we have to inverse a large matrix, say $\mathcal{A}$, of size $2N \times 2N$. Moreover, $\mathcal{A}$ is a wide banded sparse matrix.

Newton-Decoupling Method (NDM). We follow the strategy, developed in [1,4] to decouple the two equations from the system and to solve them separately, in order to avoid a large matrix inversion. Now, we need to inverse one $N \times N$ three-diagonal matrices and the solution $\widehat{V}_i$ from the second equation $L^0(\widehat{U}_i, \widehat{V}_i) = 0$ in (9) can be computed by explicit formula:

$$L^p(\widehat{U}_i, \widehat{V}_i) = L^p(\widehat{U}_i^{(k)}, \widehat{V}_i^{(k)}) + \frac{\partial L^p}{\partial U}(\widehat{U}_i^{(k)}, \widehat{V}_i^{(k)}) \triangle_{U_i}^{(k+1)},$$

$$L^0(\widehat{U}_i, \widehat{V}_i) = L^0(\widehat{U}_i^{(k+1)}, \widehat{V}_i^{(k)}) + \frac{\partial L^0}{\partial V}(\widehat{U}_i^{(k+1)}, \widehat{V}_i^{(k)}) \triangle_{V_i}^{(k+1)}.$$

Note that with Newton-like methods, the coefficient matrix $\mathcal{A}$ updates during the iteration process and we perform the matrix inversion at each iteration. *Picard-Decoupling Method* (PDM).

$$L^p(\widehat{U}_i, \widehat{V}_i) = U_{t_i}^{(k+1)} - \frac{1}{2}\sigma^2 S_i^2 \widehat{U}_{\overline{S}S,i}^{(k+1)} + ae^{\widehat{U}_i^{(k)}} e^{-\widehat{V}_i^{(k)}} - b = 0,$$

$$L^0(\widehat{U}_i, \widehat{V}_i) = V_{t_i}^{(k+1)} + ce^{-\widehat{U}_i^{(k+1)}} e^{\widehat{V}_i^{(k)}} - c = 0.$$

With this algorithm, just as for NDM, we solve both equations separately, upgrading the value $\widehat{U}_i^{(k+1)}$ in the second equation $L^0(\widehat{U}_i, \widehat{V}_i) = 0$. The coefficient matrix $\mathcal{A}$ is one and the same at each iteration and we perform a matrix inversion only ones for the whole computation process. Typically, Picard method requires more iterations than Newton-like methods in order to achieve one and the same precision.

The implementation of initial and boundary conditions in NM, NDM and PDM is standard.

5   Numerical Experiments

The aim of this section is to verify the order of convergence and efficiency of the presented methods NM, NDM and PDM. We provide experiments both with exact and original solution. For exact solution test, we add residual terms in the right-hand side of Eqs. (4),(8), incorporate appropriate initial conditions to obtain exact solutions $u_{\text{ex}}(x, \tau) = e^\tau \cos(\pi x)$ and $v_{\text{ex}}(x, \tau) = e^{\tau/2} \cos(\pi x)$.

Accuracy in maximal discrete norm ($\| \cdot \|_\infty$) and convergence rate are computed at final time T, using two consecutive meshes with formulas

$$CR^W = \log_2 \frac{E_{N/2}^W}{E_N^W}, \quad \text{where } W = \{U, V\}, \ E_N^U = \|u_{\text{ex}} - U\|_\infty, \ E_N^V = \|v_{\text{ex}} - V\|_\infty.$$

To test the order of convergence of the original system (4)–(8) at final time T, we use the numerical solution $\mathcal{W}_N := [W_0, \ldots, W_N]^T$ at three consecutive meshes

$$CR^W = \log_2 \frac{\|\mathcal{W}_{N/2} - \mathcal{W}_N\|_\infty}{\|\mathcal{W}_N - \mathcal{W}_{2N}\|_\infty}.$$

Table 1. Results for E_N^W, CR^W and CPU time, T=0.5, Example 1

	U		V		CPU		
N	E_N^U	CR^U	E_N^V	CR^U	NM	NDM	PDM
20	3.96355e-3		3.49357e-3		0.484	0.391	0.968
40	9.95449e-4	1.9934	8.76557e-4	1.9948	4.859	2.484	3.671
80	2.49283e-4	1.9976	2.19595e-4	1.9970	14.956	11.297	15.140
160	6.22588e-5	2.0014	5.48240e-5	2.0020	72.531	70.719	73.093
320	1.55645e-5	2.0000	1.37057e-5	2.0000	497.969	484.703	407.021
640	3.89120e-6	2.0000	3.42661e-6	1.9999	3 523.141	2 568.781	2 990.156
1280	9.72792e-7	2.0000	8.56635e-7	2.0000	26 791.103	22 154.631	19 179.441

Model parameters are: $\sigma = 0.2$, $\gamma = 1$, $\mu = 0.06$, $\nu_{01} = 1$, $\nu_{10} = 12$ (i.e. the liquidity shocks occur at a rate of once per year and last an average of one month [5]).

The iteration process continue until the difference between two consecutive iteration (in max discrete norm) become less than a given tolerance $tol = 1.e-10$.

Example 1 (Exact Solution Test). Let $L = 2$, $\triangle\tau = h^2$. The precision obtained by NM, NDM and PDM with one and the same model and mesh parameters differs insignificantly. In Tables 1, 2 we give the results (for both solutions U and V) from computations with exact solution: CPU time, errors and convergence rates in maximal discrete norm for each method at time $T = 0.5$ and $T = 2$, respectively. Tables 1, 2 illustrate second order convergence rate in space of the numerical solutions U and V. For long time computations, starting with *smooth initial data*, we observe better efficiency of PDM than NM and NDM.

Example 2 (Original Solution Test). Now we give the results from the computation of the difference scheme (9) with initial function (6). In Table 3 we list convergence rates of the solutions U and V, computed by NDM in maximal discrete norm, also solutions at strike $S = E$ i.e. $W(T, E)$, the corresponding absolute value of the difference (diff_W) between the solutions W at consecutive embedded meshes and convergence at strike point (CR_E^W) at final time $T = 0.5$

Table 2. Results for E_N^W, CR^W and CPU time, $T = 2$, Example 1

	U		V		CPU		
N	E_N^U	CR^U	E_N^V	CR^U	NM	NDM	PDM
160	5.76624e-4		5.76154e-4		332.065	312.641	346.765
320	1.44180e-4	1.9998	1.44062e-4	1.9998	1 954.487	2 028.421	1 942.031
640	3.60435e-5	2.0001	3.60140e-5	2.0001	13 172.906	13 706. 680	13 074.935

Table 3. Results for CR^W, $W(T, E)$, diff$_W$, CR_E^W, T=0.5, Example 2

N	Max. norm		Results at strike point $S = E$					
	CR^U	CR^V	$U(T, E)$	diff$_U$	CR_E^U	$V(T, E)$	diff$_V$	CR_E^V
40			0.283350			0.251367		
80	2.0617	2.0678	0.290271	6.9207e-3	2.0211	0.260459	9.0920e-3	2.0484
160	2.0129	2.0185	0.291929	1.6577e-3	2.0492	0.262628	2.1687e-3	2.0392
320	2.0031	2.0047	0.292340	4.1075e-4	2.0031	0.263163	5.3526e-4	2.0047
640	2.0008	2.0012	0.292442	1.0247e-4	2.0008	0.263296	1.3338e-4	2.0012
1280	2.0002	2.0003	0.292468	2.5603e-5	2.0002	0.263330	3.3317e-5	2.0003
2560			0.292474	6.4000e-6		0.263338	8.3276e-6	

Table 4. CPU time for one and the same precision, obtained by NM, NDM, PDM at different time levels, Example 2

N	$T = 0.5$			$T = 2$			$T = 5$		
	NM	NDM	PDM	NM	NDM	PDM	NM	NDM	PDM
160	1.781	1.266	6.500	4.859	4.203	24.296	11.625	10.422	61.453
320	8.609	5.828	50.125	30.453	21.891	192.859	73.828	55.611	478.640
640	55.953	40.625	545.812	210.359	156.719	2 096.109	519.344	387.766	5 031.251

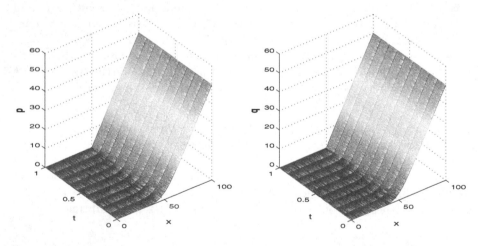

Fig. 1. Numerical solutions p (*left*) and q (*right*), $0 \leq t \leq 1$, Example 2

for $L = 10$, $E = 5$ and $\triangle \tau = h^2$. The results for other iteration methods - NM and PDM are the same. Again, we observe second order convergence rate in space. In Table 4 we show computer time, requited to obtain one and the same precision of the numerical solution, computed by NM, NDM, PDM for different time. It become clear that in the case, when the iteration process starts with *non-smooth initial data*, NDM is more efficient than NM and PDM.

On Fig. 1 we plot evolution graphics of numerical solutions p and q for $L = 100$, $E = 50$, $T = 1$, $N = 300$, $\triangle\tau = h^2$, to illustrate the non-negativity of the prices.

6 Conclusions

In this paper we developed second-order in space Newton-like and Picard-like methods for solving the non-linear system of algebraic equations resulting from fully implicit finite difference approximation of the parabolic-ODE system of European options. Convergence and discrete comparison principle are established.

Numerical tests show that when we deal with smooth initial data, for long time computations, the more effective is PDM. So, if we smooth the initial function (it would be necessary if we construct high-order scheme, see [2]) the best choice for the computations is PDM. In our case, we not need to smooth the initial data, as we construct second-order scheme and the methods are effective even for non-smooth initial function. Thus the best option for the computations is NDM, because of its fast performance. The reason is that in this case, PDM requires significant number of iterations, especially at first time levels, which leads to delay in the computations and can not be compensated by the absence of matrix inversion at each time level (which was pointed as a main advantage of PDM).

We have solved numerically the option buyer's price model (4.8) derived in [5]. One can similarly consider the writer's indifference price model (4.9) in [5].

Acknowledgement. This research is supported by the European Union under Grant Agreement number 304617 (FP7 Marie Curie Action Project Multi-ITN STRIKE - Novel Methods in Computational Finance) and Bulgarian National Fund of Science under Project I02/20-2014.

References

1. Dremkova, E., Ehrhardt, M.: A high-order compact method for nonlinear Black-Scholes option pricing equations of American options. Int. J. Comput. Math **88**(13), 2782–2797 (2011)
2. Düring, B., Heuer, C.: High-order compact schemes for parabolic problems with mixed derivatives in multiple space dimensions. Submitted for Publication, 27 June 2014. Available at SSRN: http://ssrn.com/abstract=2459861
3. Gyulov, T.B., Vulkov, L.G.: Well-posedness and compaison principle for option pricing with switching liquidity. arXiv: 1502.07622v1 (q-fin.MF)
4. Liao, W., Khaliq, A.Q.M.: High order compact scheme for solving nonlinear Black-Scholes equation with transaction cost. Int. J. Comp. Math. **86**(6), 1009–1023 (2009)
5. Ludkovski, M., Shen, Q.: European option pricing with liquidity shocks. Int. J. Theor. Appl. Financ. **16**(7), 30 (Article ID: 1350043) (2013)
6. Windcliff, H., Wang, J., Forsyth, P.A., Vetzal, K.R.: Hedging with a correlated asset: solution of a nonlinear pricing PDE. J. Comp. Appl. Math. **200**(1), 86–115 (2007)

7. Mudzimbabwe, W.: Numerical solution of a stochastic control problem of option pricing for a liquidity switching market. Acta Mathematica Universitatis Comenianae (2015, in press)
8. Mudzimbabwe, W., Vulkov, L.G.: IMEX schemes for a parabolic-ODE system of European options with liquidity shocks. arXiv Preprint arXiv:1503.09008
9. Samarskii, A.A.: The Theory of Difference Schemes. Marcel Dekker Inc, New York (2001)

Thermoelectrical Tick Removal Process Modeling

Nikola Kosturski, Ivan Lirkov$^{(\boxtimes)}$, Svetozar Margenov, and Yavor Vutov

Institute of Information and Communication Technologies,
Bulgarian Academy of Sciences, Acad. G. Bonchev Bl. 25A, 1113 Sofia, Bulgaria
ivan@parallel.bas.bg

Abstract. Ticks are widespread ectoparasites. They feed on blood of animals like birds and mammals, including humans. They are carriers and transmitters of pathogens, which cause many diseases, including *tick-borne meningoencephalitis, lyme borreliosis, typhus* to name few. The best way to prevent infection is to remove the ticks from the host as soon as possible. The removal usually is performed mechanically by pulling the tick. This however is a risky process. Tick irritation or injury may result in it vomiting infective fluids.

On a quest of creating of a portable device, which utilizes radio-frequency alternating current for contact-less tick removal, we simulate the thermo-electrical processes of the device application. We use the finite element method, to obtain both the current density inside the host and the tick, and the created temperature field. The computational domain consists of the host's skin, the tick, the electrodes, and air.

Experiments on nested grids are performed to ensure numerical correctness of the obtained solutions. Various electrode configurations are investigated. The goal is to find suitable working parameters – applied power, duration, position for the procedure.

1 Introduction

We are on a quest of developing a contact-less tick removal apparatus. Ticks spread a wide variety of diseases and their removal without disturbance is of a great importance. Ticks are not to be disturbed, because they could vomit potentially contaminated fluids into their host.

The apparatus under development consists of two electrodes. They are applied to the skin of the host, in the vicinity of the tick bite. Then radio-frequency alternating current is started through the electrodes. It is expected that weak electro and thermal stimulation can discomfort the tick and make it leave the host.

The rest of the paper is organized as follows: In Sect. 1 the model and its mathematical treatment are presented. The setup and the results of the performed numerical experiments are presented in Sect. 2. Some discussion and concluding remarks are given at the end.

© Springer International Publishing Switzerland 2015
I. Lirkov et al. (Eds.): LSSC 2015, LNCS 9374, pp. 369–376, 2015.
DOI: 10.1007/978-3-319-26520-9_41

2   The Model, Space, and Time Discretization

We consider the bio-heat equation in the following form [1]:

$$\rho c \frac{\partial T}{\partial t} = \nabla \cdot k \nabla T + J \cdot E \tag{1}$$

where the thermal energy arising from the current flow is described by $J \cdot E$. Here, a simplified model is considered, where metabolic heat production, and air convection effects are ignored. The initial and boundary conditions which are used in this approach are as follows:

$$
\begin{align}
T = 37°C \qquad & \text{when } t = 0 \text{ at } \Omega_{skin} \cup \Omega_{tick} \cup \Omega_{tick\,mouth} \tag{2a}\\
T = 20°C \qquad & \text{when } t = 0 \text{ at } \Omega_{air} \cup \Omega_{El1} \cup \Omega_{El2}, \tag{2b}\\
T = 37°C \qquad & \text{when } t \geq 0 \text{ at } \Gamma_{bottom}, \tag{2c}\\
T = 20°C \qquad & \text{when } t \geq 0 \text{ at } \Gamma_{top}, \tag{2d}\\
k\nabla T \cdot \mathbf{n} = 0 \qquad & \text{when } t > 0 \text{ at } \Gamma_{side}. \tag{2e}
\end{align}
$$

The notations which are used in (1) and (2) are given below:

- Ω – the entire domain of the model;
- $\Omega_{El1}, \Omega_{El2}$ – the domains of the two electrodes;
- $\Omega_{tick}, \Omega_{tick\,mouth}$ – the domains of the body and mouth of the tick;
- $\Omega_{skin} = \Omega_{gel} \cup \Omega_{epidermis} \cup \Omega_{dermis} \cup \Omega_{subcutis}$ – the domains of various skin tissues and the applied electro-conducting gel;
- Γ_{top} – the top boundary of Ω_{air};
- Γ_{bottom} – the bottom boundary of Ω_{skin};
- Γ_{side} – The boundary on the side of Ω;
- ρ – density [kg/m^3];
- c – specific heat [J/kg K];
- k – thermal conductivity [W/m K];
- J – current density [A/m];
- E – electric field intensity [V/m];
- t – time [s];
- T – temperature [°C];
- $\mathbf{n}$ – outward facing normal to the boundary.

The computational domain Ω is sketched on Fig. 1. It consists of air, two steel electrodes, electro-conducting gel, and several skin layers – epidermis, dermis, and subcutis. Both tick's mouth and body are present in the domain. The goal is not to model the exact tick geometry, but rather to have a distinct domain in the geometry (especially for the tick mouth) so we can analyze the temperature field there. The outermost epidermis layer – stratum corneum has high electrical impedance, which also varies a lot. To overcome this, we apply electro-conductive gel on top of the skin. In the presented algorithm the bio-heat problem (1) is solved in two steps (see [6] for more details):

1. Finding a steady state heat source $J \cdot E$ using that: (a) $E = -\nabla V$ (V is the electric potential in the computational domain Ω), and (b) $J = \sigma E$, where σ is the electric conductivity [S/m];

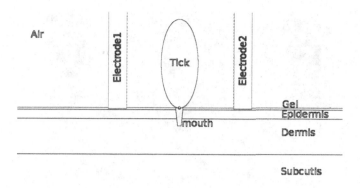

Fig. 1. Computational geometry

2. Finding the temperature T by solving the heat transfer Eq. (1) using the heat source $J \cdot E$ obtained in the first step.

At first step, in order to determine the heat source $J \cdot E$, we have to find the distribution of the electric potential V. It is known that we can neglect the contribution from the magnetic field in Maxwell's equation if the size of the computational domain is much smaller than the wavelength. We use computer simulation of the electromagnetic processes in a model domain which is a 5 cm × 5 cm × 5 cm cube for the electric current with frequency up to 900 kHz. Comsol Multiphysics module "AC/DC module — Electric and induction currents" (see [11]) is applied to solve the Maxwell's equation in the time-harmonic case. The numerical results show that the dependency of the distribution of the electric potential V on the magnetic field is unseen. Thus, the distribution of the electric potential V is found by solving the Poisson's equation:

$$\nabla \cdot \sigma \nabla V = 0, \text{ in } \Omega, \tag{3}$$

with boundary conditions

$$\nabla V \cdot \mathbf{n} = 0 \text{ at } \partial\Omega \setminus \partial\Omega_{El1} \setminus \partial\Omega_{El2}, \tag{4a}$$

$$V = 0 \text{ at } \partial\Omega_{El1} \cap \partial\Omega, \tag{4b}$$

$$V = V_0 \text{ at } \partial\Omega_{El2} \cap \partial\Omega. \tag{4c}$$

The following notations are used in the above equations:

- V – Electric potential in Ω;
- σ – electric conductivity [S/m];
- V_0 – applied voltage;
- $\partial\Omega_{El1}$ – boundary of electrode 1;

- $\partial\Omega_{El2}$ – boundary of electrode 2;

- $\partial\Omega \setminus \partial\Omega_{El1} \setminus \partial\Omega_{El2}$ – the rest of the domain surface.

Fig. 2. Computational coarse mesh.

After determining the potential distribution, the electric field intensity and the current density is computed from

$$E = -\nabla V, \qquad J = \sigma E.$$

It is more useful to have the output power as an input parameter, than the actual voltage. We have to find the potential V_0 for the last boundary condition of (3) that will yield a given electrical output power P [W]. To do this, the Poisson's equation is initially solved with an arbitrary nonzero boundary condition $V = V_0^*$ at $\partial\Omega_{El2}$. Then, E^* and J^* are obtained from the solution and the corresponding electrical power P^* can be computed as $P^* = \int_\Omega E^* \cdot J^* d\mathbf{x}$. Since the solution and all the components of E and J are proportional to the value of V_0 we can scale the obtained solution, instead of recomputing it, in the following way:

$$V_0 = \lambda V_0^*, \quad E = \lambda E^*, \quad J = \lambda J^*, \quad \text{where} \quad \lambda = \sqrt{P/P^*}.$$

Let us note that this adjustment is performed only once at the beginning of the simulation. The obtained potential V_0 remains constant during the procedure.

For the numerical solution of problems (1)–(2) and (3)–(4) the finite element method in space is used [4]. We use *linear conforming tetrahedral finite elements*. Our unstructured grid parallel solver [7] is adapted for this problem. The element matrices are directly defined on the tetrahedrons of the used unstructured mesh (see Fig. 2). The meshing is done with the Netgen mesher [8]. An *algebraic multigrid* (AMG) preconditioner is used [3] in the PCG solution of the arising linear systems. The PCG and multigrid implementation from the Hypre library [2] are used. The time derivative can be discretized via finite differences and both the *backward Euler* and the *Crank-Nicolson* schemes can be used [5]. Let the matrices K and M be the stiffness and mass matrices from the finite element discretization of (1):

$$K = \left[\int_\Omega k\nabla\Phi_i \cdot \nabla\Phi_j d\mathbf{x} \right]_{i,j=1}^N, \qquad M = \left[\int_\Omega \rho c\Phi_i\Phi_j d\mathbf{x} \right]_{i,j=1}^N.$$

The electric field intensity is given by:

$$\mathbf{F} = \left[\int_{\Omega} J \cdot E\Phi_i d\mathbf{x} \right]_{i=1}^{N},$$ (5)

Then, the spatially discretized parabolic Eq. (1) can be written in matrix form as:

$$M\frac{\partial \mathbf{T}}{\partial t} + K\mathbf{T} = \mathbf{F}.$$ (6)

The time discretization for both backward Euler method and the Crank-Nicolson one can be written in the form

$$(M + \tau^n \theta K)\mathbf{T}^{n+1} = (M - \tau^n(1-\theta)K)\mathbf{T}^n + \tau^n \mathbf{F},$$ (7)

where the current (n-th) time-step is denoted with τ^n, the unknown solution at the next time step – with $\mathbf{T}^{n+1}$, and the solution at the current time step – with $\mathbf{T}^n$. If we set the parameter $\theta = 1$, (7) gives a system for the backward Euler discretization. When $\theta = 0.5$ (7) becomes Crank-Nicolson one.

3   Experiments

The material properties of the skin are taken from [9, 10]. The tick material properties are chosen to be close to the ones of the skin. All material properties used in the experiments are collected in Table 1. The entire domain is a cylinder with diameter 100 mm and height 100 mm. The thickness of the air layer is 50 mm, the one of the gel layer—0.1 mm, all layers of the skin—49.9 mm. The thickness of epidermis is 0.5 mm. The dermis is 2 mm thick. The rest of the tissue is considered subcutis.

In all experiments the applied power P is set to 0.5 W. Four different geometries with varying diameter of the electrodes $\varnothing$ and distance between them l are studied: $\varnothing = 1.5$ mm, $l = 5$ mm for Geometry 1; $\varnothing = 1.0$ mm, $l = 5$ mm for

Table 1. Thermal and Electrical Properties of the Materials

Material	ρ (kg/m^3)	c (J/kg K)	k (W/m K)	σ (S/m)
Stainless steel	21 500	132	71	4×10^8
Gel	1 060	3 473.55	0.512	16
Epidermis	1 050	3 473.55	0.419	0.1
Dermis	1 050	3 473.55	0.314	0.6
Subcutis	1 050	3 473.55	0.209	0.6
Air	1	1	0.025	10^{-3}
Tick's body	1 050	3 473.55	0.512	1
Tick's mouth	1 050	3 473.55	0.512	0.6

Geometry 2; $\varnothing = 1.5$ mm, $l = 7$ mm for Geometry 3 and $\varnothing = 1.0$ mm, $l = 7$ mm for Geometry 4. The meshes obtained from Netgen are refined uniformly by dividing each tetrahedron into 8 smaller ones. This refinement process is performed twice. We shall refer to the three different levels of refinement as *coarse, medium, and fine*. The backward Euler method is used in all numerical experiments. In the first set of experiments we are interested in numerical accuracy and stability of the discretization of the problem. We performed experiments on Geometry 2. On the coarse mesh time-step $\tau = 1$ is used, on the medium mesh $- \tau = 0.25$, and on the fine mesh $- \tau = 0.0625$. On Fig. 3, the averaged temperatures at tick's mouth T^{coarse} for the coarse mesh, T for the medium mesh, and T^{fine} for the fine mesh are compared. As we can see, there is no significant difference between the two finer meshes. Therefore, for the rest of the experiments we use a time-step of 0.25 and the medium meshes. Then we performed experiments with all four geometries. The averaged temperatures at tick's mouth are depicted on Fig. 4. Two cross-sections for the experiments on each mesh are

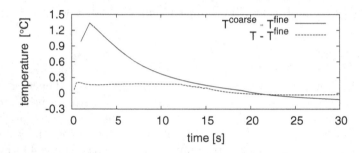

Fig. 3. Average temperature at tick's mouth, Geometry 2, compared to the ones on the finest mesh.

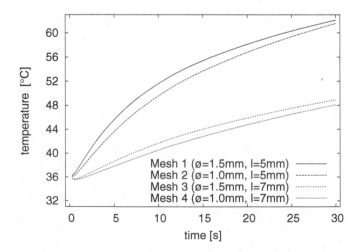

Fig. 4. Average temperature at tick's mouth, for different meshes.

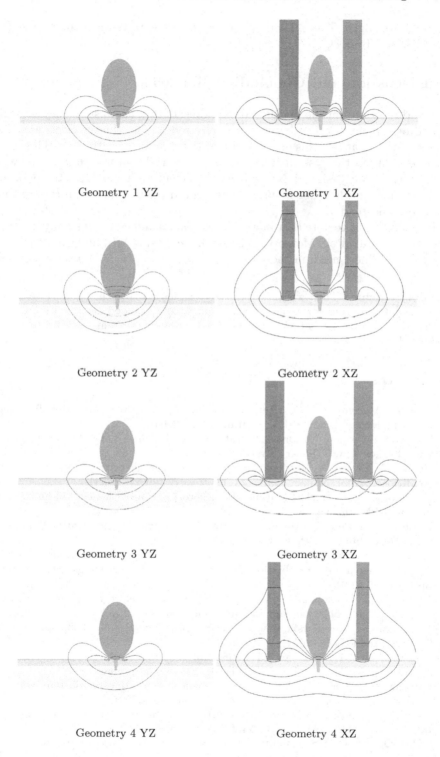

Geometry 1 YZ   Geometry 1 XZ

Geometry 2 YZ   Geometry 2 XZ

Geometry 3 YZ   Geometry 3 XZ

Geometry 4 YZ   Geometry 4 XZ

Fig. 5. Isolines of the temperature: T=40°C, T=45°C, T=50°C at time t=30 s

presented on Fig. 5. There, isolines connect points with temperatures T=40°C, T=45°C, and T=50°C.

4 Discussion and Concluding Remarks

It is readily seen from the performed experiments that the results vary strongly depending on the distance between the electrodes. The thicker electrodes also produce more heat but this effect becomes less significant with time. This leads to the conclusion that the distance between electrodes in the actual apparatus should be carefully designed. Nevertheless in all four geometries, the temperature around tick's mouth is above 45°C. The applied power and/or procedure time can also be decreased.

It is important to note that our solver is general and can handle any unstructured tetrahedral mesh. Moreover it is parallelized using MPI and can handle big meshes – the finest mesh in these experiments has ≈ 144 million tetrahedrons.

Acknowledgment. This work is partially supported by the project AComIn "Advanced Computing for Innovation", grant 316087, funded by the FP7 Capacity Program. The collaboration with the development team from AMET Ltd is also acknowledged.

References

1. Chang, I.A., Nguyen, U.D.: Thermal modeling of lesion growth with radiofrequency ablation devices. BioMed. Eng. OnLine **3**, 27 (2004)
2. Lawrence Livermore National Laboratory Scalable Linear Solvers Project. http://www.llnl.gov/CASC/linear_solvers/
3. Henson, V., Yang, U.: BoomerAMG: a parallel algebraic multigrid solver and preconditioner. Appl. Numer. Math. **41**(1), 155–177 (2002). Elsevier
4. Brenner, S., Scott, L.: The Mathematical Theory of Finite Element Methods. Texts in Applied Mathematics, vol. 15. Springer, New York (1994)
5. Iserles, A.: A First Course in the Numerical Analysis of Differential Equations. Cambridge University Press, Cambridge (2009)
6. Kosturski, N., Margenov, S., Vutov, Y.: Supercomputer simulation of radiofrequency hepatic tumor ablation. In: AMiTaNS 2012 Proceedings, AIP CP 1487, pp. 120–126 (2012)
7. Georgiev, K., Kosturski, N., Vutov, Y.: On the adaptive time-stepping in radiofrequency liver ablation simulation: some preliminary results. In: Lirkov, I., Margenov, S., Waśniewski, J. (eds.) LSSC 2013. LNCS, vol. 8353, pp. 397–404. Springer, Heidelberg (2014)
8. Schöberl, J.: NETGEN an advancing front 2D/3D-mesh generator based on abstract rules. Comput. Vis. Sci. **1**(1), 41–52 (1997). Springer
9. Gurung, D.: Thermoregulation through skin at low atmospheric temperatures. Kathmandu Univ. J. Sci. Eng. Technol. **5**(I), 14–22 (2009)
10. Gabriel, C., Peyman, A., Grant, E.: Electrical conductivity of tissue at frequencies below 1 MHz. Phys. Med. Biol. **54**(16), 4863–4878 (2009)
11. COMSOL Inc: COMSOL multiphysics: AC/DC module user's guide (2008)

Performance Analysis of Block AMG Preconditioning of Poroelasticity Equations

Nikola Kosturski[1]($\boxtimes$), Svetozar Margenov[1], Peter Popov[1,2],
Nikola Simeonov[2], and Yavor Vutov[1]

[1] Institute of Information and Communication Technologies,
Bulgarian Academy of Sciences, Sofia, Bulgaria
kosturski@parallel.bas.bg
[2] PPResearch Ltd, Sofia, Bulgaria

Abstract. The goal of this study is to develop, analyze, and implement efficient numerical algorithms for equations of linear poroelasticity, a macroscopically diphasic description of coupled flow and mechanics. We suppose that the solid phase is governed by the linearized constitutive relationship of Hooke's law. Assuming in addition a quasi-steady regime of the fluid structure interaction, the media is described by the Biot's system of equations for the unknown displacements and pressure $(\mathbf{u}, p)$. A mixed Finite Element Method (FEM) is applied for discretization. Linear conforming elements are used for the displacements. Following the approach of Arnold-Brezzi, non-conforming FEM approximation is applied for the pressure where bubble terms are added to guarantee a local mass conservation. Block-diagonal preconditioners are used for iterative solution of the arising saddle-point linear algebraic system. The BiCGStab and GMRES are the basic iterative schemes, while algebraic multigrid (AMG) is utilized for approximation of the diagonal blocks. The HYPRE implementations of BiCGStab, GMRES and AMG (BoomerAMG, [6]) are used in the presented numerical tests. The aim of the performance analysis is to improve both: (i) the convergence rate of the solvers measured by the iteration counts, and (ii) the CPU time to solve the problem. The reported results demonstrate some advantages of GMRES for the considered real-life, large-scale, and strongly heterogeneous test problems. Significant improvement is observed due to tuning of the Boomer-AMG settings.

1 Introduction

In classical linear poroelasticity it is assumed that the solid is governed by the constitutive relationship $\sigma = \mathcal{L}\mathbf{e}$, where σ is the stress tensor, $\mathbf{e}(\mathbf{u}) = \frac{1}{2}(\nabla\mathbf{u} + \nabla\mathbf{u}^\mathbf{T})$ is the strain tensor and $\mathcal{L}$ stands for the elasticity tensor. Then the media is described by the Biot law (c.f. [4,9], see also [10]):

$$\nabla \cdot (\mathcal{L}\mathbf{e}(\mathbf{u}) - \mathbf{A}p_0) = \mathbf{0}, \tag{1}$$

$$\nabla \cdot (\mathbf{K}\nabla p) = \nabla \cdot \frac{\partial \mathbf{u}}{\partial t} + \mathbf{A} : \mathbf{e}\left(\frac{\partial \mathbf{u}}{\partial t}\right) + \beta\frac{\partial p}{\partial t}. \tag{2}$$

© Springer International Publishing Switzerland 2015
I. Lirkov et al. (Eds.): LSSC 2015, LNCS 9374, pp. 377–384, 2015.
DOI: 10.1007/978-3-319-26520-9_42

Here p and $\mathbf{u}$ are the unknown pressure and displacement vector. $\mathbf{K}$ stands for the permeability tensor. The Biot coefficient (tensor) $\mathbf{A}$ takes into account the contribution of the fluid pressure p into the momentum Eq. (1), as well as the pore volume change term $\mathbf{A} : \mathbf{e}\left(\frac{\partial \mathbf{u}}{\partial t}\right)$ in the balance of mass (1), due to the displacements $\mathbf{u}$. The pore volume change in (2) due to p is captured by β, the coefficient of apparent rock compressibility due macroscopic fluid pressure.

The mesh methods provide computational technology for efficient discretization of the problem (1–2). Among others, we would mention the Galerkin finite element method (FEM) and the mixed FEM. The choice of method depends on the features of the considered class of problems. Here, conforming linear FEs are applied for approximation of the displacements in the elasticity terms of (1–2). For applications related to flows in highly heterogeneous porous media, the mixed finite element methods have proven to be accurate and locally mass conservative. While applying the mixed FEM to fluid subproblem, the continuity of the velocity normal to the boundary between two adjacent finite elements could be enforced by Lagrange multipliers. The relationship between the mixed and non-conforming FEM has been studied and simplified for various finite element spaces (see, e.g. [2]). In [3] Arnold and Brezzi have demonstrated that after the elimination of the unknowns representing the pressure and the velocity from the algebraic system the resulting Schur system for the Lagrange multipliers is equivalent to a discretization by Galerkin method using linear non-conforming finite elements. Namely, in [3] it is shown that the lowest-order Raviart-Thomas mixed finite element approximations are equivalent to the usual Crouzeix-Raviart non-conforming linear finite element approximations when the non-conforming space is augmented with quadratic bubbles. We use such kind of augmented Crouzeix-Raviart linear elements for approximation of the pressure. Implicit backward Euler method is applied for time discretization.

The rest of the paper is devoted to the solution of the linear algebraic systems arising at each time step. It is organized as follows. In Sect. 2, we consider the preconditioning algorithms for the related saddle-point problems. Section 3 contains the key results. A performance analysis based on tuning of a set of preconditioning parameters is presented here. Some concluding remarks are given at the end.

2   Preconditioning

The saddle-point poroelasticity problem corresponding to a displacements-pressure two by two block splitting (see Fig. 1(b)) is written in the form:

$$A^{Biot} = \begin{bmatrix} A_{\mathbf{uu}} & A_{\mathbf{u}p} \\ A_{p\mathbf{u}} & A_{pp} \end{bmatrix}$$

To get a symmetric matrix A^{Biot}, the pressure Eq. (2) is multiplied by -1.

For large-scale FEM systems, the advantages of the iterative solution methods are well known. There are efficient Krylov subspace methods designed for

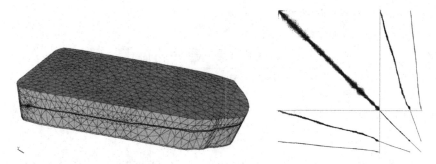

Fig. 1. (a) Example of strongly anisotropic unstructured FEM mesh in reservoir simulation; (b) Structure of the nonzero entries of the saddle-point matrix A^{Biot}

solving symmetric indefinite problems. In this study, the behaviour of the biconjugate gradient stabilized method BiCGStab and the generalized minimal residual method GMRES are studied. The potentially increasing memory consumption of GMRES is controlled with restarts.

A crucial ingredient for success of the Krylov subspace methods is the preconditioning. In this work, we use a block-diagonal preconditioning. Some related results for poroelasticity FEM systems are available in some recent papers (c.f. e.g. [1,5] and the references there in), where the media is ether homogeneous or not strongly heterogeneous. The robustness for heterogeneous problems of high-contrast is a hot topic. The derivation of uniform inf-sup estimate with respect to the coefficient jumps (c.f. e.g. [7]) is still a challenging issue for poroelasticity FEM systems.

The BoomerAMG preconditioner [6] is used as approximation of the diagonal blocks A_{uu} and A_{pp}. The resulting linear preconditioner of the Biot matrix is in the form:

$$C^{Biot} = \begin{bmatrix} A_{uu}^{AMG} & \\ & A_{pp}^{AMG} \end{bmatrix}$$

The following 3D Test Problems (TP) concerning simulation of flows in deformable porous media as appear in petroleum engineering (see Fig. 1(a)) are considered in the next section: (i) TP1: $N = N_u + N_p = 3 \times 2\,291 + 9\,592 = 16\,465$; the media is modestly heterogeneous; (ii) TP2: $N = N_u + N_p = 3 \times 150\,871 + 186\,669 = 639\,282$; the media is strongly heterogeneous; the mesh is strongly anisotropic. Here N stands for the total number of degrees of freedom, while N_u and N_p stand for the related numbers of displacements and pressures. Unstructured grids are used in both, TP1 and TP2.

The large condition number of TP2 is due to the following complementary factors: the elasticity modulus range is $E \in (0.5, 30)$ [GPa]; the Poisson ratio range is $\nu \in (0.3, 0.48)$; the permeability variation is of order of 5×10^5; the mesh anisotropy ratio of order of 6×10^3.

The distribution of nonzero entries of the matrix corresponding to TP2 is presented in Fig. 1(b). The structure of A^{Biot} corresponds to the following numbering: (i) pointwise ordering in A_{uu} (we don't assume separable displacements); (ii) the unknowns corresponding to the bubbles in A_{pp} are at the end.

Table 1. Performance of BiCGStab iterative solver

ϵ	TP1		TP2	
	N_{it}^{BiCG}	T[s]	N_{it}^{BiCG}	T[s]
10^{-6}	7	0.23	18	26.48
10^{-9}	10	0.28	63	79.67
10^{-12}	14	0.36	430	517.30

3   Performance Analysis: Tuning of the BoomerAMG Parameters

The numerical tests are performed on a 3.4 GHz Intel Core i7 CPU. The following notations are used: ϵ - relative stopping criteria for both BiCGStab and GMRES; N_{it}^{BiCG} - number of BiCGStab iterations; N_{it}^{GMRES} - number of GMRES iterations; T[s] - CPU time in seconds. The performance analysis is started with the default settings of BoomerAMG: Falgout coarsening, hybrid symmetric Gauss-Seidel relaxation. classical modified interpolation, and a Strong Threshold equal to 0.75. The results for BiCGStab are given in Table 1.

The influence of the size of Krylov subspace before restart, $Kdim$, is examined additionally when the GMRES performance is studied. In Table 2, we see how the number of iterations N_{it}^{GMRES} decreases with the increase of Kdim. The CPU time is almost always smaller for largest values of Kdim. GMRES is faster, even though the count of BiCG iterations is smaller.

In the following performance analysis, the numerical tests are only for the larger problem TP2, where the heterogeneity and mesh anisotropy are

Table 2. Performance of GMRES iterative solver

TP	ϵ	Kdim	4	8	16	32	64	128	256	512
1	10^{-6}	N_{it}^{GMRES}	24	13	11					
		T[s]	0.38	0.25	0.22					
1	10^{-9}	N_{it}^{GMRES}	39	21	16					
		T[s]	0.53	0.32	0.28					
1	10^{-12}	N_{it}^{GMRES}	47	28	24	21				
		T[s]	0.62	0.40	0.35	0.31				
2	10^{-6}	N_{it}^{GMRES}		62	44	28				
		T[s]		46.75	33.28	22.81				
2	10^{-9}	N_{it}^{GMRES}		151	128	92	88	84		
		T[s]		105.94	86.84	63.37	61.69	60.44		
2	10^{-12}	N_{it}^{GMRES}			893	617	468	372	362	351
		T[s]			574.08	396.13	307.16	259.86	275.30	302.09

Table 3. Tuning of the Coarsening: TP2, $\epsilon = 10^{-6}$, $Kdim = 32$; Relaxation - 1; Interpolation - 0

	BiCGStab		GMRES	
Coarsening	N_{it}	T[s]	N_{it}	T[s]
21	17	17.02	28	15.80
22	17	17.08	28	15.83
3	17	17.16	28	15.87
0	17	18.44	28	16.44
6	17	19.05	29	16.73

very strong. Comprehensive tests for BiCGStab ($\epsilon = 10^{-6}$) are performed at the next step, varying the following parameters: (i) Coarsening: 0 - CLJP-coarsening; 3 - Ruge-Stueben coarsening; 6 - Falgout coarsening; 21 - CGC coarsening; 22 - CGC-E coarsening; (ii) Relaxation: 1 - Gauss-Seidel, sequential; 3 - hybrid Gauss-Seidel forward solve; 4 - hybrid Gauss-Seidel backward solve; 5 - hybrid chaotic Gauss-Seidel; 6 - hybrid symmetric Gauss-Seidel; (iii) Interpolation: 0 - classical modified interpolation; 4 - multipass interpolation; 6 - extended+i interpolation; 7 - extended+i (if no common C neighbor) interpolation; 8 - standard interpolation; 12 - FF interpolation; 13 - FF1 interpolation; 14 - extended interpolation. The best result is obtained for the setting: Coarsening - 21; Relaxation - 1; Interpolation - 0. This variant is selected as default for the next parameter by parameter tunings. The comparative results for both, BiCGStab and GMRES ($Kdim = 32$), are shown in Tables 3, 4, 5.

Now, we analyze the influence of the so called Strong Threshold which (according to HYPRE documentation) is to be chosen in the interval $(0, 1)$, depending on the particular problem. The results are given in Table 6. What we observe is the monotone increasing of the number of iterations, pursued with a monotone decreasing of the time for both, BiCGStab and GMRES.

The last step of the presented performance analysis is devoted to tuning the BoomerAMG parameters separately for each of the blocks A_{uu} and A_{pp} where

Table 4. Tuning of the Relaxation: TP2, $\epsilon = 10^{-6}$, $Kdim = 32$; Coarsening - 21; Interpolation - 0

	BiCGStab		GMRES	
Relaxation	N_{it}	T[s]	N_{it}	T[s]
5	17	17.05	28	15.87
3	17	17.15	28	15.98
1	17	17.21	28	15.85
4	17	19.16	28	17.57
6	18	25.98	27	21.86

a stronger stopping criteria of $\epsilon = 10^{-12}$ is applied. Some selected best settings are given in the Tables 7–8.

The final result of this experimental study is given in Table 9.

Table 5. Tuning of the Interpolation: TP2, $\epsilon = 10^{-6}$, $Kdim = 32$; Coarsening - 21; Relaxation - 1

	BiCGStab		GMRES	
Interpolation	N_{it}	T[s]	N_{it}	T[s]
7	17	16.98	28	15.90
4	17	17.05	28	15.95
12	17	17.08	28	15.94
0	17	17.09	28	15.77
13	17	17.16	28	15.87
14	17	17.29	28	16.01
6	20	19.52	28	15.98
8	21	20.35	28	16.14

Table 6. Tuning of the Strong Threshold: TP2, $\epsilon = 10^{-6}$, $Kdim = 32$; Coarsening - 21; Relaxation - 1

	BiCGStab		GMRES	
ST	N_{it}	T[s]	N_{it}	T[s]
0.05	16	50.98	21	44.14
0.10	15	42.79	20	36.52
0.15	14	37.47	20	32.50
0.20	13	33.00	21	30.43
0.30	14	29.35	21	25.86
0.40	16	27.30	23	23.09
0.50	16	23.45	23	19.63
0.60	16	19.99	25	17.98
0.70	18	19.00	28	16.85
0.75	17	17.00	28	15.91
0.80	18	16.81	29	15.31
0.85	21	17.83	30	14.77
0.90	20	16.35	31	14.55
0.95	20	15.59	31	13.82

Table 7. Tuning of the BoomerAMG parameters for the block A_{uu}: TP2, $\epsilon = 10^{-12}$

Coarsening	Relaxation	Interpolation	N_{it}	T[s]
21	5	12	50	21.37
21	5	0	50	21.42
3	5	12	50	21.42
22	5	4	50	21.42
22	1	13	50	21.44
3	5	7	50	21.44

Table 8. Tuning of the BoomerAMG parameters for the block A_{pp}: TP2, $\epsilon = 10^{-12}$

Coarsening	Relaxation	Interpolation	N_{it}	T[s]
0	6	13	789	33.37
6	6	6	794	33.39
6	6	12	794	33.39
6	6	0	794	33.39
6	6	14	794	33.40
0	6	12	789	33.43

Table 9. Behavior of the solvers of the coupled system after tuning the BoomerAMG parameters: TP2, Kdim=32

	BiCGStab		GMRES	
ϵ	N_{it}	T[s]	N_{it}	T[s]
10^{-6}	14	15.91	22	13.94
10^{-9}	54	45.27	68	34.56
10^{-12}	279	213.36	282	170.98

4   Concluding Remarks

Block-diagonal preconditioning of the mixed FEM Biot system is studied. One commonly used technique is based on inner iterations for the related elliptic blocks. Let us note that the related preconditioners are not linear. Numerical tests illustrating the efficiency of this approach for modestly heterogeneous problems can be found in [1,5]. Our study is focussed on problems with strong heterogeneity and strong mesh anisotropy (see the related details for TP2). For this class of problems, we don't observe any advantages of inner iterations. This is the reason to concentrate on linear preconditioners where AMG approximation of the diagonal blocks is used. The presented results are of strongly expressed experimental nature. The performance analysis shows a serious potential for improvement of the computational efficiency. Due to tuning of the BoomerAMG parameters, the achieved decrease of the CPU times T[s] for TP2 for $\epsilon = 10^{-12}$ are as follows: (i) BiCGStab - from 517.30 to 213.36, that is a reduction factor of 2.42; (ii) GMRES (Kdim=32) - from 396.13 to 170.98, or a reduction factor of 2.32. The better performance of GMRES is clearly visible. In this respect we have to remember that the required memory increases with Kdim which could be a restriction for more large scale applications.

Acknowledgments. The research is partly supported by the project AComIn "Advanced Computing for Innovation", grant 316087, funded by the FP7 Capacity Program.

References

1. Arbenz, P., Turan, E.: Preconditioning for large scale micro finite element analyses of 3D poroelasticity. In: Manninen, P., Öster, P. (eds.) PARA. LNCS, vol. 7782, pp. 361–374. Springer, Heidelberg (2013)
2. Arbogast, T., Chen, Z.: On the implementation of mixed methods as nonconforming methods for second order elliptic problems. Math. Comp. **64**, 943–972 (1995)
3. Arnold, D.N., Brezzi, F.: Mixed and nonconforming finite element methods: implementation, postprocessing and error estimates, RAIRO. Model. Math. Anal. Numer. **19**, 7–32 (1985)
4. Biot, M.A.: General theory of three dimensional consolidation. J. Appl. Phys. **12**, 155–164 (1941)
5. Haga, J.B., Osnes, H., Langtangen, H.P.: A parallel preconditioner for large-scale poroelasticity with highly heterogeneous material parameters. Comput. Geosci. **16**, 723–734 (2012)
6. Henson, V.E., Yang, U.M.: BoomerAMG: a parallel algebraic multigrid solver and preconditioner. Appl. Num. Math. **41**(5), 155–177 (2002)
7. Kraus, J., Lazarov, R., Limbery, M., Margenov, S., Zikatanov, L.: Preconditioning of weighted H(div)-norm and applications to numerical simulation of highly heterogeneous media. Cornell University Library (2014). arXiv:1406.4455
8. Kraus, J., Margenov, S.: Robust Algebraic Multilevel Methods and Algorithms. Radon Series on Computational and Applied Mathematics, vol. 5. de Gruyter, Berlin (2009)
9. Lee, C.K., Mei, C.C.: Re-examination of the equations of poroelasticity. Int. J. Eng. Sci. **35**, 329–352 (1997)
10. Popov, P.: Upscaling of deformable porous media with applications to bone modelling. In: Annual Meeting of the Bulgarian Section of SIAM, BGSIAM09 Proceedings, pp. 105–110 (2010)

Surface Constructions on Irregular Grids

Arne Lakså[✉] and Børre Bang

Narvik University College, P.O.Box 385, 8505 Narvik, Norway
arne.laksa@hin.no
http://www.hin.no/Simulations

Abstract. "Big" surfaces defined on domains that can not be modeled on a single regular grid is typically made by joining several surfaces together with the aid of fillet surfaces or by intersecting the surfaces and joining them after trimming.

In computations on geometry and geometric modeling in general surface modeling is a key issue. The most important type of surfaces are tensor product spline surfaces. They are in general based on regular griding, i.e. knot vectors are the same for all "lines" or "columns". Examples are B-spline surfaces, Hermite-spline surfaces and Expo-rational B-spline surfaces. Surfaces constructions that in some way handle "irregular grids" has been developed. We find them in for example T-splines, LR B-splines, Truncated Hierarchical B-splines and PHT-splines. In general, surfaces based on irregular grids can be regarded as a collection of surfaces on regular grids that are connected at the edges and the corners in a smooth, but irregular way. This involves T-junctions and star-junctions.

To investigate a surface construction based on blending of local "small" patches into a "big" surface with arbitrary topology also requires that we can deal with T-junctions and star junctions.

Here we investigate use of blending technique at T- and star-junctions. We look at special blending surfaces between regular patches, and reparametrization to obtain a correct orientation and a better mapping in the parameter plane. The focus is on smoothness of the resulting surface.

Keywords: Surface · Spline · Junction · Blending

1 Introduction

Boundary representation, abbreviated as B-rep, is a common method for describing objects. An object is described as a collection of contiguous surfaces that represent boundaries of an object or boundaries between different materials. It follows that it is important to be able to model complex surfaces, surfaces with different genius and with varying degrees of complexity. This means that we must be able to cope with "irregular" geometry. Today this is mainly done by using trimmed surfaces with trimming curves that are computed using Boolean operations and thus surface intersections. See [1] and [2].

An important reason to handle irregularities without using trimming is the introduction of isogeometric analysis. This means to use a common function

© Springer International Publishing Switzerland 2015
I. Lirkov et al. (Eds.): LSSC 2015, LNCS 9374, pp. 385–393, 2015.
DOI: 10.1007/978-3-319-26520-9_43

space for both computation and shape description. It follows that the traditional Finite Element function-spaces can be replaced by spline-spaces, see [3].

Both, T-splines, [4,5], LR-splines, [6] and PHT-splines [7] are surface descriptions that modify B-splines to handle irregularities. Therefor, we introduce a related solution for blending-splines surfaces, GERBS, described in [8] and [9].

The most flexible solution is to use non-regular connected triangular surfaces. However, tensor product surfaces are the most used surface construction because of the simple construction, the smoothness control and that they are well known in industrial systems. Therefore, we are looking at tensor product surfaces that are joined together in a smooth, but irregular way, see [10].

This imply investigating T-junctions and star-junctions, special blending surfaces divided in sub-surfaces, and re-parametrization.

2   Surfaces, Blending, and Grid

A tensor product B-spline surface is defined by

$$S(u,v) = \sum_{i=1}^{n_u} \sum_{j=1}^{n_v} c_{i,j}\, b_{d_v,j}(v) b_{d_u,i}(u),$$

where $c_{i,j}$ is the control points of the surface, $b_{d_v,j}(v)$ and $b_{d_u,i}(u)$ are defined by the two knot-vectors, $\bar{u} = \{u_0, u_1, ..., u_{n_u+d_u}\}$ and $\bar{v} = \{v_0, v_1, ..., v_{n_v+d_v}\}$, where n_u and n_v are the number of basis-functions and d_u, d_v are the polynomial degree in respective u- and v-direction. The recursive definition of the basis functions are:

$$b_{d,k}(t) = w_{d,k}(t)\, b_{d-1,k}(t) + (1 - w_{d,k+1}(t))\, b_{d-1,k+1}(t) \tag{1}$$

where

$$w_{d,i}(t) = \frac{t - t_i}{t_{i+d} - t_i}, \tag{2}$$

terminating with

$$b_{0,i}(t) = \begin{cases} 1; \text{if} \ \ t_i \leq t < t_{i+1}; \\ 0; \text{otherwise}, \end{cases} \quad i = k, ..., k + d. \tag{3}$$

The domain of a tensor product B-spline surface is closed and rectangular. The two knot-vectors divide the domain in a regular net of sub-rectangles, and the control polygon is organized in a regular net.

2.1   B-Functions for Blending

A B-function (blending function) is a smooth monotone permutation function, $B : [0,1] \rightarrow [0,1]$. An example of a B-function is the Expo-Rational B-function inducing C^∞-smoothness,

$$B(t) = S_d \int_0^t \phi(s)\, ds, \quad t \in [0,1] \tag{4}$$

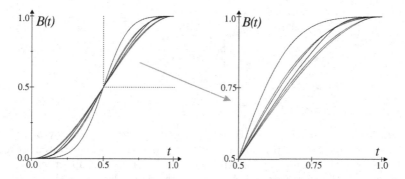

Fig. 1. Six examples of B-functions. On the right side is the upper right corner of the plot shown. At $B(t) = 0.75$ we can see from left to right, $B_2(t) = \frac{t^3}{(1-t)^3+t^3}$, $B_1(t) = \frac{t^2}{(1-t)^2+t^2}$, $B_2(t) = 6t^5 - 15t^4 + 10t^3$, $B_\infty(t)$ - ERBS, $B_1(t) = \sin^2 \frac{\pi}{2}t$ and $B_1(t) = 3t^2 - 2t^3$. The index in the B-function B_j means it induces C^j-smoothness.

where $\phi(s) = e^{-\frac{\left(s-\frac{1}{2}\right)^2}{s(1-s)}}$ and $S_d = \left[\int_0^1 \phi(s)ds\right]^{-1}$. A plot of $B(t)$ from expression (4), together with five other examples of B-functions, can be seen in Fig. 1. It follows that the Expo-Rational B-function has the following properties:

$$\begin{aligned} \textbf{At start } & B(0) = 0, \quad \text{and all derivatives } B^{(d)}(0) = 0, \quad d = 1,2,3,... \\ \textbf{At end } \quad & B(1) = 1, \quad \text{and all derivatives } B^{(d)}(1) = 0, \quad d = 1,2,3,... \end{aligned} \qquad (5)$$

2.2 Blending Surfaces

If we adjust the formula (1) for B-splines with a B-function, we get:

$$B_{d,k}(t) = B \circ w_{d,k}(t)\, B_{d-1,k}(t) + (1 - B \circ w_{d,k+1}(t))\, B_{d-1,k+1}(t), \qquad (6)$$

where B is a B-function, $w_{d,i}(t)$ is defined in (2) and $B_{0,i} = b_{0,i}$ defined in (3).

A blending ERBS surface is a 1st degree tensor product B-spline surface adjusted with an ERBS B-function as described in (6). It is defined by two knot-vectors, $\bar{u} = \{u_0, u_1, ..., u_{n_u+1}\}$ and $\bar{v} = \{v_0, v_1, ..., v_{n_v+1}\}$, where n_u and n_v is the number of basis-functions in respective u- and v-direction. The formula is:

$$S(u,v) = \sum_{i=1}^{n_u} \sum_{j=1}^{n_v} s_{i,j}(u,v)\, B_{1,j}(v)B_{1,i}(u),$$

where $s_{i,j}(u,v)$ is a net of local patches.

Thus, tensor product ERBS blending surface is a surface that also can be regarded to be constructed on a "regular grid", where the grid is defined by the net of local patches, see Fig. 2.

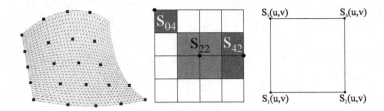

Fig. 2. On left hand side is an ERBS tensor product surface made by 5×5 "local patches". In the middle the parameter plane of the surface is illustrated as a grid. Three examples of the domain of local patches are illustrated. In upper left corner we see surface S_{04}, in the center we see S_{22}, and on the right edge S_{42} is shown. On right hand side one partition is shown. Local patches, connected to each corner, are marked.

2.3 Sub-surfaces

A 1st degree B-spline and thus an ERBS tensor product basis function has minimum support, over two knot intervals. The domain of an ERBS-surface is partitioned by the knot vectors, see Fig. 2. The parameter lines are crossing each other where the vertices of the surface are located. These vertices are the points where the local patches interpolates the ERBS-surface. It follows that the support of the local patch is the sub-partitions in the domain that is surrounding the related vertex. For all internal vertices this is four squared partitions, for vertices on the edges it is two partitions, and for vertices in the corners it is only one partition. This is illustrated in Fig. 2.

On each partition we define sub-surfaces of the four local patches covering the partition and thus also the ERBS-surface. It follows that because of the reparametrization in (6) we can regard the local domain of each sub-surface to be the unit square $[0, 1] \times [0, 1]$.

3   Properties of Surface Blending

We now look at surfaces with domain $[0, 1] \times [0, 1]$, made by blending four patches connected to the corners, see on right hand side in Fig. 2. The formula is:

$$
\begin{aligned}
S(u, v) &= (1 - B(v))((1 - B(u))s_1(u, v) + B(u)s_2(u, v)) \\
&\quad + B(v)((1 - B(u))s_3(u, v) + B(u)s_4(u, v)), \\
&= s_1(u, v) + B(u)(s_2(u, v) - s_1(u, v)) + B(v)(s_3(u, v) - s_1(u, v)) \\
&\quad + B(u)B(v)(s_4(u, v) - s_3(u, v) - s_2(u, v) + s_1(u, v))
\end{aligned}
$$

In the following we skip the parameters (u, v) for the surfaces in the expressions.

$$
S = s_1 + B(u)(s_2 - s_1) + B(v)(s_3 - s_1) + B(u)B(v)(s_4 - s_3 - s_2 + s_1) \quad (7)
$$

To investigate the behavior on the edges, we just look at the edge on the left side. The other edges will have similar behavior. Thus, the function value is

$$
S(0, v) = s_1 + B(v)(s_3 - s_1), \quad (8)
$$

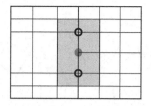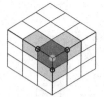

Fig. 3. On left hand side is a T-junction and the two termination points marked. The seven sub-surfaces involved in the irregular blending are marked with light gray. To the right is a Star-junction and three termination points marked. The sub-surfaces involved are, three marked dark gray and six marked light gray.

and the first and second order partial derivatives are

$$
\begin{aligned}
S_u(0,v) &= s_{1u} + B(v)(s_{3u} - s_{1u}), \\
S_v(0,v) &= s_{1v} + B(v)(s_{3v} - s_{1v}) + B'(v)(s_3 - s_1), \\
S_{uu}(0,v) &= s_{1uu} + B(v)(s_{3uu} - s_{1uu}), \\
S_{uv}(0,v) &= s_{1uv} + B'(v)(s_{3u} - s_{1u}) + B(v)(s_{3uv} - s_{1uv}), \\
S_{vv}(0,v) &= s_{1vv} + 2B'(v)(s_{3v} - s_{1v}) + B(v)(s_{3vv} - s_{1vv}).
\end{aligned}
\tag{9}
$$

The following lemma states the interpolation properties on the boundary of the blending surfaces made by blending four surfaces connected to each corners. Thus, the blended surface inherits some of its behavior from its "local patches".

Lemma 1. *At the four corners we get the following properties*

Lower left corner	$S(0,0)$	$S \equiv s_1$ including all its derivatives
Lower right corner	$S(1,0)$	$S = s_2$ including all its derivatives
Upper left corner	$S(0,1)$	$S \equiv s_3$ including all its derivatives
Upper right corner	$S(1,1)$	$S = s_4$ including all its derivatives

At the four edges we get the following properties

Left edge	$S(0,v) = s_1 + B(v)(s_3 - s_1)$	only depend on s_1 and s_3
Right edge	$S(1,v) = s_2 + B(v)(s_4 - s_2)$	only depend on s_2 and s_4
Lower edge	$S(u,0) = s_1 + B(u)(s_2 - s_1)$	only depend on s_1 and s_2
Upper edge	$S(u,1) = s_3 + B(u)(s_4 - s_3)$	only depend on s_3 and s_4

Proof. The proof follows from B-function property (5) and (7), (8) and (9). □

4   Surfaces on Irregular Grid

To connect several surfaces into one common surface, we use specialized local patches. These local patches must cover the nearest neighborhood (according to the grid) on all surfaces to be connected. In Fig. 3 there is on left hand side a grid description of a T-junction, and on right hand side a grid description of a Star-junction.

We also define points that ends irregular areas as termination points.

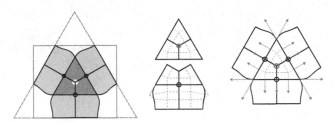

Fig. 4. On left hand side is, for a Star-junction shown the parameter plane of a surface and the sub-surfaces (tensor product surface or Bézier-triangle). In the middle we see a control polygon of a Bézier patch used in the re-parametrization. To the right is a Bézier triangle used. It is divided in 3 by re-parametrization using Bézier patches. In addition is 3 surfaces based on Hermite blended curves connected to each edges.eps

4.1   T-Junctions

Here we define, that in an irregular grid a T-junction is a line ending on a orthogonal line. In Fig. 3 this is illustrated. T-junctions typically occurs on the edge between two surfaces when joining them to one surface. How to handle T-junctions to achieve smooth surfaces, is given by the following theorem.

Theorem 1. *To get a C^k-smooth ($k > 0$) blending on a T-junction or a collection of connected T-junctions, the local patches to these T-junctions and to the termination points of this collection, must be parts of a common surface.*

Proof. From Lemma 1 it follows that, at the vertices, the surface is identical to the local patches connected to the respective vertices. If there is a T-junction, the T-junction is a vertex on two neighboring sub-surfaces on a common local patch. On the other side of the T-junction is another sub-surface from the same local patch. On this sub-surface the T-junction is not a vertex. To interpolate a local patch at an internal point on an edge (with all its derivatives), it follows from Lemma 1 that the two patches connected to the two vertices defining the edge must be parts of the same surface, and that the patch connected to the T-junction also must be part of the same surface.   □

To the left in Fig. 3 a T-junction is highlighted in solid gray and two termination points marked with a black ring. The three local patches connected to the marked vertices and covering the irregularities, are marked light gray and are divided into seven sub-surfaces.

4.2   Star-Junctions

We define Star-junction as a point where several "parameter" lines meet in a non-orthogonal way. An example of a Star-junction is given on the right hand side in Fig. 3, where three grid-lines meet in one point. Star-junctions are difficult to handle. Problems related to Star-junction appears clearly when we look at

Fig. 3. Lines in the parameter plane where only one parameter varies get a kink when they pass an edge on their way out of a Star-junction.

In the following theorem we show how to handle Star-junction in blending.

Theorem 2. *To get a smooth surface over a Star-junction, the local patch con-. nected to the Star-junction point and the local patches connected to the termination points must be sub-surfaces of one common surface. The local patches may be individually translated (not rotated, scaled,..).*

The smoothness over the edges from the Star-junction to its termination points will be C^∞, but the smoothness over the edge orthogonal to the termination points (see Fig. 4) will depend on the re-parametrization that is used.

Proof. Because the local patches are parts of the same surface (only individually translated), derivatives of all order are the same at an edge shared by two local patches. It follows from Lemma 1 that the value and all derivatives on the resulting surface must converge towards the same when we approach the edge from either side. Thus, the edge between the Star-junction and its termination points will be C^∞-smooth. At the edges perpendicular to the star-junction lines at the termination points (see Fig. 4), the smoothness will only depend on the smoothness of the re-parametrization in the parameter plane of the local surface that divide it into sub-surfaces. Thus, the resulting surface can only be as smooth as the re-parametrization in the parameter plane (see Fig. 4).   □

On left hand side of Fig. 4 an example of a parameter plane to a surface that covers an irregular T-junction area is shown. There are nine sub-surfaces (dark and light gray), which together define four local patches. The upper figure in the middle of Fig. 4 shows the local surface attached to the T-junction, below we see one of the three local patches that are connected to the termination points. Note that the upper part of the lower surface is the same as the lower part of the upper surface.

In the upper figure in the middle of Fig. 4 the re-parametrization is done using a Bézier map, $\omega : \mathbb{R}^2 \to \mathbb{R}^2$. There are three sub-surfaces in the central surface. They are each parameterized using a Bézier map of degree 2 in both directions, with a 3×3 control polygon, in a "symmetric" way.

We then look at the right side of the sub-surface below and divide it in the upper part and the lower part. The upper part is equal to the central local patch. The lower part is a Bézier map of degree 2 in one direction and degree 3 in the other direction, i.e. a 3×4 control polygon. If the control points, named $c_{i,j}$, $i = 1, 2, 3$ and $j = 1, 2, ..., 6$ are used by the upper patch $c_{i,j}$, $i = 1, 2, 3$ and $j = 1, 2, 3$ and the lower patch $c_{i,j}$, $i = 1, 2, 3$ and $j = 3, 4, 5, 6$ then the local patch is continuous. If in addition $c_{i,4} - c_{i,3} = c_{i,3} - c_{i,2}$ then the local patch is C^1-smooth over the edge.

On right hand side of Fig. 4 it is used a Bézier triangular map expanded by Hermite blended curves and vector valued functions connected to each of the edges of the triangle. The curves on the other side can be arbitrary chosen. The number of derivative functions used in the Hermite blending determines the continuity.

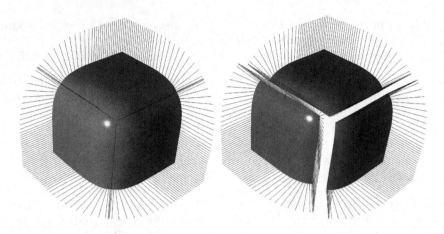

Fig. 5. A smooth surface with a Star-point. The normals on the edges are plotted and on right hand side the three parts are separated. The surface is made by blending of planar surfaces and one curved Bézier triangle expanded by Hermite blended curves.

5   Concluding Remarks

Figure 5 shows a surface with a Star-point, The map explained on the right hand side in Fig. 4 is used in the construction. The Surface is C^1-smooth, but the smoothness can be increased. The two maps that are described on the left hand side of Fig. 4 gave both the same result. Thus, the described methods works.

References

1. Mäntylä, M.: An Introduction to Solid Modeling, vol. 13. Computer Science Press, Incorporated, New York (1988)
2. Hoffmann, C.M.: Geometric and Solid Modeling: An Introduction. Morgan Kaufmann Publishers Inc., San Francisco (1989). ISBN 1-55860-067-1
3. Cottrell, J.A., Hughes, T.J.R., Bazilevs, Y.: Isogeometric Analysis: Toward Integration of CAD and FEA, 1st edn. Wiley, Chichester (2009). ISBN 0470748737, 9780470748732
4. Sederberg, T.W., Zheng, J., Bakenow, A., Nasri, A.: T-splines and T-nurces. ACM Trans. Graph. **22**, 477–484 (2003)
5. Sederberg, T.W., Cardon, D.L., Finnigan, D.G., Zheng, J., Lyche, T.: T-spline simplification and local refinement. ACM Trans. Graph. **23**, 276–283 (2004)
6. Dokken, T., Lyche, T., Pettersen, K.F.: Polynomial splines over locally refined box-partitions. Comput. Aided Geom. Des. **30**, 331–356 (2013)
7. Wang, P., Xu, J., Deng, J., Chen, F.: Adaptive isogeometric analysis using rational PHT-splines. Comput. Aided Des. **43**(11), 1438–1448 (2011). ISSN 0010–4485, Solid and Physical Modeling (2011)

8. Lakså, A., Bang, B., Dechevsky, L.T.: Exploring expo-rational B-splines for curves and surfaces. In: Dæhlen, M., Mørken, K., Schumaker, L. (eds.) Mathematical Methods for Curves and Surfaces, pp. 253–262. Nashboro Press, Brentwood (2005)
9. Lakså, A.: Basic properties of Expo-Rational B-splines and practical use in computer aided geometric design. unipubavhandlinger 606, Unipub, Oslo (2007)
10. Lakså, A.: ERBS-surface construction on irregular grids. In: Pasheva, V., Venkov, G. (eds.) 39th International Conference Applications of Mathematics in Engineering and Economics AMEE13, pp. 113–120. American Institute of Physics (AIP) (2013)

Spline Representation of Connected Surfaces with Custom-Shaped Holes

Aleksander Pedersen[✉], Jostein Bratlie, and Rune Dalmo

R & D Group in Mathematical and Geometrical Modeling,
Numerical Simulations, Programming and Visualization,
Narvik University College, PO Box 385, 8505 Narvik, Norway
aleksanderpedersen90@gmail.com
http://www.hin.no/Simulations

Abstract. Compact surfaces possessing a finite number of boundaries are important to isogeometric analysis (IGA). Generalized expo-rational B-Splines (GERBS) is a blending type spline construction where local functions associated with each knot are blended by C^k-smooth basis functions. Modeling of surfaces with custom-shaped boundaries, or holes, can be achieved by using certain features and properties of the blending type spline construction, including local refinement and insertion of multiple inner knots. In this paper we investigate representation of arbitrary inner boundaries on parametric surfaces by using the above mentioned blending type spline construction.

Keywords: Isogeometric analysis · Splines · Blending methods · Boundary representations

1 Introduction

The field of isogeometric analysis (IGA), introduced by Hughes et al. in [11], is attempting to integrate finite element analysis (FEA) and computer aided design (CAD).

Polynomials on Bernstein form, B-splines in particular, are numerically more stable than polynomials on monomial form [9]. They can provide exact representations of elementary curves, surfaces and volumes. The non-uniform rational B-splines (NURBS) have been incorporated into the initial graphics exchange specification (IGES) and standard for the exchange of product model data (STEP) industry standards used in CAD. Spline based methods is one of the current approaches on IGA, attempting to bridge the gap from CAD to FEA. Some notable variants of spline based IGA include T-splines [16], PHT-splines [7], locally refined B-splines (LR B-splines) [8] and hierarchical B-splines [10,17].

This study constitutes one part in a series of attempts to explore the use of blending type splines in construction and analysis of iso geometry. In this article we address the use of tensor product boolean sum surfaces of Coons type [3] as local surface patches for the spline construction. Our main motivation is to

© Springer International Publishing Switzerland 2015
I. Lirkov et al. (Eds.): LSSC 2015, LNCS 9374, pp. 394–400, 2015.
DOI: 10.1007/978-3-319-26520-9_44

explore how well we can control the shape of an internal hole, represented by a boundary curve, in a C^k smooth surface.

The following sections provide a brief overview of the GERBS surface construction, Coons patch and boolean sum surfaces followed by a description of our method before we present our results and finally give our concluding remarks.

2   The Blending-Type Spline Construction

Expo-rational B-splines (ERBS) were first introduced under this name in [6,12]. The generalization of expo-rational B-splines (GERBS) appeared in [5]. Later, in [13,14], the ERBS blending construction was presented in the framework of the B-spline recursion formula associated with the knots $(t_i)_{i=0}^{k+d}$:

$$B_{d,k}(t) = B \circ \omega_{d,k}(t)B_{d-1,k}(t) + (1 - B \circ \omega_{d,k+1}(t))B_{d-1,k+1}(t), \qquad (1)$$

where $\omega_{d,i}(t) = \frac{t-t_i}{t_{i+d}-t_i}, B_{0,i}(t) = \begin{cases} 1; & \text{if } t_i \le t < t_{i+1}. \\ 0; & \text{otherwise.} \end{cases}$ The degree $d = 1$ in

the case of GERBS; moreover, B is a C^k-smooth blending function possessing the following set of properties:

1. $B : I \to I$ ($I = [0,1] \subset \mathbb{R}$),
2. $B(0) = 0$,
3. $B(1) = 1$,
4. $B'(t) >= 0, t \in I$.
5. $B(t) + B(1 - t) = 1, t \in I$.

The last property is optional and specifies point symmetry around the point $(0.5, 0.5)$, however, we assume this property in the present study.

In the present work we consider tensor product surfaces defined as

$$S(u,v) = \sum_{i=1}^{n} \sum_{j=1}^{m} \ell_{i,j}(u,v)B_{1,i}(u)B_{1,j}(v), \qquad (2)$$

where $\ell_{i,j}(u,v)$ are *local surface patches* which are blended together by the C^k-smooth blending functions $B_{1,i}$ and $B_{1,j}$.

Local refinement by knot insertion for the blending construction was addressed in [2,4]. Splitting of the geometry, which can be achieved by using multiple inner knots in this case, is one form of editing which was first investigated in [1] as a technique for artistic editing where a method for creating holes in the geometry was presented.

One consequence of splitting by using multiple knots is that affine transforms can be applied to the local surface patches associated with the knots. Separating two connected double knots gives a geometric G^0 continuity of the construction, in the corresponding parametric direction, while maintaining the C^k mathematical smoothness. This follows as a consequence of the transfinite

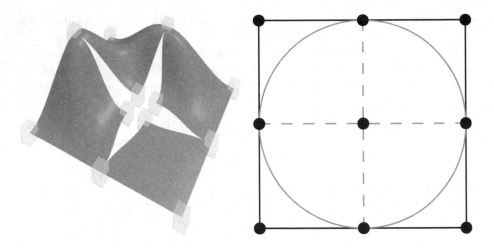

Fig. 1. A blending-type spline surface with multiple knots which are separated to generate a split (to the left) and the parametric representation of the same surface (to the right).

Hermite interpolation property at the knots, which in turn is due to the vanishing derivatives of the basis function at the knots.

Figure 1 shows an example of how the blending construction can be splitted by using multiple knots which are moved away from each other, and how the parametric representation is still continuous. Using double knots in both parametric directions splits the surface into four equal pieces. Each quadrangle has four local patches, pointing at a single patch, which is representing a corner in the final surface. A geometric hole appears in the center of the surface since the four local patches in the center is pointing to a respective patch in a corner.

3   Boolean Sum Surfaces

The Coons surface patch [3] is a surface segment bounded by four spatial curves which it interpolates to. Coons proposed using blending surfaces in [3]. Two ruled surfaces, where each of them interpolates two of the four boundary curves, are added together with this construction. A slope correction surface is subtracted to compensate for the curves which the ruled surfaces fail to interpolate.

Figure 2 illustrates the components of a bilinear Coons patch defined by the boolean sum surface

$$\ell(u,v) = L_1(u,v) + L_2(u,v) - L_3(u,v), \tag{3}$$

where the ruled surface $L_1(u,v)$ interpolates the two boundary curves $c_1(u), c_2(u)$ but is linear along v, $L_2(u,v)$ interpolates $g_1(v), g_2(v)$ but is linear along u, and $L_3(u,v)$ is a bilinear slope correction surface matching the linear components of both $L_1(u,v)$ and $L_2(u,v)$.

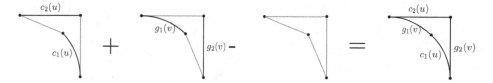

Fig. 2. The surfaces involved in a bilinear Coons patch construction. The red lines indicate linear segments between two nodes, and the black lines indicate the interpolated curves. Adding two ruled surfaces and subtracting a bilinear one yields the final Coons patch surface (right). From left to right: $L_1 + L_2 - L_3 = \ell$

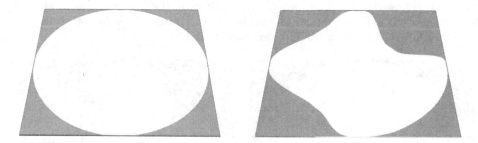

Fig. 3. The illustration shows the shaded version of the two example constructions. The surface on the left has a circular inner boundary while the one on the right has a free form ERBS-based inner boundary. Both inner boundary curves interpolate the outer boundary on the middle of the four edges.

4   Coons Patches as Local Geometry

We propose using Coons patches as local surfaces in connection with double knots to obtain splits, or holes, with arbitrary shape. The desired shape for the inner boundary can be described by curves which are interpolated by the Coons patch local surfaces. The only requirement for each quadrant is that they interpolate in the parametric central knot of every edge.

The initial shape of the resulting surface (before separating the multiple knots) will be preserved since the local patches which are blended together have exactly the same shape and since the desired geometric hole is constructed using the inner boundary curve.

As an experiment and proof of concept we provide two surfaces. They are constructed from an inner and outer boundary curve. In both cases the outer boundary curve is a closed parametric rectangular curve. Both examples have uniform 3×3 knots vectors, where the middle knot is a double knot, in both parametric directions.

In the first example, shown on the left in Fig. 3, the inner boundary is a circular curve interpolating the outer boundary at the middle double knots. In the second example, shown on the right in Fig. 3, the inner boundary is a parametric ERBS-based free form blending-type spline curve interpolating the outer boundary curve at the middle knots.

Fig. 4. The illustration shows a shaded version of the two example surface constructions, $S(u,v)$ where $u, v \in [0,2]$, with a curve plotted in the parametric plane along v for a fixed $u = 0.5$. The surface on the left has a circular inner boundary while the one on the right has a free form ERBS-based inner boundary.

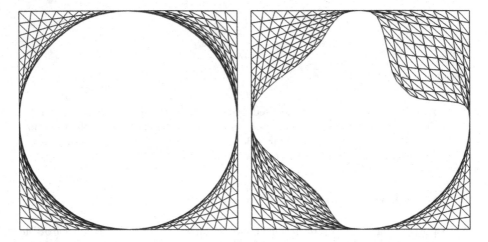

Fig. 5. The illustrations shows the tensor-product parameter grid distribution as wireframes of the two example constructions. The surface on the left has a circular inner boundary while the one on the right has a free form ERBS-based inner boundary. Both surfaces is sampled at 20×20 samples.

A curve evaluated on the boundary of the two example surfaces is shown in Fig. 4 and a wireframe representation of the examples, showing the distribution of the parametric domain, is provided in Fig. 5. On both surfaces, $s(u,v)$ where $u, v \in [0,2]$, the surface curve is evaluated along v for a fixed $u = 0.5$.

5   Concluding Remarks

In both examples the middle knots on the boundary is equally distanced from both sides. We note that this is not a general requirement. Moreover, in both examples, a compact C^k-smooth tensor-product blending-type spline surface construction is created where the inner boundary curve limits a natural hole in the surface construction. Both constructions are exact constructions in the sense that the inner and outer boundary curves are interpolated. The construction is C^k everywhere and at the knots the smoothness depends on the local geometry. In the case where bilinear Coons patches is used as representation of the local geometry the smoothness is C^1. We note that in the case where bicubic blending would be used the construction would be C^2 at the knots. Further exploration is a topic for future work. All the while the construction has a naturally inner hole represented by the inner boundary curve and is therefore G^0. The construction is also evaluable everywhere on the tensor-product parametric domain.

One expansion of this construction could be such that the inner and outer boundary curves do not interpolate at the boundary knots. Another interesting expansion could be to expand the construction into a three-tensor volumetric representation where the inner boundary would be a parametric surface.

References

1. Andresen, K.M.: Brukergrensesnitt for kunstnerisk design ved bruk av ERBS. Master's thesis, Narvik University College (2008)
2. Bratlie, J.: Local refinement of GERBS surfaces with applications to interactive geometric modeling. In: Pasheva, V., Venkov, G. [15], pp. 18–25
3. Coons, S.A.: Surfaces for computer-aided design of space forms. Project MAC-TR 41, Massachusetts Institute of Technology, Massachusetts, USA (1967)
4. Dalmo, R.: Local refinement of ERBS curves. In: Pasheva, V., Venkov, G. [15], pp. 204–211
5. Dechevsky, L.T., Bang, B., Lakså, A.: Generalized expo-rational B-splines. Int. J. Pure Appl. Math. **57**(6), 833–872 (2009)
6. Dechevsky, L.T., Lakså, A., Bang, B.: Expo-rational B-splines. Int. J. Pure Appl. Math. **27**(3), 319–362 (2006)
7. Deng, J., Chen, F., Feng, Y.: Dimensions of spline spaces over T-meshes. J. Comput. Appl. Math. **194**(2), 267–283 (2006). http://dx.doi.org/10.1016/j.cam.2005.07.009
8. Dokken, T., Lyche, T., Pettersen, K.F.: Polynomial splines over locally refined box-partitions. Comput. Aided Geom. Des. **30**(3), 331–356 (2013). http://www.sciencedirect.com/science/article/pii/S0167839613000113
9. Farouki, R.T., Rajan, V.T.: On the numerical condition of polynomials in Bernstein form. Comput. Aided Geom. Des. **4**(3), 191–216 (1987)
10. Giannelli, C., Jüttler, B., Speleers, H.: THB-splines: the truncated basis for hierarchical splines. Comput. Aided Geom. Des. **29**(7), 485–498 (2012)
11. Hughes, T.J.R., Cottrell, J.A., Bazilevs, Y.: Isogeometric analysis: CAD, finite elements, NURBS, exact geometry and mesh refinement. Comput. Meth. Appl. Mech. Eng. **194**(39–41), 4135–4195 (2005). http://dx.doi.org/10.1016/j.cma.2004.10.008
12. Lakså, A., Bang, B., Dechevsky, L.T.: Exploring expo-rational B-splines for curves and surfaces. In: Dæhlen, M., Mørken, K., Schumaker, L. (eds.) Mathematical Methods for Curves and Surfaces, pp. 253–262. Nashboro Press, Brentwood (2005)
13. Lakså, A.: ERBS-surface construction on irregular grids. In: Pasheva, V., Venkov, G. [15], pp. 113–120

14. Lakså, A.: Construction and properties of non-polynomial spline curves. In: Sivasundaram, S. (ed.) ICNPAA 2014 World Congress: 10th International Conference on Mathematical Problems in Engineering, Aerospace and Sciences. AIP Conference Proceedings, vol. 1637, pp. 545–554. AIP Publishing (2014)

15. Pasheva, V., Venkov, G. (eds.): 39th International Conference Applications of Mathematics in Engineering and Economics AMEE13, AIP Conference Proceedings, vol. 1570. AIP Publishing (2013)

16. Sederberg, T.W., Zheng, J., Bakenov, A., Nasri, A.: T-splines and T-NURCCs. In: ACM SIGGRAPH 2003 Papers, pp. 477–484. SIGGRAPH 2003, ACM, New York, NY, USA (2003). http://doi.acm.org/10.1145/1201775.882295

17. Vuong, A.V., Giannelli, C., Jüttler, B., Simeon, B.: A hierarchical approach to adaptive local refinement in isogeometric analysis. Comput. Meth. Appl. Mech. Eng. 200(49–52), 3554–3567 (2011)

Scalability of Shooting Method for Nonlinear Dynamical Systems

Stanislav Stoykov$^{(\boxtimes)}$ and Svetozar Margenov

Institute of Information and Communication Technologies,
Bulgarian Academy of Sciences, Acad. G. Bonchev Street,
Bl. 25A, 1113 Sofia, Bulgaria
{stoykov,margenov}@parallel.bas.bg

Abstract. The computation of periodic solutions of nonlinear dynamical systems is essential step for their analysis. The variation of the steady-state periodic responses of elastic structures with the frequency of vibration or with the excitation frequency provides valuable information about the dynamical behavior of the structure. Shooting method computes iteratively the periodic solutions of dynamical systems. In the current paper a parallel implementation of the shooting method is presented. The nonlinear equation of motion of Bernoulli-Euler beam is used as a model equation. Large-scale system of ordinary differential equations is generated by applying the finite element method. The speedup and efficiency of the method are studied and presented.

1 Introduction

In the last decades many researchers have been involved in development of numerical methods and tools for investigating the dynamical characteristics of nonlinear systems. The knowledge of the steady-state response of elastic structures due to external excitations with different frequencies presents valuable information which is used by engineers for design and maintenance of the structures. The nonlinear frequency-response function, together with determination of bifurcation points, secondary branches and stability provides such information about the dynamics of the structure.

Shooting method is an iterative method for computing periodic responses [6]. In a combination with the arc-length continuation method, it allows one to compute complete branch of solutions from the bifurcation diagram [5].

Due to complexity of the structures three-dimensional finite elements become more appropriate for space discretization [8]. This leads to systems with large number of degrees of freedom. The complete dynamical analysis becomes computationally burdensome and time consuming process when it is applied to large-scale systems. Thus, the parallel computation of the numerical methods used in the analysis becomes unavoidable part when one investigates real-life structures.

In the present work, the nonlinear equation of motion of Bernoulli-Euler beam is used for investigating the scalability of the shooting method. A parallel

© Springer International Publishing Switzerland 2015
I. Lirkov et al. (Eds.): LSSC 2015, LNCS 9374, pp. 401–408, 2015.
DOI: 10.1007/978-3-319-26520-9_45

implementation of the complete process of computing the nonlinear frequency-response function is presented and its efficiency is studied. The process involves algebraic operations of sparse and dense matrices and also solutions of linear and nonlinear systems with both types of matrices. Appropriate libraries are used for these operations and in a compliance with the parallel realization of the method.

2   Nonlinear Equation of Motion of Bernoulli-Euler Beam

The nonlinear equation of motion of Bernoulli-Euler beam is used as a model equation for investigating the efficiency of the parallel implementation of the shooting method. The equation has the following form:

$$\rho A \frac{\partial^2 w(x,t)}{\partial t^2} + EI \frac{\partial^4 w(x,t)}{\partial x^4} - \frac{3}{2} EA \left(\frac{\partial w(x,t)}{\partial x} \right)^2 \left(\frac{\partial^2 w(x,t)}{\partial x^2} \right) = f(x,t) \quad (1)$$

where $w(x,t)$ is the transverse displacement of the beam, ρ is the density of the beam, A is the cross sectional area, E is the elastic modulus, I is the second moment of area and $f(x,t)$ is the applied external force. The beam is assumed to have length l, $x \in [-l/2, l/2]$. A clamped-clamped boundary conditions are considered, that are given by:

$$w(-l/2, t) = 0, \left. \frac{\partial w(x,t)}{\partial x} \right|_{x=-l/2} = 0 \qquad (2)$$

$$w(l/2, t) = 0, \left. \frac{\partial w(x,t)}{\partial x} \right|_{x=l/2} = 0 \qquad (3)$$

Details about the derivation of the equation of motion are given in [6].

The equation of motion (1) is written in its variational form, integration by parts is applied and using the boundary conditions (2) and (3), the following weak formulation is obtained:

$$\rho A \int_\Omega \frac{\partial^2 w(x,t)}{\partial t^2} \psi(x) dx + EI \int_\Omega \frac{\partial^2 w(x,t)}{\partial x^2} \frac{\partial^2 \psi(x)}{\partial x^2} dx$$
$$+ \frac{1}{2} EA \int_\Omega \left(\frac{\partial w(x,t)}{\partial x} \right)^3 \frac{\partial \psi(x)}{\partial x} dx = \int_\Omega f(x,t)\psi(x) dx, \qquad (4)$$

where $\psi_i(x)$ are the trial functions. The transverse displacement is expressed by shape functions and generalized coordinates in a local coordinate system:

$$w^e(\xi, t) = \mathbf{N}^T(\xi) \mathbf{q}^e(t) \qquad (5)$$

where $\mathbf{N}(\xi)$ is the vector of shape functions, Hermite cubic polynomial functions are used (Fig. 1), $\mathbf{q}^e(t)$ is the local vector of generalized coordinates and ξ is the local coordinate, $\xi \in [-1, 1]$.

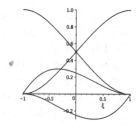

Fig. 1. Hermite cubic polynomial functions used as shape functions for the beam elements, ξ is the local coordinate.

The local mass and stiffness matrices are given in the following form:

$$\mathbf{M}^e = \rho A \frac{l}{2} \int_{-1}^{1} \mathbf{N}(\xi) \mathbf{N}^T(\xi) d\xi \tag{6}$$

$$\mathbf{K_L}^e = EI \left(\frac{2}{l}\right)^3 \int_{-1}^{1} \frac{\partial^2 \mathbf{N}(\xi)}{\partial \xi^2} \frac{\partial^2 \mathbf{N}^T(\xi)}{\partial \xi^2} d\xi \tag{7}$$

$$\mathbf{K_{NL}}^e(\mathbf{q}^e(t)) = EA \frac{4}{l^3} \int_{-1}^{1} \frac{\partial \mathbf{N}(\xi)}{\partial \xi} \frac{\partial \mathbf{N}^T(\xi)}{\partial \xi} \left(\frac{\partial \mathbf{N}^T(\xi)}{\partial \xi} \mathbf{q}^e(t)\right)^2 d\xi \tag{8}$$

$$\mathbf{F}^e(t) = \frac{l}{2} \int_{-1}^{1} f(\xi, t) \mathbf{N}(\xi) d\xi \tag{9}$$

where superscript ^{e} denotes the element matrix. $\mathbf{M}$ represents the mass matrix, $\mathbf{K_L}$ - the stiffness matrix of constant terms, $\mathbf{K_{NL}}(\mathbf{q}(t))$ - the stiffness matrix that depends on the transverse displacement $w(\xi, t)$ and consequently on the vector of generalized coordinates $\mathbf{q}(t)$, $\mathbf{F}(t)$ is the generalized vector of external forces.

After assembling the matrices of the elements and introducing a damping matrix, the following nonlinear system of ordinary differential equations is obtained:

$$\mathbf{M}\ddot{\mathbf{q}}(t) + \mathbf{C}\dot{\mathbf{q}}(t) + \mathbf{K_L}\mathbf{q}(t) + \mathbf{K_{NL}}(\mathbf{q}(t))\mathbf{q}(t) = \mathbf{F}(t) \tag{10}$$

where $\mathbf{C}$ is the damping matrix. It is proportional to the mass matrix and depends on the frequency of vibration. The damping matrix is expressed as $\mathbf{C} = 0.01 \frac{\omega_l^2}{\omega} \mathbf{M}$, where ω_l is the fundamental natural frequency and ω is the excitation frequency. The total number of degrees of freedom of system (10) is denoted by N.

3   Computation of Periodic Responses by Parallel Algorithms

Periodic responses are of interest, thus one needs to compute the initial conditions which lead to periodic oscillations. At this point it is assumed that the period of vibration is $T = 2\pi/\omega$. It is pointed that, because the system is nonlinear, a period-multiplying bifurcation point can exist, and the period of vibration T can become an integer multiplied by $2\pi/\omega$.

The response of the system depends on the time t but also on the initial conditions, thus the response and velocity at time t due to initial conditions $\mathbf{q}_0$ and $\dot{\mathbf{q}}_0$ are written by $\mathbf{q}(t, \mathbf{q}_0, \dot{\mathbf{q}}_0)$ and $\dot{\mathbf{q}}(t, \mathbf{q}_0, \dot{\mathbf{q}}_0)$. Dot is used to denote time derivative. The initial conditions $\mathbf{q}_0$ and $\dot{\mathbf{q}}_0$ are of interest, they satisfy the equations:

$$\mathbf{q}(T, \mathbf{q}_0, \dot{\mathbf{q}}_0) = \mathbf{q}_0, \tag{11}$$

$$\dot{\mathbf{q}}(T, \mathbf{q}_0, \dot{\mathbf{q}}_0) = \dot{\mathbf{q}}_0. \tag{12}$$

Since the initial conditions that lead to periodic response are not known preliminary, the shooting method computes them iteratively by predictor-corrector scheme. Details of the shooting method for systems of second order ODE are given in [7]. The main characteristics of the method and its parallel implementation are summarized here.

The corrections $\delta\mathbf{q}_0$ and $\delta\dot{\mathbf{q}}_0$ of the initial conditions are obtained by solving the linear system:

$$\begin{bmatrix} \mathbf{Q_d}(T) - \mathbf{I} & \mathbf{Q_v} \\ \dot{\mathbf{Q}}_\mathbf{d}(T) & \dot{\mathbf{Q}}_\mathbf{v}(T) - \mathbf{I} \end{bmatrix} \begin{Bmatrix} \delta\mathbf{q}_0 \\ \delta\dot{\mathbf{q}}_0 \end{Bmatrix} = \begin{Bmatrix} \mathbf{q}_0 - \mathbf{q}(T, \mathbf{q}_0, \dot{\mathbf{q}}_0) \\ \dot{\mathbf{q}}_0 - \dot{\mathbf{q}}(T, \mathbf{q}_0, \dot{\mathbf{q}}_0) \end{Bmatrix} \tag{13}$$

where $\mathbf{Q_d}(T)$ and $\dot{\mathbf{Q}}_\mathbf{d}(T)$ are solutions at time T of system:

$$\mathbf{M}\ddot{\mathbf{Q}}(t) + \mathbf{C}\dot{\mathbf{Q}}(t) + \mathbf{K_L}\mathbf{Q}(t) + \mathbf{J}(\mathbf{q}(t))\mathbf{Q}(t) = \mathbf{0} \tag{14}$$

due to initial conditions $\mathbf{Q_d}(0) = \mathbf{I}$ and $\dot{\mathbf{Q}}_\mathbf{d}(0) = \mathbf{0}$. $\mathbf{Q_v}(T)$ and $\dot{\mathbf{Q}}_\mathbf{v}(T)$ are solutions at time T of the system (14) due to initial conditions $\mathbf{Q_v}(0) = \mathbf{0}$ and $\dot{\mathbf{Q}}_\mathbf{v}(0) = \mathbf{I}$. System (14) needs to be solved for $2N$ independent initial conditions. $\mathbf{J}(\mathbf{q}(t))$ is the Jacobian of the nonlinear terms of the equation of motion (10), i.e.

$$\mathbf{J}(\mathbf{q}(t)) = \frac{\partial \mathbf{K_{NL}}(\mathbf{q}(t))\mathbf{q}(t)}{\partial \mathbf{q}(t)} \tag{15}$$

When the Jacobian $\mathbf{J}(\mathbf{q}(t))$ is computed, (14) becomes a linear system of ordinary differential equations. Thus, the solution vectors $\mathbf{q}(t)$ and $\dot{\mathbf{q}}(t)$, and the solution matrices $\mathbf{Q_d}(t)$, $\dot{\mathbf{Q}}_\mathbf{d}(t)$, $\mathbf{Q_v}(t)$ and $\dot{\mathbf{Q}}_\mathbf{v}(t)$ are computed simultaneously on each time step by Newmarks's method [2]. Newton-Raphson's method [2] is used for the resulting algebraic nonlinear system which is a consequence of the application of Newmark's method to system (10).

The parallel realization of the algorithm is presented here. It is noted that system (14) presents $2N$ independent linear systems of ODE. The matrices $\mathbf{Q_d}(t)$ and $\dot{\mathbf{Q}}_\mathbf{d}(t)$, $\mathbf{Q_v}(t)$ and $\dot{\mathbf{Q}}_\mathbf{v}(t)$ are dense matrices, consequently system (13) is a dense linear system. The proposed parallel implementation of the method consists of separation of the matrices $\mathbf{Q_d}(t)$ and $\dot{\mathbf{Q}}_\mathbf{d}(t)$, $\mathbf{Q_v}(t)$ and $\dot{\mathbf{Q}}_\mathbf{v}(t)$ into blocks of vector-columns (Fig. 2). The solution in the time domain of each block of vector-columns is computed independently. Let Δt be the time integration step, $a_0 = \frac{1}{\beta \Delta t^2}$, $a_1 = \frac{\gamma}{\beta \Delta t}$, $a_2 = \frac{1}{\beta \Delta t}$, $a_3 = \frac{1}{2\beta} - 1$, $a_4 = \frac{\gamma}{\beta} - 1$, and $a_5 = \frac{\Delta t \gamma}{2\beta} - 1$ are the constants of Newmark's method, β and γ are parameters that can be determined to obtain integration accuracy and stability. The UMFPACK library [4], which is a direct solver, is used for solving the sparse systems (10) and (14). ScaLAPACK [3] library, which is library of high-performance linear algebra routines for parallel distributed memory machines, is used for solving the dense system (13).

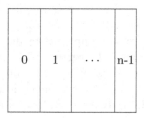

Fig. 2. Partition of dense matrices among n processes.

The algorithm is summarized here by the following pseudo-code. The subscripts t and $t+\Delta t$ denote the values of the vectors and the matrices at these time steps.

1. Define predictions to the initial conditions $\mathbf{q}_0$ and $\dot{\mathbf{q}}_0$ for chosen excitation frequency ω.
2. Start time integration from 0 to T. At time step $t + \Delta t$ do
   (a) Solve the nonlinear system
   $$\left(a_0\mathbf{M} + a_1\mathbf{C} + \mathbf{K_L} + \mathbf{K_{NL}}(\mathbf{q}_{t+\Delta t})\right)\mathbf{q}_{t+\Delta t} =$$
   $$= \mathbf{F}_{t+\Delta t} + (a_0\mathbf{M} + a_1\mathbf{C})\mathbf{q}_t + (a_2\mathbf{M} + a_4\mathbf{C})\dot{\mathbf{q}}_t + (a_3\mathbf{M} + a_5\mathbf{C})\ddot{\mathbf{q}}_t \text{ by}$$
   Newton-Raphson method and UMFPACK and compute $\mathbf{J}(\mathbf{q}_{t+\Delta t})$.
   (b) Compute the factorization of the matrix
   $$\mathbf{M_E} = a_0\mathbf{M} + a_1\mathbf{C} + \mathbf{K_L} + \mathbf{J}(\mathbf{q}_{t+\Delta t})$$
   (c) Compute matrix-matrix products and sums by parallel algorithms
   $$\mathbf{R_D} = (a_0\mathbf{M} + a_1\mathbf{C})\mathbf{Q_{d_t}} + (a_2\mathbf{M} + a_4\mathbf{C})\dot{\mathbf{Q}}_{\mathbf{d_t}} + (a_3\mathbf{M} + a_5\mathbf{C})\ddot{\mathbf{Q}}_{\mathbf{d_t}}$$
   $$\mathbf{R_V} = (a_0\mathbf{M} + a_1\mathbf{C})\mathbf{Q_{v_t}} + (a_2\mathbf{M} + a_4\mathbf{C})\dot{\mathbf{Q}}_{\mathbf{v_t}} + (a_3\mathbf{M} + a_5\mathbf{C})\ddot{\mathbf{Q}}_{\mathbf{v_t}}$$
   (d) Solve the $2N$ systems by parallel algorithms and UMFPACK
   $$\mathbf{M_E}\mathbf{Q}_{\mathbf{d}_{t+\Delta t}} = \mathbf{R_D}$$
   $$\mathbf{M_E}\mathbf{Q}_{\mathbf{v}_{t+\Delta t}} = \mathbf{R_V}$$
3. Compute the corrections $\delta\mathbf{q}_0$ and $\delta\dot{\mathbf{q}}_0$ of the initial conditions. Use parallel solvers for the dense matrix - ScaLAPACK.
   $$\begin{bmatrix} \mathbf{Q_d}(T) - \mathbf{I} & \mathbf{Q_v} \\ \dot{\mathbf{Q}}_{\mathbf{d}}(T) & \dot{\mathbf{Q}}_{\mathbf{v}}(T) - \mathbf{I} \end{bmatrix} \begin{Bmatrix} \delta\mathbf{q}_0 \\ \delta\dot{\mathbf{q}}_0 \end{Bmatrix} = \begin{Bmatrix} \mathbf{q}_0 - \mathbf{q}(T) \\ \dot{\mathbf{q}}_0 - \dot{\mathbf{q}}(T) \end{Bmatrix}$$
4. Check for convergence, i.e. if
   $$\|\mathbf{q}(T, \mathbf{q}_0 + \delta\mathbf{q}_0, \dot{\mathbf{q}}_0 + \delta\dot{\mathbf{q}}_0) - \mathbf{q}_0 - \delta\mathbf{q}_0\| < \epsilon$$
   $$\|\dot{\mathbf{q}}(T, \mathbf{q}_0 + \delta\mathbf{q}_0, \dot{\mathbf{q}}_0 + \delta\dot{\mathbf{q}}_0) - \dot{\mathbf{q}}_0 - \delta\dot{\mathbf{q}}_0\| < \epsilon$$

5. If convergence is achieved, define new prediction of q_0, $\dot{q}_0$ and ω and compute the next point from the frequency-response function, i.e. go to 1. Else, further improve the initial conditions by repeating the process from step 2.

4   Scalability

The speedup and efficiency of the parallel implementation of the shooting method applied to the nonlinear equation of motion of Bernoulli-Euler beam (1) is studied in this section. A mesh with 4097 finite elements is used for the numerical experiments. The total number of degrees of freedom of the nonlinear system of ODE (10) is 8192. The dimension of the dense matrix, used to correct the initial conditions is 16384. The numerical computations are carried out on HPCG cluster [1] located at IICT-BAS. The cluster has 576 computing cores (Intel Xeon X5560 @ 2.8 GHz) organized in blade system with 36 blades BL 280c. Each blade has 24 GB RAM. Non-blocking DDR interconnection via Voltaire Grid director 2004 with latency 2.5 μs and bandwidth 20 Gbps is used to connect the blades.

Table 1. Strong scalability of time integration part of shooting method (step 2).

P	CPU (s)	Speed up	Efficiency %
4	3393.18	-	-
8	1710.07	1.98	99.58
16	860.92	3.94	98.53
32	445.25	7.62	95.26
64	230.22	14.74	92.12

The process of a single shooting iteration is divided into two parts: the first one presents the time integration, i.e. solving the nonlinear system (10) for the time integration interval and computing the dense matrices which come from (14). This part presents step 2 from the pseudo-code. The second part is the gathering process of the dense matrices which define system (13), distributing the dense matrix by performing two-dimensional block cyclic distribution and computing its solution by ScaLAPACK. This part is denoted as step 3 from the pseudo-code.

The speedup and efficiency of both parts are presented in Tables 1 and 2. Very good efficiency of the parallel computations is achieved for the time integration process. The most time consuming operations are the matrix-matrix products and the solutions of the systems, i.e. steps 2c and d from the pseudo-code. These operations are realized on parallel processors. Steps 2a and b from the pseudo-code are not implemented for parallel computations. Parallel implementation of these steps can be considered for future improvement of the method.

Subroutine PDGESV from ScaLAPACK library is used to solve the dense system. Scalability of this subroutine was shown to perform reasonably well when

Table 2. Strong scalability results of generating, distributing and solving the dense system by ScaLAPACK (step 3).

P	CPU (s)	Speed up	Efficiency %
4	478.16	-	-
8	253.22	1.89	94.42
16	127.68	3.74	93.62
32	80.11	5.97	74.61
64	51.79	9.23	57.71

Table 3. Strong scalability results of one complete shooting iteration (steps 2 4).

P	CPU (s)	Speed up	Efficiency %
4	3871.33	-	-
8	1963.29	1.97	98.59
16	988.60	3.92	97.90
32	525.36	7.37	92.11
64	282.01	13.73	85.80

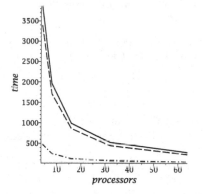 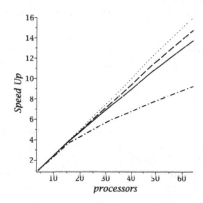

Fig. 3. (a) Variation of CPU time with the number of processors, (b) Scalability of shooting method, − − − time integration part of shooting method, −·−·−·− distributing the dense matrix and solving the dense system, ——complete shooting iteration, · · · · · optimal (linear) performance.

appropriate parameters are used [3]. The decrease of the efficiency, which can be noted in Table 2, is due to the additional communications between the processors. These communications gather the separate parts of the dense matrices $\mathbf{Q_d}(T)$, $\dot{\mathbf{Q}}_\mathbf{d}(T)$, $\mathbf{Q_v}(T)$ and $\dot{\mathbf{Q}}_\mathbf{v}(T)$ and distribute the global dense matrix from (13) among the processors by two-dimensional block cyclic distribution. Because of these communications, the efficiency of the second part of the shooting method differs from optimal.

Scalability of the whole shooting iteration is presented in Table 3. The results demonstrate that the most time consuming operations were successfully computed by parallel algorithms. Figure 3 presents the CPU time versus processors and the scalability of the parallel shooting method. The results confirm that the shooting method was parallelized successfully and the scalability is close to optimal.

5   Conclusion

The shooting method, used to compute periodic responses of elastic structures and applied to systems of second order ordinary differential equations, was realized by parallel algorithms for distributed memory machines. The method consists of multiple computations of solutions of systems composed of sparse and dense matrices, as well of algebraic operations between both types of matrices. Its parallel realization simultaneously involves efficient algorithms for sparse and dense matrices.

The scalability of the method was studied on the nonlinear equation of motion of Bernoulli-Euler beam. The results show that the shooting method, implemented for parallel computations, has the potential to be used for dynamical analysis of engineering applications. The future work will consider its implementation of elastic structures discretized by three-dimensional finite elements.

Acknowledgments. This work was supported by the project AComIn "Advanced Computing for Innovation", grant 316087, funded by the FP7 Capacity Programme.

References

1. Atanassov, E., Gurov, T., Karaivanova, A., Ivanovska, S., Durchova, M., Georgiev, D., Dimitrov, D.: Tuning for scalability on hybrid HPC cluster. Math. Ind. **224**, 64–77 (2014)
2. Bathe, K.J.: Finite Element Procedures. Prentice Hall, Englewood Cliffs (1996)
3. Blackford, L.S., Choi, J., Cleary, A., D'Azevedo, E., Demmel, J., Dhillon, I., Dongarra, J., Hammarling, S., Henry, G., Petitet, A., Stanley, K., Walker, D., Whaley, R.C.: ScaLAPACK Users' Guide. Society for Industrial and Applied Mathematics, Philadelphia (1997)
4. Davis, T.: Algorithm 832: UMFPACK, an unsymmetric-pattern multifrontal method. ACM Trans. Math. Softw. **30**(2), 196–199 (2004)
5. Kerschen, G., Peeters, M., Golinval, J., Vakakis, A.: Nonlinear normal modes, part I: a useful framework for the structural dynamicist. Mech. Syst. Sig. Process. **23**(1), 170–194 (2009)
6. Nayfeh, A., Pai, P.: Linear and Nonlinear Structural Mechanics. Wiley, New York (2004)
7. Stoykov, S., Margenov, S.: Numerical computation of periodic responses of nonlinear large-scale systems by shooting method. Comput. Math. Appl. **67**(12), 2257–2267 (2014)
8. Zienkiewicz, O., Taylor, R., Zhu, J.: The Finite Element Method: Its Basis and Fundamentals. Elsevier, Oxford (2005)

Selecting Explicit Runge-Kutta Methods
with Improved Stability Properties

Zahari Zlatev[1], Krassimir Georgiev[2](✉), and Ivan Dimov[2]

[1] Department of Environmental Science, Aarhus University,
Frederiksborgvej 399, P.O. 358, 4000 Roskilde, Denmark
[2] Institute of Information and Communication Technologies,
BAS, Acad. G. Bonchev Str., Bl. 25-A, 1113 Sofia, Bulgaria
georgiev@parallel.bas.bg

Abstract. Explicit Runge-Kutta methods can efficiently be used in the numerical integration of initial value problems for non-stiff systems of ordinary differential equations (ODEs). Let m and p be the number of stages and the order of a given explicit Runge-Kutta method. We have proved in a previous paper [8] that the combination of any explicit Runge-Kutta method with $m = p$ and the Richardson Extrapolation leads always to a considerable improvement of the absolute stability properties. We have shown in [7] (talk presented at the NM&A14 conference in Borovets, Bulgaria, August 2014) that the absolute stability regions can be further increased when $p < m$ is assumed. For two particular cases, $p = 3 \wedge m = 4$ and $p = 4 \wedge m = 6$ it is demonstrated that
(a) the absolute stability regions of the new methods are larger than those of the corresponding explicit Runge-Kutta methods with $p = m$, and
(b) these regions are becoming much bigger when the Richardson extrapolation is additionally applied.
The explicit Runge-Kutta methods, which have optimal absolute stability regions, form two large classes of numerical algorithms (each member of any of these classes having the same absolute stability region as all the others). Rather complicated order conditions have to be derived and used in the efforts to obtain some special methods within each of the two classes.

We selected two particular methods within these two classes and tested them by using appropriate numerical examples.

Keywords: Ordinary differential equations · Explicit Runge-Kutta methods · Richardson extrapolation · Absolute stability regions · Order conditions

1 Statement of the Problem

Consider the classical initial value problem for non-linear systems of ordinary differential equations (ODEs) defined by:

$$\frac{dy}{dt} = f(t, y), \ t \in [a, b], \ a < b, \ y \in R^s, \ s \geq 1, \ f \in D \subset R \times R^s, \ y(a) = \eta, \quad (1)$$

© Springer International Publishing Switzerland 2015
I. Lirkov et al. (Eds.): LSSC 2015, LNCS 9374, pp. 409–416, 2015.
DOI: 10.1007/978-3-319-26520-9_46

An m-stage explicit Runge-Kutta method is based on the following formula:

$$y_n = y_{n-1} + h \sum_{i=1}^{m} c_i k_i^n, \tag{2}$$

The coefficients c_i are given constants, while at time-step n the stages k_i^n are defined by

$$
\begin{aligned}
k_1^n &= f(t_{n-1}, y_{n-1}), \\
k_i^n &= f\left(t_{n-1} + h a_i, y_{n-1} + h \sum_{j=1}^{i-1} b_{ij} k_j^n\right), & i = 2, 3, \ldots, m \\
a_i &= \sum_{j=1}^{i-1} b_{ij}, & i = 2, 3, \ldots, m
\end{aligned}
\tag{3}
$$

where b_{ij} are some constants depending on the particular numerical method. The order of accuracy of the approximation $y_n \approx y(t_n)$ will be denoted by p and mainly the case $p < m$ will be considered in this paper.

Assume that: (a) the order of method (2) and (3) is p and (b) two approximations z_n and w_n are calculated by using the starting value y_{n-1} as well as stepsizes h and $0.5h$ respectively. Then the following two relationships can be obtained:

$$y(t_n) - z_n = h^p K + O(h^{p+1}), \tag{4}$$
$$y(t_n) - w_n = (0.5h)^p K + O(h^{p+1}). \tag{5}$$

The quantity K that participates in the right-hand-side of both (4) and (5) depends on the numerical method applied in the calculation of z_n and w_n and on the particular problem (1) that is handled. However, this quantity does not depend on the time-stepsize h. Let us now eliminate K from (4) and (5). After some obvious manipulations, the following relationship can be obtained:

$$y(t_n) - \frac{2^p w_n - z_n}{2^p - 1} = O(h^{p+1}) \tag{6}$$

Let us introduce the notation:

$$y_n = \frac{2^p w_n - z_n}{2^p - 1}. \tag{7}$$

The following relationship can be obtained by applying (7) in (6):

$$y(t_n) - y_n = O(h^{p+1}). \tag{8}$$

The method for calculating the approximation (7) is called Richardson extrapolation, [4], and it is seen from (8) that its order of accuracy is $p + 1$, while each of the approximations z_n and w_n is of order of accuracy p. This means that the Richardson extrapolation can be used to increase the accuracy of the numerical solution. It can also be shown that this device can be used to control the stepsize too (see [6,7]). It should be emphasized, however, that the Richardson extrapolation is computationally more expensive. Moreover, sometimes it also affects the stability of the computational process, [5].

If the problem (1) is definitely non-stiff, then the application of the Richardson extrapolation is very efficient. The same or even better accuracy can be achieved by using larger stepsizes. This compensates the drawback caused by the fact that the Richardson extrapolations is about three times more expensive than the direct use of (2) and (3) if the same stepsize is used with both methods. The situation changes drastically when the problem (1) is slightly stiff and even more when it is moderately stiff, because the requirement for stability of the computational process puts restrictions on the choice of larger stepsizes. Therefore, it is necessary to design numerical methods for which the use of Richardson extrapolation results in better stability properties. We shall present such methods in the remaining part of this paper.

2   Stability of the Computational Process

The stability of the calculations with the m-stage explicit Runge-Kutta methods of order p is normally investigated, [2], by using the polynomial:

$$R(v) = 1 + v + \frac{(v)^2}{2!} + \frac{(v)^3}{3!} + \ldots + \frac{(v)^p}{p!} + \frac{(v)^{p+1}}{\gamma_{p+1}^{(m,p)}(p+1)!} + \ldots + \frac{(v)^m}{\gamma_{p+1}^{(m,p)}(m)!}, \quad (9)$$

which can be obtained when the scalar equation $y' - \lambda y$ with $\lambda \in C^-$ is solved instead of the non-linear system (1). This implies that $v = \lambda h$ is a point in the complex plane located to the left of the imaginary axis and it is said that the method given by (2) and (3) is absolutely stable for a given value of $v \in C^-$ when $|R(v)| \leq 1$. All points $v \in C^-$, for which $|R(v)| \leq 1$ is satisfied, form the absolute stability region of the method defined by (2) and (3). It is assumed in this paper that $p < m$. This means that the parameters λ in the last $m - p$ terms in the right-hand-side of (9), are free parameters that could be used in the efforts to improve the stability properties of the methods (note that all these terms vanish when $p = m$ and, thus, no optimization with regard to the absolute stability can be achieved in this case). If the selected explicit Runge-Kutta method is combined with the Richardson extrapolation, then the stability polynomial $R(v)$ from (9) should be replaced by

$$R(v) = \frac{2^p \left[R\left(\frac{v}{2}\right) \right]^2 - R(v)}{2^p - 1}. \quad (10)$$

Two special cases were selected in this study: $p = 3 \wedge m = 4$ and $p = 4 \wedge m = 6$. Numerical methods with optimal stability properties were found by applying an optimization procedure. In the first case, the methods depend on parameter $\gamma_4^{(4,3)}$ and the best one was found when $\gamma_4^{(4,3)} = 2.4$. In the second case, the search for largest absolute stability regions depended on two parameters and the best choice was found when $\gamma_5^{(6,4)} = 1.42 \wedge \gamma_6^{(6,4)} = 4.86$ were selected. Much more details about the stability polynomials and the absolute stability regions can be found in [2, 6, 7]. Some information about the lengths of the stability intervals along the

Table 1. Lengths of the absolute stability intervals on the negative real axis of four explicit Runge-Kutta methods and their combinations with the Richardson extrapolation. It is clearly seen that the optimized methods have better stability properties than the classical ones and, furthermore, the combinations with the Richardson extrapolation are always giving better results.

Numerical method	Direct implementation	Combined with Richardson extrapolation
$p = m = 3$	2.51	4.02
$p = 3$ and $m = 4$	3.65	8.93
$p = m = 4$	2.70	6.40
$p = 4$ and $m = 6$	5.81	16.28

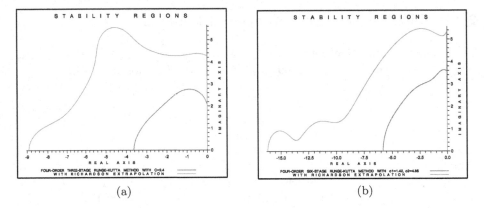

(a)   (b)

Fig. 1. (a) Stability regions of any representative of the class of explicit third-order four-stage Runge-Kutta(**ERK43**) methods with $\gamma_4^{(4,3)}$ and its combination with the Richardson extrapolation; (b) Stability regions of any representative of the class of explicit Runge-Kutta methods determined with $p = 4$, $m = 6$, $\gamma_5^{(6,4)} = 1.42$ and $\gamma_6^{(6,4)} = 4.86$ together with its combination with the Richardson extrapolation

negative real axis is presented in Table 1 for four explicit Runge-Kutta methods and their combinations with the Richardson extrapolation. The absolute stability regions are shown in Fig. 1.

3   Selecting Appropriate Numerical Methods

The explicit Runge-Kutta methods with optimal absolute stability regions, which were discussed in the previous section, are forming two large classes of numerical algorithms. It is necessary to select appropriate particular representatives in each of these classes. A rather complicated computation shows that the conditions (mainly order conditions), which have to be satisfied, lead to the solution of a non-linear algebraic system of 8 equations and 13 unknowns in the case $p = 3 \wedge m = 4$. The case $p = 4 \wedge m = 6$ is much more difficult. Very long calculations show that a six-stage method of order four can be obtained by solving a non-linear system of 15 equations and 26 unknowns for the coefficients in (2) and (3). Both the derivation of the order conditions and the calculations of the coefficients needed for this study can be found on the web-site http://parallel.bas.bg/dpa/EN/publications_2015.htm

4   Numerical Experiments

Consider the system of ordinary differential equations (ODEs) defined by:

$$\frac{dy}{dt} - Ay, \ t \in [0, 13.1072] \ y = (y_1, y_2, y_3)^T, \ y(0) = (1, 0, 2)^T, \ A \in R^{3 \times 3}, \quad (11)$$

where $A = (a_{ij})_{i=1,3}^{j=1,3}$ and

$$
\begin{array}{llll}
a_{11} = & 741.4, & a_{12} = & 749.7, & a_{13} = -741.7, & (12) \\
a_{21} = -765.7, & a_{22} = -758.0, & a_{23} = & 757.7, & (13) \\
a_{31} = & 725.7, & a_{32} = & 741.7, & a_{33} = -734.0. & (14)
\end{array}
$$

The three components of the exact solution of the problem defined above) are given by:

$$y_1(t) = e^{-0.3t} \sin 8t + e^{-750t} \quad (15)$$
$$y_2(t) = e^{-0.3t} \cos 8t - e^{-750t} \quad (16)$$
$$y_3(t) = e^{-0.3t}(\sin 8t + \cos 8t) + e^{-750t}. \quad (17)$$

The eigenvalues of matrix A from (11) are given by:

$$\mu_1 = -750, \qquad \mu_2 = -0.3i \qquad \mu_3 = -0.3 - 8i. \quad (18)$$

The components of the solution of this example presented above are shown in Fig. 2. Some results are given in Tables 2 and 3 bellow.

We can draw three main conclusions from the numerical experiments:

(a) the Richardson extrapolation is improving (by order one) the accuracy of the calculated approximations (in many of the cases the results obtained by the Richardson extrapolation are the most accurate),

(b) the optimized numerical methods derived in this study have better stability properties than the classical methods and, therefore, can be used with larger stepsizes (including also the case when the problem solved is moderately stiff), and

(c) the speed of convergence (given in brackets in Tables 2 and 3) corresponds precisely to the order of accuracy of the tested methods.

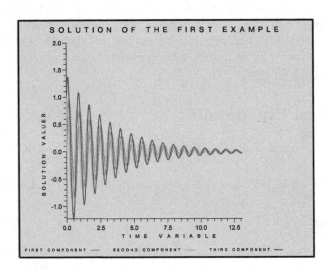

Fig. 2. Plots of the three components of the solution of the system of ODEs defined by (11)–(14).

Table 2. Comparison of the third-order four stages explicit Runge-Kutta (ERK43) method and its combination with the Richardson extrapolation (ERK43 | RE) with the traditionally used third-order three-stages and fourth-order four-stages explicit Runge-Kutta methods (ERK33 and ERK44). The largest errors made during the solution and the speeds of convergence (in brackets) are listed in this table. "N.S" means that the method is not stable (the computations are declared as unstable and stopped when the norm of the calculated solution becomes greater than 1.0E+07).

Stepsize	ERK33	ERK44	ERK43	ERK43+RE
0.02048	N.S	N.S	N.S	N.S
0.01024	N.S	N.S	N.S	N.S
0.00512	N.S	N.S	8.43E-03	4.86E-08
0.00256	5.97E-06	2.46E-08	3.26E-06	3.04E-09(15.99)
0.00128	7.46E-07(8.00)	1.54E-09(15.97)	4.07E-07(8.01)	1.90E-10(16.00)
0.00064	9.33E-08(8.00)	9.62E-12(16.00)	5.09E-08(8.00)	1.19E-11(15.97)
0.00032	1.17E-08(7.97)	6.01E-12(16.01)	6.36E-09(8.00)	7.42E-13(16.04)
0.00016	1.46E-09(8.01)	3.76E-13(15.98)	7.95E-10(8.00)	4.64E-14(15.99)
0.00008	1.82E-10(8.02)	2.35E-14(16.00)	9.94E-11(8.00)	2.90E-15(16.00)
0.00004	2.28E-11(7.98)	1.47E-15(15.99)	1.24E-11(8.02)	1.81E-16(16.02)
0.00002	2.85E-12(8.00)	9.18E-17(16.01)	1.55E-12(8.00)	1.13E-17(16.02)
0.00001	3.56E-13(8.01)	5.74E-18(15.99)	1.94E-13(7.99)	7.08E-19(15.96)

Table 3. Comparison of the first fourth-order six stages explicit Runge-Kutta (ERK64) method and its combination with the Richardson extrapolation (ERK64+RE) with the classical fourth-order four stages explicit Runge-Kutta (ERK44) method and the fifth-order six stages (ERK65B and ERK65F) Runge-Kutta methods proposed respectively by Butcher in [1] and by Fehlberg in [3]. The largest errors made during the solution and the speeds of convergence (in brackets) are listed in this table. "N.S" means that the method is not stable (the computations are declared as unstable and stopped when the norm of the calculated solution becomes greater than 1.0E+07).

Stepsize	ERK44	ERK65B	ERK65F	ERK64	ERK64+RE
0.02048	N.S.	N.S.	N.S.	N.S.	9.00E-08
0.01024	N.S.	N.S.	N.S.	N.S.	1.93E-04
0.00512	N.S.	1.18E-09	N.S.	1.16E-07	8.82E-11
0.00256	2.46E-08	3.69E-11(31.97)	5.51E-11	7.28E-09(15.93)	2.76E-12(31.96)
0.00128	1.54E-09(15.97)	1.15E-12(32.09)	1.72E-12(32.03)	4.55E-10(16.00)	8.62E-14(32.02)
0.00064	9.62E-11(16.00)	3.61E-14(31.86)	5.39E-14(31.91)	2.85E-11(15.96)	2.69E-15(32.04)
0.00032	6.01E-12(16.01)	1.13E-15(31.95)	1.68E-15(32.08)	1.78E-12(16.01)	8.42E-17(31.95)
0.00016	3.76E-13(15.98)	3.52E-17(32.10)	5.26E-17(31.94)	1.11E-13(16.04)	2.63E-18(32.01)
0.00008	2.35E-14(16.00)	1.10E-18(32.00)	1.64E-18(32.07)	6.95E-15(15.97)	8.22E-20(32.00)
0.00004	1.47E-15(15.99)	3.44E-20(31.98)	5.14E-20(31.91)	4.34E-16(16.01)	2.57E-21(31.98)
0.00002	9.18E-17(16.01)	9.18E-21(32.15)	1.61E-21(31.93)	2.71E-17(16.01)	8.03E-23(32.00)
0.00001	5.74E-18(15.00)	5.74E-23(31.85)	5.02E-23(32.07)	1.70E-18(15.94)	2.51E-24(31.99)

5   Plans for Future Research

It is necessary to carry out more numerical experiments, also by using non-linear systems of ordinary differential equations.

The procedure of selecting particular explicit Runge-Kutta methods within the classes with the same absolute stability regions can be improved (by imposing an extra requirement to minimize the coefficients in the leading terms of the truncation error).

Acknowledgments. This research is supported in part by Grants DFNI I-01/5 and DFNI I-02/20 from the Bulgarian National Science Found. The authors thanks Center of Scientific Computing at Technical University of Denmark for giving access to their computers for making computations.

References

1. Butcher, J.C.: Numerical Methods for Ordinary Differential Equations, 2nd edn. Wiley, New York (2003)
2. Fehlberg, E.: New high-order Runge-Kutta formulas with an arbitrary small truncation error. Z. Angew. Math. Mech. **46**, 1–15 (1966)
3. Lambert, J.D.: Numerical Methods for Ordinary Differential Equations. Wiley, New York (1991)

4. Richardson, L.F.: The deferred approach to the limit. I-Single Lattice Philos. Trans. R. Soc. Lond., Ser. A **226**, 299–349 (1927)
5. Zlatev, Z., Farago, I., Havasi, A.: Stability of the Richardson extrapolation applied together with the θ - method. J. Comput. Appl. Math. **235**(2), 507–520 (2010)
6. Zlatev, Z., Georgiev, K., Dimov, I.: Influence of climatic changes on pollution levels in the Balkan Peninsula. Comput. Math. Appl. **65**(3), 544–562 (2013)
7. Zlatev, Z., Georgiev, K., Dimov, I.: Improving the absolute stability properties of some explicit Runge-Kutta methods and their combinations with Richardson extrapolation, Talk presented at the NM&A14 Conference, Borovets, Bulgaria, August (2014)
8. Zlatev, Z., Georgiev, K., Dimov, I.: Studying absolute stability properties of the Richardson extrapolation combined with explicit Runge-Kutta methods. Comput. Math. Appl. **67**, 2294–2307 (2014)

Contributed Papers

Schur Complement Matrix and Its (Elementwise) Approximation: A Spectral Analysis Based on GLT Sequences

Ali Dorostkar[1]($\boxtimes$), Maya Neytcheva[1], and Stefano Serra-Capizzano[1,2]

[1] Department of Information Technology, Uppsala University, Uppsala, Sweden
{ali.dorostkar,maya.neytcheva,stefano.serra}@it.uu.se
[2] Department of Science and High Technology, Insubria University, Como, Italy
stefano.serrac@uninsubria.it

Abstract. Using the notion of the so-called *spectral symbol* in the Generalized Locally Toeplitz (GLT) setting, we derive the GLT symbol of the sequence of matrices $\{A_n\}$ approximating the elasticity equations. Further, as the GLT class defines an algebra of matrix sequences and Schur complements are obtained via elementary algebraic operation on the blocks of A_n, we derive the symbol f^S of the associated sequences of Schur complements $\{S_n\}$ and that of its element-wise approximation.

1 Introduction and Preliminaries

In this paper, the notions of (block)-Toeplitz matrices and related notations are used in their broadly accepted conventional meaning. We refer to [5] for details and include only some definitions for clarity and self-consistency of this paper.

Definition 1 (Generating function of Toeplitz sequences). *Denote by $f(\theta_1, \cdots, \theta_d)$ a d-variate complex-valued integrable function, defined over the domain $Q^d = [-\pi, \pi]^d, d \geq 1$. Denote by f_k the Fourier coefficients of f,*

$$f_k = \frac{1}{m\{Q^d\}} \int_{Q^d} f(\theta) e^{-i(k,\theta)} \, d\theta, \ k = (k_1, \cdots, k_d) \in \mathbb{Z}^d, \ i^2 = -1,$$

where $(k, \theta) = \sum_{j=1}^{d} k_j \theta_j$, $n = (n_1, \cdots, n_d)$, and $N(n) = n_1 \cdots n_d$. Following the multi-index notation in [15], with each f we can associate a sequence of Toeplitz matrices $\{T_n\}$, $T_n = \{f_{k-\ell}\}_{k,\ell=e^T}^n \in \mathbb{C}^{N(n) \times N(n)}$, $\mathbf{e} = [1, 1, \cdots, 1] \in \mathbb{N}^d$.

The function f is referred to as the generating function (or the symbol of) T_n. Using a more compact notation, we say that the function f is the generating function of the whole sequence $\{T_n\}$ and we write $T_n = T_n(f)$.

Definition 2 (Spectral symbol of a matrix). *Given a sequence $\{A_n\}$, A_n of size n, we say that g is the (spectral) symbol of $\{A_n\}$ if all the eigenvalues of A_n are given, up to a small error and for large n, by an evaluation of g over a equispaced grid in the definition set of g.*

A noteworthy example is the Topliz case where the spectral symbol of $\{T_n(f)\}$ is exactly the generating function f: in that case, for $d = 1$, the possible grid is given by $\{x_j^{(n)}\}$, $x_j^{(n)} = -\pi + \frac{2\pi j}{n}$, $j = 1, \ldots, n$.

© Springer International Publishing Switzerland 2015
I. Lirkov et al. (Eds.): LSSC 2015, LNCS 9374, pp. 419–426, 2015.
DOI: 10.1007/978-3-319-26520-9_47

1.1 Toeplitz Matrices in the Context of Discrete PDEs

Consider a differential boundary value problem of the general form $\mathcal{L}u = f$ on Ω, complemented with proper boundary conditions, where $\mathcal{L}$ is a given differential operator and $\Omega \subset \mathbb{R}^d$, $d \geq 1$ is some open, bounded, connected domain.

The techniques to approximate partial differential equations (PDEs) by local methods such as the Finite Element method (FEM) [4] lead to sequences of matrices that admit a Toeplitz structure. When discretizing this problem for a sequence of discretization parameters h_n we obtain a corresponding sequence of matrices $\{A_n\}$ of size n that grows to infinity as the approximation error tends to zero. The study of the spectrum of A_n for fixed dimension and its behavior in an asymptotic sense is often a prerequisite for designing efficient solvers and preconditioners.

Most often, the theoretical results in the related literature (as well as all derivations in this paper) are done for PDEs with constant coefficients, square domains and uniform grids. At first glance, the limitations on square domains, uniform meshes, and constant coefficients seem quite strong. However, substantial steps for overcoming these limiting factors to variable coefficients, domains of arbitrary shape, nonequidistant discretization meshes, and preconditioning, have been done by Tilli [14] and by the third author [12,13]. There, the definition of Generalized Locally Toeplitz (GLT) sequences is introduced and characterized as follows.

GLT1. Each GLT sequence has a symbol (f).
GLT2. The set of GLT sequences form a $*$-algebra that is close under linear combinations, conjugation, products, inversion. Hence, the sequence obtained via algebraic operations on a finite set of GLT sequences is still a GLT sequence and its symbol is obtained by the same algebraic manipulations on the corresponding symbols of the input GLT sequences.
GLT3. Every Toeplitz sequence generated by an L^1 function f is a GLT sequence and its symbol is f, possessing the properties from **GLT1**.
GLT4. The approximation of PDEs with non-constant coefficients, general domains, nonuniform gridding by local methods (FDM, FEM, etc.), under very mild assumptions leads also to GLT sequences (see [3,7,12–14].

The paper is organized as follows. Section 2 introduces the target problem, its discrete formulation and a preconditioner of interest. In Sect. 3, we use the GLT machinery to derive the corresponding symbols of the arising matrices, the exact Schur complement and its approximation.

2 Target Problem, Preconditioning

We simulate the so-called Glacial Isostatic Adjustment (GIA) model. It describes the response of the solid Earth to redistribution of mass due to alternating glaciation and deglaciation periods and is characterized by the coupled system

$$-\nabla \cdot (2\mu\varepsilon(\mathbf{u})) - \nabla(\mathbf{u} \cdot \mathbf{b}) + (\nabla \cdot \mathbf{u})\mathbf{c} - \mu\nabla p = \mathbf{f} \quad \text{in } \Omega, \tag{1a}$$

$$\mu\nabla \cdot \mathbf{u} - \frac{\mu^2}{\lambda}p = 0 \quad \text{in } \Omega, \tag{1b}$$

with $\mathbf{u}$ - the displacement vector, $\varepsilon(\mathbf{u}) = \frac{1}{2}\left(\nabla\mathbf{u} + \nabla\mathbf{u}^T\right)$, λ and μ - the Lamé (material) coefficients. It is assumed that $\mathbf{b} = \{b_i\}$, $\mathbf{c} = \{c_i\}$, $i = 1, 2$ are some given vectors, for simplicity with constant coefficients. We note that the Lamé coefficients μ and λ depend on the material properties and can vary through the domain. Remarkably, the GLT machinery works also in presence of variable coefficients as already mentioned in **GLT4** and all the results, derived here, hold for variable problem parameters. System (1) is first formulated in variational terms and discretized with a stable pair of finite element spaces that satisfy the Ladyzhenskaya-Babuška-Brezzi (LBB) stability condition. We consider below the so-called Modified Taylor-Hood elements (Q1isoQ1, cf. [4]). The target geometry of the problem is rectangular, therefore a discretization with a square or a rectangular mesh is the natural choice. We use square grid and a lexicographical ordering of the node points. The variational setting and the discretization of (1) lead to the algebraic system of equations to be solved,

$$\mathcal{A}\begin{bmatrix}\mathbf{u}_h \\ \mathbf{p}_h\end{bmatrix} = \begin{bmatrix}\mathbf{f} \\ \mathbf{g}\end{bmatrix} \text{ where } \mathcal{A} = \begin{bmatrix}K & B^T \\ B & -\rho M\end{bmatrix} = \begin{bmatrix}K_{11} & K_{12} & B_1^T \\ K_{21} & K_{22} & B_2^T \\ B_1 & B_2 & -\rho M\end{bmatrix} \begin{matrix}\}\text{displ. in } x \\ \}\text{displ. in } y \\ \}\text{pressure}\end{matrix}. \tag{2}$$

Here M is the pressure mass matrix; $\rho = \frac{\mu^2}{\lambda} \neq 0$ for compressible materials, $\rho = 0$ for purely incompressible materials and $\rho \to 0$ in the nearly incompressible case. The block K is symmetric and positive definite when $\mathbf{b} = \mathbf{c} = \mathbf{0}$, otherwise it is nonsymmetric. The blocks B and B^T correspond to discrete divergence and gradient operators, correspondingly. Imposing separate displacement ordering (SDO) for $\mathbf{u}$, i.e., ordering first the displacements in x-direction and then the displacements in y-direction, we induce a two-by-two block structure of the block K and on B as $B = \begin{bmatrix}B_1 & B_2\end{bmatrix}$. The system matrix is depicted in (2), right.

To solve systems with $\mathcal{A}$ we consider preconditioned Krylov subspace iterative solution methods for general matrices, that are suitable for variable preconditioning schemes. We consider a preconditioner $\mathcal{B} = \begin{bmatrix}A_{11} & 0 \\ A_{21} & S\end{bmatrix}$, known to be very efficient, provided that S is a high quality approximation of the exact Schur complement $S_{\mathcal{A}}$ of $\mathcal{A}$, $S_{\mathcal{A}} = A_{22} - A_{21}A_{11}^{-1}A_{12}$, cf. [2] and systems with A_{11} are solved accurately enough.

Various studies have shown (cf. [1, 8–11]) that one particular approximation of $S_{\mathcal{A}}$, obtainable in the finite element context, is very efficient for the target problem, namely, the so-called element-wise Schur complement. To briefly describe it, we assume that the spatial discretization is done by the FEM method on some mesh with characteristic mesh-size h, denoted by $\mathcal{T}_h = \{\tau_\ell^e\}, \ell = 1, \cdots, L$, where τ_ℓ^e denote the individual elements (triangles, quadrilaterals, bricks etc.) and L is the number of the finite (macro-)elements in the discretization mesh.

It has been observed that the matrix $\mathcal{A}$ can be assembled based on local matrices that have the same structure as $\mathcal{A}$, namely, $\mathcal{A} = \sum_{\ell=1}^{L} R^{(\ell)^T} A^{(\ell)} R^{(\ell)}$, $\mathcal{A} \in \mathbb{R}^{N \times N}$, $A^{(\ell)} \in \mathbb{R}^{n \times n}$, where

$$A^{(\ell)} = \begin{bmatrix} A_{11}^{(\ell)} & A_{12}^{(\ell)} \\ A_{21}^{(\ell)} & A_{22}^{(\ell)} \end{bmatrix} \begin{matrix} \}n_1 \\ \}n_2 \end{matrix}, \qquad S = \sum_{\ell=1}^{L} R_2^{(\ell)^T} S^{(\ell)} R_2^{(\ell)}. \tag{3}$$

Here $n = n_1 + n_2$, $\ell = 1, \cdots, L$. The matrices $R^{(\ell)} \in \mathbb{R}^{n \times N}$ are the standard Boolean matrices which provide the local-to-global correspondence of the numbering of the degrees of freedom.

Based on (3) (left) we can compute the local Schur complements exactly and assemble those into a global matrix that is then used as an approximation of S ((3) (right)), where $S^{(\ell)} = A_{22}^{(\ell)} - A_{21}^{(\ell)} A_{11}^{(\ell)^{-1}} A_{12}^{(\ell)}$ and $R_2^{(\ell)}$ are the parts of $R^{(\ell)}$ corresponding to the degrees of freedom in A_{22}. The matrix S in (3) (right) is referred to as the element-wise Schur complement approximation. Without loss of generality we assume that all $A_{11}^{(\ell)}$ are invertible. Otherwise we add a diagonal perturbation of order h^2, where h is the characteristic discretization parameter.

For coupled systems of equations of the form (2) that are discretized with mixed finite elements, the macroelement is tightly related to the choice of the stable finite element pair of spaces we use. For the Q1isoQ1 case we have two meshes, based on one consecutive regular refinement, characterized by a mesh size H and $h = H/2$. Using Linear Algebra tools it is possible to explain the experimentally observed high qualities of the element-wise Schur complement for the case when $\mathcal{A}$ is symmetric and positive definite as well as when it is symmetric indefinite and A_{11} is positive semi-definite. Those tools and the available results for Schur complements are not applicable for both definite or indefinite nonsymmetric matrices. Therefore, to get a better insight in the above, we apply the GLT framework.

3   The Symbols of $\mathcal{A}$, $\mathcal{S_A}$ and $\mathcal{S}$

The matrix $\mathcal{A}$ in (2) can be seen as a generalized block Toeplitz matrix. Note that here we deal with matrices which are Toeplitz up to low rank corrections E_n, i.e., these can be written as $T_n(f) + E_n$ for some function f, where E_n is a low rank perturbation matrix. If the matrices are unilevel then rank(E_n) is bounded by a constant independent of n. Therefore by **GLT2**, the whole sequence $\{T_n(f) + E_n\}$ is a GLT sequence with the same symbol as $\{T_n(f)\}$. Hence, again by **GLT2**, we deduce that the symbol of $\{T_n(f) + E_n\}$ is the generating function of Toeplitz part i.e. the function f.

We next show the related symbols for the blocks and for the whole matrix in (2). Under the lexicographical ordering, all matrix blocks can be seen as stencil-based. All stencils and the detailed derivation of the symbols can be found in [5].

The mass matrix M is block-tridiagonal and each block has a tridiagonal structure. The block-symbol of M, $f^M(\theta_1, \theta_2)$ is $f^M(\theta_1, \theta_2) = 4(2 + \cos(\theta_1))(2 + \cos(\theta_2))$, where θ_1 and θ_2 are generic angles between 0 and π.

Symbols of K, B and the Schur Complement for Q1isoQ1. The symbols for K_{11}, K_{22} and K_{12} read as follows:

$$f^{K_{11}}(\theta_1, \theta_2) = 4 - 2\cos(\theta_1)(1 + \cos(\theta_2)),$$
$$f^{K_{22}}(\theta_1, \theta_2) = 4 - 2(1 + \cos(\theta_1))\cos(\theta_2), \qquad (4)$$
$$f^{K_{12}}(\theta_1, \theta_2) = 4\sin(\theta_1)\sin(\theta_2).$$

Correspondingly, the symbol of the block K has the matrix form

$$f^K = \mu \begin{bmatrix} 4 - 2\cos(\theta_1)(1 + \cos(\theta_2)) & \sin(\theta_1)\sin(\theta_2) \\ \sin(\theta_1)\sin(\theta_2) & 4 - 2(1 + \cos(\theta_1))\cos(\theta_2) \end{bmatrix}. \qquad (5)$$

The derivation of the symbols of the blocks B_1 and B_2 deserves a special attention as these blocks are rectangular. For the case of Q1isoQ1, the blocks B_ℓ^T, $\ell = 1, 2$ are of size $n^2 \times m^2$, where m and n are the number of mesh points in one direction, on two consecutive meshes, i.e., $n = 2(m - 1) + 1$. As the symbol can be related only to square matrices, in order to use the technique, we represent B_ℓ as a result of *downsampling* of larger square matrices $\tilde{B}_\ell$ of size $n \times n$, namely, $B_\ell(n, m) = \tilde{B}_\ell(n, n)H(n, m)$, where H has a particular structure used in various contexts, including multigrid methods, cf., e.g., [6], where it referred to as the *cutting* matrix. For the considered discretization and ordering, $\tilde{B}_\ell$ are five-diagonal block matrices, where each block is itself five-diagonal of size (n, n). The term *downsampling* describes a particular size reduction of a square matrix (of odd size), obtained by deleting each second column, deleting every second block column, or both. More details on how the sampling matrices work can be found in [5]. The corresponding symbols of $\tilde{B}$ are found to be $f^{\tilde{B}_1}(\theta_1, \theta_2) = -4i\phi(\theta_1)\psi(\theta_2)$, $f^{\tilde{B}_2}(\theta_1, \theta_2) = -4i\psi(\theta_1)\phi(\theta_2)$, where $\phi(\theta) = 2\sin(\theta) + \sin(2\theta)$ and $\psi(\theta) = 5 + 6\cos(\theta) + \cos(2\theta)$.

Having constructed all the symbols, using symbolic computations, we compute the symbol of $\tilde{B}K^{-1}\tilde{B}^T$ as $G = f^{\tilde{B}K^{-1}\tilde{B}^T} = v^*(f^K)^{-1}v$ with the vector v such that $v_1 = f^{\tilde{B}_1}$ and $v_2 = f^{\tilde{B}_2}$.

Finally we consider the effect of H and H^T on the underlying symbol. The symbol of the exact Schur f^S is computed by the formula below

$$f^S(\theta_1, \theta_2) = f^M(\theta_1, \theta_2) + \frac{1}{4}\left(\sum_{l=0}^{1}\sum_{m=0}^{1} G\left(\frac{\theta_1}{2} + l\pi, \frac{\theta_2}{2} + m\pi\right)\right). \qquad (6)$$

As already mentioned, the detailed derivation is shown in [5].

Next we deal with the advection term in the 11-block of the matrix $\mathcal{A}$. We consider only a term of the form $\nabla(\mathbf{b} \cdot \mathbf{u})$, with an advection vector $\mathbf{b} = [b_1, b_2]$. We denote the matrix, arising from the discretization of $\nabla(\mathbf{b} \cdot \mathbf{u})$ by A. Similarly to K, SDO induces a two-by-two structure on A, where the blocks $A_{k,\ell}$,

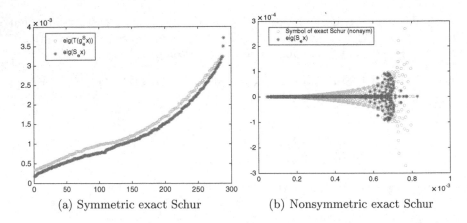

(a) Symmetric exact Schur   (b) Nonsymmetric exact Schur

Fig. 1. The spectrum of the symmetric and nonsymmetric $\mathcal{S}$ (stars) vs sampling of their symbols (circles) (Color figure online)

$k, \ell = 1, 2$ are block-tridiagonal and each block is also block tridiagonal. The symbol of the block A is found to be

$$f^A = -4i \begin{bmatrix} b_1 \sin(\theta_1)(2 + \cos(\theta_2)) & b_2 \sin(\theta_1)(2 + \cos(\theta_2)) \\ b_1 \sin(\theta_2)(2 + \cos(\theta_1)) & b_2 \sin(\theta_2)(2 + \cos(\theta_1)) \end{bmatrix}. \tag{7}$$

In an analogous way we can derive the symbol of the matrix, arising from the term $(\nabla \cdot \mathbf{u}) \mathbf{c}$ with $\mathbf{c} = [c_1, c_2]$. The symbol of the nonsymmetric Schur complement is obtained in the same way as in the symmetric case. To illustrates how well the symbols describe the spectral properties of the corresponding matrices, in Fig. 1 we show the spectrum of the exact symmetric and the nonsymmetric ($\mathbf{b} = [0, 1]$) Schur complements and the sampled symbols.

The Symbol of the Element-Wise Schur Complement Approximation. Using exactly the same machinery, we derive the symbols of the elementwise Schur complement approximation S, both in the symmetric and nonsymmetric case. In contrast to the exact Schur complement, the symbol of S depends on h as we add a diagonal perturbation of order h^2 to A_{11}^ℓ in order to invert them. When constructing the symbol, we use the matrices, corresponding to seven refinements ($h = 0.002$). The symbols read as follows

$$f^{S_{sym}} = \left(0.7145 + 0.3766 \cos(\theta_1)\right) + 2 \cos(\theta_2)\left(0.1883 + 0.0996 \cos(\theta_1)\right)$$
$$f^{S_{nonsym}} = \left(0.7145 + 0.3765 \cos(\theta_1) + 0.0001i \sin(\theta_1)\right)$$
$$+ \left(0.1882 + 0.0996 \cos(\theta_1)\right)\left(\cos(\theta_2) + i \sin(\theta_2)\right)$$
$$+ \left(0.7145 + 0.0995 \cos(\theta_1) - i0.0001 \sin(\theta_1)\right)\left(\cos(\theta_2) - i \sin(\theta_2)\right).$$

The nonsymmetric case is illustrated in Fig. 2.

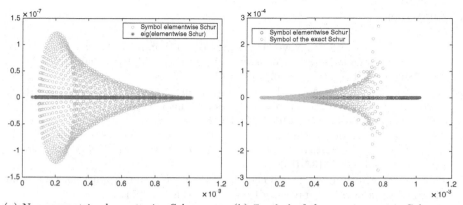

(a) Nonsymmetric element-wise Schur, symbol and eigenvalues

(b) Symbol of the nonsymmetric Schur, exact vs elementwise

Fig. 2. Four refinements (Color figure online)

4   Conclusions and Open Problems

In this work, using the notion of the so-called *spectral symbol* in the Generalized Locally Toeplitz (GLT) setting, we identify the GLT symbol of the sequence of matrices $\{\mathcal{A}_n\}$ approximating the elasticity equations. Further, by exploiting the property that the GLT class defines an algebra of matrix sequences and the fact that Schur complements are obtained via elementary operation on the blocks of $\mathcal{A}_n$, we derived the symbols g_s of the associated sequences of Schur complements $\{\mathcal{S}_n\}$. As a consequence of the GLT theory, the eigenvalues of $\mathcal{S}_n$ for large n are described by a sampling of g_s on a uniform grid of its domain of definition.

We derive the symbols of $\mathcal{A}_n$ and $\mathcal{S}_n$ for the Q1isoQ1 stable FEM pair and the corresponding symbols for the case where the PDE problem includes an advection term and the corresponding system matrix, and respectively, the Schur complement matrix are nonsymmetric. Further, we derive the symbol of the elementwise Schur complement approximation and visually compare it with that of the exact Schur complement. One unexpected result of the study is that the elementwise Schur complement approximation for the considered problem converges to a symmetric matrix when $h \to 0$.

All numerical experiments show that, for the studied discrete problems, the sampling of the symbol agrees very well with the computed spectrum even for a relatively small-sized matrices.

Acknowledgements. The work of the third author is partly supported by Donation **KAW 2013.0341** from the Knut & Alice Wallenberg Foundation in collaboration with the Royal Swedish Academy of Sciences, supporting Swedish research in mathematics.

References

1. Axelsson, O., Blaheta, R., Neytcheva, M.: Preconditioning of boundary value problems using elementwise Schur complements. SIAM J. Matrix Anal. Appl. **31**(2), 767–789 (2009). http://dx.doi.org/10.1137/070679673
2. Axelsson, O., Neytcheva, M.: Preconditioning methods for linear systems arising in constrained optimization problems. Numer. Linear Algebra Appl. **10**(1–2), 3–31 (2003). http://dx.doi.org/10.1002/nla.310
3. Beckermann, B., Serra Capizzano, S.: On the asymptotic spectrum of finite element matrix sequences. SIAM J. Numer. Anal. **45**(2), 746–769 (2007). http://dx.doi.org/10.1137/05063533X
4. Braess, D.: Finite Elements: Theory, Fast Solvers, and Applications in Solid Mechanics. Cambridge University Press, Cambridge (2007)
5. Dorostkar, A., Neytcheva, M., Serra Capizzano, S.: Spectral analysis of coupled PDEs and of their Schur complements via the notion of generalized locally Toeplitz sequences. Technical report 2015–008, Numerical Analysis, Uppsala University (2015)
6. Fiorentino, G., Serra Capizzano, S.: Multigrid methods for Toeplitz matrices. CALCOLO **28**(3–4), 283–305 (1991). http://dx.doi.org/10.1007/BF02575816
7. Garoni, C., Serra Capizzano, S., Sesana, D.: Spectral analysis and spectral symbol of d-variate $\mathbb{Q}_p$ Lagrangian FEM stiffness matrices. SIAM J. Sci. Comput. **36**(3), 1100–1128 (2015). http://dx.doi.org/10.1137/140976480
8. Kraus, J.: Algebraic multilevel preconditioning of finite element matrices using local Schur complements. Numer. Linear Algebra Appl. **13**(1), 49–70 (2006). http://dx.doi.org/10.1002/nla.462
9. Kraus, J.: Additive Schur complement approximation and application to multilevel preconditioning. SIAM J. Sci. Comput. **34**(6), A2872–A2895 (2012). http://dx.doi.org/10.1137/110845082
10. Neytcheva, M., Bängtsson, E.: Preconditioning of nonsymmetric saddle point systems as arising in modelling of viscoelastic problems. Electron. Trans. Numer. Anal. (ETNA) **29**, 193–211 (2007). http://eudml.org/doc/130827
11. Neytcheva, M., Do-Quang, M., He, X.: Element-by-element Schur complement approximations for general nonsymmetric matrices of two-by-two block form. In: Lirkov, I., Margenov, S., Waśniewski, J. (eds.) LSSC 2009. LNCS, vol. 5910, pp. 108–115. Springer, Heidelberg (2010)
12. Serra Capizzano, S.: Generalized locally Toeplitz sequences: spectral analysis and applications to discretized partial differential equations. Linear Algebra Appl. **366**, 371–402 (2003). http://www.sciencedirect.com/science/article/pii/S00243795 02005049, special issue on Structured Matrices: Analysis, Algorithms and Applications
13. Serra Capizzano, S.: The GLT class as a generalized Fourier analysis and applications. Linear Algebra Appl. **419**(1), 180–233 (2006). http://www.sciencedirect.com/science/article/pii/S0024379506002175
14. Tilli, P.: Locally Toeplitz sequences: spectral properties and applications. Linear Algebra Appl. **278**(1–3), 91–120 (1998). http://www.sciencedirect.com/science/article/pii/S0024379597100799
15. Tyrtyshnikov, E.E.: A unifying approach to some old and new theorems on distribution and clustering. Linear Algebra Appl. **232**, 1–43 (1996). http://www.sciencedirect.com/science/article/pii/0024379594000255

An Iterative Process for the Solution of Semi-Linear Elliptic Equations with Discontinuous Coefficients and Solution

Aigul Manapova[✉]

Bashkir State University, Zaki Validi Street,
32, Ufa, Republic of Bashkortostan, Russia
aygulrm@yahoo.com

Abstract. We consider and investigate boundary value problems (BVPs) for semi-linear elliptic equations with discontinuous coefficients and solutions (with imperfect contact matching conditions). Finite difference approximations of these problems are constructed. An iterative method for solving difference BVPs of contact for semi-linear elliptic equations with iterations on the inner boundary where the coefficients and solutions are discontinuous is constructed and validated. The convergence rate of iterations (with calculated constants) is estimated.

Keywords: Semi-linear elliptic equations · Numerical method · Iterative method · Operator

1 Introduction and Setting of the Problem

In this article we study boundary value problems (BVPs) for semi-linear elliptic equations with variable coefficients in inhomogeneous anisotropic media, with discontinuous coefficients and solutions (with imperfect contact matching conditions) [1,2]. This kind of problems naturally appears in the mathematical modeling and solution to problems of heat transfer, diffusion, filtration, elasticity, etc., in a study of contact BVPs for equations of mathematical physics in multilayered media.

Before solving such problems numerically, they have to be approximated by problems of a simpler nature, specifically, by "finite-dimensional BVPs" (see [3]). One of the most convenient, universal, and widespread techniques for finite-dimensional approximation as applied to BVPs is the grid method [1,2,4–7].

This work can be considered as an extension of the results, obtained in [8]. The aim of the work is to construct efficient, cost-effective, high-precision approximate methods for solving contact problems for PDEs with discontinuous coefficients and solutions, as well to develop and implement finite-dimensional approximations of iterative problems at each iteration step.

This work was supported by a grant of the President of the Russian Federation for state support of young Russian scientists and PhDs, number MK 4147.2015.1.

© Springer International Publishing Switzerland 2015
I. Lirkov et al. (Eds.): LSSC 2015, LNCS 9374, pp. 427–434, 2015.
DOI: 10.1007/978-3-319-26520-9_48

We develop and validate an iterative method for solving BVPs of contact for semi-linear elliptic equations with discontinuous coefficients and solutions and estimate the convergence rate of iterations (with calculated constants). Note also that while proving the iterative process convergence, we develop an idea of the work [9], where the linear setting of BVP is studied, and approximation issues are not considered. As a result, the numerical solution of these problems can be effectively implemented on the basis of the developed iterative method (with iterations on the inner boundary where the coefficients and solutions are discontinuous) in combination, for example, with the difference method for solving some already traditional "independent" BVPs arising in each contacting subdomain inside the composite integration domain.

Now consider the following problem. Let $\Omega = \{r = (r_1, r_2) \in \mathbf{R}^2 : 0 \leq r_\alpha \leq l_\alpha, \alpha = 1, 2\}$ be a rectangle in $\mathbf{R}^2$ with a boundary $\partial\Omega = \Gamma$. The domain Ω is divided by the line $r_1 = \xi$, where $0 < \xi < l_1$ (by the internal interface $\overline{S} = \{r_1 = \xi, \ 0 \leq r_2 \leq l_2\}$, where $0 < \xi < l_1$) into the left $\Omega_1 \equiv \Omega^- = \{0 < r_1 < \xi, \ 0 < r_2 < l_2\}$ and right $\Omega_2 \equiv \Omega^+ = \{\xi < r_1 < l_1, \ 0 < r_2 < l_2\}$ subdomains with boundaries $\partial\Omega_1 \equiv \partial\Omega^-$ and $\partial\Omega_2 \equiv \partial\Omega^+$. Thus, $\Omega = \Omega_1 \cup \Omega_2 \cup S$, while $\partial\Omega$ is the outer boundary of Ω. Let $\overline{\Gamma}_k$ denote the boundaries of Ω_k without S, $k = 1, 2$. Therefore $\partial\Omega_k = \overline{\Gamma}_k \cup S$, where Γ_k, $k = 1, 2$ are open nonempty subsets of $\partial\Omega_k$, $k = 1, 2$; and $\overline{\Gamma}_1 \cup \overline{\Gamma}_2 = \partial\Omega = \Gamma$. Let n_α, $\alpha = 1, 2$ denote the outward normal to the boundary $\partial\Omega_\alpha$ of Ω_α, $\alpha = 1, 2$. Let $n = n(x)$ be a unit normal to S at a point $x \in S$, directed, for example, so that n is the outward normal on S with respect to Ω_1; i.e., n is directed inside Ω_2. While formulating BVPs for states of control processes below, we assume that S is a straight line across which the coefficients and solutions of the problems are discontinuous, while being smooth within Ω_1 and Ω_2.

Assume that the conditions imposed on a controlled physical process are such that it can be modeled in the domain Ω by the following Dirichlet problem for a semi-linear elliptic equation with discontinuous coefficients and solution: Find a function $u(x)$, defined on $\overline{\Omega}$ that satisfies in Ω_1 and Ω_2 the equations:

$$Lu(x) = -\sum_{\alpha=1}^{2} \frac{\partial}{\partial x_\alpha}\left(k_\alpha(x)\frac{\partial u}{\partial x_\alpha}\right) + d(x)q(u) = f(x), \quad x \in \Omega_1 \cup \Omega_2, \quad (1)$$

and the conditions $u(x) = 0, \quad x \in \partial\Omega = \overline{\Gamma}_1 \cup \overline{\Gamma}_2$,

$$\left[k_1(x)\frac{\partial u}{\partial x_1}\right] = 0, \quad G(x) = \left(k_1(x)\frac{\partial u}{\partial x_1}\right) = \theta(x_2)[u], \quad x \in S,$$

where $u(x) = \begin{cases} u_1(x), x \in \Omega_1; \\ u_2(x), x \in \Omega_2, \end{cases}$ $q(\xi) = \begin{cases} q_1(\xi_1), \xi_1 \in \mathbf{R}; \\ q_2(\xi_2), \xi_2 \in \mathbf{R}, \end{cases}$

$$k_\alpha(x), d(x), f(x) = \begin{cases} k_\alpha^{(1)}(x), d_1(x), f_1(x), x \in \Omega_1; \\ k_\alpha^{(2)}(x), d_2(x), f_2(x), x \in \Omega_2, \alpha = 1, 2. \end{cases}$$

Here $[u] = u_2(x) - u_1(x)$ is the jump in $u(x)$ across S; $k_\alpha(x)$, $\alpha = 1, 2$, $f(x)$, $d(x)$ are given functions that are defined variously in Ω_1 and Ω_2, and have a jump discontinuity on S; $q_\alpha(\xi_\alpha)$, $\alpha = 1, 2$, are given functions defined for $\xi_\alpha \in \mathbf{R}$, $\alpha = 1, 2$; $\theta(x_2)$, $x_2 \in S$, is a given function. The given functions are assumed to satisfy the following conditions: $k_\alpha(x) \in W_\infty^1(\Omega_1) \times W_\infty^1(\Omega_2)$, $\alpha = 1, 2$, $d(x) \in L_\infty(\Omega_1) \times L_\infty(\Omega_2)$, $f(x) \in L_2(\Omega_1) \times L_2(\Omega_2)$ $\theta(x_2) \in L_\infty(S)$, $0 < \nu \le k_\alpha(x) \le \overline{\nu}$, $\alpha = 1, 2$, $0 \le d_0 \le d(x) \le \overline{d}_0$ for $x \in \Omega_1 \cup \Omega_2$ and $0 < \theta_0 \le \theta(x_2) \le \overline{\theta}_0$ for $x \in S$, where $\nu, \overline{\nu}, d_0, \overline{d}_0, \theta_0, \overline{\theta}_0$ are given constants; and the functions $q_\alpha(\zeta_\alpha)$ satisfy the conditions: $q_\alpha(0) = 0$, $0 < q_0 \le (q_\alpha(\zeta_\alpha) - q_\alpha(\overline{\zeta}_\alpha))/(\zeta_\alpha - \overline{\zeta}_\alpha) \le L < \infty$, for all $\zeta_\alpha, \overline{\zeta}_\alpha \in \mathbf{R}$, $\zeta_\alpha \ne \overline{\zeta}_\alpha$.

Let $\overset{\circ}{\Gamma}_k$ be a portion of the boundary $\partial\Omega_k$. Denote by $W_2^1\left(\Omega_k; \overset{\circ}{\Gamma}_k\right)$ the closed subspace of $W_2^1(\Omega_k)$ in which the set of all functions from $C^1(\overline{\Omega}_k)$ vanishing near $\overset{\circ}{\Gamma}_k \subset \partial\Omega_k$, $k = 1, 2$ in a dense set. We introduce the space $\overset{\circ}{V}_{\Gamma_1,\Gamma_2}(\Omega^{(1,2)})$ of pairs $u = (u_1, u_2)$: $\overset{\circ}{V}_{\Gamma_1,\Gamma_2}(\Omega^{(1,2)}) = \{u = (u_1, u_2) \in W_2^1(\Omega_1; \Gamma_1) \times W_2^1(\Omega_2; \Gamma_2)\}$ with the norm $\|u\|_{\overset{\circ}{V}_{\Gamma_1,\Gamma_2}}^2 = \sum_{k=1}^2 \int_{\Omega_k} \sum_{\alpha=1}^2 \left(\frac{\partial u_k}{\partial x_\alpha}\right)^2 d\Omega_k + \int_S [u]^2 dS$.

We say that a function $u(g) \in \overset{\circ}{V}_{\Gamma_1,\Gamma_2}(\Omega^{(1,2)})$ is a generalized solution to the problem (1), satisfying the identity:

$$\int_{\Omega_1 \cup \Omega_2} \left[\sum_{\alpha=1}^2 k_\alpha(x)\frac{\partial u(x)}{\partial x_\alpha}\frac{\partial \vartheta(x)}{\partial x_\alpha} + d(x)\,q(u)\,\vartheta(x)\right] d\Omega_0$$
$$+ \int_S \theta(x)[u][\vartheta]dS = \int_{\Omega_1 \cup \Omega_2} f(x)\vartheta(x)d\Omega_0, \quad \text{for all } \vartheta \in \overset{\circ}{V}_{\Gamma_1,\Gamma_2}(\Omega^{(1,2)}). \tag{2}$$

2 Difference Approximation of the BVP

For the numerical solution of boundary-value problems (1) we approximate them based on the grid method [1,5]. We introduce one-dimensional nonuniform grids in x_1 and x_2: $\widehat{\omega}_\alpha = \{x_\alpha^{(i_\alpha)} \in [0, l_\alpha] : i_\alpha = \overline{0, N_\alpha}, x_\alpha^{(0)} = 0, x_\alpha^{(N_\alpha)} = l_\alpha, h_{\alpha i_\alpha} = x_\alpha^{(i_\alpha)} - x_\alpha^{(i_\alpha-1)}, i_\alpha = \overline{1, N_\alpha}\}$, $\alpha = 1, 2$. Additionally, a nonuniform grid in x_1 and x_2 is introduced in $\overline{\Omega} = \overline{\Omega}_1 \cup \overline{\Omega}_2$: $\widehat{\omega} = \widehat{\omega}_1 \times \widehat{\omega}_2$. Obviously, we can always construct a grid $\widehat{\omega}_1$ on $[0, l_1]$ such that the point $x_1 = \xi$ is one of its nodes. In applications, it is reasonable to set uniform mesh sizes $h_1^{(1)}$ and $h_1^{(2)}$ in $\overline{\Omega}_1$ and $\overline{\Omega}_2$, respectively. Then, based on the location of the point $x_1 = \xi$, the number of nodes is determined by the assumption $h_1^{(1)} \approx h_1^{(2)}$. We set $x_1^{(i_1)} - x_1^{(i_1-1)} = h_1$ for $i_1 = \overline{1, N_1}$ and $x_2^{(i_2)} - x_2^{(i_2-1)} = h_2$ for $i_2 = \overline{1, N_2}$. The value x_1 at the point $x_1 = \xi$ is denoted by x_ξ, and the corresponding node index is $N_{1\xi}$, $1 < N_{1\xi} < N_1 - 1$. We introduce the following grids of nodes: $\overline{\omega}_1^{(1)} = \{x_1^{(i_1)} = i_1 h_1 \in [0, \xi] : i_1 = \overline{0, N_{1\xi}}, N_{1\xi}h_1 = \xi\}$, $\overline{\omega}_1^{(2)} = \{x_1^{(i_1)} = i_1 h_1 \in [\xi, l_1] : i_1 = \overline{N_{1\xi}, N_1}, N_1 h_1 = l_1\}$, $\omega_1^{(1)} = \overline{\omega}_1^{(1)} \setminus \{x_1 = 0, x_1 = \xi\}$, $\omega_1^{(2)} = \overline{\omega}_1^{(2)} \setminus \{x_1 = \xi, x_1 = l_1\}$; $\overline{\omega}_2 = \{x_2^{(i_2)} = i_2 h_2 \in [0, l_2] : i_2 = \overline{0, N_2}, N_2 h_2 = l_2\}$, $\omega_2 = \overline{\omega}_2 \setminus \{x_2 = 0, x_2 = l_2\}$; $\overline{\omega}_1 = $

$\overline{\omega}_1^{(1)} \cup \overline{\omega}_1^{(2)}$; $\omega_1 = \omega_1^{(1)} \cup \omega_1^{(2)}$; $\overline{\omega}^{(1)} = \overline{\omega}_1^{(1)} \times \overline{\omega}_2$; $\overline{\omega}^{(2)} = \overline{\omega}_1^{(2)} \times \overline{\omega}_2$; $\omega^{(1)} = \omega_1^{(1)} \times \omega_2$;
$\omega^{(2)} = \omega_1^{(2)} \times \omega_2$; $\overline{\omega} \equiv \overline{\omega}^{(1,2)} = \overline{\omega}^{(1)} \cup \overline{\omega}^{(2)} = (\overline{\omega}_1^{(1)} \cup \overline{\omega}_1^{(2)}) \times \overline{\omega}_2 = \{x_1^{(i_1)} =$
$i_1 h_1$, $i_1 = \overline{0, N_1}$, $N_{1\xi} h_1 = \xi$, $(N_1 - N_{1\xi}) h_1 = l_1 - \xi$, $1 < N_{1\xi} < N_1 - 1\} \times \overline{\omega}_2$, $\omega \equiv$
$\omega^{(1,2)} = \omega^{(1)} \times \omega^{(2)}$; $\omega_1^{(1)+} = \overline{\omega}_1^{(1)} \cap (0, \xi]$, $\omega_1^{(1)-} = \overline{\omega}_1^{(1)} \cap [0, \xi)$, $\omega_1^{(2)-} = \overline{\omega}_1^{(2)} \cap [\xi, l_1)$,
$\omega^{(1)(+1)} = \omega_1^{(1)+} \times \overline{\omega}_2$; $\gamma_S = \{x_1 = \xi,\ x_2 = h_2, 2h_2, \ldots, (N_2 - 1)h_2\} = \{x_1 =$
$\xi,\ x_2^{(i_2)} = i_2 h_2,\ i_2 = \overline{1, N_2 - 1}\}$; $\gamma^{(k)} = \partial\omega^{(k)} \setminus \gamma_S$; $\omega_1^{(1)+} \times \omega_2 = \omega^{(1)} \cup \gamma_S =$
$\overline{\omega}^{(1)} \setminus \gamma^{(1)}$; and $\partial\omega^{(k)} = \overline{\omega}^{(k)} \setminus \omega^{(k)}$ is the set of boundary grid nodes of the grid
$\overline{\omega}^{(k)}$, $k = 1, 2$. In the sequel we need the inner products, norms, and seminorms of
grid functions defined on various grids (see [7]). Particularly, let $W_2^1(\overline{\omega}^{(k)}; \gamma^{(k)})$
denote the subspace of grid functions from $W_2^1(\overline{\omega}^{(k)})$ that vanish on $\gamma^{(k)}$, $k = 1, 2$.
The spaces $\overset{\circ}{V}_{\gamma^{(1)}\gamma^{(2)}}(\overline{\omega}^{(1,2)})$ of pairs of grid functions $y(x) = (y_1(x), y_2(x))$ are
defined as: $\overset{\circ}{V}_{\gamma^{(1)}\gamma^{(2)}}(\overline{\omega}^{(1,2)}) = \{y = (y_1, y_2) \in W_2^1(\overline{\omega}^{(1)}; \gamma^{(1)}) \times W_2^1(\overline{\omega}^{(2)}; \gamma^{(2)})\}$,
with the norm $\|y\|_{\overset{\circ}{V}_{\gamma^{(1)}\gamma^{(2)}}}^2 = \|\nabla y_k\|^2 + \|[y]\|_{L_2(\gamma_S)}^2$.

BVP (1) is associated with the following difference BVP. Namely, the grid
function $y \in \overset{\circ}{V}_{\gamma^{(1)}\gamma^{(2)}}(\overline{\omega}^{(1,2)})$, which is the solution of the difference BVP for
problem (1), satisfies, for $\forall v \in \overset{\circ}{V}_{\gamma^{(1)}\gamma^{(2)}}(\overline{\omega}^{(1,2)})$, the summation identity

$$
\begin{aligned}
&\left\{\sum_{\omega_1^{(1)+}} \sum_{\omega_2} a_{1h}^{(1)} y_{1\overline{x}_1} v_{1\overline{x}_1} h_1 h_2 + \left(\sum_{\omega_1^{(1)}} \sum_{\omega_2^+} a_{2h}^{(1)} y_{1\overline{x}_2} v_{1\overline{x}_2} h_1 h_2 \right.\right. \\
&\left.\left. + \frac{1}{2} \sum_{\omega_2^+} a_{2h}^{(1)}(\xi, x_2) y_{1\overline{x}_2}(\xi, x_2) v_{1\overline{x}_2}(\xi, x_2) h_1 h_2 \right)\right\} \\
&+ \left\{\sum_{\omega_1^{(2)+}} \sum_{\omega_2} a_{1h}^{(2)} y_{2\overline{x}_1} v_{2\overline{x}_1} h_1 h_2 + \left(\sum_{\omega_1^{(2)}} \sum_{\omega_2^+} a_{2h}^{(2)} y_{2\overline{x}_2} v_{2\overline{x}_2} h_1 h_2 \right.\right. \\
&\left.\left. + \frac{1}{2} \sum_{\omega_2^+} a_{2h}^{(2)}(\xi, x_2) y_{2\overline{x}_2}(\xi, x_2) v_{2\overline{x}_2}(\xi, x_2) h_1 h_2 \right)\right\} \\
&+ \sum_{\omega_2} \theta_h(x_2)[y][v](\xi, x_2) h_2 + \left\{\left(\sum_{\omega^{(1)}} d_{1h}(x) q_1(y_1) v_1(x) h_1 h_2\right.\right. \\
&\left. + \frac{1}{2} \sum_{\omega_2} d_{1h}(\xi, x_2) q_1(y_1(\xi, x_2) v_1(\xi, x_2) h_1 h_2 \right) \\
&+ \left(\sum_{\omega^{(2)}} d_{2h}(x) q_2(y_2(x)) v_2(x) h_1 h_2\right. \\
&\left.\left. + \frac{1}{2} \sum_{\omega_2} d_{2h}(\xi, x_2) q_2(y_2(\xi, x_2)) v_2(\xi, x_2) h_1 h_2 \right)\right\} \\
&= \left\{\left(\sum_{\omega^{(1)}} \Phi_{1h}(x) v_1(x) h_1 h_2 + \frac{1}{2} \sum_{\omega_2} \Phi_{1h}(\xi, x_2) v_1(\xi, x_2) h_1 h_2\right)\right. \\
&\left. + \left(\sum_{\omega^{(2)}} \Phi_{2h}(x) v_2(x) h_1 h_2 + \frac{1}{2} \sum_{\omega_2} \Phi_{2h}(\xi, x_2) v_2(\xi, x_2) h_1 h_2\right)\right\}.
\end{aligned}
\tag{3}
$$

Here, $a_{\alpha h}^{(1)}(x)$, $a_{\alpha h}^{(2)}(x)$, $\Phi_{\alpha h}(x)$ and $d_{\alpha h}(x)$, $\alpha = 1, 2$, $\theta_h(x_2)$, and $u_{0h}^{(1)}(x)$ are grid approximations of the functions $k_{\alpha}^{(1)}(r)$, $k_{\alpha}^{(2)}(r)$, $f_{\alpha}(r)$ and $d_{\alpha}(r)$, $\alpha = 1, 2$, $\theta(r_2)$, and $u_0^{(1)}(r)$ defined via Steklov averages (see [6, 7]).

Problem (3) is a grid analogue of the original problem for state (1) with discontinuous coefficients and a discontinuous solution (state).

Now we explicitly write difference scheme (3) at nodes of the grid $\overline{\omega} = \overline{\omega}_1 \cup \overline{\omega}_2 = \overline{\omega}^{(1,2)}$. The search for the solution of the difference problem is reduced to the following problem: Find a function: $y(x) = (y_1(x), y_2(x))$, defined on $\overline{\omega} = \overline{\omega}_1 \cup \overline{\omega}_2 = \overline{\omega}^{(1,2)}$, where $y(x) = y_1(x)$ for $x \in \overline{\omega}^{(1)}$, $y(x) = y_2(x)$ for $x \in \overline{\omega}^{(2)}$, and the components $y_1(x)$ and $y_2(x)$ satisfy the following conditions:

(1) The grid function $y_1(x)$, $x \in \overline{\omega}^{(1)} = \omega^{(1)} \cup \partial\omega^{(1)}$, satisfies the equation

$$- \left(a_{1h}^{(1)}(x)y_{1\overline{x}_1} \right)_{x_1} - \left(a_{2h}^{(1)}(x)y_{1\overline{x}_2} \right)_{x_2} + d_{1h}(x)q_1(y_1) = \Phi_{1h}(x), \quad x \in \omega^{(1)}, \quad (4)$$

and, on the boundary $\gamma^{(1)} = \partial\omega^{(1)} \setminus \gamma_S$, obeys the condition $y_1(x) = 0$, $x \in \gamma^{(1)}$;

(2) The grid function $y_2(x)$, $x \in \overline{\omega}^{(2)} = \omega^{(2)} \cup \partial\omega^{(2)}$, satisfies the equation

$$- \left(a_{1h}^{(2)}(x)y_{2\overline{x}_1} \right)_{x_1} - \left(a_{2h}^{(2)}(x)y_{2\overline{x}_2} \right)_{x_2} + d_{2h}(x)q_2(y_2) = \Phi_{2h}(x), \quad x \in \omega^{(2)}, \quad (5)$$

and, on the boundary $\gamma^{(2)} = \partial\omega^{(2)} \setminus \gamma_S$, obeys the condition $y_2(x) = 0$, $x \in \gamma^{(2)}$;

(3) The sought functions $y_1(x)$ and $y_2(x)$ are related by additional matching conditions on $\gamma_S = \{x_1 = \xi, x_2 \in \omega_2\}$, namely:

$$\frac{2}{h_1} \left[a_{1h}^{(1)}(\xi, x_2)y_{1\overline{x}_1}(\xi, x_2) + \theta_h(x_2)y_1(\xi, x_2) \right] + d_{1h}(\xi, x_2)q_1(y_1(\xi, x_2))$$
$$- \left(a_{2h}^{(1)}(\xi, x_2)y_{1\overline{x}_2}(\xi, x_2) \right)_{x_2} = \Phi_{1h}(\xi, x_2) + \frac{2}{h_1}\theta_h(x_2)y_2(\xi, x_2), \quad (6)$$

$$-\frac{2}{h_1} \left[a_{1h}^{(2)}(\xi + h_1, x_2)y_{2x_1}(\xi, x_2) - \theta_h(x_2)y_2(\xi, x_2) \right] + d_{2h}(\xi, x_2)q_2(y_2)$$
$$- \left(a_{2h}^{(2)}(\xi, x_2)y_{2\overline{x}_2}(\xi, x_2) \right)_{x_2} = \Phi_{2h}(\xi, x_2) + \frac{2}{h_1}\theta_h(x_2)y_1(\xi, x_2). \quad (7)$$

3   An Iterative Process and Its Convergence

The question we are going to discuss in this section is to construct an effective convergent iterative process (with calculated constants) for solving the grid problem (4)–(7). The problem (4)–(7) is associated with the following iterative process with iterations on the inner boundary $\gamma_S = \{x_1 = \xi, x_2 \in \omega_2\}$:

$$- \left(a_{1h}^{(1)}(x)y_{1\overline{x}_1}^n \right)_{x_1} - \left(a_{2h}^{(1)}(x)y_{1\overline{x}_2}^n \right)_{x_2} + d_{1h}(x)q_1(y_1^n) = \Phi_{1h}(x), \quad x \in \omega^{(1)},$$
$$y_1^n(x) = 0, \quad \gamma^{(1)} = \partial\omega^{(1)} \setminus \gamma_S; \quad (8)$$

$$\frac{2}{h_1} \left[a_{1h}^{(1)}(\xi, x_2)y_{1\overline{x}_1}^n(\xi, x_2) + \theta_h(x_2)y_1^n(\xi, x_2) \right] + d_{1h}(\xi, x_2)q_1(y_1^n(\xi, x_2))$$
$$- \left(a_{2h}^{(1)}(\xi, x_2)y_{1\overline{x}_2}^n(\xi, x_2) \right)_{x_2} = \Phi_{1h}(\xi, x_2) + \frac{2}{h_1}\theta_h(x_2)y_2^{n-1}(\xi, x_2), \quad x \in \gamma_S, \quad (9)$$

$$- \left(a_{1h}^{(2)}(x)y_{2\overline{x}_1}^n\right)_{x_1} - \left(a_{2h}^{(2)}(x)y_{2\overline{x}_2}^n\right)_{x_2} + d_{2h}(x)q_2(y_2^n) = \Phi_{2h}(x), \quad x \in \omega^{(2)}, \tag{10}$$
$$y_2^n(x) = 0, \quad x \in \gamma^{(2)} = \partial\omega^{(2)} \setminus \gamma_S;$$

$$-\frac{2}{h_1}\left[a_{1h}^{(2)}(\xi+h_1,x_2)y_{2x_1}^n(\xi,x_2) - \theta_h(x_2)y_2^n(\xi,x_2)\right] + d_{2h}(\xi,x_2)q_2(y_2^n(\xi,x_2))$$
$$- \left(a_{2h}^{(2)}(\xi,x_2)y_{2\overline{x}_2}^n(\xi,x_2)\right)_{x_2} = \Phi_{2h}(\xi,x_2) + \frac{2}{h_1}\theta_h(x_2)y_1^n(\xi,x_2), \quad x \in \gamma_S. \tag{11}$$

where $n = 1, 2, \ldots$; $y_2^0(x)$ is an initial approximation.

Thus, the iterative process $(8)-(11)$ reduces the solution of the original BVP $(4)-(7)$ with discontinuous coefficients and solution to the solution of two BVPs $(8)-(9)$ and $(10)-(11)$ in subdomains Ω_1 and Ω_2 at each iteration n, respectively.

In a generalized statement the iterative process with respect to functions y_1^n and y_2^n is to find a sequence of pairs of functions $\{y^n\} = \left\{(y_1^n, y_2^n)\right\}_{n=1}^{\infty}$, such that $y_k^n \in W_2^1(\omega^{(k)}; \gamma^k)$, $k = 1, 2$ and satisfy the summation identities:

$$\sum_{\omega_1^{(1)}+\omega_2}\sum a_{1h}^{(1)}y_{1\overline{x}_1}^n v_{1\overline{x}_1}h_1h_2 + \sum_{\omega_1^{(1)}}\sum_{\omega_2^+}a_{2h}^{(1)}y_{1\overline{x}_2}^n v_{1\overline{x}_2}h_1h_2$$
$$+\frac{1}{2}\sum_{\omega_2^+}a_{2h}^{(1)}(\xi,x_2)y_{1\overline{x}_2}^n(\xi,x_2)v_{1\overline{x}_2}(\xi,x_2)h_1h_2$$
$$+\sum_{\omega^{(1)}}d_{1h}(x)\,q_1(y_1^n)\,v_1(x)h_1h_2 + \frac{1}{2}\sum_{\omega_2}d_{1h}(\xi,x_2)\,q_1(y_1^n)\,v_1(\xi,x_2)h_1h_2 \tag{12}$$
$$+\sum_{\omega_2}\theta_h(x_2)y_1^n(\xi,x_2)\,v_1(\xi,x_2)h_2 = \sum_{\omega^{(1)}}\Phi_{1h}(x)v_1(x)h_1h_2$$
$$+\frac{1}{2}\sum_{\omega_2}\Phi_{1h}(\xi,x_2)v_1(\xi,x_2)h_1h_2 + \sum_{\omega_2}\theta_h(x_2)y_2^{n-1}(\xi,x_2)\,v_1(\xi,x_2)h_2,$$
$$\forall v_1(x) \in W_2^1(\overline{\omega}^{(1)}; \gamma^{(1)});$$

$$\sum_{\omega_1^{(2)}+\omega_2}\sum a_{1h}^{(2)}y_{2\overline{x}_1}^n v_{2\overline{x}_1}h_1h_2 + \sum_{\omega_1^{(2)}}\sum_{\omega_2^+}a_{2h}^{(2)}y_{2\overline{x}_2}^n v_{2\overline{x}_2}h_1h_2$$
$$+\frac{1}{2}\sum_{\omega_2^+}a_{2h}^{(2)}(\xi,x_2)y_{2\overline{x}_2}^n(\xi,x_2)v_{2\overline{x}_2}(\xi,x_2)h_1h_2$$
$$+\sum_{\omega^{(2)}}d_{2h}(x)\,q_2(y_2^n)\,v_2(x)h_1h_2 + \frac{1}{2}\sum_{\omega_2}d_{2h}(\xi,x_2)\,q_2(y_2^n)\,v_2(\xi,x_2)h_1h_2 \tag{13}$$
$$+\sum_{\omega_2}\theta_h(x_2)y_2^n(\xi,x_2)\,v_2(\xi,x_2)h_2 = \sum_{\omega^{(2)}}\Phi_{2h}(x)v_2(x)h_1h_2$$
$$+\frac{1}{2}\sum_{\omega_2}\Phi_{2h}(\xi,x_2)v_2(\xi,x_2)h_1h_2 + \sum_{\omega_2}\theta_h(x_2)y_1^n(\xi,x_2)\,v_2(\xi,x_2)h_2,$$

$\forall v_2(x) \in W_2^1(\overline{\omega}^{(2)}; \gamma^{(2)})$; $n = 1, 2, \ldots$; $y_2^0(x)$ is an initial approximation.
For our further analysis, we use the following results.

Theorem 1. *The problems of finding a solution to difference scheme (12) and (13) with any fixed control $\Phi_{\alpha h} \in U_{\alpha h}$, $\alpha = 1, 2$ are equivalent to the* operator equations $A_{\alpha h} y_\alpha = F_{\alpha h}$, $\alpha = 1, 2$, where the difference operators $A_{\alpha h}$ from $W_2^1(\overline{\omega}^{(\alpha)}; \gamma^{(\alpha)})$ to $W_2^1(\overline{\omega}^{(\alpha)}; \gamma^{(\alpha)})$, $\alpha = 1, 2$, and the grid functions $F_{\alpha h} \in W_2^1(\overline{\omega}^{(\alpha)}; \gamma^{(\alpha)})$, $\alpha = 1, 2$, are defined by the relations

$$\left(A_{\alpha h} y_\alpha, v_\alpha\right)_{W_2^1(\overline{\omega}^{(\alpha)}; \gamma^{(\alpha)})} = Q_{\alpha h}(y_\alpha, v_\alpha), \quad \left(F_{\alpha h}, v_\alpha\right)_{W_2^1(\overline{\omega}^{(\alpha)}; \gamma^{(\alpha)})} = l_{\alpha h}(v_\alpha),$$
$$\forall y_\alpha, v_\alpha \in W_2^1(\overline{\omega}^{(\alpha)}; \gamma^{(\alpha)}), \quad \alpha = 1, 2.$$

Problems (difference schemes) (12) and (13) are uniquely solvable; moreover,

$$\|y_\alpha\|_{W_2^1(\overline{\omega}^{(\alpha)}; \gamma^{(\alpha)})} \leq M_\alpha \|\Phi_{\alpha h}\|_{L_2(\omega^{(\alpha)} \cup \gamma_S)}, \quad \alpha = 1, 2.$$

Lemma 1. *For any functions $v_1 \in W_2^1(\omega^{(1)})$ and $v_2 \in W_2^1(\omega^{(2)})$ the inequalities*

$$\|v_1\|_{L_2(\gamma_S)}^2 \leq \frac{2}{\xi} \|v_1\|_{L_2(\omega^{(1)})}^2 + 2\xi \|v_{1\overline{x}_1}\|_{L_2(\omega^{(1)+})}^2,$$

$$\|v_2\|_{L_2(\gamma_S)}^2 \leq \frac{2}{l_1 - \xi} \|v_2\|_{L_2(\omega^{(2)})}^2 + 2(l_1 - \xi) \|v_{2\overline{x}_2}\|_{L_2(\omega^{(2)+})}^2,$$

are valid.

Lemma 2. *For any functions $v_1 \in W_2^1(\omega^{(1)}; \gamma^{(1)})$ and $v_2 \in W_2^1(\omega^{(2)}; \gamma^{(2)})$ we have the estimates*

$$\|v_1\|_{L_2(\omega^{(1)})}^2 \leq \max\{\xi^2; l_2^2\} \left\{ \|v_{1\overline{x}_1}\|_{L_2(\omega_1^{(1)+} \times \omega_2)}^2 + \|v_{1\overline{x}_2}\|_{L_2(\omega_1^{(1)} \times \omega_2^+)}^2 \right\},$$

$$\|v_2\|_{L_2(\omega^{(2)})}^2 \leq \max\{(l_1 - \xi)^2; l_2^2\} \left\{ \|v_{2\overline{x}_1}\|_{L_2(\omega_1^{(2)+} \times \omega_2)}^2 + \|v_{2\overline{x}_2}\|_{L_2(\omega_1^{(2)} \times \omega_2^+)}^2 \right\}.$$

The following theorem proves the convergence of the iterative process $(8)-(11)$ (in a generalized statement the convergence of the iterative process (12), (13)).

Theorem 2. *Suppose that the condition $q = q_1 q_2 < 1$ holds true, where*

$$q_1^2 = \frac{1}{\nu} \|\theta_h\|_{L_\infty(\gamma_S)} \left(\frac{l_1 - \xi}{2} + \frac{M_2^2}{2(l_1 - \xi)} \right), \quad q_2^2 = \frac{1}{\nu} \|\theta_h\|_{L_\infty(\gamma_S)} \left(\frac{\xi}{2} + \frac{M_1^2}{2\xi} \right),$$
$$M_1^2 = \max\{\xi^2; l_2^2\}, \quad M_2^2 = \max\{(l_1 - \xi)^2; l_2^2\}.$$

Then the iterative process $(8)-(11)$ converges in the norm

$$\|v\|_{\overset{\circ}{V}_{\gamma_1, \gamma_2}(\omega^{(1,2)})}^2 = \sum_{k=1}^2 \|\nabla v_k\|^2 + \|[v]\|_{L_2(\gamma_S)}^2,$$

where $\|[v]\|_{L_2(\gamma_S)}^2 = \sum_{\omega_2} (v_2(\xi, x_2) - v_1(\xi, x_2))^2 h_2$;

$$\|\nabla v_k\|^2 = |v_k|_{W_2^1(\omega^{(k)})}^2 = \sum_{\overline{\omega}_1^{(k)+} \times \overline{\omega}_2} v_{k\overline{x}_1}^2 h_1 h_2 + \sum_{\overline{\omega}_1^{(k)} \times \overline{\omega}_2^+} v_{k\overline{x}_2}^2 h_1 h_2, \quad k = 1, 2,$$

(and, therefore, in the norm $\|v\|_{V(\overline{\omega}^{(1,2)})} = \sum_{k=1}^2 \|v_k\|_{W_2^1(\overline{\omega}^{(k)})}^2 = \sum_{k=1}^2 (\|\nabla v_k\|^2 + \|v_k\|_{L_2(\overline{\omega}^{(k)})}^2)$, $\|v_k\|_{L_2(\overline{\omega}^{(k)})}^2 = \sum_{\overline{\omega}^{(k)}} v_k^2(x) h_1 h_2$, because of their equivalence) to a

unique solution of the problem (4)–(7) for any initial approximation $y_2^{(0)} \in W_2^1(\overline{\omega}^{(2)}; \gamma^{(2)})$, *and we have convergence rate estimates:*

$$\begin{cases} |z_1^{(n)}|_{W_2^1(\omega^{(1)})} \le q_1 |z_2^{(n-1)}|_{W_2^1(\omega^{(2)})}, & |z_2^{(n)}|_{W_2^1(\omega^{(2)})} \le q_2 |z_1^{(n)}|_{W_2^1(\omega^{(1)})}; \\ |z_2^{(n)}|_{W_2^1(\omega^{(2)})} \le q_1 q_2 |z_2^{(n-1)}|_{W_2^1(\omega^{(2)})}, & n = 1, 2, \ldots; \end{cases}$$

$$\begin{cases} |z_2^{(n)}|_{W_2^1(\omega^{(2)})} \le q^n |z_2^{(0)}|_{W_2^1(\omega^{(2)})}, & |z_1^{(n)}|_{W_2^1(\omega^{(1)})} \le q_1 q^{n-1} |z_2^{(0)}|_{W_2^1(\omega^{(2)})}; \\ \|z_1^{(n)}\|_{L_2(\omega^{(1)})} \le M_1 q_1 q^{n-1} |z_2^{(0)}|_{W_2^1(\omega^{(2)})}, & \|z_2^{(n)}\|_{L_2(\omega^{(2)})} \le M_2 q^n |z_2^{(0)}|_{W_2^1(\omega^{(2)})}, \\ n = 1, 2, \ldots; \end{cases}$$

$$\begin{cases} \|z_1^{(n)}\|_{L_2(\gamma_S)} \le \left(\frac{2}{\xi} M_1^2 + 2\xi \right)^{1/2} q_1 q^{n-1} |z_2^{(0)}|_{W_2^1(\omega^{(2)})}, & n = 1, 2, \ldots; \\ \|z_2^{(n)}\|_{L_2(\gamma_S)} \le \left(\frac{2}{l_1 - \xi} M_2^2 + 2(l_1 - \xi) \right)^{1/2} q^n |z_2^{(0)}|_{W_2^1(\omega^{(2)})}, & n = 1, 2, \ldots; \\ \|[z^{(n)}]\|_{L_2(\gamma_S)}^2 \le 2 \left\{ \left(\frac{2}{\xi_1} M_1^2 + 2\xi \right) (q_1 q^{n-1})^2 + \left(\frac{2}{l_1 - \xi} M_2^2 + \right. \\ \left. + 2(l_1 - \xi) \right) (q^n)^2 \right\} |z_2^{(0)}|_{W_2^1(\omega^{(2)})}, & n = 1, 2, \ldots. \end{cases}$$

References

1. Samarskii, A.A., Andreev, V.B.: Difference Methods for Elliptic Equations. Nauka, Moscow (1976). In Russian
2. Samarskii, A.A.: The Theory of Difference Schemes. Marcel Dekker, New York (2001). Nauka, Moscow (1989)
3. Vasil'ev, F.P.: Optimization Methods. Faktorial, Moscow (2002). In Russian
4. Samarskii, A.A., Vabishchevich, P.N.: Computational Heat Transfer. Wiley, New York (1996). Librokom, Moscow (2009)
5. Samarskii, A.A., Lazarov, R.D., Makarov, V.L.: Difference Schemes for Differential Equations with Weak Solutions. Vysshaya Shkola, Moscow (1987). In Russian
6. Lubyshev, F.V., Manapova, A.R.: On some optimal control problems and their finite difference approximations and regularization for quasilinear elliptic equations with controls in the coefficients. Comput. Math. Math. Phys. **47**(3), 361–380 (2007)
7. Lubyshev, F.V., Manapova, A.R., Fairuzov, M.E.: Approximations of optimal control problems for semilinear elliptic equations with discontinuous coefficients and solutions and with control in matching boundary conditions. Comput. Math. Math. Phys. **54**(11), 1700–1724 (2014)
8. Manapova, A.R., Lubyshev, F.V.: Accuracy estimate with respect to state of finite-dimensional approximations for optimization problems for semi-linear elliptic equations with discontinuous coefficients and solutions. Ufimsk. Mat. Zh. **6**(3), 72–87 (2014)
9. Lubyshev, F.V., Fairuzov, M.E.: Some iterative processes of solution of elliptic equations with discontinuous coefficients and solutions with a design estimates of the rate of convergence of iterations. J. Middle Volga Math. **16**(1), 89–107 (2014). In Russian

Extremal Interpolation of Convex Scattered Data in $\mathbb{R}^3$ Using Tensor Product Bézier Surfaces

Krassimira Vlachkova[✉]

Faculty of Mathematics and Informatics, Sofia University "St. Kliment Ohridski",
Blvd. James Bourchier 5, 1164 Sofia, Bulgaria
krassivl@fmi.uni-sofia.bg

Abstract. We consider the problem of extremal interpolation of convex scattered data in $\mathbb{R}^3$ and propose a feasible solution. Using our previous work on edge convex minimum L_p-norm interpolation curve networks, $1 < p \leq \infty$, we construct a bivariate interpolant F with the following properties:

(i) F is G^1-continuous;
(ii) F consists of tensor product Bézier surfaces (patches) of degree (n, n) where $n \in \mathbb{N}, n \geq 4$, is priorly chosen;
(iii) The boundary curves of each patch are convex;
(iv) Each Bézier patch satisfies the tetra-harmonic equation $\Delta^4 F = 0$. Hence F is an extremum to the corresponding energy functional.

1 Introduction

Scattered data interpolation is a fundamental problem in approximation theory and finds applications in various areas including geology, meteorology, cartography, medicine, computer graphics, geometric modeling, etc. Different methods for solving this problem were applied and reported, excellent surveys are [4,5,8,9].

The problem can be formulated as follows: Given *scattered data* $\mathbf{d}_i = (x_i, y_i, z_i) \in \mathbb{R}^3$, $i = 1, \ldots, N$, that is points $\mathbf{v}_i = (x_i, y_i)$ are different and non-collinear, find a bivariate function F defined in a certain domain D containing points $\mathbf{v}_i$, such that F possesses continuous partial derivatives up to a given order and $F(x_i, y_i) = z_i$. One of the possible approaches to solving the problem is due to Nielson [14]. The method consists of the following three steps:

Step 1. Triangulation. Construct a triangulation T of $\mathbf{v}_i$, $i = 1, \ldots N$.

Step 2. Minimum Norm Network (MNN). The interpolant F and its first order partial derivatives are defined on the edges of T so as to satisfy an extremal property.

Step 3. Interpolation Surface. The obtained network is extended to F by an appropriate *blending method*.

In [1] Andersson et al. paid special attention to the second step of the above method, i.e. the construction of the MNN. The authors applied a novel approach and gave an alternative proof of Nielson's result. Their method allows to consider and handle the case where the data are convex and a convex interpolant

© Springer International Publishing Switzerland 2015
I. Lirkov et al. (Eds.): LSSC 2015, LNCS 9374, pp. 435–442, 2015.
DOI: 10.1007/978-3-319-26520-9_49

is sought. Andersson et al. formulated the corresponding extremal constrained interpolation problem of finding an MNN that is convex along the edges of the triangulation. The extremal network was characterized as a solution to a non-linear system of equations and a Newton-type algorithm for solving this type of systems was proposed. The results from [1] are extended in [16] to the class of L_p-norms for $1 < p \leq \infty$. The validity and convergence of the Newton-type algorithm for $1 < p \leq \infty$ were studied further in [17]. We note that the edge convex MNN may not be globally convex and hence a convex interpolation surface may not exist at all. Moreover, even in the case where the edge convex MNN is globally convex, Nielson's blending method may produce non-convex surface.

In this paper we propose the following solution to the convex scattered interpolation problem. Instead of triangulation we construct a rectilinear quadrangulation Q having points $\mathbf{v}_i = (x_i, y_i), i = 1, \ldots, N$, as its vertices, see Fig. 1. We define suitable z-values for different from $\mathbf{v}_i$ vertices of Q (if any) and add the new points to our data. Then we compute the edge convex minimum L_p-norm network for $p = \frac{n-1}{n-2}$ where $n \in \mathbb{N}, n \geq 3$, is chosen in advance. Hereafter we assume that n is part of our input data. The obtained edge convex MNN on every edge of Q is either a polynomial of degree n or a C^1-continuous spline with one inner knot consisting of a linear function plus a polynomial of degree n, see [16]. Moreover, the obtained network is not only edge convex but also it is convex on every whole row or column of Q. This is one of the reasons we use rectilinear quadrangulation instead of triangulation. Despite that, in general the edge convex MNN still may not be globally convex. For that reason we are seeking to construct an interpolation surface that is computationally simple and minimizes some appropriately chosen energy functional. Surfaces with such properties tend to preserve convexity of the input data. Nielson's blending method [13, 14] produces an interpolant which is a rational function on every triangle in T and consecutively may have large values in terms of energy. So, our idea is as follows. First, we slightly modify the edge-convex MNN on the edges where it is a spline. The modified MNN is C^1-continuous with the same tangent planes at the vertices of Q, consists of edge convex Bézier curves of degree n, and is convex on every row or column of Q. Next, we find a piecewise polynomial surface that interpolates the modified MNN and minimizes an appropriate energy functional. Although the modified MNN is C^1-continuous at the vertices of T, it is preferable and more appropriate to require G^1-continuity for the interpolant instead of C^1-continuity since the latter is parametrization dependent. We recall that two surfaces with a common boundary curve are G^1-*continuous* if they have a continuously varying tangent plane along that boundary curve.

Let D be the union of all quadrangles in Q. For simplicity we assume that D contains no holes. We construct a surface $F(u, v)$ defined on D that interpolates the modified MNN and has the following properties:

(i) F consists of tensor product Bézier surfaces (patches) of degree (n, n). Each patch is defined on a quadrangle of the mesh;
(ii) F is G^1-continuous;
(iii) The boundary curves of each patch are convex;

(iv) F satisfies the tetra-harmonic equation $\Delta^4 F = 0$ a.e. for $(u,v) \in D$ where $\Delta = \frac{\partial^2}{\partial u^2} + \frac{\partial^2}{\partial v^2}$ is the Laplace operator. Hence F is a solution to the extremal problem

$$\mathrm{argmin}_{\mathbf{x} \in \mathcal{F}} \int_D \|\Delta^4 \mathbf{x}\| du dv, \tag{1}$$

where $\mathcal{F} := \{\mathbf{x}(u,v) : \mathbf{x}(\mathbf{v}_i) = z_i, \ i = 1, \ldots, N, \ \mathbf{x} \in W_2^8(D)\}$, and $W_2^8(D)$ is the corresponding Sobolev space. Then F is an extremum to the corresponding energy functional.

The harmonic and bi-harmonic Bézier surfaces were studied by Monterde and Ugail [11]. Their method was extended to general 4th-order PDE Bézier surfaces in [12]. Here we use a result by Centella et al. [2] to generate tetra-harmonic tensor product Bézier surfaces from given boundary curves and tangent conditions along them. The corresponding unconstrained problem for scattered data interpolation in $\mathbb{R}^3$ is considered and solved in [18] using a rectangular quadrangulation.

The paper is organised as follows. In Sect. 2 we introduce the notation, present some related results from [1,16], and propose our Algorithm 1 for solving the convex scattered data interpolation problem. In Sect. 3 we discuss the construction of the surface F.

2   Preliminaries and Description of the Algorithm

A *quadrangulation* of given points in $\mathbb{R}^2$ is a collection of non-overlapping, non-degenerate closed quadrangles such that the set of the vertices of the quadrangles coincides with the set of the points. We shall assume that our points $\mathbf{v}_i, i = 1, \ldots, N$, are vertices of a quadrangulation Q that is homeomorphic to a rectilinear quadrangulation where vertical (horizontal) lines are not necessarily parallel. Given set of points in a general position in $\mathbb{R}^2$ one can not construct a rectilinear quadrangulation having these points as its vertices. However, we can construct it so that all of our points are among its vertices. The remaining vertices are added to the given points, see Fig. 1. An obvious way is to draw vertical and horizontal lines through each of our points. We can also choose appropriately two directions in the plane and draw lines parallel to the chosen directions through each of $\mathbf{v}_i, i = 1, \ldots, N$. Clearly this approach does not lead to unique quadrangulation. It is an open question to find the dependance of the input surface on the initial choice of the quadrangulation. Furthermore, the above approach has a drawback that the number of the new vertices is quadratic in terms of N. Our method works directly in the case where Q is a rectilinear quadrangulation, see Fig. 1. The benefit of

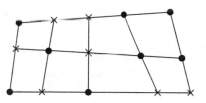

Fig. 1. Rectilinear quadrangulation of the projection points $\mathbf{v}_i, i = 1, \ldots, N$, where $\bullet$ denotes old (given) points, and $\times$ denotes new (added) points.

constructing a rectilinear quadrangulation so that points $\mathbf{v}_{i,}$, $i = 1, \ldots, N$, are its vertices is that the number of the new (added) vertices would be reduced considerably although in the general case their number still would be quadratic in terms of N. Given scattered data finding an optimal in terms of size rectilinear quadrangulation is beyond the scope of this paper.

We define z_i-values for the new points so that the data are in a convex position. The latter is possible due to the following lemma from [1].

Lemma 1. ([1]) *If the data are convex (strictly convex) then there exists a convex (strictly convex) function $\psi \in C^\infty(\mathbb{R}^2)$ interpolating the points $\mathbf{d}_i$, $i = 1, \ldots, N$.*

The proof of Lemma 1 is constructive and the function ψ is constructed in a polynomial time. We choose the z_i-values for the new points $\mathbf{d}_i$ so that $\psi(x_i, y_i) = z_i$. Hereafter we suppose that our data are strictly convex.

The union of all quadrangles in Q is the domain D. The set of the edges of the quadrangles in Q is denoted by E. If there is an edge between $\mathbf{v}_i$ and $\mathbf{v}_j$ in E, it will be referred to by e_{ij} or simply by e if no ambiguity arises. A *curve network* is a collection of real-valued univariate functions $\{f_e\}_{e \in E}$ defined on the edges in E. With any real-valued bivariate function F defined on D we naturally associate the curve network defined as the restriction of F on the edges in E, i.e. for $e = e_{ij} \in E$,

$$f_e(t) := F\left(\left(1 - \frac{t}{\|e\|}\right)x_i + \frac{t}{\|e\|}x_j, \left(1 - \frac{t}{\|e\|}\right)y_i + \frac{t}{\|e\|}y_j\right), \qquad (2)$$

$$\text{where } 0 \le t \le \|e\| \text{ and } \|e\| = \sqrt{(x_i - x_j)^2 + (y_i - y_j)^2}.$$

In our presentation, according to the context, F will denote either a real-valued bivariate function or a curve network defined by (2). Let $1 < p < \infty$. We introduce the following class of functions defined on D

$$\mathcal{F}_p := \big\{ F(x, y) \in C^1(D) : F(x_i, y_i) = z_i, \ i = 1, \ldots, N,$$

$$f_e' \in AC[0, \|e\|], \ f_e'' \in L^p[0, \|e\|], \ e \in E \big\},$$

and the corresponding class of *smooth interpolation edge convex curve networks*

$$\mathcal{C}_p(E) := \big\{ F|_E = \{f_e\}_{e \in E} : F(x, y) \in \mathcal{F}_p, \ f_e'' \ge 0, \ e \in E \big\}.$$

For $F \in \mathcal{C}_p(E)$ we denote the curve network of second derivatives of F by $F'' := \{f_e''\}_{e \in E}$. The L_p-norm of F'' is defined by

$$\|F''\|_p := \left(\sum_{e \in E} \int_0^{\|e\|} |f_e''(t)|^p dt\right)^{1/p}.$$

We consider the following extremal problem.

$$(\mathbf{P}_p) \qquad \text{Find } F^* \in \mathcal{C}_p(E) \text{ such that } \|F^{*''}\|_p = \inf_{F \in \mathcal{C}_p(E)} \|F''\|_p.$$

The degree of all inner vertices in Q, i.e. the number of the edges in E incident to each inner vertex, is four. Let $\{e_{ii_1}, \ldots, e_{ii_4}\}$ be the edges incident to the inner vertex $\mathbf{v}_i$ listed in clockwise order around $\mathbf{v}_i$. A *basic curve network* B_{is} is defined on E for $s = 1, 2$ as follows.

$$B_{is} := \begin{cases} 1 - \dfrac{t}{\|e_{ii_{s+r}}\|} & \text{on } e_{ii_{s+r}}, \; 0 \le t \le \|e_{ii_{s+r}}\|, \; r = 0, 2, \\ 0 & \text{on the other edges of } E. \end{cases}$$

Note that basic curve networks are associated with vertices that have at least two collinear edges incident to them. Thus, one basic curve network is associated with each vertex on the boundary of Q except the four corner vertices. We denote by N_B the set of pairs of indices is for which a basic curve network is defined. With each basic curve network B_{is} for $is \in N_B$ we associate a number d_{is} defined by $d_{is} = (z_{i_s} - z_i)/\|e_{ii_s}\| + (z_{i_{s+2}} - z_i)/\|e_{ii_{s+2}}\|$.

The next theorem characterizes the solution to problem $(\mathbf{P}_p)$.

Theorem 1 *([1, 16]). In the case of strictly convex data the problem $(\mathbf{P}_p)$, $1 < p < \infty$, has a unique solution F^*. The second derivative of the solution $F^{*\prime\prime}$ has the form*

$$F^{*\prime\prime} = \left(\sum_{is \in N_B} \alpha_{is} B_{is} \right)_+^{q-1}$$

where $1/p + 1/q = 1$, $(x)_+ := \max(x, 0)$ and the coefficients α_{is} satisfy the following nonlinear system of equations

$$\int_E \left(\sum_{is \in N_B} \alpha_{is} B_{is} \right)_+^{q-1} B_{kl} \, dt = d_{kl}, \; \text{for } kl \in N_B. \tag{3}$$

The basic curve networks B_{is} are the univariate basic B-splines defined along every row and column of the quadrangulation Q and the numbers d_{is} are the univariate second-order divided differences. Our data are strictly convex which guarantees that d_{is} are strictly positive and therefore Theorem 1 applies. The solution to $(\mathbf{P}_p)$ decomposes to $n_1 + n_2$ solutions to the problem in the univariate case along every row and column of Q, where n_1, n_2 are the numbers of the rows and columns of Q respectively, and $n_1 n_2 = N$. In the univariate case the problem of finding a convex function which interpolates given convex data and minimises the energy functional is considered e.g. by Hornung [6] for $p = 2$, and for $1 < p < \infty$ by Iliev and Pollul [7], Micchelli et al. [10].

It follows from Theorem 1 that in the case where $q \in \mathbb{N}, q > 1$, then F^* is a C^1-continuous polynomial network and the degree of the polynomials is $n = q + 1$. Hence to obtain a polynomial solution to $(\mathbf{P}_p)$ of degree n we solve the nonlinear system (3) for $p = \frac{q}{q-1} = \frac{n-1}{n-2}$ using the Newton-type algorithm [17]. Note that although Theorem 1 holds for $1 < p < \infty$, we apply it only for $1 < p \le 2$ since $n \ge 3$. Further on, we consider the polynomials in its Bézier form and propose Algorithm 1 for solving the convex scattered data extremal interpolation problem. Step 5 and Step 6 of Algorithm 1 are similar to the corresponding steps of the algorithm for the unconstrained case proposed in [18] where they are considered in detail. In the next Sect. 3 we focus mainly on Step 4 - the construction of the modified edge convex MNN.

Algorithm 1. Extremal Convex Scattered Data Interpolation

Input: Strictly convex scattered data $\mathbf{d}_i = (x_i, y_i, z_i) \in \mathbb{R}^3$, $i = 1, \ldots, N$;
$n \in \mathbb{N}$, $n \geq 3$

Output: Interpolation surface F with certain extremal property

Step 1. Construct quadrangulation Q of the projection points $\mathbf{v}_i = (x_i, y_i)$,
$i = 1, \ldots, N$, using straight lines through them

Step 2. Add new input points to the data if necessary

Step 3. Solve $(\mathbf{P}_p)$ for $p = \frac{n-1}{n-2}$

Step 4. Construct the modified edge convex MNN
4.1 Compute the control points of the modified curves (if any)
4.2 Degree elevate all curves to curves of degree $n + 1$

Step 5. For each quadrangle in Q find nearest to the boundary control points
that satisfy G^1 continuity conditions

Step 6. Find the remaining inner control points so that the tensor product Bézier
surface for each quadrangle satisfies the tetra-harmonic equation $\Delta^4 F = 0$

---

3   Construction of the Bézier Patches

Let B_1 and B_2 be tensor product Bézier patches whose common boundary is the polynomial $q(t)$ of degree $n, n \in \mathbb{N}$. First, we consider sufficient conditions for G^1 continuity between B_1 and B_2. Let $q(t) = \sum_{i=0}^{n} \mathbf{q}_i B_i^n(t)$ where $\mathbf{q}_i, i = 0, \ldots, n$, are the control points of $q(t)$, and $B_i^n(t) := \binom{n}{i} t^i (1-t)^{n-i}$, $i = 0, \ldots, n$, are the Bernstein polynomials defined for $0 \leq t \leq 1$. We degree elevate $q(t)$ to a polynomial of degree $n+1$. Then $q(t) = \sum_{i=0}^{n+1} \hat{\mathbf{q}}_i B_i^{n+1}(t)$ where $\hat{\mathbf{q}}_i, i = 0, \ldots, n+1$, are the degree elevated control points. Let $\mathbf{p}_i$ and $\mathbf{r}_i, i = 0, \ldots, n$, be nearest to the boundary control points of B_1 and B_2, respectively. Farin [3] proposed the following sufficient conditions for G^1 continuity between B_1 and B_2:

$$\frac{i}{n+1} d_{i,n+1} + \left(1 - \frac{i}{n+1}\right) d_{i,0} = 0, \; i = 0, \ldots, n+1, \text{ where} \qquad (4)$$

$$d_{i,0} = \alpha_0 \mathbf{p}_i + (1 - \alpha_0)\mathbf{r}_i - \left(\beta_0 \hat{\mathbf{q}}_i + (1 - \beta_0)\hat{\mathbf{q}}_{i+1}\right),$$
$$d_{i,n+1} = \alpha_1 \mathbf{p}_{i-1} + (1 - \alpha_1)\mathbf{r}_{i-1} - \left(\beta_1 \hat{\mathbf{q}}_{i-1} + (1 - \beta_1)\hat{\mathbf{q}}_i\right),$$

and $0 < \alpha_0, \alpha_1 < 1$. The coefficients α_0 and α_1 are uniquely determined by the intersection point of segments $\mathbf{p}_0 \mathbf{r}_0$, $\hat{\mathbf{q}}_0 \hat{\mathbf{q}}_1$, and $\mathbf{p}_n \mathbf{r}_n$, $\hat{\mathbf{q}}_n \hat{\mathbf{q}}_{n+1}$, respectively. In [18] it is shown that in the case where $\alpha_0 = \alpha_1$, system (4) always has a solution. The *vertex enclosure problem* is also solved since we use a rectilinear quadrangulation, see [15, 18] for details.

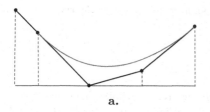
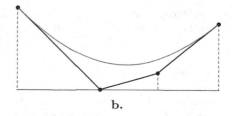

a.   **b.**

Fig. 2. Case $n = 3$: **a.** Convex C^1-continuous spline with one inner knot and its control polygon. The spline consists of a linear function plus a cubic function. **b.** The modified cubic convex Bézier curve and its control polygon.

To construct an interpolating Bézier patch, the boundary curves need to be polynomial curves. For that reason, we first modify the edge convex MNN so that all edge curves comprising it are Bézier curves of degree n. In the case where f_e^* for some $e \in E$ is a spline (see Fig. 2a for $n = 3$) we modify it to the Bézier curve of degree n whose Bézier polygon has the same graph as the Bézier polygon of the spline, see Fig. 2b for $n = 3$. The modified curve is convex and has the same tangents at the endpoints as the spline.

Next, consider a pair of Bézier curves defined on same line neighbouring edges of Q, see Fig. 3. Let $\mathbf{d}_i$ be their common point and s_1, s_2 be the two segments of their Bézier polygons with common endpoint $\mathbf{d}_i$. By construction s_1 and s_2 are collinear. We shorten the longer segment by moving its endpoint towards $\mathbf{d}_i$ so that the new segments s_1 and s_2 become equal. Then we replace the corresponding Bézier polygon by the modified one. In this way we ensure that $\alpha_0 = \alpha_1$ for every edge of Q and hence system (4) can be solved. Then we compute nearest to the boundary control points, see [18] for details.

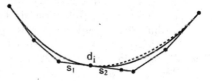

Fig. 3. Pair of cubic Bézier curves defined on same line neighbouring edges of Q. The right curve has been modified. The modified curve is shown dashed.

To compute the rest of the control points for each tensor product Bézier patch B_i we use a result by Centella et al. [2]. It states that given the boundary control points and those adjacent to them of an $(n+1) \times (n+1)$ net there exists a unique tetra-harmonic Bézier surface whose control net has those points as boundary control points and those adjacent to them. Finally, using Algorithm 1 we construct surface $F(u, v)$ defined on D which consists of tensor product Bézier patches of degree (n, n). The surface F interpolates the data since it interpolates the modified edge convex MNN. The next theorem states main properties of F.

Theorem 2. *The interpolant $F(u, v)$ is G^1-continuous and is a solution to the extremal problem* (1).

Acknowledgments. This work was partially supported by the Bulgarian National Science Fund under Grant No. DFNI-T01/0001.

References

1. Andersson, L.-E., Elfving, T., Iliev, G., Vlachkova, K.: Interpolation of convex scattered data in $\mathbb{R}^3$ based upon an edge convex minimum norm network. J. Approx. Theory **80**(3), 299–320 (1995)
2. Centella, P., Monterde, J., Moreno, E., Oset, R.: Two C^1-methods to generate Bézier surfaces from the boundary. Comput. Aided Geom. Des. **26**, 152–173 (2009)
3. Farin, G.: Curves and Surfaces for CAGD: A Practical Guide, 5th edn. Morgan-Kaufmann, San Francisco (2002)
4. Foley, T.A., Hagen, H.: Advances in scattered data interpolation. Surv. Math. Ind. **4**, 71–84 (1994)
5. Franke, R., Nielson, G.M.: Scattered data interpolation and applications: a tutorial and survey. In: Hagen, H., Roller, D. (eds.) Geometric Modeling, pp. 131–160. Springer, Berlin (1991)
6. Hornung, U.: Interpolation by smooth functions under restrictions on the derivatives. J. Approx. Theory **28**, 227–237 (1980)
7. Iliev, G., Pollul, W.: Convex interpolation by functions with minimal L_p-norm $(1 < p < \infty)$ of the k-th derivative. In: Proceedings of the 13 Spring Conference of the Union of the Bulgarian Mathematicians, pp. 31–42 (1984)
8. Lodha, S.K., Franke, K.: Scattered data techniques for surfaces. In: Hagen, H., Nielson, G.M., Post, F. (eds.) Proceedings of Dagstuhl Conference on Scientific Visualization, pp. 182–222. IEEE Computer Society Press, Washington (1997)
9. Mann, S., Loop, C., Lounsbery, M., Meyers, D., Painter, J., DeRose, T., Sloan, K.: A survey of parametric scattered data fitting using triangular interpolants. In: Hagen, H. (ed.) Curve and Surface Design, pp. 145–172. SIAM, Philadelphia (1992)
10. Micchelli, C.A., Smith, P.W., Swetits, J., Ward, J.D.: Constrained L_p approximation. Constr. Approx. **1**, 93–102 (1985)
11. Monterde, J., Ugail, H.: On harmonic and biharmonic Bézier surfaces. Comput. Aided Geom. Des. **21**, 697–715 (2004)
12. Monterde, J., Ugail, H.: A general 4th-order PDE method to generate Bézier surfaces from the boundary. Comput. Aided Geom. Des. **23**, 208–225 (2006)
13. Nielson, G.M.: Minimum norm interpolation in triangles. SIAM J. Numer. Anal. **17**(1), 44–62 (1980)
14. Nielson, G.M.: A method for interpolating scattered data based upon a minimum norm network. Math. Comput. **40**(161), 253–271 (1983)
15. Peters, J.: Smooth interpolation of a mesh of curves. Constr. Approx. **7**(1), 221–246 (1991)
16. Vlachkova, K.: Interpolation of convex scattered data in $\mathbb{R}^3$ based upon a convex minimum L_p-norm network. C. R. Acad. Bulg. Sci. **45**, 13–15 (1992)
17. Vlachkova, K.: A Newton-type algorithm for solving an extremal constrained interpolation problem. Num. Linear Algebra Appl. **7**(3), 133–146 (2000)
18. Vlachkova, K.: Extremal scattered data interpolation using tensor product Bézier surfaces. In: Ivanov, K., Nikolov, G., Uluchev, R. (eds.) Constructive Theory of Functions Sozopol 2013, pp. 253–264. Marin Drinov Academic Publishing House, Sofia (2014)

Author Index

Printed in the United States
By Bookmasters

Printed in the United States
By Bookmasters